Betty Haines, Roger Haines & Andrew May

Mathematics A Level

$$\int \frac{x^4}{1+x^2}\,dx$$

CW01474849

MACMILLAN

Acknowledgements

To our children and grandchildren

© Roger & Betty Haines 1986, 1991
© Roger & Betty Haines & Andrew May 1996

All rights reserved. No reproduction, copy or transmission of
this publication may be made without written permission.

No paragraph of this publication may be reproduced, copied or
transmitted save with written permission or in accordance with
the provisions of the Copyright, Designs and Patents Act 1988,
or under the terms of any licence permitting limited copying
issued by the Copyright Licensing Agency, 90 Tottenham Court
Road, London W1P 9HE.

Any person who does any unauthorised act in relation to this
publication may be liable to criminal prosecution and civil
claims for damages.

First published 1996 by
MACMILLAN PRESS LTD
Houndmills, Basingstoke, Hampshire RG21 6XS
and London
Companies and representatives
throughout the world

ISBN 0–333–64383–6

A catalogue record for this book is available
from the British Library.

10 9 8 7 6 5 4 3 2 1
05 04 03 02 01 00 99 98 97

Printed in Hong Kong

Over many years the questions set by the various Examination Boards have stimulated and enhanced the teaching of mathematics throughout education. Everyone involved in mathematics, both the teachers and the taught, owes a debt to the Boards for the ever-present challenge that new examination questions bring to mathematics education.

Once again our thanks go to everyone who has helped with the preparation of this book, especially to the staff of Macmillan Press.

We shall be greatly indebted to anyone notifying us of any errors.

The authors and publishers wish to thank the following for permission to use copyright material: The Associated Examining Board, the Midland Examining Group, the Northern Examinations and Assessment Board (incorporating Northern Examining Association and the Joint Matriculation Board), Northern Ireland Council for the Curriculum, Examinations and Assessment 1995, the University of Cambridge Local Examinations Syndicate (incorporating the University of Oxford Delegacy of Local Examinations and the Oxford and Cambridge Schools Examination Board), University of London Examinations and Assessment Council, and the Welsh Joint Education Committee for questions from past examination papers.

Every effort has been made to trace all the copyright holders but if any have been inadvertently overlooked the publishers will be pleased to make the necessary arrangement at the first opportunity.

Good design
for effective revision

2

∫ Basic algebra and arithmetic

Start and completion columns Keep tabs on your progress – see at a glance which areas still need to be worked through

Syllabus analysis Ensures you only do the topics you need – no more, no less

AS Level			A Level			Topic	Date attempted	Date completed	Self-assessment
CORE	MODULAR	TRADITIONAL	CORE	MODULAR	TRADITIONAL				
✓	✓	✓	✓	✓	✓	Numbers			
✓	✓	✓	✓	✓	✓	Indices			
✓	✓	✓	✓	✓	✓	Logarithms			
✓	✓	✓	✓	✓	✓	Derivation of equations			

Chapter breakdowns Shows you in detail what is covered in the chapters

Self assessment For you to note how well you've done or areas which need to be improved

Contents

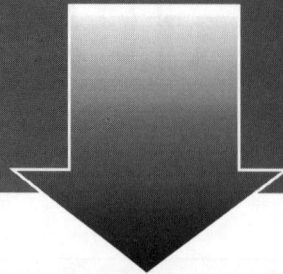

Exam Boards Addresses

For syllabuses and past papers contact the Publications Office at the following addresses:

The Associated Examining Board   (AEB)
Publications Department
Stag Hill House
GUILDFORD
Surrey
GU2 5XJ
Tel. 01483 302302

University of Cambridge Local Examinations
    Syndicate   (UCLES)
1 Hills Road
CAMBRIDGE
CB1 2EU
Tel. 01223 553311

Joint Matriculation Board   (JMB)
78 Park Road
ALTRINCHAM
Cheshire
WA14 5QQ
Tel. 0161 953 1180

University of London Examinations and
    Assessment Council   (L)
Stewart House
32 Russell Square
LONDON
WC1B 5DN
Tel. 0171 331 4000

Mathematics in Education and Industry (MEI)
Monkton Coombe
BATH
BA2 7HG

Northern Examinations and Assessment
    Board   (NEAB)
12 Harter Street
MANCHESTER
M1 6HL
Tel. 0161 953 1170
(Also shop at the above address)

Northern Ireland Council for the Curriculum,
    Examinations and Assessment   (NICCEA)
Beechill House
42 Beechill Road
BELFAST
BT8 4RS
Tel. 01232 704666

University of Oxford Delegacy of Local
  Examinations   (OLE)
Unit 23
Monks Brook Industrial Park
Chandler's Ford
EASTLEIGH
Hants
SO5 3RA
Tel. 01703 266232

Oxford and Cambridge Schools Examination
  Board   (O&C)
Purbeck House
Purbeck Road
CAMBRIDGE
CB2 2PU
Tel. 01223 411211

Welsh Joint Education Committee   (WJEC)
245 Western Avenue
Llandaff
CARDIFF
CF5 2YX
Tel. 01222 265000

Scottish Examination Board   (SEB)
Ironmills Road
DALKEITH
Midlothian
EH22 1LE
Tel. 0131 663 6601

Remember to check your syllabus number
with your teacher!

Introduction

The books in the Work Out series are not 'just textbooks'. They are based on the revision needs of AS and A Level students and have been designed to help students obtain the best possible grades in their examinations.

Work Out Mathematics covers AS and A Level syllabuses in Pure Mathematics, Statistics and Mechanics. Additionally, students entering the Foundation Year of Degree Courses requiring a mathematical background, such as science, engineering, business or management studies, will find that the book provides a bridge which allows them to rapidly familiarize themselves with new topics.

Mathematics Beyond GCSE

Everyone involved in mathematics beyond GCSE, be they students, teachers, lecturers or examiners have common aims and objectives.

Aims

Our (and that includes *your*) aim is to further develop your understanding of mathematics and mathematical processes wherever they are applied. This may be in purely mathematical situations or it may be in the analysis (modelling) of any problem, whatever the subject.

This development should be a positive thing, encouraging confidence and enjoyment. We have to become familiar with mathematical skills and techniques so that we can think clearly, work carefully and communicate ideas successfully. We have to try to appreciate how mathematical ideas can be applied in the everyday world; to develop the ability to change a problem into its mathematical expressions and then, after doing the 'sums', we have to interpret the solution in the context of the original problem.

Objectives

In order to see how successful students and teachers have been in fulfilling the aims, Examination Boards set assessment objectives. These are broadly common to all Examination Boards.

Candidates should be able to demonstrate that they are able to:

(a) recall, select and use their knowledge of appropriate mathematical facts, concepts and techniques in a variety of contexts

(b) construct rigorous mathematical arguments through the appropriate use of precise statements, logical deduction and inferences and by the manipulation of mathematical expressions

(c) evaluate mathematical models, including an appreciation of the assumptions made; and interpret, justify and present the results from a mathematical analysis in a form relevant to the original problem

(d) organize information in tabular, graphical and other modes of presentation

(e) communicate mathematical ideas clearly, using tabular, graphical or other methods

(f) appreciate how to use appropriate technology, such as computers and calculators, as a mathematical tool, and have an awareness of the limitations of this technology

(g) comprehend and investigate the application of mathematics.

Assessment

Different Examination Boards have various ways of assessing how well you have achieved the above objectives. Your teachers or lecturers will inform you of the name of your Examination Board and of their method of assessment. Traditionally assessment was by examination, one three-hour examination at AS Level and two at A Level.

Some courses are now modularized. At AS Level there may be between two and five modules, with the number doubling at A Level. Typically, modules are examined at intervals throughout the course. If you are a private candidate you will have to make the appropriate arrangements at a local centre (possibly a school or college) for your chosen Examination Board. Your Board will provide you with the names of centres in your locality.

Modularized courses with their frequent assessments require you, whether school or college based or private, to work steadily and regularly throughout the *whole* course.

Many Examination Boards now include a Coursework and/or a Project component in the course. Your teachers will inform you if there is any such requirement in your course. In the event of any query you can always obtain the appropriate syllabus and regulations from your Examination Board. (Addresses are given on pages v and vi)

How to Use the Book

(a) By repeated use and practice, endeavour to become familiar with the frequently used formulae listed in the fact sheets.

(b) Practice in answering examination questions is important. Open the book at a definite topic, choose a question and cover up the solution until you have tried to do it by yourself. If you get really stuck, your mind will be receptive when you uncover the solution.

(c) The methods used in the book are not necessarily the shortest. Always be on the lookout for shorter and neater solutions. 'There is always a shorter way' is an excellent maxim to adopt throughout mathematics.

(d) Be 'calculator confident'. Your graphic calculator is great! Learn how to use it fluently. Chapter 1 is devoted to things it can do (with your help!). Throughout the book you will encounter calculator symbols:

1.2

In these places use your graphic calculator. The number(s) in boxes beside the calculator symbols indicate which section of Chapter 1 you should refer to for help with using that particular calculator function.

Use it to solve equations, plot graphs, differentiate, integrate, check answers, etc., etc.

(e) Syllabus content varies between Examination Boards and between various courses examined by the same Board. You do not need to study all the topics in the book. If you are in doubt about a topic check your syllabus with your teacher or lecturer.

(f)   Throughout the book you will find hints from the authors in thin, italic type. These will help you understand a working or a rule or help you to work through an example.

Coursework and Projects

If your course contains one or more Coursework or Project components the marks for that component may be up to 20% of the total available mark. (The percentage may vary between Boards.) If you do your Coursework and/or Project well, and obtain a good mark, then you are well on the way to a good overall grade. (Done badly and handed in late it becomes a millstone, dragging down not only your Mathematics grade but your overall ambience.) If you are set Coursework or a Project, *do it*, do it *well* and hand it in *on time*.

Typically you may have to submit a portfolio of work or a report, demonstrating an in-depth study based on a topic or combination of topics on the syllabus. In marking the work your teacher, and maybe an external moderator, will consider how well you have satisfied various criteria. These include:

(a)   researching, processing and evaluating relevant information
(b)   analysing the problem, designing or formulating a mathematical model
(c)   solving the model using appropriate techniques which may include graphic calculators and computers
(d)   interpreting the model solution, leading to conclusions
(e)   communicating the work precisely and effectively.

Mathematical Modelling

There is an increasing emphasis on the idea of a 'mathematical model' and also on the process of 'mathematical modelling'. To some extent these terms are little more than new labels for things that a mathematician would always have used when tackling a real life problem. One advantage of having this new terminology, however, is that it provides a clearly defined framework to work within which helps mathematicians solve real life problems.

Mathematical Model

A mathematical model is a mathematical description of some aspect of the real world. It comprises a general relationship between two or more quantities – typically it may take the form of a formula or a graph.

The Modelling Process

There are three parts to this:

(a)   **Formulating** the mathematical model
(b)   **Solving** the model
(c)   **Interpreting and evaluating/validating** the model.

If the evaluation/validation indicates that an improved model is called for then the modelling process is repeated until a satisfactory model is arrived at. This idea of repeating the process gives rise to the notion of a 'modelling loop', illustrated by this diagram:

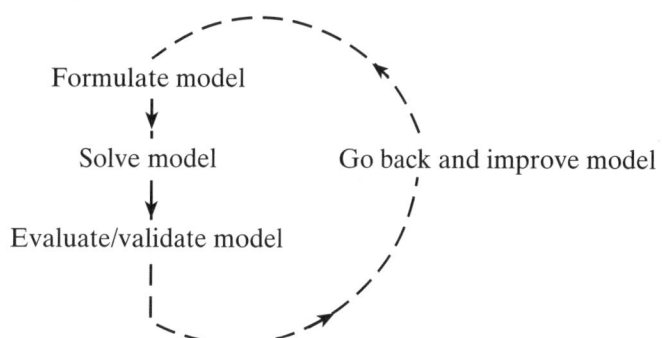

Formulate model
↓
Solve model   Go back and improve model
↓
Evaluate/validate model

Formulating the Mathematical Model

Define the problem

A problem whose solution is required is the motivation for mathematical modelling in the first place. Problems can vary from complicated, for example wanting to know which number the ball will end up in on the roulette wheel, to very simple, for example wanting to know the relationship between distance and time when a ball rolls down a slope. Your problem should be stated clearly and unambiguously. Seek information about it from books and by talking to people. This is background research.

Design the model

This is where aspects of the exact real life situation are replaced with approximations to them. Although the approximations will tend to vary in their extent, they all have the shared aim of simplifying the mathematical analysis which follows. This simplification can be either in order to avoid a mathematical analysis which is impossibly difficult or to eliminate unwanted, insignificant detail which would only serve to clutter the analysis.

Examples of assumptions include 'the object behaves as a particle', 'the string is light (massless)', 'air resistance is negligible', 'the population is Normally distributed', 'there is no resistance to the motion', 'there is an equal chance of the coin landing heads or tails'.

This is where the problem, having been suitably simplified, is translated into symbols and numbers. Units should be stated. For example, 'initial speed of ball = u m s^{-1}', 'acceleration due to gravity = 9.8 m s^{-2}', 'total time for the descent = T seconds', 'number of people = N'.

Solving the Mathematical Model

This is where the mathematical analysis takes place. Very frequently use will need to be made of various Laws. For example, in mechanics, most problems will involve Newton's 1st or 2nd Law. Laws such as these can themselves be regarded as mathematical models that have been so well tested and have such wide-ranging use that they are referred to as Laws (they are sometimes referred to as 'standard models'). The mathematical analysis will often involve the solution of one or more equations, either exactly or by using some approximate numerical method, maybe using a graphic calculator or a computer. The result of the analysis constitutes the mathematical model itself.

Interpreting and Evaluating/Validating the Mathematical Model

Interpreting the model

Some form of clear explanation of what the model implies should always be included. A graph with clearly labelled axes and a title can be very effective; a written interpretation may also be included. For example, 'the time taken is directly proportional to the distance moved, so if the distance doubles then so does the time taken', or 'the height of the object above its equilibrium position is a sinusoidal function of time with amplitude 3 metres and period 2.4 seconds'. The interpretation should also include making certain predictions on the basis of the model; for example, 'if the ball were rolled down a slope of angle 30° it would take 2 seconds to travel 8 metres'.

Evaluating/validating the model

It is essential to evaluate how well the mathematical model describes the real life situation that it is supposed to apply to. The predictions that have been made are therefore tested. In the light of the accuracy of the predictions the mathematical model can be appraised and some idea gained of how valid the mathematical model is.

Generally the question of how valid a model may be is not clear-cut; typically the circumstances in which a model applies are limited. In practice, a model may therefore be regarded as a good approximation to reality in some circumstances, reasonable in some others, and rather poor in others. This is true even of most of the 'standard models' or 'Laws': Hooke's Law is fairly accurate so long as the spring is not near its elastic limit; Newton's 2nd Law is very accurate indeed unless the speeds involved start to approach the speed of light.

In any A Level piece of modelling the limitations of a mathematical model are likely to be significant so they need to be discussed carefully and it should be made clear under what circumstances the model is thought to be valid. Reasons why the model is limited should also be put forward. A main source of limitation will often be the original assumptions on which the model is based; for example, it may have been assumed that the object could be treated as a particle where in fact the object is a hollow sphere which is rolling. Also, certain factors ruled out altogether by the original assumptions may in fact play a significant role. Having identified limitations, suggestions for the refinement of the model can be made. Actually carrying out some degree of refinement by going back and modifying the set of assumptions constitutes the 'modelling loop' referred to earlier.

Coursework and Project Hints

(a) When your topic is decided, start a file or note book and keep all the work together. Throughout the work, remember to note the names of people, books and other sources of information.

(b) Your Coursework/Project report is a communication document. It should be *clear*, *unambiguous*, and use *standard notation* and *defined symbols*.

(c) Tables, graphs and diagrams should be clearly labelled, including units and dimensions.

(d) Don't be afraid to be critical (but not too critical!) of the assumptions you have made. Suggest refinements or possible alternative approaches. Could your method be used for similar problems?

(e) If appropriate write a summary. Write it *last*, put it *first*. Three sentences may suffice: (i) the problem, (ii) your solution, (iii) what it means.

Revision Hints

(a) Your school or college should be able to supply you with a syllabus and typical examination papers. If not, you should write to the Secretary of the Examination Board. A list of addresses is given on pages v and vi.

(b) Use the book to revise topics before trying past papers.

(c) Familiarize yourself with the contents of your formula booklet (if one is provided) well before the examination. There are always some candidates who fail to answer questions because they are not aware that a vital formula which escapes them at that instant is to be found in the booklet so thoughtfully provided by the Board!

(d) Develop a good examination technique. Try papers 'to time' under examination conditions (quiet, no cups of coffee!). Don't be too depressed by your first efforts; practice can make perfect.

(e) Get into the habit of writing solutions tidily first time. It makes it so much easier on 'the day' if you don't have to think about legibility.

Examination Hints

(a) Spend a few minutes reading the paper and select questions in order of priority.

(b)   Do the 'easy' questions at the start of the examination – this boosts your confidence.

(c)   If you get stuck on a question and cannot see an alternative approach, cut your losses and go on to another question. Examinations have been failed by good candidates who spent too long stuck on one or two questions.

(d)   Never cross out. You may be crossing out marks.

(e)   Never walk out of an examination. Reread the questions right through – even if you cannot do the first part of a question, there may be parts you can try.

(f)   Good luck.

Graphic calculators

1

Chapter Contents

1 INTRODUCTION

In the last few years the use of graphic calculators within mathematics A Level courses has increased enormously. Many Examination Boards now positively encourage their use in the examination and in fact some Boards already consider them to be essential. Whatever the attitude of your particular Board, however, there is little doubt that a graphic calculator can, with a little practice, become an extremely effective tool in tackling a very wide range of A Level questions as well as serving as a powerful learning aid throughout an A Level course. Apart from the facility to graph functions, which a 'traditional' calculator cannot do of course, the range of graphic calculators' capabilities includes manipulation of functions and the analysis of their graphs, performing repeated or complex calculations very efficiently, carrying out one-variable and two-variable statistical calculations, working with matrices, solving equations and generating and summing sequences.

All graphic calculators can be programmed, although any stored programs must be cleared before starting an examination. It is, however, a good idea to become familiar with the basics of programming your calculator – writing and experimenting with programs can be a very good way of developing an understanding of certain areas of an A Level course as well as being useful in encouraging and practising logical thought. Very short programs written quickly during an examination can also prove effective.

This chapter includes detailed information on how graphic calculators can be used to help solve or check the solutions of A Level problems. Section 3 provides a summary of the range of important features of the most popular graphic calculators that are used by A Level students. It is followed in Section 4 by a description of how a graphic calculator can be used to its best advantage within different parts of the syllabus. Specific parts of Section 4 are referred to throughout the book at points where a graphic calculator could be helpful – look for the calculator icon together with a reference to a specific technique.

The particular models of graphic calculator that will be referred to are the Texas TI-82, the Casio fx-7700GE (and its predecessor the fx-7700GB), the Casio fx-9700GE and the Sharp EL-9300/9200. Each of these can perform a vast range of mathematical techniques, many of which require several keys to be pressed in sequence in order to carry them out. The manual that is supplied with the calculator, at least 150 pages in length, describes these techniques and their associated key sequences in fine detail. It is the aim in Section 4 of this chapter to cover only the most useful things for A Level mathematics that a graphic calculator is capable of doing. The level of detail included in the descriptions assumes you are already familiar with the basic operation of your particular model.

2 USING A GRAPHIC CALCULATOR IN THE A LEVEL EXAMINATION

- At the start of the examination you are likely to be required to 'Reset' your calculator, which clears all the programs and any other data that may be stored and returns the calculator to its initial settings. It is therefore worth making absolutely sure, by trying it beforehand, that you know what these settings are (for example does it reset to degrees or to radians?) and how to alter them in the examination to what you need.

- Where no particular method is specified in the question then a solution using the graphic calculator is perfectly acceptable (but note that if you write an answer down with no working then if there is even a very small error in the answer no marks will be awarded).

- Most Examination Boards' papers include a statement in the rubric at the beginning of the paper to the effect that working should be shown. Expressions such as 'all necessary working', 'sufficient detail of the working' are used frequently. It is obviously always going to be a grey area as to exactly what constitutes 'sufficient' or 'necessary' in this context. It should, however, be borne in mind that in the past it has been standard practice for most Examination Boards to award full marks for correct answers written down with no accompanying working or explanation.

- Every graphic calculator has over 20 value memories. Make sure that you know how to use these and are used to using them in the course of longer questions where intermediate answers may be needed later on in the question. Using the stored values not only saves time and avoids inputting errors when copying from your script but it also ensures that rounding errors are kept to a minimum. (See '1.1 – Approximate Answers' in Section 4 of this chapter for more details.)

- When an exact answer is required then it is obviously preferable to tackle the question by conventional means, and then possibly to check the solution using the graphic calculator. If you do not know how to tackle it by conventional means then it can be worth finding an answer on the calculator, which is very likely to be approximate, and then make an educated guess at what the exact answer is likely to be. (See '1.2 – Exact Answers' in Section 4 of this chapter for more details.)

3 IMPORTANT COMMANDS AND FEATURES OF INDIVIDUAL CALCULATORS

This section contains a summary, for each calculator, of the main features that can be useful in tackling A

Level problems. Details of exactly how to apply the
calculator commands are given in Section 4 of this
chapter. All the different models of graphic calculator
provide the standard range of mathematical functions
such as the trig functions (sin, cos, tan), logarithms, the
exponential function, powers, roots, reciprocal and so
on. They are not included in the following summary as
it is expected that you will already be entirely familiar
with them.

  Bold italic type is used to denote specific in-built
commands and menus of each calculator.

Texas TI-82

Memory

32K total consisting of:

- 27 value memories
- Automatic storage of last answer (***ANS***)
- Automatic storage of last 128 bytes of commands
  (typically equivalent to the last 10 to 20 commands)
- Up to 10 cartesian, 6 parametric, 6 polar and
  2 sequence functions
- Up to 6 lists (up to 99 members each); these are in
  effect extra value memories
- Statistical data
- Matrices
- Programs
- Up to 6 Graph databases
- Up to 6 Pictures

Graphs

(a) Graph Plotting

There are 4 graphing modes – cartesian, parametric,
polar and sequence.

When ***GRAPH*** is pressed all the stored functions in
the current graphing mode which are 'switched on'
(their '=' sign being highlighted) are plotted. Any
statistics graphs which are 'on' are also plotted.

(b) Graphical Function Analysis

Notes
(i)   ***ZOOM*** refers to changing the scale at which the
      graph is viewed – this can be specified numerically
      as a scale factor; alternatively an area of the graph
      to be magnified can be marked with the cursor.
(ii) ***TRACE*** refers to moving the cursor along the
      curve and its coordinates being displayed.
(iii) All of these commands, except ***TRACE*** and
      ZOOM, are on the ***CALC*** menu.
(iv) Finding roots, maxima, minima, intersections and
      function values can be done with patient use of
      TRACE and ***ZOOM*** but it is a great deal quicker
      to use the 'tailor-made' commands listed below.

TRACE
traces any of the plotted functions.

ZOOM
zooms in on any of the plotted functions.

Root
finds (real) roots of cartesian functions. This means in
effect that equations of the form $f(x) = 0$ can be
solved.

Maximum
locates maximum turning point (cartesian graph).

Minimum
locates minimum turning point (cartesian graph).

Intersect
locates intersections of two cartesian functions.

Value
evaluates the function at a chosen value of x, θ or t
(depending on the graphing mode).

∫f(x)dx
calculates and shades the definite integral between two
points specified using trace.

dy/dx
calculates the gradient at any point specified using
trace.

dr/dθ
calculates its value at any point, specified by trace, on a
polar curve.

dy/dt
calculates its value at any point, specified by trace, on a
parametric curve.

dx/dt
calculates its value at any point, specified by trace, on a
parametric curve.

(c) Graph Transformations

A stored function can be defined in terms of another
stored function in order to produce a transformation of
the original graph:

$Y_2 = Y_1 + a$	Y_2 will be a translation of Y_1 by a units in the y-direction.
$Y_2 = aY_1$	Y_2 will be a stretch of Y_1 in the y-direction with a scale factor of a.
$Y_2 = Y_1(aX)$	Y_2 will be a stretch of Y_1 in the x-direction with a scale factor of $1/a$.
$Y_2 = Y_1(X + a)$	Y_2 will be a translation of Y_1 by $-a$ units in the x-direction.

(d) Inequality Graphs

A function defined as $f(x) > g(x)$ takes the value of 1
when the inequality is satisfied and 0 when it is not.
(Works the same with '<'.) For example, if Y_1 is

defined as $X^3 > X^2$ then Y_1 will be equal to 1 for $x > 1$ and it will be 0 elsewhere; it can of course be plotted and traced like any other function.

$f(x) > g(x)$ can also be solved by plotting both functions and finding their points of intersection (by using **Intersect**).

Functions

(a) Function Storage

Up to 10 cartesian, 6 parametric, 6 polar, and 2 sequence functions can be stored. Each of these can be defined to be 'on' or 'off'. Several related functions can all be stored in one function memory by using { }. For example $Y_1 = \sin\{X, 2X, 3X\}$ will store the three functions $\sin X$, $\sin 2X$ and $\sin 3X$.

(b) Evaluating Functions

Tabulate using **TABLE** command. This enables the values of a function to be tabulated for a set of values of x. **TblSet** enables the minimum value of x and the incremental change in the value of x to be input. All the stored functions in the current graphing mode which are 'on' are tabulated.

Enter the stored function name and the value of the variable, e.g. $Y_1(5)$.

Enter the stored function name, e.g. Y_1, to evaluate at the current value of X.

Enter the function itself to evaluate at the current value of X.

(c) Composite Functions

Stored functions can be defined as the composite of other functions, including other stored functions, for example:

- If $Y_1 = \sin X$ and $Y_2 = X^2$, then $Y_2(Y_1)$ will be the composite function $(\sin X)^2$.
- $Y_1(X + 5)$ will be the composite function $\sin (X + 5)$.

Calculus

nDeriv(
enables a function to be defined as the derivative of another, which can then be plotted and analysed graphically like any other function. It also enables the first derivative of a function to be calculated at a particular value of the variable.

fnInt(
enables a function to be defined as the integral of another, which can then be plotted and analysed

graphically like any other function. It also enables a definite integral to be calculated between any two points.

Notes
(i) These can all be applied in a similar way to polar and parametric functions.
(ii) See 'Graphical Function Analysis' for more calculus commands.

Solving Equations

(a) Numerical Methods

Solve(
finds the real roots of $f(x) = 0$. For every root to be found an initial guess is required. For example, **Solve(** $3 \sin X + 2 \cos X - 1.5, X, 90)$ employs an initial guess of $90°$ to solve the equation $3 \sin x + 2 \cos x = 1.5$.

To solve an equation of the form $x = f(x)$ use an iterative formula with **ANS** representing the successive approximations to the solution.

A set of n linear equations in n unknowns can be solved using matrices.

(b) Graphical Methods

Root
finds roots of $f(x) = 0$ having plotted $y = f(x)$.

Intersect
solves 2 simultaneous equations in 2 unknowns by locating the points of intersection of $y = f(x)$ and $y = g(x)$.

ZOOM and **TRACE**
can be used to achieve the same final result as 'Root' or 'Intersect'.

Sequences and Series

Sequence facilities on the calculator enable:

- General term to be input.
- First order inductive definition to be input.
- Terms to be stored as a list (up to 99 terms).
- Sum of sequence to be calculated.
- Terms of sequence to be tabulated.
- Graph of sequence to be plotted.

A sequence defined inductively can be generated by using **ANS** to represent the ith term of the sequence.

Fractions

Fractions have to be done in terms of division (e.g. 3/4) and answers are given as decimals.

▶Frac

This is an extremely useful command that converts a decimal to its equivalent fraction, provided the number is rational and the denominator is no greater than 4 digits.

Statistics

Statistical data can be stored in any of the 6 lists – the numbers in any of the lists can represent frequencies but these must be non-negative integers less than 100.

Single-variable statistics (x)

The following quantities are automatically calculated: $x, \Sigma x, \Sigma x^2, \sigma x$ (standard deviation), Sx (unbiased estimate of population standard deviation), n (number of data values), Min x (minimum value of x), Max x (maximum value of x), Q_1 (lower quartile), Med (median), Q_3 (upper quartile).

Single-variable graphics

Box and whisker plot, bar chart.

Two-variable statistics (x, y)

All the quantities that are calculated for single-variable statistics, apart from Q_1, Med and Q_3, are calculated for both the x and the y data. In addition Σxy is calculated.

Four of the 9 regression models available use the method of least squares: linear $y = ax + b$ (or $y = a + bx$), logarithmic $y = a + b \ln x$, exponential $y = ab^x$, and power $y = ax^b$. For each of these the values of a, b and r, the correlation coefficient, are calculated.

Two-variable graphics

Scatter diagram and line (or curve) of best fit.

Programming

Program control commands include:

If
conditional test.

If-Then
conditional test.

If-Then-Else
conditional test.

Lbl
define a label.

Goto
go to a label.

While
create a conditional loop.

Repeat
create a conditional loop.

Pause
pause execution of program.

IS>(
increment a variable by 1 and either skip or execute the next command according to the value of the variable.

DS<(
decrement a variable by 1 and either skip or execute the next command according to the value of the variable.

Menu(
branch to one of a number of labels.

prgm
branch to another program.

Return
branch back to original program.

All calculator facilities can be used within programs.

Casio fx-7700GE

The fx-7700GE operates in many ways exactly the same as its predecessor the fx-7700GB so much of this section is relevant to both models. The main changes that have been introduced on the fx-7700GE are an icon-based menu system to select the different modes of operation of which there are two extra: 'GRAPH' mode and 'EQUATION' mode.

Memory

4K total including:

– 28 value memories (this number can be increased)
– Automatic storage of last answer (**Ans**)
– Automatic storage of last command
– Up to 6 cartesian functions in the 'Function Memory'
– Up to 20 functions in the 'Graphic Function Memory' (up to a total of 127 bytes). The functions can be cartesian, polar, parametric or inequality
– Statistical data
– Matrices
– Programs

Graphs

(a) Graph Plotting

There are 4 main graph types – cartesian, parametric, polar and inequality.

In GRAPH mode: when **DRW** is pressed all the stored functions in the 'Graphic Function Memory' (Y1, Y2, etc.) which are 'switched on' (their '=' sign being highlighted) are plotted. Any functions which are stored in the 'Function Memory' (f_1, f_2, etc.) are not plotted.

In COMP mode: each graph requires a separate command (**Graph Y=**) to plot it.

(b) Graphical Function Analysis

Notes
(i) **Zoom** refers to changing the scale at which the graph is viewed – this can be specified numerically as a scale factor; alternatively an area of the graph to be magnified can be marked with the cursor.
(ii) **Trace** refers to moving the cursor along the curve and its coordinates being displayed.

Trace
In GRAPH mode any of the plotted functions can be traced.
In COMP mode only the graph corresponding to the most recent **Graph Y=** command can be traced.

Zoom
In GRAPH mode any of the plotted functions can be zoomed.
In COMP mode only the graph corresponding to the most recent **Graph Y=** command can be zoomed.

G-∫dx
calculates and shades the definite integral between any two points.

(c) Graph Transformations

A stored function (either in the 'Graphic Function Memory' or in the 'Function Memory') can be defined in terms of another stored function in order to produce a transformation of the original graph:

Y2 = Y1 + a	Y2 will be a translation of Y1 by a units in the y-direction.
Y2 = aY1	Y2 will be a stretch of Y1 in the y-direction with a scale factor of a.
Y2 = Y1(aX)	not recognized in the usual sense.
Y2 = Y1($X + a$)	not recognized in the usual sense.

(d) Inequality Graphs

A function can be defined using any of the 4 basic inequalities and the region satisfied will be shaded. For example, $Y > X^2$ will shade the region above the curve $y = x^2$.

$f(x) > g(x)$ can be solved by plotting both functions and finding their points of intersection (by using **Zoom** and **Trace**).

Functions
(a) Function Storage

Storage of up to 6 cartesian functions (f_1, f_2, etc.) in the 'Function Memory'.

In GRAPH mode up to 20 functions (Y1, Y2, etc.) can be stored in the 'Graphic Function Memory', each of which can be 'selected' or 'cancelled'.

(b) Evaluating Functions

Enter the stored function name, e.g. **Y**1 or f_1, to evaluate at the current value of X.

Enter the function itself to evaluate at the current value of X.

The multistatement '? $\rightarrow$ X: *function of x*' is very efficient for evaluating a function at several values of x in succession.

Calculus

d/dx
enables a function to be defined as the derivative of another, which can then be plotted and analysed graphically like any other function. It also enables the first derivative of a function to be calculated at a particular value of the variable.

∫dx
definite integral calculated between any two points.

G-∫dx
definite integral calculated between any two points and the corresponding area shaded on the graph.

Solving Equations
(a) Numerical Methods

To solve an equation of the form $x = f(x)$ use an iterative formula with **Ans** representing the successive approximations to the solution.

A set of n linear equations in n unknowns can be solved using matrices.

In EQUATION mode a set of 2 linear simultaneous equations in 2 unknowns and 3 in 3 unknowns can be solved. Quadratic equations can also be solved, complex solutions included.

(b) Graphical Methods

Zoom and **Trace**
can be used to find the roots of $f(x) = 0$ having plotted $y = f(x)$.

Sequences and Series

A sequence defined inductively can be generated by using **Ans** to represent the ith term of the sequence.

Fractions

Fractions can be entered directly, including ones with a preceding whole number (e.g. $2\frac{3}{4}$). The answer is given in the same form.

Statistics

Statistical data can be stored as a frequency distribution. Since the frequencies need not be positive integers, probability distributions can also be analysed.

Single-variable statistics (x)

The following quantities are automatically calculated: x, Σx, Σx^2, $x\sigma_n$ (standard deviation), $x\sigma_{n-1}$ (unbiased estimate of population standard deviation), n (number of data values), Min (minimum value of x), Max (maximum value of x), Med (median), Mod (mode). Areas under the Normal curve can also be calculated.

Single-variable graphics

Bar chart, the Normal curve.

Two-variable statistics (x, y)

All the quantities that are calculated for single-variable statistics, apart from Min, Max, Med and Mod, are calculated for both the x and the y data. In addition Σxy is calculated.

There are 4 regression models which use the method of least squares: linear $y = A + Bx$, logarithmic $y = A + B \ln x$, exponential $y = Ae^{Bx}$, and power $y = Ax^B$. For each of these the values of A, B and r, the correlation coefficient, are calculated.

Two-variable graphics

Scatter diagram and line (or curve) of best fit.

Programming

Program control commands include:

If-Then-Else
conditional test (the words 'If-Then-Else' are not actually used but the effect is the same).

Lbl
define a label.

Goto
go to a label.

Isz
increment a variable by 1 and either skip or execute the next command according to the value of the variable.

Dsz
decrement a variable by 1 and either skip or execute the next command according to the value of the variable.

prgm
branch to another program (returns from it automatically).

All calculator facilities except 'Trace' can be used within programs.

Programming commands can be used outside programs (e.g. ? $\rightarrow$ X:).

Casio fx-9700GE

The fx-9700GE operates in a virtually identical way to the Casio fx-7700GE. There are some extra features on the fx-9700GE including 32K of memory and two extra modes of operation, 'DYNAMIC GRAPH' and 'TABLE'. Only the important features that do not appear on the fx-7700GE are detailed in this section.

Graphs

(a) Graphical Function Analysis

All of the following are on the **G-SOLV** menu, which is only available in GRAPH mode.

ROOT
finds (real) roots of cartesian functions. This means in effect that equations of the form f$(x) = 0$ can be solved.

MAX
locates maximum turning point (cartesian graph).

MIN
locates minimum turning point (cartesian graph).

Y-ICPT
locates the intercept on the y-axis.

ISCT
locates intersections of two cartesian functions.

Y:CAL
evaluates the function at a chosen value of x.

X:CAL
evaluates x at a chosen value of y.

d/dx
for cartesian and parametric curves this calculates the

gradient at any point specified using trace. For polar curves this calculates the value of $dr/d\theta$ at any point specified using trace.

(b) Graph Transformations

In DYNAMIC GRAPH mode, functions can be defined which include a number of parameters (A, B, C, D and so on). One of the parameters whose effect on the graph of the function is to be observed is then selected. Its minimum and maximum values and its incremental change between these values are input. The calculator will then display in succession the individual graphs that correspond to each of the changing values of the parameter – the effect being to 'animate' the display and make the relationship between the particular parameter and the graph of the function very clear.

Functions

(a) Evaluating Functions

In TABLE mode, the values of a function for a set of values of x are tabulated. The values of x are input as a start value, an end value and also a value of 'pitch' (which is the incremental change in value of x between the start and end values). The function can be specified as one of those stored in the 'Graphic Function Memory'.

It is also possible to tabulate recursion formulae – see 'Sequences and Series'.

Solving Equations

(a) Numerical Methods

In EQUATION mode a set of up to 6 linear simultaneous equations in 6 unknowns can be solved. Quadratic and cubic equations can also be solved, complex solutions included.

(b) Graphical Methods

ROOT
finds roots of $f(x) = 0$ having plotted $y = f(x)$.

ISCT
solves 2 simultaneous equations in 2 unknowns by locating the points of intersection of $y = f(x)$ and $y = g(x)$.

X:CAL
solves $f(x) = k$. Plot $y = f(x)$, and enter k as the y-coordinate.

Sequences and Series

In TABLE mode, sequence facilities enable:

– General term to be input.
– Inductive definition to be input (first order linear or second order linear).
– Sequence terms and sums to be tabulated.
– Sequence terms and sums to be graphed.

In COMP mode a sequence defined using a general term can be summed.

Sharp EL-9300 and EL-9200

The main differences between these calculators are that the EL-9300 has an equation-solving mode and it can evaluate functions when applied to complex numbers.

Memory

32K (EL-9300) or 8K (EL-9200) total including:

– 28 value memories
– Automatic storage of last answer (***Ans***)
– Automatic storage of last command
– Up to 4 cartesian, 2 parametric and 2 polar functions
– Statistical data
– Matrices
– Programs
– Equations

Graphs

(a) Graph Plotting

Takes place in GRAPH mode. There are 3 main graph types – cartesian, parametric and polar. Only one type can be active at a time; this is selected on the ***SET UP*** menu.

When the graph key is pressed all the stored functions of the current graph type which are 'switched on' (their '=' sign being highlighted) are plotted.

(b) Graphical Function Analysis

Notes
(i) ***ZOOM*** refers to changing the scale at which the graph is viewed – this can be specified numerically as a scale factor; alternatively an area of the graph to be magnified can be marked with the cursor.
(ii) ***TRACE*** refers to moving the cursor along the curve and its coordinates being displayed. There is no key labelled 'Trace' – it is activated by pressing the left or right arrow keys when there are one or more graphs displayed.
(iii) All of these commands, except ***TRACE***, ***ZOOM*** and ***Y'*** are on the ***JUMP*** menu.
(iv) Finding roots, maxima, minima, intersections and intercepts can be done with patient use of ***TRACE***

and ***ZOOM*** but it is a great deal quicker to use the 'tailor-made' commands listed below.

TRACE
traces any of the plotted functions.

ZOOM
zooms in on any of the plotted functions.

INTERSEC
locates intersections of two cartesian functions.

MIN
locates minimum turning point (cartesian graph).

MAX
locates maximum turning point (cartesian graph).

X INCPT
locates intersections with the *x*-axis (cartesian graph).

Y INCPT
locates intersections with the *y*-axis (cartesian graph).

Y'
displays the value of the first derivative when a graph is being traced.

(c) Inequalities

Regions above or below a function can be shaded.

$f(x) > g(x)$ can be solved by plotting both functions and finding their points of intersection.

Functions

(a) Function Storage

Up to 4 cartesian, 2 parametric and 2 polar functions which can be defined to be 'on' or 'off'.

(b) Evaluating Functions

Enter the function itself to evaluate at the current value of *X*.

Calculus

d/dx(
enables a function to be defined as the derivative of another, which can then be plotted and analysed graphically like any other function. It also enables the first derivative of a function to be calculated at a particular value of the variable.

∫dx
definite integral calculated between any two points.

Solving Equations

(a) Numerical Methods

To solve an equation of the form $x = f(x)$ use an iterative formula with ***Ans*** representing the successive approximations to the solution.

A set of *n* linear equations in *n* unknowns can be solved using matrices.

In SOLVER mode (EL-9300 only) the calculator will find real solutions of equations. If possible it will use direct algebraic rearrangement, otherwise it will use an iterative method.

(b) Graphical Methods

X INCPT
locates intersections of a function with the *x*-axis.

Sequences and Series

A sequence defined inductively can be generated by using ***Ans*** to represent the *i*th term of the sequence.

Fractions

Fractions can be entered directly, and the answer is given in the same form.

Statistics

Statistical data is stored on 'cards'. Each card occupies the whole screen and contains a value of *x*, a value of *y* and a weighting, *w*. The set of data cards therefore comprise a frequency distribution for either single-variable or paired-variable data. Since the frequencies (*w*) need not be positive integers, probability distributions can also be analysed.

Single-variable statistics (x)

The following quantities are automatically calculated: x, Σx, Σx^2, σx (standard deviation), sx (unbiased estimate of population standard deviation), n (number of data values), xmi (minimum value of x), xma (maximum value of x).

Single-variable graphics

Bar graph, broken line graph, cumulative frequency curve, normal distribution, box chart.

Two-variable statistics (x, y)

All the quantities that are calculated for single-variable statistics are calculated for both the x and the y data. In addition Σxy is calculated.

There are 6 regression models which use the method of least squares: linear $y = a + bx$, logarithmic

$y = a + b \ln x$, logarithmic $y = a + b \log x$, exponential $y = ae^{bx}$, power $y = ax^b$ and inverse $y = a + bx^{-1}$. For each of these the values of a, b and r, the correlation coefficient, are calculated.

Two-variable graphics

Scatter diagram and line (or curve) of best fit.

Programming

Program control commands include:

If
conditional test.

Label
define a label.

Goto
go to a label.

Gosub
go to a subroutine.

Return
returns from a subroutine.

4 TECHNIQUES AND APPLICATIONS

1 Approximate and Exact Answers

As you will be aware, numerical answers are sometimes required to be approximate and stated to a specified level of accuracy (e.g. 3 significant figures) and sometimes they are required to be given exactly (e.g. $\sqrt{5}$, $\pi/7$, $3e^2$). An approximate answer is of course what is generally obtained on a calculator; if this is what is wanted then all you have to ensure is that your answer is correct to the specified number of significant figures or decimal places. Exact solutions are generally harder to obtain from a calculator since it can only ever work internally to a finite number of decimal places. The two following subsections contain precautions and techniques when using the graphic calculator to obtain approximate answers and also some techniques for arriving at exact answers.

1.1 Approximate Answers

(i) Using value memories

In any problem that involves a number of calculations which are interdependent it is important to avoid a build-up of rounding errors which occurs when an intermediate result is rounded off before using it in subsequent calculations. It is worth developing the

habit of storing in a 'value memory' any result that is to be used again and recall it from there when needed. Every graphic calculator has over 20 value memories so there are always enough! Rounding off an answer should be done as the very last step.

(ii) Iterative methods

When using an iterative formula (see 2.2(a)(ii)) look carefully at the successive values that appear each time 'Execute' is pressed. If, for example, the answer is required to 4 decimal place accuracy, then keep going until it is absolutely clear that the 4th decimal place has stopped changing (this will certainly be so if the 5th decimal place has also stabilized).

1.2 Exact Answers

(i) Identifying an answer in decimal form as the root of a rational number

Raise the number to the power of 2, then 3, then 4, then 5, and so on until the answer is either an integer or a terminating decimal or a recurring decimal. (Converting from a decimal to a fraction is covered in (ii).) The quickest way of carrying out this repeated trial is to use the last command memory after each trial so that the only change you need to make each time is to increase the power of the number by one.

Example
Starting with the decimal 1.801 983 127
1.801 983 127^2 EXE   3.247 143 191
1.801 983 127^3 EXE   5.851 297 239
1.801 983 127^4 EXE   10.543 938 9
1.801 983 127^5 EXE   18.99999998
Therefore $1.801\ 983\ 127 = \sqrt[5]{19}$

(ii) Identifying an answer in decimal form as a rational number

Multiply the number by successive integers 2, 3, 4, 5 and so on until the result is an integer (or a very close approximation to one).

Example
Starting with the number 0.272 727 273 4 it is found that

$$0.272\ 727\ 273\ 4 \times 11 = 3.000\ 000\ 007$$

It is very likely that the exact answer that is required is $\frac{3}{11}$.

Note: The Texas TI-82 has the command ►*Frac* (on *MATH* menu) which automatically converts, where possible, a decimal to a fraction.

(iii) Identifying an answer in decimal form as an exact multiple of a given constant

The 'given constant' is frequently π or e or the root of an integer. The technique is to divide the answer by the given constant and then if necessary apply (i) or (ii) to the result.

(iv) Identifying an answer in decimal form as a natural logarithm

If a question is given in terms of e then it is possible that the exact answer can be expressed as the natural logarithm of a 'simple' number. To test this, raise e to the power of the answer.

Example
Find the exact solutions of the equation

$$2e^{2x} - 11e^x + 15 = 0$$

Using a calculator to solve the equation numerically the answers are 0.916 290 731 9 and 1.098 612 289. Raising e to each of these in turn gives the results 2.5 and 3 respectively, which means that the exact answers are ln 2.5 and ln 3.

2 Equations

2.1 Solving Equations Graphically

(a) One Equation in One Unknown

Method 1   Solving f(x) = 0

The principle used here is that the solutions of the equation $f(x) = 0$ are given by the values of x at which the graph of $y = f(x)$ intersects the x-axis, i.e. the 'roots' of f(x).

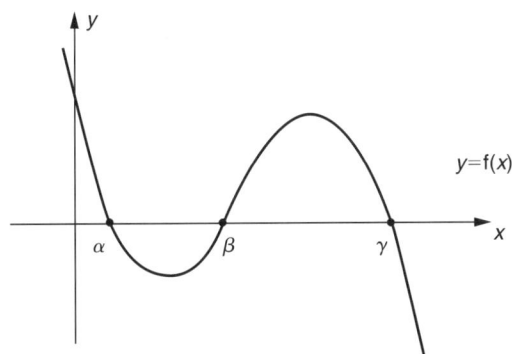

α, β and γ are the solutions of the equation $f(x) = 0$.

In practice you therefore rearrange an equation into the form $f(x) = 0$, plot the graph of $y = f(x)$ and locate the roots.

Texas TI-82: Use **Root** command (on **CALC** menu). Mark a lower bound and an upper bound with the cursor, then make a guess at the root by positioning the cursor between the two bounds or on one of the bounds.

Casio fx-7700GE: Use **Zoom** and **Trace** for each

root. For an immediate return to the whole graph following each root, use **ORG** (on **Zoom** menu).

Casio fx-9700GE: Use **ROOT** command (on **G-SOLV** menu). The calculator locates the lowest root first then moves to the next root to the right of it when the right-hand arrow is pressed.

Sharp EL-9300/9200: Use **X INCPT** command (on **JUMP** menu). The calculator locates the lowest root first or, if the cursor is displayed, the first one to the right of it.

Sharp EL-9300: Use SOLVER mode and select **GRAPHIC** method. Enter a lower and upper bound for the search for solutions and an initial guess for each one.

Method 2   Solving f(x) = g(x) directly

The solutions will be the x-coordinates of the points of intersection of the graphs of $y = f(x)$ and $y = g(x)$. See the following section for details of finding solutions of two equations in two unknowns.

(b) Two Equations in Two Unknowns – 'Simultaneous Equations'

The principle here is that each solution of the two equations

$$y = f(x)$$
$$y = g(x)$$

is given by the x and y coordinates of the points of intersection of the two graphs $y = f(x)$ and $y = g(x)$.

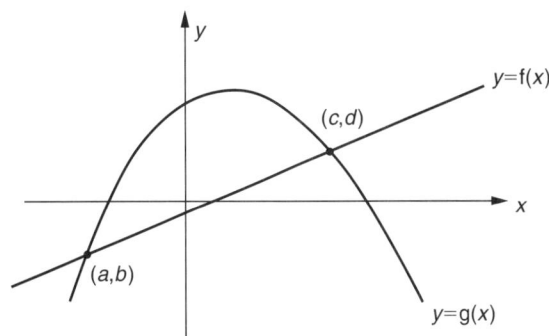

$x = a, y = b$ and $x = c, y = d$ are the solutions of the simultaneous equations $y = f(x)$ and $y = g(x)$.

So, in practice, rearrange each equation to the form $y = function\ of\ x$ then plot both curves and find their points of intersection.

Texas TI-82: Use **Intersect** command (on **CALC** menu). Mark each of the curves with the cursor then make a guess at the point of intersection by positioning the cursor near to it.

Casio fx-7700GE: Use **Zoom** and **Trace** for point of intersection. For an immediate return to the whole graph following each point of intersection, use **ORG** (on **Zoom** menu).

Casio fx-9700GE: Use **ISCT** command (on **G-SOLV** menu). Select the two curves using the up and down arrows. The calculator then finds the coordinates of the point of intersection furthest to the left. Move to the next point of intersection by pressing the right hand arrow.

Sharp EL-9300/9200: Use **INTERSEC** command (on **JUMP** menu). The calculator locates the intersection furthest to the left on the screen or, if the cursor is displayed, the first one to the right of it on the particular curve that the cursor is on.

2.2 Solving Equations Numerically

(a) One Equation in One Unknown

(i) Use a specific calculator command

Texas TI-82: Rearrange the equation to the form f(x) = 0 then use **Solve(** command (on **MATH** menu). In full the command is **Solve(** f(X), X, a) where a is a guess at the solution.

Casio fx-7700GE: For quadratic equations use EQUATION mode. Select **PLY**, enter the coefficients a, b and c then select **SOL** which produces the two solutions (which can be complex numbers).

Casio fx-9700GE: For quadratic or cubic equations use EQUATION mode. Select **POLY**, select **2** or **3** (for quadratic or cubic), enter the coefficients a, b, c (and d if cubic) then select **SOLV** which produces the solutions (which can be complex numbers).

Sharp EL-9300: Select SOLVER mode. Enter the equation in any form (it need not have zero on one side), assign values to all but one of the 'variables' in the equation. The calculator will then, if possible, solve the equation by simple algebraic rearrangement, or if it cannot do this it will use an iterative method which requires a starting point and a step size to be entered.

(ii) Use an iterative formula

Find an iterative formula x_{i+1} = f(x_i) either by direct algebraic arrangement of the given equation, for example

$$x^3 + x - 3 = 0 \text{ gives } x_{i+1} = \sqrt[3]{(3 - x_i)}$$

or by using the Newton–Raphson method, for example applying it to the equation $x^3 + x - 3 = 0$ gives the iterative formula

$$x_{i+1} = (2x_i^3 + 3)/(3x_i^2 + 1)$$

In practice, the procedure is carried out by inputting a number into the 'Ans' memory (by typing a number and pressing the 'ENTER' or 'EXE' key) then typing the right-hand side of the iterative formula with **Ans** taking the place of x_i. This command is then executed repeatedly until the sequence of numbers converges to a sufficiently good approximation to the solution. In the case of the iterative formula $x_{i+1} = (2x_i^3 + 3)/(3x_i^2 + 1)$ if a starting value of 1.5 is used then the procedure would be:

1.5	EXE
$(2\mathbf{Ans}^3 + 3)/(3\mathbf{Ans}^2 + 1)$	EXE
	EXE and so on.

(iii) Quadratic equations ($ax^2 + bx + c = 0$)

As well as the specific calculator commands already mentioned that could be used to solve a quadratic equation, there is of course the quadratic formula. On any calculator the coefficients a, b and c can be stored in memories A, B and C. Each (real) solution is produced by a single command which is the quadratic formula itself.

Example

$3x^2 - 4x - 9 = 0$	
$3 \to A$	EXE
$-4 \to B$	EXE
$-9 \to C$	EXE
$(-B + \sqrt{(B^2 - 4AC)})/(2A)$	EXE

which gives the answer 2.522 588 121.

The second solution is quickly obtained by recalling the last command, changing the '+' to '−' then executing it. If the formula is entered correctly then the possibility of making an error, which is commonly a sign error, when evaluating the formula is eliminated. On the Texas TI-82 or any of the Casio models, the formula can be entered into a function memory – then whenever a quadratic equation is to be solved the function name itself is recalled then executed as a command to produce the answer.

(b) Simultaneous Equations

(i) Use a specific calculator command

Casio fx-7700GE: Use EQUATION mode to solve 2 or 3 linear simultaneous equations. Select **S12** or **S13**, enter the coefficients and the right-hand sides of each equation then select **SOL** to produce the solutions.

Casio fx-9700GE: Use EQUATION mode to solve up to 6 linear simultaneous equations. Select **SIML** followed by the number of equations to be solved. Enter the coefficients and the right-hand sides of each equation then select **SOLV** to produce the solutions.

(ii) Use matrices to solve *n* linear simultaneous equations

A set of *n* simultaneous equations can be represented in matrix form as $[A][X] = [B]$ where $[A]$ is an $n \times n$ square matrix with each row consisting of the coefficients of one of the equations, $[X]$ is an $n \times 1$ matrix consisting of the *n* variables, and $[B]$ is an $n \times 1$ matrix consisting of the constants from the right-hand side of each equation. Pre-multiplying both sides by $[A]^{-1}$, the inverse of $[A]$, gives $[X] = [A]^{-1}[B]$. The technique is therefore to enter the matrices $[A]$ and $[B]$ then multiply the inverse of matrix $[A]$ by matrix $[B]$ to give the solution.

Example
Find the equation of the plane which contains the points $(-3, -1, -3)$, $(1, 1, -1)$ and $(2, 5, 1)$.

Solution
Each point must satisfy the equation $ax + by + cz = 1$. This produces the 3 equations

$$-3a - b - 3c = 1$$
$$1a + 1b - 1c = 1$$
$$2a + 5b + 1c = 1$$

The matrix $[A]$ is therefore $\begin{bmatrix} -3 & -1 & -3 \\ 1 & 1 & -1 \\ 2 & 5 & 1 \end{bmatrix}$ and

the matrix $[B]$ is $\begin{bmatrix} 1 \\ 1 \\ 1 \end{bmatrix}$.

The matrix product $[A]^{-1}[B]$ produces the matrix

$\begin{bmatrix} 1/6 \\ 1/4 \\ -7/12 \end{bmatrix}$ which are the values of a, b and c.

Multiplying through by 12 gives the equation of the plane as

$$2x + 3y - 7z = 12$$

Texas TI-82: Define the dimensions and enter the elements of matrices $[A]$ and $[B]$ via the **MATRIX − EDIT** menu. The calculation is entered as $[A]^{-1}[B]$, each matrix name being obtained from the **MATRIX − NAMES** menu.

Casio fx-7700GE/9700GE: Use MATRIX mode. Select **LIST** then **DIM** to define the dimensions of each matrix. Select **EDIT** to enter the elements of each matrix. The calculation is entered as

Mat A^{-1} Mat B, '**Mat**' being obtained from the main matrix menu and A and B from the keyboard in the usual way.

Sharp EL-9300/9200: Use MATRIX mode. Define the dimensions and enter the elements of matrices $[A]$ and $[B]$ via the **MATRIX − EDIT** menu. The calculation is entered as mat A^{-1} mat B, all of which can be obtained directly from the keyboard.

2.3 The Roots of Quadratic Equations

The essential point to appreciate is that a quadratic equation can be written in terms of its two roots as follows:

$$x^2 - (\text{sum of roots})x + (\text{product of roots}) = 0$$

Example
The equation $x^2 + 7x + 11 = 0$ has roots α and β. Find a quadratic equation with integer coefficients having roots $\alpha^2 + 3\beta$ and $\beta^2 + 3\alpha$.
(See Chapter 4, Worked Example 7.)

Solution
Find the two roots of $x^2 + 7x + 11 = 0$ and store them in memories A and B. Evaluate $A^2 + 3B$ and $B^2 + 3A$ and store these values in memories C and D. The required equation will be $x^2 - (C + D)x + CD = 0$.

Note
(i) If the appropriate routines are used on any of the calculators (see above) for obtaining the roots of the original equation, the values of $C + D$ and of CD are obtained exactly. Using zoom and trace to find the original roots will not be accurate enough for this sort of question.
(ii) On all but the Texas TI-82 a mode change is necessary before the values of α and β can be accessed.

2.4 Demonstrating the Existence of Solutions of One Equation in One Unknown

(a) Graphically

Rearrange the equation into the form $f(x) = g(x)$ such that both $f(x)$ and $g(x)$ are functions whose graphs are reasonably 'standard'. Solutions will correspond to the x-coordinates of any points of intersection of the graphs.

Example
This graph demonstrates that the equation $x - 2 \sin x = 0$ has only one positive solution, α.

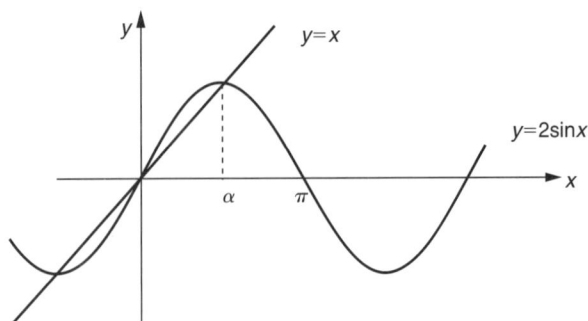

(b) Numerically

To show that the equation $f(x) = 0$ has a solution between $x = a$ and $x = b$ it is necessary to show that $f(a)$ and $f(b)$ have opposite signs. If the values of a and b are given in the question it is simply a matter of evaluating the function at the given values. If you are first required to find the values of a and b (often successive integers) between which a root lies then either locate the root graphically (see 2.1), from which the two values are easily found, or use your calculator to tabulate the function (Texas TI-82 or Casio fx-9700GE only – see details under 'Functions' in Section 3 of this chapter).

2.5 Solving an Equation Expressed in Terms of Transformed Functions

Note: This can only be done directly on the Texas TI-82.

Example
Solve the equation $fg(x) = 2f(x)$ given that

$$f(x) = 4x^2 - 1 \text{ and } g(x) = \sqrt{(x + 6)}$$

(See Chapter 5, Exercise 7(c).)

Solution
Define the functions $Y_1 = 4x^2 - 1$, $Y_2 = \sqrt{(x + 6)}$ and $Y_3 = Y_1(Y_2) - 2Y_1$ then find the required root of Y_3. (Plot the graph of Y_3 then use the **Root** command – see 2.1.)

3 Inequalities

3.1 Solving Inequalities

There are three main alternatives:

(a) Express the inequality in the form $f(x) > 0$ (or $<$ or $\geq$ or $\leq$) and find the roots of $f(x)$ graphically (see 2.1(a)), from which the solution of the inequality is easy.
(b) Express the inequality in the form $f(x) > g(x)$ (or $<$ or $\geq$ or $\leq$), plot both functions and find the x-coordinates of their points of intersection

graphically (see 2.1(b)). The solution of the inequality can then be written down.
(c) Use a calculator command specific to inequalities. For details see Section 3 of this chapter.

4 Algebra

4.1 Factorizing Polynomials

This is based on the factor theorem which states that if $f(a) = 0$ then $(x - a)$ is a factor of $f(x)$, and if $f'(a) = 0$ also then $(x - a)$ is a repeated factor of $f(x)$ (see Chapter 3, Fact Sheet). The technique therefore is to find all the roots graphically (see 2.1(a)), which may be repeated roots of the function, and hence write down the factors. (For details of repeated roots see Chapter 3, Fact Sheet.) It may be necessary to multiply one or more of the factors by an appropriate constant in order to make the factorized form equal the original function.

Example
Factorize $3x^3 + x^2 - 38x + 24$.

Solution
The roots of the function are found to be $x = 3, \frac{2}{3}$ and -4. Therefore, by the factor theorem $(x - 3)$, $(x - \frac{2}{3})$ and $(x + 4)$ are factors. By considering the coefficient of x^3 it is easy to see that the second of these needs to be multiplied by 3, giving

$$3x^3 + x^2 - 38x + 24 = (x - 3)(3x - 2)(x + 4)$$

4.2 Completing the Square

If the vertex of the graph of the quadratic function $ax^2 + bx + c$ has coordinates (p, q) then the completed square form of the function is $a(x - p)^2 + q$.

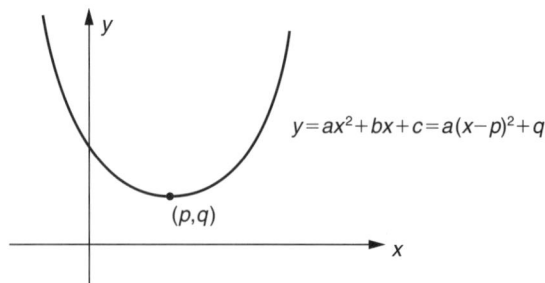

The method therefore is to plot the quadratic function and find the coordinates of the vertex. The Texas TI-82, Sharp EL-9300 and 9200 and the Casio fx-9700GE have commands to do this automatically (see 5.3).

4.3 Partial Fractions

Easy to *check* by plotting two graphs – firstly the expression to be split into partial fractions and secondly the partial fractions themselves (as a single function). If the graphs appear to be indistinguishable then it is almost certain that the partial fractions are correct.

5 Differentiation

5.1 Checking a Derived Function

It is easy to check whether you have differentiated a given function correctly by plotting two graphs – firstly the calculator's version of the derived function (which it calculates numerically on the basis of the given function) and secondly the derived function that you have found by differentiating the given function. If the two graphs are indistinguishable then it is almost certain that you are correct.

Suppose the derivative of a function f(x) is to be checked and f'(x) is what you believe to be the correct answer.

Texas TI-82: Use **nDeriv(** command (on **MATH** menu). Define Y_1 = **nDeriv(** f(X), X, X) and Y_2 = f'(X).

Casio fx-7700GE and fx-9700GE: Use **d/dx** command. Define Y1 = **d/dx(** f(X), X) and Y2 = f'(X).

Sharp EL-9300/9200: Use **d/dx(** command (on **MATH − CALC** menu). Define Y1 = **d/dx(** f(X)) and Y2 = f'(X).

5.2 Evaluating the Derived Function/Calculating the Gradient at a Point

Texas TI-82:
(i) **nDeriv(** f(X), X, a) evaluates the derivative of f(x) at $x = a$.
(ii) **dy/dx** (on **CALC** menu) evaluates the derivative at any point on a curve specified by **TRACE**.

Casio fx-7700GE:
(i) **d/dx(** f(X), a) evaluates the derivative of f(x) at $x = a$.

Casio fx-9700GE:
(i) **d/dx(** f(X), a) evaluates the derivative of f(x) at $x = a$.
(ii) **d/dx** (on **G-SOLV** menu) evaluates the derivative at any point on a curve specified by **Trace**.

Sharp EL-9300/9200:
(i) **d/dx(** f(X), a) evaluates the derivative of f(x) at $x = a$.
(ii) **Y' ON** (on GRAPH mode menu) evaluates the derivative at any point on a curve when it is traced.

5.3 Turning Points

Maxima and minima can be found on any graphic calculator with patient use of 'trace' and 'zoom' but it is a great deal quicker to use the specific commands listed here.

Texas TI-82: Use **Maximum** and **Minimum** (on **CALC** menu). For each one to be found mark, with the cursor, a lower bound and an upper bound on the curve then make a guess at the position of the turning point by positioning the cursor between the two bounds or on one of the bounds.

Casio fx-9700GE: Use **MAX** and **MIN** (on the **G-SOLV** menu). The right and left arrow keys move the cursor to the adjacent turning point of the same type.

Sharp EL-9300/9200: Use **MAX** and **MIN** (on **JUMP** menu).

5.4 Parametric and Polar Curves

Texas TI-82:
Parametric
(i) dy/dt and dx/dt can be calculated at a chosen value of t using the **nDeriv(** command. Specifically, to evaluate the derivative with respect to t of the parametric function f(t) at the point where $t = a$ the command is **nDeriv(** f(T), T, a). Combining the results for dy/dt and dx/dt in accordance with the chain rule enables the value of dy/dx at any value of t to be calculated.
(ii) **dy/dx**, **dy/dt** and **dx/dt** (all on the **CALC** menu) will calculate their respective values at points specified by **TRACE**.

Polar
(i) Differentiation of a polar function can be checked in a similar way to that of a cartesian function by using the **nDeriv(** command (on **MATH** menu). Suppose the derivative of a function f(θ) is to be checked and f'(θ) is what you believe to be the correct answer, then define r_1 = **nDeriv(** f(θ), θ, θ) and r_2 = f'(θ).
(ii) dr/dθ can be calculated at a chosen value of θ

using the **nDeriv(** command. Specifically, to evaluate the derivative with respect to θ of the polar function $f(\theta)$ at the point where $\theta = a$ the command is **nDeriv(** $f(\theta), \theta, a$).

(iii) **dy/dx** and **dr/dθ** (all on the **CALC** menu) will calculate their respective values at points specified by **TRACE**.

Casio fx-9700GE:
Parametric
d/dx (on the **G-SOLV** menu) enables the value of dy/dx to be displayed when the function is traced.

Polar
d/dx (on the **G-SOLV** menu) enables the value of $dr/d\theta$ to be displayed when the function is traced.

6 Integration

6.1 Checking an Integral Function

It is easy to check whether you have integrated a given function correctly by exploiting the fact that integration is the reverse of differentiation. Suppose you wish to check that your answer, $g(x)$ say, is the correct integral of a given function $f(x)$. If it is correct then the derived function of $g(x)$ will be precisely the function $f(x)$. The technique therefore is to plot two graphs – firstly the calculator's version of the derived function of your answer and secondly the function that you were originally given to integrate (see 5.1). If the two graphs are indistinguishable then it is almost certain that your answer is correct. For example, to check that $\sin 2x + 2x$ is the correct integral (arbitrary constant excepted) of $4 \cos^2 x$ the two graphs to plot would be the calculator's version of the derived function of $\sin 2x + 2x$ and the graph of $4 \cos^2 x$.

6.2 Evaluating a Definite Integral

(i) Using a specific calculator command (answers are approximate in general)

Texas TI-82:
(i) **fnInt(** $f(X), X, a, b$) evaluates the definite integral of $f(x)$ between limits a and b. (**fnInt(** is on the **MATH** menu.)
(ii) **∫f(x)dx** (on **CALC** menu) calculates and shades the definite integral between two points specified using trace.

Casio fx-7700GE and fx-9700GE:
(i) **∫dx(** $f(X), a, b$) evaluates the definite integral of $f(x)$ between limits a and b.
(ii) **G-∫dx(** $f(X), a, b$) evaluates the definite

integral of $f(x)$ between limits a and b and shades the relevant area on a graph.

Sharp EL-9300/9200:
∫f(X), a, b dx evaluates the definite integral of $f(x)$ between limits a and b. (**∫** and **dx** are on the **MATH − CALC** menu.)

(ii) Evaluating integral functions accurately

Many questions involving a definite integral will either require an exact answer and/or will want the final answer to be obtained in the 'traditional' way so obliging you to find an integral function and then substitute the limits into it to obtain the answer. Having obtained the correct integral function (for checking it see 6.1), it is important not to waste marks by making careless mistakes when evaluating it at the two values of x. This is where the function-handling features of your calculator can be useful.

Suppose the integral function is $g(x)$ and the limits are a and b. Then you need to find the values of $g(a)$ and $g(b)$.

Texas TI-82: Define $Y_1 = g(X)$. $g(a)$ is then evaluated by the command $Y_1(a)$. $g(b)$ is evaluated in a similar way. (Y_1 is on the **Y-VARS − Function** menu.)

Casio fx-7700GE and fx-9700GE: Define $Y1 = g(X)$. $g(a)$ is then evaluated by the command $Y1$, having entered the value a into the X memory. $g(b)$ is evaluated in a similar way. (Y is on the **VAR − GRP** menu on the fx-7700GE and on the **VAR − GRPH** menu on the fx-9700GE.)

Note: Alternative ways of evaluating functions are listed under 'Functions' for each calculator in Section 3 of this chapter.

6.3 Trapezium, Simpson's and Mid-Ordinate Rules

All of these rules involve evaluating a particular function at a number of different values of x then combining the values in the way specified by the rule. Questions on these can be done very quickly and accurately by:

(i) using the function-handling features of your calculator to evaluate the function (see 6.2)
(ii) storing each function value in a separate memory
(iii) using the stored values when applying the rule itself to give the final answer.

7 Graphs and Functions

7.1 Curve Sketching

The graphic calculator can obviously be of great use when answering any question or part of a question that requires a graph to be sketched, be it cartesian, parametric or polar. Remember though that a 'sketch' should include details of all important points on the curve, such as turning points and axis intercepts. You should therefore ensure that you are familiar with all the relevant calculator facilities – see 'Graphical Function Analysis' for your particular calculator in Section 3 of this chapter.

7.2 $y^2 = f(x)$

(i) Sketching $y^2 = f(x)$ given $y = f(x)$

To plot $y^2 = f(x)$ on a graphic calculator it is necessary to plot two separate functions, $y = +\sqrt{f(x)}$ and $y = -\sqrt{f(x)}$.

(ii) Sketching an ellipse or hyperbola

Rearrange the equation into the form $y^2 = f(x)$ then proceed as in (i).

7.3 Motion in 2 Dimensions (e.g. Projectiles)

If the x- and y-coordinates of a moving point are given as functions of time then its path can be plotted by entering the functions as a pair of parametric equations. As well as the usual x and y ranges that must be input for the graph, the range of values of t and its incremental change between those values must also be input. After plotting the path there are various details about the motion that the calculator can find (see 'Graphical Function Analysis' for your particular calculator in Section 3 of this chapter). These include:

– Position of point at any time – use the 'trace' facility.
– Direction of motion at any time: dy/dx is equal to $\tan\theta$ where θ is the angle the velocity makes with the horizontal.
– x component of velocity at any time: dx/dt.
– y component of velocity at any time: dy/dt.
– Speed at any time: $v = \sqrt{((dx/dt)^2 + (dy/dt)^2)}$.

Note: Careful choice of the t range and its increment may be needed in order to ensure that the calculator can give details of the motion at the particular times in question. For example, a range of t from 7.5 to 8 with an increment of 0.01 would be sufficient to gain details of the motion at $t = 7.5, 7.51, 7.52, 7.53,\ldots, 7.98, 7.99$ and 8.00.

7.4 Transformations of Graphs

See 'Graph Transformations' for your particular calculator in Section 3 of this chapter.

7.5 Evaluating Functions

See 'Evaluating Functions' for your particular calculator in Section 3 of this chapter.

7.6 Composite Functions

Note: Only the Texas TI-82 has an equivalent of the fg(x) notation – see 'Composite Functions' in the Texas TI-82 part of Section 3 of this chapter.

8 Sequences and Series

8.1 Checking a Polynomial Approximation to a Function (Binomial, Maclaurin, Taylor)

It is easy to check whether a polynomial approximation to a function is likely to be correct by plotting a graph for each. For a binomial or Maclaurin expansion the two graphs should appear to merge as the origin is approached; for a Taylor series they should appear to merge as they approach the value of x around which the approximation is based. For example, to check that $4x - 4x^2 + \frac{11}{3}x^3$ is the correct Maclaurin expansion up to x^3 of

$$f(x) = 2\ln(1 + x) + \frac{2x - x^2}{1 + x}$$

the following two graphs are plotted (see Chapter 10, Exercise 8(ii)):

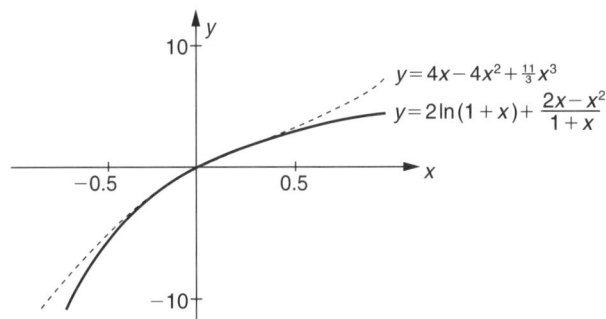

On the graphic calculator screen the two curves are indistinguishable for values of x between approximately -0.45 and $+0.5$ which suggests that the expansion is correct. Using the trace facility and switching between the two curves gives a clear numerical indication of how well matched the two functions are at different values of x.

8.2 Generating a Sequence

(i) Using the 'Answer' memory

A sequence defined inductively can be generated by using the value stored in the 'Answer' memory to represent the ith term of the sequence. For example, the sequence with $u_1 = 5$ and $u_{i+1} = 3u_i$ would require 5 to be entered into the ANS memory first then the command 3ANS to be executed repeatedly.

(ii) Using a specific calculator command

Texas TI-82:
(i) Use *seq(* command (on **LIST** menu) to generate a sequence from a general term. For example, to generate the first 12 terms of the sequence (a geometric progression in this case) with general term $2(3)^{r-1}$ the command is *seq(* $2(3)^{R-1}, R, 1, 12, 1)$. If this command is followed by **STO▶ L1** (which stores the terms in list L1) then an individual term, for example the 7th term, can be accessed with the command **L1**(7).
(ii) Use SEQUENCE mode to define inductively or from a general term or using a mixture of both. For example to generate the sequence defined as $u_n = 2u_{n-1} + 3$ with $u_1 = 4$ define the function $u_n = 2u_{n-1} + 3$ on the 'Y=' screen, then on the 'WINDOW' screen set UnStart = 4. The terms of the sequence can then be listed using the **TABLE** command.

Casio fx-9700GE:
Use TABLE mode to define a sequence either from a general term or from a first-order or second-order linear inductive relationship. For example, to generate the first 12 terms of the sequence (a geometric progression in this case) with general term $2(3)^{n-1}$, first specify the type of definition by selecting **F1** from the **TYPE** menu. Secondly input the general term (the *n* is on function key **F3**). Finally select **TABL** (on function key **F1**) which produces three columns, values of n (from 1 to 12), the terms themselves, and the sum of all the terms up to that point. A sequence defined inductively is input in a similar manner except that the first term (or first two terms for a second-order inductive definition) must also be input.

8.3 Summing a Series

Texas TI-82:
Use *sum seq(* command. (*sum* is on **LIST − MATH** menu, *seq(* is on **LIST − OPS** menu.) For example, to sum the first 12 terms of the series (a geometric

progression in this case) with general term $2(3)^{r-1}$ the command is *sum seq(* $2(3)^{R-1}, R, 1, 12, 1)$.

Casio fx-9700GE:
(i) Use **Σ(** command (on **MATH − PRB** menu). For example, to sum the first 12 terms of the series (a geometric progression in this case) with general term $2(3)^{r-1}$ the command is **Σ(** $2(3)^{R-1}, R, 1, 12)$.
(ii) In TABLE mode the sum of all the terms up to each term of a series is automatically displayed in each row of the table (see 8.2).

9 Trigonometry

9.1 Solving Trigonometrical Equations

If approximate answers are required (e.g. 3 significant figures, 2 decimal places) then the techniques described in 2.1(a) and 2.2(a) can be used directly. If exact answers are required the same techniques can be used but the answers given by the calculator will, unless already exact, need to be 'adjusted'. The following points are worth noting in connection with this:

(a) Nearly all answers in trigonometrical equations on A Level papers, when expressed in degrees, are either integers or involve very simple fractions of a degree. The calculators which have a specific routine for locating roots will either find them exactly, or at worst get close enough for the answer to be obvious (e.g. the calculator may give $133.499\,99°$ as an answer – the exact answer is almost certainly $133.5°$)
(b) If exact answers are required in radians 'in terms of π' then it is probably easier to let the calculator find each answer exactly in degrees, as described in (a), then convert to an exact multiple of π as follows.

Suppose the answer is $x = 25°$; then this becomes $\frac{25}{180}\pi$ radians. The fraction should be simplified if possible – all the Casio calculators will do this when $\frac{25}{180}$ is entered using the fraction key and executed.

In the case of the Texas TI-82, exact answers in terms of π can be found easily without resorting to working in degrees. Find the root graphically as above, divide by π, then use the 'convert to fraction' command (on the **MATH** menu).

Example
Find the first solution in the range 0 to 2π of the equation
$$\cos 2\theta + \sqrt{3} \sin 2\theta = 2 \cos \theta$$
(see Chapter 6, Worked Example 14).

Solution

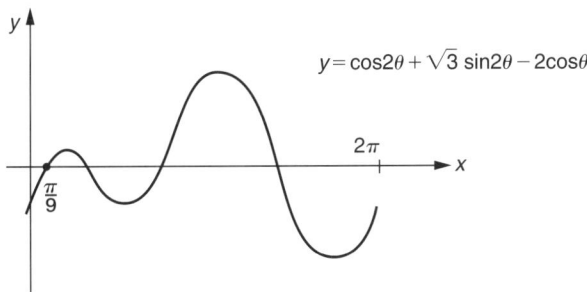

$$y = \cos2\theta + \sqrt{3}\,\sin2\theta - 2\cos\theta$$

Root	gives the answer	$X = .349\,065\,85$
$X/\pi \blacktriangleright \textbf{Frac}$	gives the answer	1/9

So the first solution in the range is $\pi/9$.

9.2 R, α Method

It follows from the general results about transformations of $\sin\theta$ and $\cos\theta$ (Chapter 6, Fact Sheet) that the graph of $R\sin(x + \alpha)$ has a maximum value of R and is a translation of $\sin x$ by $-\alpha$ in the x-direction and that the graph of $R\cos(x + \alpha)$ has a maximum value of R and is a translation of $\cos x$ by $-\alpha$ in the x-direction. Therefore, the procedure for converting a function of the form $a\sin x \pm b\cos x$ to the form $R\sin(x + \alpha)$ or $R\cos(x + \alpha)$ is: plot the graph of $y = a\sin x \pm b\cos x$, find its maximum value (see 5.3), which is the value of R, and find a root (see 2.1(a) and 2.2(a)) from which the value of α can easily be calculated.

Example
Express $3\sin x + 4\cos x$ in the form

(i) $R\sin(x + \alpha)$
(ii) $R\cos(x + \alpha)$

where $R > 0$ and α is an acute angle.

Solution

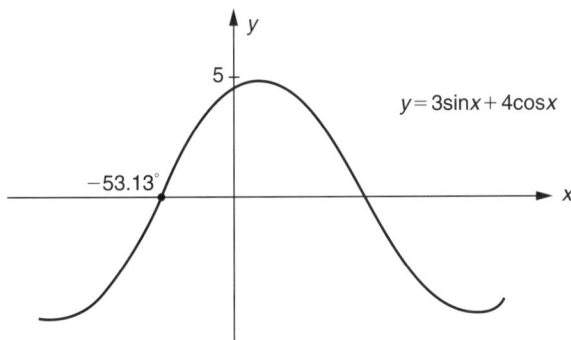

$$y = 3\sin x + 4\cos x$$

The maximum value is found to be 5 and the relevant root is $-53.13°$. This means that the graph can be regarded as a translation of $5\sin x$ by $-53.13°$ in the x-direction or as a translation of $5\cos x$ by $+36.87°$ in the x-direction ($5\cos x$ cuts the x-axis at $-90°$). The solutions are therefore

(i) $5\sin(x + 53.13°)$
(ii) $5\cos(x - 36.87°)$.

10 Statistics

10.1 Single-Variable Statistics

For details of which quantities are evaluated automatically by your particular calculator see Section 3 of this chapter. The present section describes how to input data, how to output the calculated values and how to display statistical graphics.

Texas TI-82:
Input
Select *Edit* (on *STAT* − *EDIT* menu) and input data into one of the lists. Use another list to represent frequencies if required.
Output
Select *SetUp* (on *STAT* − *EDIT* menu). Under '1-Var Stats' define which list is to be analysed and which list, if any, is to represent frequencies. Select *1-Var Stats* (on *STAT* − *EDIT* menu) to output all calculated values.
Graphics
Select type of statistical plot from the *STAT PLOT* menu then press *GRAPH*.

Casio fx-7700GE and fx-9700GE: Use SD mode.
Input
Use *DT* and *;*. For example 3 *;* 5 *DT* inputs 3 with a frequency of 5. *EDIT* allows the frequency distribution to be viewed and then edited using *INS* or *DEL* and the 4 arrow keys.
Output
All calculated quantities are output using *DEV* or Σ (both on the main SD menu).
Graphics
Use the command *Graph Y=* to draw a histogram.

Sharp EL-9300 and EL-9200: Use STATISTICS mode.
Input
Select the type of data (on *DATA FORMAT* menu) then input.
Output
Select *STAT − X-VARS* on main statistics menu to output all calculated quantities.
Graphics
Use STATISTICS GRAPH mode and select which type of plot from the main menu.

10.2 The Normal Distribution

Texas TI-82:
An area under the Normal curve can be found using the *fnInt(* command (on *MATH* menu). If the integral is stored as a function:
$Y_1 = 1/\sqrt{(2\pi)} * \textbf{fnInt(}\, e^{(-X/2)}, X, -5, X)$ then the

area between $x = a$ and $x = b$ can be found directly using the command $Y_1(b) - Y_1(a)$.

Casio fx-7700GE and fx-9700GE: Use SD mode. Probabilities associated with areas under the Normal curve are calculated when a standardized value (z) is input. There are 3 choices: *P(* z gives the area up to the chosen value of z, *Q(* z gives the area between $z = 0$ and the chosen value of z, *R(* z gives the area under the curve for values of z greater than the chosen value. The area corresponding to any of these probabilities can be shaded by preceding the command with *Graph Y=*.

Sharp EL-9300 and EL-9200: Use REAL mode. An area under the Normal curve can be found using the $\int$ and *dx* commands (on *MATH − CALC* menu.) The area between $x = a$ and $x = b$ would be found with the command
$1/\surd(2\pi) * \int e^{(-X/2)}, -5, X\ dx$.

10.3 Linear Regression

For details of which quantities are evaluated automatically by your particular calculator see Section 3 of this chapter. The present section describes how to input data, how to output the calculated values and how to display statistical graphics.

Texas TI-82:
Input
Same as for single-variable except input values of x and y into 2 separate lists and their frequencies (if not all equal to 1) into a third list.

Output
Select *SetUp* (on *STAT − EDIT* menu). Under '2-Var Stats' define which list is to be analysed and which list, if any, is to represent frequencies. Select *2-Var Stats* (on *STAT − EDIT* menu) to output all calculated values. The required regression model is selected from the same menu; on making the selection the values of a, b and r are output.
Graphics
Same as for single-variable. The equation of the line (or curve) of best fit is accessed from the *VARS-Statistics-EQ* menu.

Casio fx-7700GE and fx-9700GE: Use REG mode.
Input
Same procedure as for single-variable.
Output
Select required regression model from *SET UP* menu. All calculated quantities are output using *DEV* or Σ or *REG* (all on the main *REG* menu).
Graphics
Use the command *Graph Y = Line* 1 to draw a scatter diagram and line (or curve) of best fit.

Sharp EL-9300 and EL-9200:
Input
Use STATISTICS mode. Select the type of data (on *DATA FORMAT* menu) then input.
Output
Select *STAT − X-VARS* or *STAT − Y-VARS* or *STAT-REG* (all on main *Statistics* menu) to output all calculated quantities.
Graphics
Use STATISTICS GRAPH mode. Scatter diagram and line (or curve) of best fit are drawn when the required regression model to be graphed is selected (from *REG* menu). When the graph is complete, pressing *MENU* then selecting *REG* outputs the values of a, b and r for the chosen regression model.

Basic algebra and arithmetic

2

AS Level			A Level			Topic	Date attempted	Date completed	Self-assessment
CORE	MODULAR	TRADITIONAL	CORE	MODULAR	TRADITIONAL				
✓	✓	✓	✓	✓	✓	**Numbers**			
✓	✓	✓	✓	✓	✓	**Indices**			
✓	✓	✓	✓	✓	✓	**Logarithms**			
✓	✓	✓	✓	✓	✓	**Derivation of equations**			

Fact Sheet

Numbers

- Natural $\mathbb{N}$ $\{1, 2, 3, \ldots\}$
- Integers $\mathbb{Z}$ $\{0, \pm 1, \pm 2, \ldots\}$
- Rational $\mathbb{Q}$ any number which can be expressed as a fraction, for example $\{\frac{3}{7}, -\frac{2}{3}, \frac{9}{5}, \frac{8}{2}\}$
- Irrational   a real number which cannot be expressed as a fraction, for example $\{\pi, \text{e}, \sqrt{3}\}$
- Real $\mathbb{R}$ any element of the sets Rational and Irrational
- Imaginary $\mathbb{I}$ square roots of negative numbers, for example $a\text{i}$ or $a\text{j}$ where i and/or j represent $\sqrt{-1}$ and $a \in \mathbb{R}$
- Complex $\mathbb{C}$ numbers of the form $a + b\text{i}$ or $a + b\text{j}$

Indices or Powers

$2^{-8}, 3^4, 5^{0.3}$ are numbers expressed in index form. The indices are $-8, 4, 0.3$, the base numbers are $2, 3, 5$, respectively.

Rules of Indices

- $a^m \times a^n = a^{m+n}$
- $a^m / a^n = a^{m-n}$
- $a^{-m} = 1/a^m$
- $a^0 = 1$
- $a^{m/n} = \sqrt[n]{a^m}$
- $(a^m)^n = a^{mn} = (a^n)^m$
- $a^{1/n} = \sqrt[n]{a}$
- $(ab)^n = a^n b^n$

Numbers represented in index form are called **exponentials**. 2^x is an exponential function; e^x is *the* exponential function.

Logarithms

Logarithms are the inverse of exponentials.
$y = 2^x \Rightarrow x = \log_2 y$
$y = a^x \Rightarrow \log_b y = x \log_b a$

Rules of Logarithms

(i) $\log_a bc = \log_a b + \log_a c$
(ii) $\log_a (b/c) = \log_a b - \log_a c$
(iii) $\log_a b^c = c \log_a b$

(iv) $\log_a b = \dfrac{\log_c b}{\log_c a}$

(v) $\log_a b = \dfrac{1}{\log_b a}$

Surds

$\sqrt{a} \times \sqrt{b} = \sqrt{ab}$ or $a^{\frac{1}{2}} \times b^{\frac{1}{2}} = (ab)^{\frac{1}{2}}$

$\sqrt{\dfrac{a}{b}} = \dfrac{\sqrt{a}}{\sqrt{b}}$

$\dfrac{1}{\sqrt{a}} = \dfrac{\sqrt{a}}{a}$

Proportion

- Direct proportion $x = ky$

- Inverse proportion $\quad x = \dfrac{k}{y}$

- Inverse square law $\quad x = \dfrac{k}{y^2}$

where k is the constant of proportionality

Standard Form

Any real number can be expressed as $a \times 10^n$ where $1 \leqslant a < 10$ and n is an integer. For example, $42\,000 = 4.2 \times 10^4$, $\quad 0.000\,071\,6 = 7.16 \times 10^{-5}$.

Worked Examples

1 Show that $17\left(1 - \dfrac{1}{17^2}\right)^{\frac{1}{2}} = n\sqrt{2}$ where n is an

integer.

- $17\left(1 - \dfrac{1}{17^2}\right)^{\frac{1}{2}}$ $= (17^2 - 1)^{\frac{1}{2}}$ Using $a\sqrt{b} = \sqrt{a^2 b}$

$= (288)^{\frac{1}{2}}$
$= (144 \times 2)^{\frac{1}{2}}$
$= 12\sqrt{2}$ $\boxed{1.2}$

2 Simplify $\sqrt{3} + \sqrt{12} + \sqrt{108} - \sqrt{75}$.

- $\sqrt{3} + \sqrt{12} + \sqrt{108} - \sqrt{75}$
$= \sqrt{3} + \sqrt{4 \times 3} + \sqrt{36 \times 3} - \sqrt{25 \times 3}$
$= \sqrt{3} + 2\sqrt{3} + 6\sqrt{3} - 5\sqrt{3}$
$= 4\sqrt{3}$ $\boxed{1.2}$

3 Find the exact value of x without the use of a calculator given that

$\dfrac{2^{x+5}}{8^x} = \dfrac{4^{x-1}}{2^{2x-1}}$.

- 4 and 8 can be expressed as powers of 2
$4^{x-1} = 2^{2x-2}, 8^x = 2^{3x}$.

Hence $\dfrac{2^{x+5}}{2^{3x}} = \dfrac{2^{2x-2}}{2^{2x-1}}$ Subtract the indices for division

$2^{5-2x} = 2^{-1}$ Equate the indices
$5 - 2x = -1$
$x = 3$ $\boxed{\text{2.1a and 2.2a}}$

4 Solve $27^x = 9^{x-1}$
   (a) by indices
   (b) by logarithms   (L, 1992)

- (a) $27 = 3^3$ and $9 = 3^2 \Rightarrow 3^{3x} = 3^{2x-2}$ Equate
$3x = 2x - 2 \Rightarrow x = -2.$ the indices

   (b) Take logs of both sides
$\log 27^x = \log 9^{x-1} \Rightarrow x \log 27 = (x - 1) \log 9^*$
Either use your calculator here, though using log or ln can be inaccurate, or use log theory to simplify the equation

$x(1.4314) = (x - 1)(0.9542)$
$x(1.4314 - 0.9542) = -0.9542$
$x = -1.9996 = -2$ to 3 d.p.

* $or \log 27 = \log 3^3 = 3 \log 3$
$\log 9 = \log 3^2 = 2 \log 3$
Hence $3x \log 3 = 2(x - 1) \log 3$
$3x = 2x - 2$
$x = -2$

or   logs may be taken to base 3

5 The variables x and y satisfy a relationship of the form $y = ax^b$, where a and b are constants. Given the table below, draw a suitable straight line graph and estimate the values of a and b, giving your answers to two significant figures.

x	2	3	4	5	6
y	6.32	7.24	7.98	8.60	9.12

(AEB, 1992)

- Since the relationship involves multiplication and powers and *no* additions or subtractions, take logs of both sides

$\log y = \log a + b \log x$
$Y = bX + \log a$ where $X = \log x$, $Y = \log y$

Plot the graph of Y against X.

X	0.301	0.477	0.602	0.699	0.778
Y	0.801	0.860	0.902	0.934	0.96

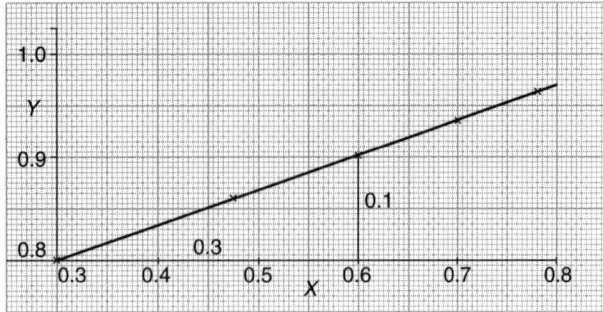

x	57.9	108.2	227.9	778.3
T	0.24	0.62	1.88	11.86

Note the false origin   GC

Gradient = b = 0.33   (2 s.f.)

   $\log a = Y - bX = 0.7$ (by using the point
                                      (0.6, 0.9) on the line)

      $a = 5.0$ (2 s.f.)

(log a would be the intercept on the Y axis if the X axis began at 0.)

Assuming a law of the form $T = Ax^n$, draw a graph of ln T against ln x. Estimate the values of A and n, giving your answers to two significant figures.

   Use your graph to estimate the approximate mean distance in millions of km of the Earth from the Sun.   (AEB, 1994)

5 Solve $3 \times 2^{x+5} = 5^{2x-1}$ to 3 significant figures.

6 Solve the equation $\log_x a + 2 \log_a x = 3$.

7 A curve with equation $ky = a^x$ passes through the points with coordinates (7, 12) and (12, 7). Find, to 2 significant figures, the values of the constants k and a.

Brief Solutions

1 (a) $\log_2 \dfrac{15 \times 40}{75} = \log_2 8 = 3.$

Exercises

1 (a) Simplify $\log_2 15 + \log_2 40 - \log_2 75$
   (b) Solve $2^{x+1} = 7$

   (c) Using the result $\log_a x = \dfrac{\log_b x}{\log_b a}$, solve the
        equation

        $\log_3 x + \log_4 x = 2$

        giving your answer to 3 significant figures.
                                           (OLE, 1993)

2 Given that $1 + \log_a (7x - 3a) = 2 \log_a x + \log_a 2$, find, in terms of a, the possible values of x.

3 (a) The variables p and q are related by the law

        $q = ap^b$ where a and b are constants

        Given that   ln p = 1.32 when ln q = 1.73
        and   ln p = 0.44 when ln q = 1.95
        find the values of b and ln a.
   (b) Given that $y = \log_2 x$ and that

        $\log_2 x - \log_x 8 + \log_2 2^k + k \log_x 4 = 0$

        prove that $y^2 + ky + (2k - 3) = 0$.
   (i) Hence deduce the set of values of k for which y is real.
   (ii) Find the values of x when $k = 1.5$. (AEB, 1986)

4 For certain planets, the approximate mean distance x, in millions of km from the centre of the Sun, and the period of the orbit T, in Earth years, are recorded.

   (b) $(x + 1) \log 2 = \log 7$

        $\Rightarrow x + 1 = \dfrac{\log 7}{\log 2} \Rightarrow x = 1.81$ to 3 s.f.

   2.1a and 2.2a

   (c) $\dfrac{\log_{10} x}{\log_{10} 3} + \dfrac{\log_{10} x}{\log_{10} 4} = 2$

        $\Rightarrow \log_{10} x \left[\dfrac{1}{\log_{10} 3} + \dfrac{1}{\log_{10} 4} \right] = 2$

        $\Rightarrow x = 3.41$ to 3 s.f.

2 $1 = \log_a \dfrac{2x^2}{7x - 3a} \Rightarrow a = \dfrac{2x^2}{7x - 3a}$

        $\Rightarrow 2x^2 - 7xa + 3a^2 = 0$

        $\Rightarrow x = 3a$ or $\dfrac{a}{2}$

3 (a) $q = ap^b \Rightarrow$ ln q = ln a + b ln p
        $\Rightarrow$ 1.73 = ln a + 1.32b
        1.95 = ln a + 0.44b

        $\Rightarrow b = \dfrac{-0.22}{0.88} = -\dfrac{1}{4}$, ln a = 2.06

   (b) $\log_x 8 = 3 \log_x 2 = \dfrac{3}{y}$,

        $\log_2 2^k = k$,

        $\log_x 4 = \dfrac{2}{y}$

Hence $y - \dfrac{3}{y} + k + \dfrac{2k}{y} = 0$,

$$y^2 + ky + (2k - 3) = 0$$

y is real if '$b^2 - 4ac \geqslant 0$', that is
$k^2 - 4(2k - 3) \geqslant 0 \Rightarrow (k - 6)(k - 2) \geqslant 0$
$\Rightarrow \quad k \leqslant 2$ or $k \geqslant 6$

When $k = 1.5$, $y = 0$ or -1.5
$\Rightarrow \quad x = 2^y = 1$ or $2^{-1.5}\ (= 0.354)$.

4 $\ln T = \ln A + n \ln x$. Gradient of the graph is n, y-intercept is $\ln A$.

When $T = 1$, $\ln x = -\dfrac{\ln A}{n} = 5.01 \quad \Rightarrow \quad x = 150$

$\therefore$ Distance of the Earth from the Sun is 150 million km (2 s.f.).

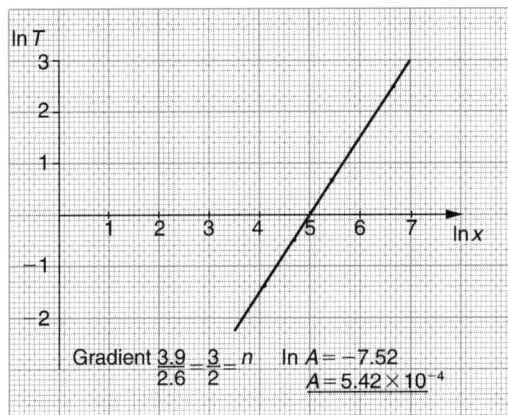

5 Take logs:
$\log 3 + (x + 5) \log 2 = (2x - 1) \log 5$

2.1a and 2.2a

$\log 3 + \log 32 + \log 5 = x(\log 25 - \log 2)$

$x = \dfrac{\log 480}{\log 12.5} = 2.44$ to 3 s.f.

6 $\log_x a = \dfrac{1}{\log_a x} \quad \Rightarrow \quad 1 + 2L^2 - 3L = 0 \quad L = \log_a x$

$\Rightarrow \quad (2L - 1)(L - 1) = 0$
$\Rightarrow \quad L = \frac{1}{2}$ or 1
$\Rightarrow \quad x = a^{\frac{1}{2}}$ or a

7 $ky = a^x \quad \Rightarrow \quad \ln k + \ln y = x \ln a$

$\Rightarrow \quad \ln k + \ln 12 = 7 \ln a$
$\quad \ln k + \ln 7 = 12 \ln a$

$\Rightarrow \quad \ln \dfrac{12}{7} = -5 \ln a$

$\Rightarrow \quad a = 0.90, k = 0.039$

3

Polynomials

AS Level			A Level			Topic	Date attempted	Date completed	Self-assessment
CORE	MODULAR	TRADITIONAL	CORE	MODULAR	TRADITIONAL				
✓	✓	✓	✓	✓	✓	**Algebra of polynomials**			
✓	✓	✓	✓	✓	✓	**Factorization**			
	✓	✓	✓	✓	✓	**Factor Theorem**			
	✓			✓	✓	**Remainder Theorem**			
	✓	✓		✓	✓	**Curve sketching of polynomials**			

Fact Sheet

A polynomial in x is an expression consisting of sums and differences of terms $a_r x^n$ where n is a positive integer or zero.

The highest value of n gives the degree of the polynomial.

- A linear polynomial can be written as $a_0 + a_1 x$ (degree 1)
- A quadratic polynomial can be written as $a_0 + a_1 x + a_2 x^2$ (degree 2)

Terms which have the same power can be added or subtracted.

Long Division

If a polynomial $f(x)$ is divided by a polynomial $g(x)$, of a lower degree than $f(x)$, the remainder is a polynomial of a lower degree than $g(x)$.

Example: If $x^5 - 3x^2 + x - 2$ (degree 5) is divided by $x^2 + 2x - 5$ (degree 2) the remainder will be of a lower degree than $x^2 + 2x - 5$, that is $ax + b$ (degree 1 or degree 0 if $a = 0$).

$$
\begin{array}{r}
x^3 - 2x^2 + 9x - 31 \\
x^2 + 2x - 5 \overline{)x^5 - 0x^4 + 0x^3 - 3x^2 + x - 2} \\
\underline{x^5 + 2x^4 - 5x^3} \\
-2x^4 + 5x^3 - 3x^2 \\
\underline{-2x^4 - 4x^3 + 10x^2} \\
9x^3 - 13x^2 + x \\
\underline{9x^3 + 18x^2 - 45x} \\
-31x^2 + 46x - 2 \\
\underline{-31x^2 - 62x + 155} \\
108x - 157
\end{array}
$$

$x^5/x^2 = x^3$

$\longleftarrow (x^2 + 2x - 5) \times x^3$

Subtract, bring down the next term and repeat the process

$(x^5 - 3x^2 + x - 2) \div (x^2 + 2x - 5) = x^3 - 2x^2 + 9x - 31$ remainder $108x - 157$

or

$$\frac{x^5 - 3x^2 + x - 2}{x^2 + 2x - 5} = x^3 - 2x^2 + 9x - 31 + \frac{108x - 157}{x^2 + 2x - 5}$$

Remainder Theorem

When a polynomial $f(x)$ is divided by $(x - a)$ the remainder is $f(a)$.
That is, $f(x) = (x - a)g(x) + f(a)$

Factor Theorem

- If $f(a) = 0$ then $(x - a)$ is a factor of $f(x)$.
- If $f(a) = 0$ and $f'(a) = 0$ then $(x - a)$ is a repeated factor of $f(x)$.

Curve Sketching of Polynomial Function $ax^n + bx^{n-1} + \ldots + c$

(i)   Factorize fully.
(ii) Each linear factor will give a point of intersection on the x-axis.
(iii) If $(x - d)^2$ is a factor of $f(x)$ then the curve will have a turning point at $x = d, y = 0$.
(iv) If $(x - d)^3$ is a factor of $f(x)$ then the curve will have a point of inflexion at $x = d, y = 0$.
(v)   If a is positive then the graph will satisfy $y \to +\infty$ as $x \to +\infty$.
(vi) If a is positive
    and (a) n is even then $y \to \infty$ as $x \to -\infty$
    or (b) n is odd then $y \to -\infty$ as $x \to -\infty$.

(vii) $f(x) = (x - a)(x - b)^2(x - c)^3$
where $a < b < c$

degree 6

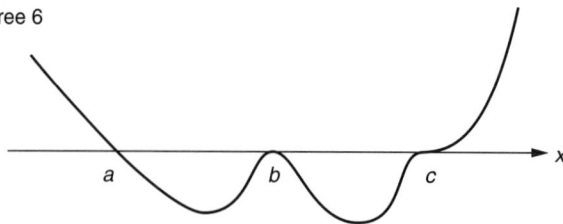

where $a < c < b$

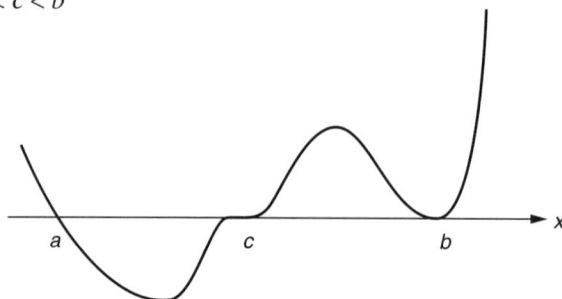

Worked Examples

1 The cubic function f is given by

$$f(x) = x^3 + ax^2 - 4x + b$$

where a, b are constants.

Given that $x - 2$ is a factor of $f(x)$ and that a remainder of 6 is obtained when $f(x)$ is divided by $(x + 1)$, find the values of a and b. (WJEC, 1992)

* $f(x) = x^3 + ax^2 - 4x + b$
  $x - 2$ is a factor $\Rightarrow$ $f(2) = 0$
  Hence $8 + 4a - 8 + b = 0$ $\Rightarrow$ $4a + b = 0$ (1)
  $f(x)$ divided by $(x + 1)$ leaves a remainder of 6
  $\Rightarrow$ $f(-1) = 6$
  Hence $-1 + a + 4 + b = 6$ $\Rightarrow$ $a + b = 3$ (2)
  Solving equations 1 and 2, $3a = -3$
  $\Rightarrow$ $\underline{a = -1, b = 4}$

2 (a) Given that $f(x) = 2x^3 - 7x^2 + x + 1$, show that $(2x - 1)$ is a factor of $f(x)$ and find the remaining quadratic factor.
Hence find the exact values of the three roots of the equation $f(x) = 0$.
(b) Solve the equation
$2 \cos^3 \theta - 7 \cos^2 \theta + \cos \theta + 1 = 0$ for $0° \leqslant \theta \leqslant 360°$, giving your answers to the nearest degree. (AEB, 1994)

* (a) $f(x) = 2x^3 - 7x^2 + x + 1$

  If $(2x - 1)$ is a factor then $f(\tfrac{1}{2}) = 0$

  $f(\tfrac{1}{2}) = 2(\tfrac{1}{2})^3 - 7(\tfrac{1}{2})^2 + (\tfrac{1}{2}) + 1$
  $\quad = \tfrac{1}{4} - \tfrac{7}{4} + \tfrac{2}{4} + 1$
  $\quad = 0$

  Hence $(2x - 1)$ is a factor.

  To find the quadratic factor use long division or inspection

$$\begin{array}{r} x^2 - 3x - 1 \\ 2x - 1 \overline{)2x^3 - 7x^2 + x + 1} \\ \underline{2x^3 - x^2} \\ -6x^2 + x \\ \underline{-6x^2 + 3x} \\ -2x + 1 \\ \underline{-2x + 1} \end{array}$$

$\dfrac{2x^3}{2x} = x^2$
$\longleftarrow (2x - 1)x^2$
Subtract and bring down the next term
Repeat the process

$f(x) = (2x - 1)(x^2 - 3x - 1)$.
When $f(x) = 0$: $2x - 1 = 0$ $\Rightarrow x = \tfrac{1}{2}$

or $x^2 - 3x - 1 = 0$ $\Rightarrow$

$$x = \frac{3 \pm \sqrt{9 + 4}}{2}$$

$$= \frac{3 \pm \sqrt{13}}{2}$$

Hence the three roots of $f(x) = 0$ are

$$\frac{1}{2}, \frac{3 + \sqrt{13}}{2} \text{ and } \frac{3 - \sqrt{13}}{2}.$$

'Exact' values were asked for so do not evaluate $\sqrt{13}$

(b) If $2 \cos^3 \theta - 7 \cos^2 \theta + \cos \theta + 1 = 0$ then $2c^3 - 7c^2 + c + 1 = 0$ where $c = \cos \theta$.

From above $\cos \theta = \tfrac{1}{2}$ or $\dfrac{3 + \sqrt{13}}{2}$ or $\dfrac{3 - \sqrt{13}}{2}$.

$\cos \theta = \tfrac{1}{2}$ $\Rightarrow$ $\theta = 60°, 300°$

$\cos \theta = \dfrac{3 + \sqrt{13}}{2}$ $\Rightarrow$ no solutions since $|\cos \theta| \leqslant 1$

9.1

$\cos \theta = \dfrac{3 - \sqrt{13}}{2}$ $\Rightarrow$ $\theta = 108°, 252°$

Solutions in the range $0 \leqslant \theta \leqslant 360°$ are $\theta = 60°$, $108°$, $252°$, $300°$.

3 Multiply $3x^2 - 7x + 2$ by $5x^3 + x^2 - 3x - 7$.

•

$$
\begin{array}{r}
5x^3 + \ x^2 - \ 3x - \ 7 \\
3x^2 - \ 7x + \ 2 \\
\hline
15x^5 + 3x^4 - \ 9x^3 - 21x^2 \\
-35x^4 - \ 7x^3 + 21x^2 + 49x \\
10x^3 + \ 2x^2 - \ 6x - 14 \\
\hline
15x^5 - 32x^4 - \ 6x^3 + \ 2x^2 + 43x - 14
\end{array}
$$

Multiplying
by $3x^2$
by $-7x$
by 2

4 Given that $(x + 1)$ and $(x - 2)$ are factors of $f(x) = x^4 - 3x^3 + ax^2 + bx + 4$ find the values of a and b and factorize completely.

• If $(x + 1)$ and $(x - 2)$ are factors, $f(-1)$ and $f(2) = 0$

$f(-1) = (-1)^4 - 3(-1)^3 + a(-1)^2 + b(-1) + 4 = 0$
$\Rightarrow \quad 8 + a - b = 0 \quad\quad\quad (1)$
$f(2) = (2)^4 - 3(2)^3 + a(2)^2 + b(2) + 4 = 0$
$\Rightarrow \quad -4 + 4a + 2b = 0 \quad\quad (2)$
$\quad\quad\quad -2 + 2a + b = 0 \quad\quad (3)$

2.1b and 2.2b

Adding equations 1 and 3:
$6 + 3a = 0 \quad \Rightarrow \quad a = -2, b = 6$

$f(x) = (x + 1)(x - 2)(\text{quadratic in } x)$

Find the quadratic in x by inspection or long division.

$x^4 - 3x^3 - 2x^2 + 6x + 4$
$= (x^2 - x - 2)(cx^2 + dx + e)$
Term in x^4 $\quad\quad 1 = c$
Constant term $\quad 4 = -2e \quad \Rightarrow \quad e = -2$
Term in x $\quad\quad 6 = -e - 2d \quad \Rightarrow \quad d = -2$
$x^4 - 3x^3 - 2x^2 + 6x + 4$
$= (x + 1)(x - 2)(x^2 - 2x - 2)$

4.1

5 When a polynomial $g(x)$ is divided by $(x + 3)$ the remainder is 8, and when $g(x)$ is divided by $(x - 2)$ the remainder is 3. Find the remainder when $g(x)$ is divided by $(x - 2)(x + 3)$.

• When a polynomial is divided by $(x - 2)(x + 3)$ the remainder will be a linear polynomial $ax + b$
Let $g(x) = (x - 2)(x + 3)f(x) + ax + b$
$\quad g(2) = 2a + b = 3$
$\quad g(-3) = -3a + b = 8$
Subtracting, $5a = -5 \quad \Rightarrow \quad a = -1, b = 5$.
Hence the remainder when the polynomial is divided by $(x - 2)(x + 3)$ is $-x + 5$.

6 The polynomial $x^5 - 3x^4 + 2x^3 - 2x^2 + 3x + 1$ is denoted by $f(x)$.
(i) Show that neither $(x - 1)$ nor $(x + 1)$ is a factor of $f(x)$.
(ii) By substituting $x = 1$ and $x = -1$ in the identity

$f(x) \equiv (x^2 - 1) q(x) + ax + b$

where $q(x)$ is a polynomial and a and b are constants, or otherwise, find the remainder when $f(x)$ is divided by $(x^2 - 1)$.
(iii) Show, by carrying out the division, or otherwise, that when $f(x)$ is divided by $(x^2 + 1)$, the remainder is $2x$.
(iv) Find all the real roots of the equation $f(x) = 2x$.

(UCLES, 1991)

• $x^5 - 3x^4 + 2x^3 - 2x^2 + 3x + 1$ is denoted by $f(x)$.
(i) $f(1) = 1 - 3 + 2 - 2 + 3 + 1 = 2$
$\therefore (x - 1)$ is not a factor of $f(x)$

7.5

$f(-1) = -1 - 3 - 2 - 2 - 3 + 1 = -10$
$\therefore (x + 1)$ is not a factor of $f(x)$
(ii) If $\quad\quad f(x) \equiv (x^2 - 1) q(x) + ax + b$
then $\quad\quad f(1) = a + b = 2 \quad\quad (1)$
$\quad\quad\quad f(-1) = -a + b = -10 \quad (2)$
Solving equations 1 and 2:
$2b = -8 \quad \Rightarrow \quad b = -4, a = 6$

$\therefore f(x) = (x^2 - 1) q(x) + 6x - 4$

When $f(x)$ is divided by $(x^2 - 1)$ the remainder is $6x - 4$.

(iii)

$$
\begin{array}{r}
x^3 - 3x^2 + \ x + 1 \\
x^2 + 1 \overline{)x^5 - 3x^4 + 2x^3 - 2x^2 + 3x + 1} \\
\underline{x^5 + 0 \ + \ x^3} \\
-3x^4 + \ x^3 - 2x^2 \\
\underline{-3x^4 + \ 0 \ - 3x^2} \\
x^3 + \ x^2 + 3x \\
\underline{x^3 + \ 0 \ + \ x} \\
x^2 + 2x + 1 \\
\underline{x^2 + 0 \ + 1} \\
2x
\end{array}
$$

You can divide by inspection or long division

$f(x) = (x^2 + 1)(x^3 - 3x^2 + x + 1)$ remainder $2x$
(iv) If $f(x) = 2x$ then
$(x^2 + 1)(x^3 - 3x^2 + x + 1) + 2x = 2x$
$\Rightarrow (x^2 + 1)(x^3 - 3x^2 + x + 1) = 0$

(a) When $x^2 + 1 = 0$ then no real roots.
(b) When $x^3 - 3x^2 + x + 1 = 0$ then, by inspection, $x = 1$ is a root.

If $g(x) = x^3 - 3x^2 + x + 1$ then
$g(x) = (x - 1)(x^2 - 2x - 1)$ by inspection
If $g(x) = 0$ then $x = 1$ or $x^2 - 2x - 1 = 0$

$$x = \frac{2 \pm \sqrt{4 + 4}}{2} = 1 \pm \sqrt{2}$$

Hence all the real roots of $f(x) = 2x$ are

$x = 1, 1 + \sqrt{2}$ and $1 - \sqrt{2}$

7 Find a and b if

$a(x + 3)^2 + b(x - 2) + 1 \equiv 3x^2 + 20x + 24$.

- Substitute values for x
  When $x = -3$,
  $-5b + 1 = 3(-3)^2 + 20(-3) + 24 = -9$
  $\Rightarrow b = 2$
  When $x = 2$,
  $25a + 1 = 3(2)^2 + 20(2) + 24 = 76$
  $\Rightarrow a = 3$
  Alternatively, expand the left-hand side of this identity:
  LHS $a(x + 3)^2 + b(x - 2) + 1$
  $\equiv a(x^2 + 6x + 9) + bx - 2b + 1$
  $\equiv ax^2 + 6ax + 9a + bx - 2b + 1$
  $\equiv ax^2 + x(6a + b) + 9a - 2b + 1$

  If $ax^2 + x(6a + b) + 9a - 2b + 1 \equiv 3x^2 + 20x + 24$ then the coefficients of like terms may be equated.
  That is, $a = 3 \quad 6a + b = 20 \quad 9a - 2b + 1 = 24$.
  Solve the first two equations for a and b, and check in the third equation
  Hence $a = 3, b = 2$.

8 Given that $(x + 1)$ and $(x - 2)$ are factors of $x^4 - 3x^3 + ax^2 + bx + 4$, find the values of a and b.

- If $(x + 1)$ and $(x - 2)$ are factors of f(x) then f(-1) and f(2) are zero.
  $f(-1) = (-1)^4 - 3(-1)^3 + a(-1)^2 + b(-1) + 4 = 0,$
  $\Rightarrow 8 + a - b = 0 \qquad (1)$
  $f(2) = (2)^4 - 3(2)^3 + a(2)^2 + b(2) + 4 = 0,$
  $\Rightarrow 4a + 2b - 4 = 0 \qquad (2)$
  $\Rightarrow 2a + b - 2 = 0 \qquad (3)$

  Adding equations 1 and 3:
  $3a + 6 = 0 \Rightarrow a = -2$

  Substitute into equation 1:
  $8 - 2 - b = 0 \Rightarrow b = 6$

9 The polynomial $x^3 + 2x^2 + ax + b$, where a, b are constants, leaves a remainder of 7 and 17 when divided by $(x - 2)$ and $(x + 3)$ respectively. Find the values of a and b and the remainder when this polynomial is divided by $(x - 4)$.

- Let $g(x) = x^3 + 2x^2 + ax + b$
  $g(2) = $ the remainder when $g(x)$ is divided by $(x - 2)$
  $g(2) = 8 + 8 + 2a + b = 7$
  $\Rightarrow 2a + b = -9 \qquad (1)$

  Similarly, $g(-3) = -27 + 18 - 3a + b = 17,$
  $\Rightarrow -3a + b = 26 \qquad (2)$

  Subtracting equation 2 from equation 1:
  $5a = -35 \Rightarrow a = -7$

  Substituting into equation 1:
  $-14 + b = -9 \Rightarrow b = 5$

Therefore $g(x) = x^3 + 2x^2 - 7x + 5$
$g(4) = 64 + 32 - 28 + 5 = 73$

The remainder when $g(x)$ is divided by $(x - 4)$ is 73.

10 When a polynomial $g(x)$ is divided by $(x + 3)$ the remainder is 8, and when $g(x)$ is divided by $(x - 2)$ the remainder is 3. Find the remainder when $g(x)$ is divided by $(x - 2)(x + 3)$.

- When a polynomial is divided by $(x - 2)(x + 3)$ (by a quadratic), the remainder is of the form $ax + b$ (linear)
  Let $g(x) = (x - 2)(x + 3) f(x) + ax + b$.
  Then $g(-3) = 0 + -3a + b = 8$
  $\Rightarrow -3a + b = 8 \qquad (1)$
  $g(2) = 0 + 2a + b = 3$
  $\Rightarrow 2a + b = 3. \qquad (2)$

  Subtracting equation 1 from equation 2,
  $5a = -5 \Rightarrow a = -1$

  Substituting in equation 2,
  $-2 + b = 3 \Rightarrow b = 5.$

  The remainder when $g(x)$ is divided by $(x - 2)(x + 3)$ is $-x + 5$.

11 For what values of x is $x^3 - 2x^2 > 5x - 6$?

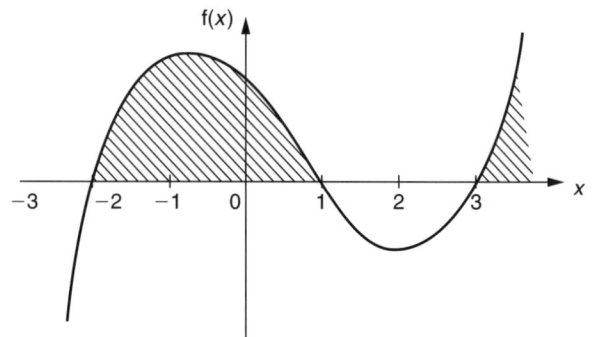

	-2	-1	0	1	2	3
$x - 1$	$-$		$-$		$+$	$+$
$x - 3$	$-$		$-$		$-$	$+$
$x + 2$	$-$		$+$		$+$	$+$
$f(x)$			$+$			$+$

- If $x^3 - 2x^2 > 5x - 6$ then $x^3 - 2x^2 - 5x + 6 > 0$.
  Let $f(x) = x^3 - 2x^2 - 5x + 6$.
  Use the factor theorem to find the factors of f(x).
  Values of x to try are the factors of 6
  $(\pm1, \pm2, \pm3, \pm6)$.
  $f(1) = 1 - 2 - 5 + 6 = 0$
  so $(x - 1)$ is a factor of f(x)
  $f(2) = 8 - 8 - 10 + 6 = -4$
  so $(x - 2)$ is not a factor.
  $f(3) = 27 - 18 - 15 + 6 = 0$

so $(x - 3)$ is a factor.
f$(-2) = -8 - 8 + 10 + 6 = 0$
so $(x + 2)$ is a factor.
f$(x) = (x - 1)(x - 3)(x + 2)$.
Check that the coefficient of x^3 is correct, since the linear factors may be multiplied by a constant
If f$(x) > 0$, $(x - 1)(x - 3)(x + 2) > 0$.
From the sketch above or by considering the number line below it,

$$x^3 - 2x^2 > 5x - 6 \text{ when } -2 < x < 1 \text{ or } x > 3.$$

Exercises

1 Given that $(x - 1)$ is a factor of $x^3 - 6x^2 + px - 6$, then

$p =$ **A:** -13 **B:** -1 **C:** 1 **D:** 11 **E:** 13
(SEB, 1993)

2 Given that $(1 + kx)^8 = 1 + 12x + px^2 + qx^3 + \ldots$ for all $x \in \mathbb{R}$
(a) Find the value of k, and the value of p, and the value of q.
(b) Using your values of k, p and q find the numerical coefficient of the x^3 term in the expansion of $(1 - x)(1 + kx)^8$.   (L, 1994)

3 Use the Remainder Theorem to find one of the factors of the cubic

$$f(x) \equiv 2x^3 - 9x^2 + 7x + 6$$

Hence factorize f(x) into its linear factors.
(WJEC, 1993)

4 Show that $x^2 + 8x + 18$ can be written in the form $(x + a)^2 + b$. Hence or otherwise find the coordinates of the turning point of the curve with equation $y = x^2 + 8x + 18$.   (SEB, 1994)

5 Show that $(x - 2)$ is a factor of $p(x) = 2x^3 + x^2 - 13x + 6$ and express $p(x)$ as the product of linear factors.
Sketch the graph of $y = p(x)$, showing clearly the coordinates of the points where the curve crosses the x-axis.

(a) Solve the inequality $2x^3 + x^2 - 13x + 6 < 0$.
(b) Find the equation of the tangent to the curve with equation $y = 2x^3 + x^2 - 13x + 6$ at the point $(2, 0)$ and determine the x-coordinate of the point where this tangent intersects the curve again.   (AEB, 1993)

6 Both $(x - 2)$ and $(x - 3)$ are factors of the cubic polynomial

$$x^3 + ax^2 + (b - 2)x + c - 1$$

Given also that $(x - 3)$ is a factor of the cubic polynomial

$$x^3 + ax^2 + bx$$

evaluate the constants a, b and c.   (NICCEA, 1993)

7 Given that $(x - 2)$ and $(x + 2)$ are factors of $x^3 + ax^2 + bx + 4$, find the value of a and the value of b.   (UCLES, 1988)

8 The function f is given by f$(x) = x^3 - 3x^2 - 2x + 6$.

(a) Use the Factor Theorem to show that $(x - 3)$ is a factor of f(x).
(b) Write f(x) in the form $(x - 3)(ax^2 + bx + c)$ giving the values of a, b and c.
(c) Hence solve f$(x) = 0$.
(d) Using your solutions to f$(x) = 0$, write down the solutions of the equation f$(x + 1) = 0$.
(NEAB, 1993)

9 If $(x - a)^2$ is a factor of the polynomial f(x), prove that both f(a) and f$'(a)$ are zero.
Find the coefficients p and q if $(x - 2)^2$ is a factor of

$$f(x) \equiv 3x^4 + px^3 + 14x^2 + qx - 8$$

Find all the roots of the equation f$(x) = 0$.
(O&C, 1991)

10 If $2x - 1$ is a factor of $2x^3 + bx^2 - 8x + 2$, b is equal to:
A 9 **B** -5 **C** 1 **D** 3 **E** 7

11 Given that $(x + 3)$ is a factor of f(x) where f$(x) \equiv 2x^3 - ax + 12$, find the constant a. Express f(x) as a product of linear factors.

12 Find the sets of values of x for which

(a) $|3x - 5| < 6$
(b) $(3x - 1)(x + 3) > 0$
(c) $|x^2 - 9| < 8$.

13 Prove that for all real x, $0 < \dfrac{1}{x^2 - 5x + 9} \leq \dfrac{4}{11}$.

Sketch the curve $y = \dfrac{1}{x^2 - 5x + 9}$.

Brief Solutions

1 f$(x) = x^3 - 6x^2 + px - 6$. $(x - 1)$ is a factor hence f$(1) = 0$.
f$(1) = 1 - 6 + p - 6 = p - 11$. Hence $p = 11$.
Answer **D**

2 $(1 + kx)^8 = 1 + 12x + px^2 + qx^3 + \ldots$ Expand by the binomial series

$$(1 + kx)^8 = 1 + 8(kx) + \frac{8.7}{1.2}(kx)^2$$
$$+ \frac{8.7.6}{1.2.3}(kx)^3 + \ldots$$

Comparing terms: $8k = 12$ $k = \frac{3}{2}$
                     $28k^2 = p$ $p = 63$
                     $56k^3 = q$ $q = 189$

$\therefore (1 + \frac{3}{2}x)^8 = 1 + 12x + 63x^2 + 189x^3 + \ldots$

$(1 - x)(1 + \frac{3}{2}x)^8 \Rightarrow$

term in x^3 is $-63x^3 + 189x^3$

The numerical coefficient of x^3 is 126.

3 $f(x) \equiv 2x^3 - 9x^2 + 7x + 6$
factors of 6 are 1, 2, 3, 6   4.1
$f(2) = 16 - 36 + 14 + 6 = 0$

$(x - 2)$ is a factor   7.5
$f(3) = 54 - 81 + 21 + 6 = 0$
$(x - 3)$ is a factor
$f(x) = (x - 2)(x - 3)(2x + 1)$
          Remaining factor of $2x^3$ is $2x$
          Remaining factor of $+6$ is $+1$

4 $x^2 + 8x + 18 = x^2 + 8x + 16 + 18 - 16$
          $= (x + 4)^2 + 2$ 4.2
The curve with equation $y = x^2 + 8x + 18$ has a minimum point when $x = -4$, $y = 2$.

5 $p(x) = 2x^3 + x^2 - 13x + 6$
$p(2) = 16 + 4 - 26 + 6 = 0 \Rightarrow (x - 2)$ is a factor
$p(x) = (x - 2)(2x^2 + 5x - 3)$ *Find the quadratic*
          $= (x - 2)(x + 3)(2x - 1)$ *factor now*
          4.1

The curve crosses the x-axis at $(-3, 0)$,
$(\frac{1}{2}, 0)$ and $(2, 0)$.   7.1

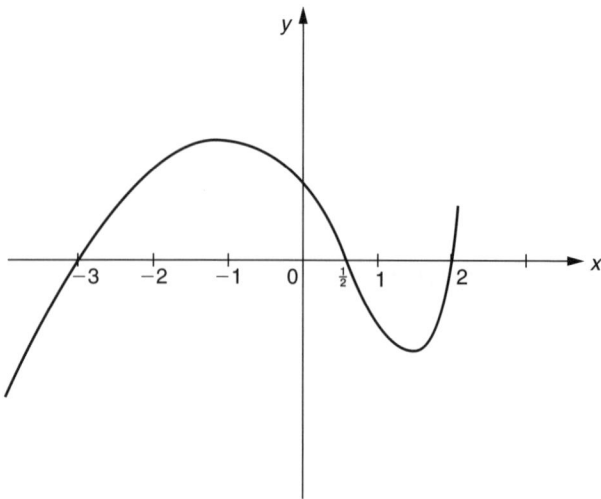

(a) $2x^3 + x^2 - 13x + 6 < 0$
    when $x < -3$ or $\frac{1}{2} < x < 2$.
(b) $p'(x) = 6x^2 + 2x - 13$
    $p'(2) = 24 + 4 - 13 = 15$
    $\therefore$ gradient of the curve at $(2, 0)$ is 15.
          5.2

The equation of the tangent at $(2, 0)$
is $y = 15x - 30$.

This meets the curve when
$2x^3 + x^2 - 13x + 6 = 15x - 30$
$2x^3 + x^2 - 28x + 36 = 0$ 2.1b and 2.2b
This will have a repeated root of $x = 2$

$2x^3 + x^2 - 28x + 36 = (x - 2)(x - 2)(2x + 9)$

The tangent will intersect the curve again when $x = -\frac{9}{2}$.

6 $x^3 + ax^2 + (b - 2)x + c - 1$ has factor $(x - 2)$
$\Rightarrow$ $f(2) = 8 + 4a + 2b - 4 + c - 1 = 0$
          $4a + 2b + c = -3$ (1)
Factor $(x - 3)$
$\Rightarrow$ $f(3) = 27 + 9a + 3b - 6 + c - 1 = 0$
          $9a + 3b + c = -20$ (2)
$x^3 + ax^2 + bx$ has factor $(x - 3)$
$\Rightarrow$ $g(3) = 27 + 9a + 3b = 0$
          $9a + 3b = -27$ (3)
From equations 2 and 3, $c = 7$
$\Rightarrow$ $4a + 2b = -10$ $2a + b = -5$
                                          $3a + b = -9$
$\Rightarrow$ $a = -4$, $b = 3$

$\therefore a = -4$, $b = 3$, $c = 7$

7 $f(x) = x^3 + ax^2 + bx + 4$
$f(2) = 8 + 4a + 2b + 4 = 0$ (1)
$\Rightarrow$ $(x - 2)$ is a factor
$f(-2) = -8 + 4a - 2b + 4 = 0$ (2)
$\Rightarrow$ $(x + 2)$ is a factor
Equation 1 + equation 2 $\Rightarrow$ $8a + 8 = 0$
                                          $\Rightarrow$ $a = -1$, $b = -4$.

8 $f(x) = x^3 - 3x^2 - 2x + 6$
$f(3) = 27 - 27 - 6 + 6 = 0$ $\Rightarrow$ $x - 3$ is a factor
$f(x) = (x - 3)(x^2 - 2)$ $\Rightarrow$ $a = 1$, $b = 0$, $c = -2$

$f(x) = 0$ $\Rightarrow$ $x = 3, \sqrt{2}$ or $-\sqrt{2}$ 2.5
If $x + 1 = y$, $f(y) = y^3 - 3y^2 - 2y + 6$ has
roots $y = 3, \sqrt{2}$ or $-\sqrt{2}$. $x = y - 1$
$\therefore x = 2, \sqrt{2} - 1, -(\sqrt{2} + 1)$
Solutions of $f(x + 1) = 0$ are $2, \sqrt{2} - 1$,
$-(\sqrt{2} + 1)$.

9 If $f(x) = (x - a)^2 g(x)$
then $f'(x) = (x - a)^2 g'(x) + 2(x - a)g(x)$
          $= (x - a)((x - a) g'(x) + 2g(x))$
When $x = a$, $f(x) = 0$ and $f'(x) = 0$.
     $f(x) = 3x^4 + px^3 + 14x^2 + qx - 8$
$\Rightarrow$ $f(2) = 48 + 8p + 56 + 2q - 8 = 0$ (1)
     $f'(x) = 12x^3 + 3px^2 + 28x + q$
$\Rightarrow$ $f'(2) = 96 + 12p + 56 + q = 0$ (2)
Simplifying equations 1 and 2: $4p + q = -48$
                                          $\underline{12p + q = -152}$
                                          $8p \quad\quad = -104$
                              $\Rightarrow$ $p = -13$, $q = 4$
$f(x) = 3x^4 - 13x^3 + 14x^2 + 4x - 8$

$$
\begin{array}{r}
3x^2 - x - 2 \\
x^2 - 4x + 4\overline{)3x^4 - 13x^3 + 14x^2 + 4x - 8} \\
\underline{3x^4 - 12x^3 + 12x^2} \\
-x^3 + 2x^2 + 4x \\
\underline{-x^3 + 4x^2 - 4x} \\
-2x^2 + 8x - 8 \\
\underline{-2x^2 + 2x - 8}
\end{array}
$$

$\therefore \quad f(x) = (x - 2)^2(3x^2 - x - 2)$
$\quad\quad f(x) = (x - 2)^2(3x + 2)(x - 1)$

When $f(x) = 0$, $x = 2$ (repeated), $-\frac{2}{3}$ or 1. | 4.1 |

10 Since $(2x - 1)$ is a factor, $f(\frac{1}{2}) = 0$.
Substitution of $x = \frac{1}{2}$ gives $b = 7$.   Answer **E**

11 $(x + 3)$ is a factor so

$$
\begin{aligned}
f(-3) = 0 \quad &\Rightarrow \quad a = 14 \\
&\Rightarrow \quad f(x) = 2x^3 - 14x + 12
\end{aligned}
$$

Trying integers gives $f(2) = 0$ and $f(1) = 0$.

$$f(x) = 2(x + 3)(x - 2)(x - 1)$$

12 (a) $|3x - 5| < 6$
$\Rightarrow 3x - 5 < 6$ or $5 - 3x < 6$
$\Rightarrow x < \frac{11}{3}$ or $x > -\frac{1}{3}$
Thus $-\frac{1}{3} < x < \frac{11}{3}$.
(b) $(3x - 1)(x + 3) > 0$.
From the sketch, $f(x) = (3x - 1)(x + 3)$,
$f(x) > 0$ when $x < -3$ or $x > \frac{1}{3}$.

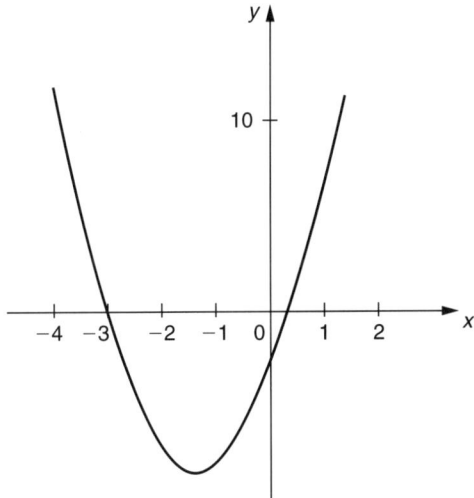

(c) $|x^2 - 9| < 8 \quad \Rightarrow \quad x^2 - 9 < 8$ or $9 - x^2 < 8$.
These give respectively $x^2 < 17$ or $x^2 > 1$.

Thus $-\sqrt{17} < x < -1$ or $1 < x < \sqrt{17}$.

13 Let $y = \dfrac{1}{x^2 - 5x + 9}$

$\Rightarrow \quad yx^2 - 5xy + 9y - 1 = 0.$ (1)

x is real, so $25y^2 \geq 4y(9y - 1) \quad \Rightarrow \quad 0 \geq y(11y - 4)$.
Thus $0 \leq y \leq \frac{4}{11}$.
If $y = 0$, equation 1 gives $-1 = 0$, which is
unacceptable, hence $y > 0$.
If $y = \frac{4}{11}$, equation 1 gives $(2x - 5)^2 = 0$
$\Rightarrow x = \frac{5}{2}$.

Hence $0 < \dfrac{1}{x^2 - 5x + 9} \leq \dfrac{4}{11}$.

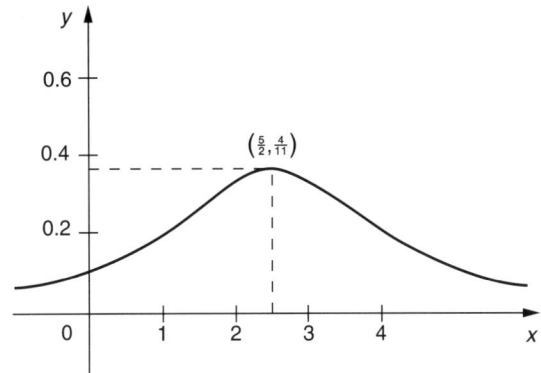

Equations

AS Level			A Level			Topic	Date attempted	Date completed	Self-assessment
CORE	MODULAR	TRADITIONAL	CORE	MODULAR	TRADITIONAL				
✓	✓	✓	✓	✓	✓	**Quadratic equations**			
✓	✓	✓	✓	✓	✓	**Simultaneous equations**			
✓	✓	✓	✓	✓	✓	**Inequalities**			
	✓	✓		✓	✓	**Partial fractions**			
✓	✓	✓	✓	✓		**Modulus function**			

Fact Sheet

Quadratic Equations

The **quadratic equation** $ax^2 + bx + c = 0$ has two roots given by

$$x = \frac{-b \pm \sqrt{(b^2 - 4ac)}}{2a}$$

Discriminant

The **discriminant** $b^2 - 4ac$ gives the nature of the roots.

- If $b^2 - 4ac > 0$, roots are real and distinct.
- If $b^2 - 4ac = 0$, roots are real and equal.
- If $b^2 - 4ac < 0$, roots are complex.
- If $b^2 - 4ac \geqslant 0$ and $ac > 0$, roots have the same sign.
- The quadratic function $f(x) = ax^2 + bx + c > 0$ for all real x if $b^2 - 4ac < 0$ and $a > 0$.

Roots

- If the roots of $ax^2 + bx + c = 0$ are α and β, then:

$$\alpha + \beta = -\frac{b}{a} \text{ and } \alpha\beta = \frac{c}{a}$$

- Given the roots of a quadratic equation, the equation is:

$$x^2 - (\text{sum of roots})\, x + \text{product of roots} = 0$$

- If coefficients a, b and c are rational then the roots occur in conjugate pairs $d \pm \sqrt{e}$.

Useful Identities

- $\alpha^2 + \beta^2 \equiv (\alpha + \beta)^2 - 2\alpha\beta \equiv (\alpha - \beta)^2 + 2\alpha\beta$.
- $(\alpha - \beta)^2 \equiv (\alpha + \beta)^2 - 4\alpha\beta$.
- $\alpha^3 + \beta^3 \equiv (\alpha + \beta)(\alpha^2 - \alpha\beta + \beta^2)$
  $\equiv (\alpha + \beta)[(\alpha + \beta)^2 - 3\alpha\beta]$.
- $\alpha^3 - \beta^3 \equiv (\alpha - \beta)(\alpha^2 + \alpha\beta + \beta^2)$
  $\equiv (\alpha - \beta)[(\alpha - \beta)^2 + 3\alpha\beta]$.

Completing the Square

- $x^2 + px$ requires $(p/2)^2$ to complete the square.
- $x^2 + px + (p/2)^2 = (x + p/2)^2$.
- If $f(x) = a(x + p)^2 + q$, and $a > 0$, then
  $f(x)$ has a least value of q, when $x = -p$ (first diagram).
- If $a < 0$, $f(x)$ has a greatest value of q, when $x = -p$ (second diagram).
- The graph of $y = f(x)$ has a line of symmetry at $x = -p$.

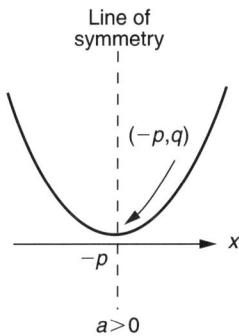

Line of symmetry

$(-p, q)$

$-p$

x

$a > 0$

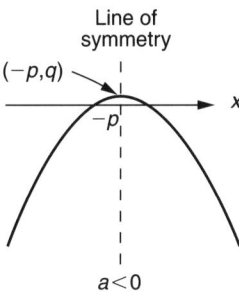

Line of symmetry

$(-p, q)$

$-p$

x

$a < 0$

Partial Fractions

If the degree of the numerator is equal to or higher than the degree of the denominator, the numerator must be divided by the denominator until the remainder is of lower degree than the denominator.

Linear Factors

$$\frac{f(x)}{(ax + b)(cx + d)(ex + f)} \equiv \frac{A}{(ax + b)} + \frac{B}{(cx + d)} + \frac{C}{(ex + f)}$$

Repeated Factors

$$\frac{f(x)}{(ax+b)(cx+d)^2} \equiv \frac{A}{(ax+b)} + \frac{B}{(cx+d)} + \frac{C}{(cx+d)^2}$$

Quadratic Factors

$$\frac{f(x)}{(ax+b)(x^2+c^2)} \equiv \frac{A}{(ax+b)} + \frac{Bx+C}{(x^2+c^2)}$$

To find the constants A, B and C: Put all the terms on the right-hand side over the same denominator as the left-hand side. Equate the numerators. Equate the coefficients of the powers of x and/or substitute well-chosen values for x.

Partial fractions are frequently combined with the binomial series (page 99), differentiation or integration (page 134).

Particular binomial series:
$(1+x)^{-1} = 1 - x + x^2 - x^3 + x^4 - \dots$ valid for $-1 < x < 1 \Leftrightarrow |x| < 1$,
$(1+x)^{-2} = 1 - 2x + 3x^2 - 4x^3 + \dots$ valid for $-1 < x < 1 \Leftrightarrow |x| < 1$.

Inequalities

Notation: Inequalities

- If a and b are numbers, represented by points on the x-axis, and $a > b$, then a is to the right of b.
- Thus $6 > 3$ and $-6 > -9$.
- If $a \geqslant b$ then a is greater than b or equal to b.

$$b \quad < \quad a$$

Modulus

- $|x|$ may be regarded as the distance from the origin to the point on the x-axis.
- $|2| = 2, |-3| = 3, |x| = 5 \Rightarrow x = 5$ or -5.
  $|x| < 5 \Rightarrow -5 < x < 5$.
- $|f(x)| = f(x)$ if $f(x)$ is positive and $-f(x)$ if $f(x)$ is negative.
- If $f(x) = x^2 - 4$ then $f(3) = 5, |f(3)| = 5$,
$$f(1) = -3, |f(1)| = 3.$$

Properties of Inequalities

- Any quantity may be added to or subtracted from both sides of an inequality:

  If $a > b$ then $a + c > b + c$ and $a - c > b - c$.

- Both sides of an inequality may be multiplied or divided by any *positive* quantity:

  If $c > 0$ and $a > b$ then $ac > bc$ and $\dfrac{a}{c} > \dfrac{b}{c}$.

- If both sides of an inequality are multiplied or divided by a *negative* quantity then the inequality is reversed:

  If $d < 0$ and $a > b$ then $ad < bd$ and $\dfrac{a}{d} < \dfrac{b}{d}$.

  (Play safe – never multiply or divide by a negative number.)
- If inequalities are of the same kind they may be added together:

  If $a > b$ and $c > d$ then $a + c > b + d$.

- **Never subtract inequalities.**
  (Example: $5 > 2$ and $4 > 0$ but $5 - 4 \not> 2 - 0$.)

Worked Examples

1 Find numbers a, b and c such that

$$2x^4 + 2x^3 + 5x^2 + 3x + 3$$
$$\equiv (x^2 + x + 1)(ax^2 + bx + c)$$

for all values of x.   (UCLES, 1989)

$\equiv$ is an identity which is true for all values of x.
Standard methods are
          (i)   substitute value of x
or   (ii)   compare coefficients of x^n
or   (iii)   long division
Decide for yourself which method you find easiest – or perhaps a mixture

• *Method (i)*
When $x = 0$, $3 = c$ (1)
When $x = 1$,
  $2 + 2 + 5 + 3 + 3 = (1 + 1 + 1)(a + b + c)$
  $\Rightarrow 15 = 3(a + b + 3)$
  $\Rightarrow a + b = 2$ (2)
When $x = -1$,
  $2 - 2 + 5 - 3 + 3 = (1 - 1 + 1)(a - b + c)$
  $\Rightarrow 5 = a - b + 3$
  $\Rightarrow 2 = a - b$ (3)
From equations 2 and 3, $a = 2$, $b = 0$.

• *Method (ii)*

$(x^2 + x + 1)(ax^2 + bx + c)$
$= ax^4 + x^3(a + b) + x^2(a + b + c) + x(b + c) + c$

Hence

$2x^4 + 2x^3 + 5x^2 + 3x + 3$
$\equiv ax^4 + (a + b)x^3 + (a + b + c)x^2 + (b + c)x + c$

From the x^4 terms, $a = 2$; from the constant terms, $c = 3$; from the x or x^3 terms, $b = 0$.

• *Method (iii)*

$$
\begin{array}{r}
2x^2 \qquad\quad + 3 \\
x^2 + x + 1 \overline{)2x^4 + 2x^3 + 5x^2 + 3x + 3} \\
\underline{2x^4 + 2x^3 + 2x^2} \\
3x^2 + 3x + 3 \\
\underline{3x^2 + 3x + 3}
\end{array}
$$

See pp. 34 and 35

Hence $a = 2$, $b = 0$, $c = 3$.

2 The sum of the roots of a quadratic equation is 9 and the product of the roots is 4. The equation could be:

A $9x^2 + 4x + 1 = 0$ **B** $4x^2 + 9x + 1 = 0$
C $x^2 + 9x - 4 = 0$ **D** $x^2 - 9x + 4 = 0$
E $x^2 - 4x + 9 = 0$

• A quadratic equation can be written as

$x^2 - (\text{sum of roots})\, x + \text{product of roots} = 0$

that is, $x^2 - 9x + 4 = 0$.   <u>Answer **D**</u>

3 A sheet of card is in the shape of the pentagon $ABCDE$ where angles ABC and BCD are right angles. The side BC is of length $4x$ cm and the sides AB, CD, DE and EA are each of length $3x$ cm. Show that the area of the pentagon is

$$x^2(12 + 2\sqrt{5})\ \text{cm}^2$$

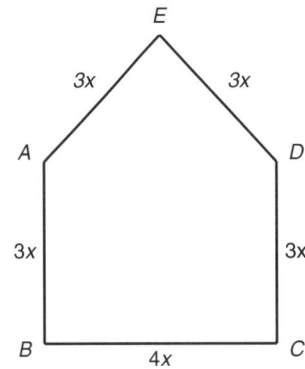

Given that the area of the card is 15.5 cm², find the value of x^2 in the form $(a + b\sqrt{5})$ where a and b are rational numbers.   (AEB, 1994)

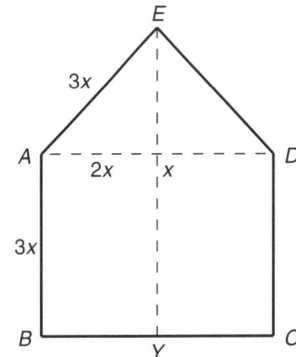

$(EX)^2 = (3x)^2 - (2x)^2$
$\qquad\quad = 5x^2$
$\therefore EX = \sqrt{5}x$

Area $ABCDE$ = area $ABCD$ + area AED
$\qquad\qquad\quad = 3x.4x + 2x.\sqrt{5}x$
$\qquad\qquad\quad = x^2(12 + 2\sqrt{5})$

If the area is 15.5 cm² then $x^2(12 + 2\sqrt{5}) = 31/2$

$$\Rightarrow x^2 = \frac{31}{2(12 + 2\sqrt{5})}$$

$$= \frac{31(12 - 2\sqrt{5})}{2(144 - 20)}$$

$$= \frac{12 - 2\sqrt{5}}{8}$$

$$= \frac{3}{2} - \frac{1}{4}\sqrt{5}$$

4 The equation of a curve is $y = ax^2 - 2bx + c$, where a, b and c are constants, with $a > 0$.

(i) Find, in terms of a, b and c, the coordinates of the turning point on the curve, and state whether it is a maximum point or a minimum point.

(ii) Given that the turning point of the curve lies on the line $y = x$, find an expression for c in terms of a and b. Show that, in this case, whatever the value of b,

$$c \geqslant -\frac{1}{4a}.$$

(iii) Find the numerical values of a, b and c when the curve passes through the point $(0, 6)$ and has a turning point at $(2, 2)$.

(UCLES, 1989)

• (i) Since $a > 0$, the turning point is a minimum.

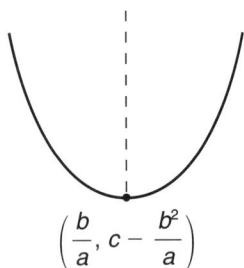

$$\left(\frac{b}{a}, c - \frac{b^2}{a} \right)$$

$$y = ax^2 - 2bx + c$$

$$= a\left(x - \frac{b}{a}\right)^2 + c - \frac{b^2}{a} \qquad \text{completing the square}$$

Minimum value of y is $c - \dfrac{b^2}{a}$ when

$$x = \frac{b}{a}.$$

(ii) If this point lies on the line $y = x$ then

$$\frac{b}{a} = c - \frac{b^2}{a} \qquad (1)$$

$$c = \frac{1}{a}(b^2 + b)$$

$$= \frac{1}{a}(b + \tfrac{1}{2})^2 - \frac{1}{4a}$$

Since $(b + \tfrac{1}{2})^2 \geqslant 0$ for all b and $a > 0$,

$$c \geqslant -\frac{1}{4a}.$$

(iii)

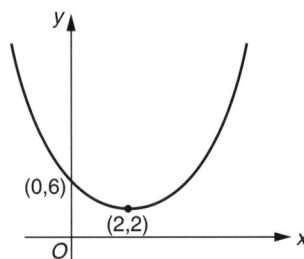

If the curve passes through $(0, 6)$, then $c = 6$ (from the equation of the curve).

From equation 1, $\dfrac{b}{a} = 6 - \dfrac{b^2}{a}$ and $\dfrac{b}{a} = 2$

from the minimum point. In equation 2, $2 = 6 - 2b \Rightarrow b = 2, a = 1, c = 6$. Hence $y = x^2 - 4x + 6$.

5 Given that $px^2 + 2px - 5 < 0$ for all real values of x, determine the set of possible values of p.

• Two conditions must be satisfied in this question:
(a) $px^2 + 2px - 5 = 0$ has no real solutions
(b) the graph of $px^2 + 2px - 5$ is entirely below the x-axis (i.e. has a maximum turning point)

(a) $px^2 + 2px - 5 = 0$ has no real roots when

$$b^2 - 4ac < 0$$
$$\Rightarrow 4p^2 - 4(p)(-5) < 0 \qquad \boxed{3.1}$$
$$4p^2 + 20p < 0$$
$$4p(p + 5) < 0$$

If $f(p) = 4p(p + 5) < 0$ then $-5 < p < 0$.

(b) $px^2 + 2px - 5$ has a maximum turning point when $p < 0$.

Combining these two conditions gives $-5 < p < 0$.

6 Solve the equation $\sqrt{(4x - 7)} + \sqrt{(2x - 4)} = 1$ (where $\sqrt{}$ means positive root only).

• In any equation with square root terms put one square root term on one side of the equation and all other terms on the other, then square both sides of the equation.

If $\sqrt{(4x - 7)} + \sqrt{(2x - 4)} = 1$ then $\sqrt{(4x - 7)} = 1 - \sqrt{(2x - 4)}$.
Square both sides:

$$4x - 7 = (1 - \sqrt{(2x - 4)})^2$$
$$= 1 + 2x - 4 - 2\sqrt{(2x - 4)}$$
$$\Rightarrow \quad 2x - 4 = -2\sqrt{(2x - 4)}$$
$$\Rightarrow \quad x - 2 = -\sqrt{(2x - 4)}$$

Square both sides:

$$x^2 - 4x + 4 = 2x - 4$$
$$x^2 - 6x + 8 = 0$$
$$(x - 2)(x - 4) = 0$$
$$x = 2 \text{ or } 4$$

2.1a and 2.2a

Check both answers in the original equation:

$$x = 2: \text{LHS} = \sqrt{1} + \sqrt{0} = 1 = \text{RHS}$$
$$x = 4: \text{LHS} = \sqrt{9} + \sqrt{4} = 5 \neq \text{RHS}$$

Hence $x = 2$ is the only solution of the original equation.

Note: $\sqrt{(4x - 7)} - \sqrt{(2x - 4)} = 1$ would have led to the same values of x: $x = 2$ or 4.

In this case both $x = 2$ and $x = 4$ would be correct solutions − moral: *always* check.

7 Find the set of values of k for which the quadratic equation

$$x^2 + kx + (2k - 3) = 0$$

has real roots.

(a) In the case when $k = 4$, find the complex roots of the equation.

(b) In the case when $k = 7$, the roots of the quadratic equation are α and β. Without finding the values of α and β, show that

$$\alpha^3 + \beta^3 = -112$$

Find a quadratic equation with integer coefficients having roots $\alpha^2 + 3\beta$ and $\beta^2 + 3\alpha$.   (AEB, 1990)

• $x^2 + kx + (2k - 3) = 0$ has real roots if '$b^2 - 4ac$' $\geqslant 0$, that is

$$k^2 - 4(2k - 3) \geqslant 0$$
$$k^2 - 8k + 12 \geqslant 0$$
$$(k - 6)(k - 2) \geqslant 0$$
$$k \leqslant 2 \quad \text{or} \quad k \geqslant 6$$

3.1

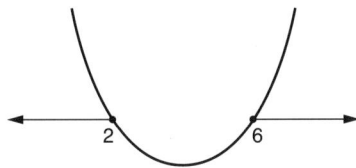

(a) When $k = 4$, $x^2 + 4x + 5 = 0$

$$x = \frac{-4 \pm \sqrt{16 - 20}}{2} = -2 \pm i$$

2.2a

Hence the complex roots are $-2 + i$ and $-2 - i$.

(b) When $k = 7$, $x^2 + 7x + 11 = 0$. If this has roots α, β, then

$$\alpha + \beta = -7 \qquad (1)$$
$$\alpha\beta = 11 \qquad (2)$$

Factorize $\alpha^3 + \beta^3$ to express it in terms of $\alpha + \beta$ and $\alpha\beta$

$$\begin{aligned}
\alpha^3 + \beta^3 &= (\alpha + \beta)(\alpha^2 - \alpha\beta + \beta^2) \\
&= (\alpha + \beta)((\alpha + \beta)^2 - 3\alpha\beta) \\
&= -7(49 - 33) \\
&= -7 \times 16 \\
&= -112
\end{aligned}$$

If a quadratic equation has roots $\alpha^2 + 3\beta$, $\beta^2 + 3\alpha$, then

$$\begin{aligned}
\text{Sum of roots} &= \alpha^2 + \beta^2 + 3(\alpha + \beta) \\
&= (\alpha + \beta)^2 + 3(\alpha + \beta) - 2\alpha\beta \\
&= 49 - 21 - 22 \\
&= 6
\end{aligned}$$

$$\begin{aligned}
\text{Product of roots} &= (\alpha^2 + 3\beta)(\beta^2 + 3\alpha) \\
&= \alpha^2\beta^2 + 3(\beta^3 + \alpha^3) + 9\alpha\beta \\
&= 121 + 3(-112) + 99 \\
&= -116
\end{aligned}$$

2.3

Hence the quadratic equation is $x^2 - 6x - 116 = 0$.

8 Express $\dfrac{3 + x}{(2 - x)(1 + 2x)}$ in partial fractions,

and hence, or otherwise, obtain the first three non-zero terms in the expansion of this expression in ascending powers of x.

State the range of values of x for which the expansion is valid.

• Look at the denominator − both factors are linear, so we need numerators, A and B

$$f(x) = \frac{3 + x}{(2 - x)(1 + 2x)} \equiv \frac{A}{2 - x} + \frac{B}{1 + 2x}$$

$$\equiv \frac{A(1 + 2x) + B(2 - x)}{(2 - x)(1 + 2x)}$$

4.3

Equating the numerators:
$$3 + x \equiv A(1 + 2x) + B(2 - x)$$
Put $x = 2$: $\qquad 5 = 5A \Rightarrow A = 1$
Put $x = -\frac{1}{2}$: $\qquad 2\frac{1}{2} = 2\frac{1}{2}B \Rightarrow B = 1$

Hence

$$f(x) \equiv \frac{3 + x}{(2 - x)(1 + 2x)} \equiv \frac{1}{2 - x} + \frac{1}{1 + 2x}$$

$$\equiv \frac{1}{2(1 - x/2)} + \frac{1}{1 + 2x}$$

$$\equiv \tfrac{1}{2}(1 - x/2)^{-1} + (1 + 2x)^{-1}$$

By the binomial expansion:

$\frac{1}{2}(1 - x/2)^{-1} = \frac{1}{2}(1 + x/2 + (x/2)^2 + (x/2)^3 + \ldots)$

  (valid for $-1 < x/2 < 1 \Rightarrow -2 < x < 2$)

$(1 + 2x)^{-1} = 1 - 2x + (2x)^2 - (2x)^3 + \ldots$

  (valid for $-1 < 2x < 1 \Rightarrow -\frac{1}{2} < x < \frac{1}{2}$)

Adding gives:

$f(x) = \frac{1}{2} + x/4 + x^2/8 + x^3/16 + 1 - 2x +$
$\qquad\qquad\qquad 4x^2 - 8x^3 + \ldots$

$\qquad = 3/2 - 7x/4 + 33x^2/8$ to the first three terms

Validity is given when both conditions are valid – the smaller interval is taken

Series is valid for $-\frac{1}{2} < x < \frac{1}{2}$.

9 Express $\dfrac{x^3 + 2x^2 - x + 3}{(x + 2)(x - 3)}$ in partial fractions.

● *Compare the orders of the numerator and denominator. Since the numerator is of order 3 and the denominator of order 2 a long division must be carried out first*

$$
\begin{array}{r}
x + 3 \\
x^2 - x - 6 \overline{)x^3 + 2x^2 - x + 3} \\
\underline{x^3 - x^2 - 6x} \\
3x^2 + 5x + 3 \\
\underline{3x^2 - 3x - 18} \\
8x + 21
\end{array}
$$

$\dfrac{x^3 + 2x^2 - x + 3}{(x + 2)(x - 3)} = x + 3 + \dfrac{8x + 21}{(x + 2)(x - 3)}$

Let $\dfrac{8x + 21}{(x + 2)(x - 3)} \equiv \dfrac{A}{x + 2} + \dfrac{B}{x - 3}$

$\qquad\qquad \equiv \dfrac{A(x - 3) + B(x + 2)}{(x + 2)(x - 3)}$ $\boxed{4.3}$

Equating the numerators:

$\qquad\qquad 8x + 21 \equiv A(x - 3) + B(x + 2)$

Put $x = 3$: $\qquad 45 = B(5) \Rightarrow B = 9$
Put $x = -2$: $\qquad 5 = A(-5) \Rightarrow A = -1$

Therefore

$\dfrac{x^3 + 2x^2 - x + 3}{(x + 2)(x - 3)} \equiv x + 3 - \dfrac{1}{x + 2} + \dfrac{9}{x - 3}.$

10 Express $f(x) = \dfrac{4x^2 - 7x + 3}{(2 - x)(1 + x^2)}$ in partial

fractions.

Expand $f(x)$ in ascending powers of x as far as, and including, the term in x^3. For what values of x is this expansion valid?

● *Look at the denominator: one factor is linear and needs a numerator A, one is quadratic and needs a numerator Bx + C*

$f(x) = \dfrac{4x^2 - 7x + 3}{(2 - x)(1 + x^2)} \equiv \dfrac{A}{2 - x} + \dfrac{Bx + C}{1 + x^2}$

$\qquad = \dfrac{A(1 + x^2) + (Bx + C)(2 - x)}{(2 - x)(1 + x^2)}$ $\boxed{4.3}$

Equating the numerators:

$\quad 4x^2 - 7x + 3 \equiv A(1 + x^2) + (Bx + C)(2 - x)$

Put $x = 2$:

$\quad 16 - 14 + 3 = A(1 + 4) \Rightarrow 5A = 5, A = 1$

Put $x = 0$: $\quad 3 = A(1) + C(2) \Rightarrow C = 1$

Put $x = 1$:

$\quad 4 - 7 + 3 = A(1 + 1) + (B + C)(1)$
$\qquad\qquad 0 = 2A + B + C \Rightarrow B = -3$

Alternatively, compare coefficients of x^2:

$4 = A - B \Rightarrow B = -3$.

Hence $f(x) \equiv \dfrac{1}{(2 - x)} + \dfrac{1 - 3x}{(1 + x^2)}$

$\dfrac{1}{2 - x} = \dfrac{1}{2(1 - x/2)} = \dfrac{1}{2}\left(1 - \dfrac{x}{2}\right)^{-1}$

$\qquad\qquad = \dfrac{1}{2}\left(1 + \dfrac{x}{2} + \dfrac{x^2}{4} + \dfrac{x^3}{8} + \ldots\right)$

  (valid $-2 < x < 2$) $\boxed{8.1}$

$\dfrac{1}{1 + x^2} = (1 + x^2)^{-1} = 1 - x^2 + x^4 - \ldots$

  (valid $-1 < x < 1$)

$f(x) = \dfrac{1}{2}\left(1 + \dfrac{x}{2} + \dfrac{x^2}{4} + \dfrac{x^3}{8} + \ldots\right)$
$\qquad + (1 - 3x)(1 - x^2 + x^4 - \ldots)$

$\qquad = \dfrac{1}{2} + \dfrac{x}{4} + \dfrac{x^2}{8} + \dfrac{x^3}{16} + 1 - 3x$

$\qquad - x^2 + 3x^3 \ldots$

$\qquad = \frac{3}{2} - \frac{11}{4}x - \frac{7}{8}x^2 + \frac{49}{16}x^3 \ldots$

The series is valid when $-2 < x < 2$ and $-1 < x < 1$. Taking the smaller range, $-1 < x < 1$.

11 Express the function $\dfrac{9x}{(1 + x)(1 - 2x)^2}$ as the

sum of three partial fractions. Hence, or otherwise, find the first three terms in the expansion of the function in ascending powers of x.

● *Notice that one of the factors is a repeated factor $\Rightarrow$ there must be three fractions, denominators $(1 + x)$, $(1 - 2x)$ and $(1 - 2x)^2$*

Let $\dfrac{9x}{(1+x)(1-2x)^2}$

$\equiv \dfrac{A}{(1+x)} + \dfrac{B}{(1-2x)} + \dfrac{C}{(1-2x)^2}$

$\equiv \dfrac{A(1-2x)^2 + B(1+x)(1-2x) + C(1+x)}{(1+x)(1-2x)^2}$

4.3

Equating the numerators:
$9x \equiv A(1-2x)^2 + B(1+x)(1-2x) + C(1+x)$
Put $x = -1$:　　$-9 = A(3)^2$　　$\Rightarrow A = -1$
Put $x = \frac{1}{2}$:　　$\frac{9}{2} = C(\frac{3}{2})$　　$\Rightarrow C = 3$
Put $x = 0$:　　$0 = A + B + C$　$\Rightarrow B = -2$

Hence $\dfrac{9x}{(1+x)(1-2x)^2}$

$\equiv \dfrac{3}{(1-2x)^2} - \dfrac{1}{(1+x)} - \dfrac{2}{(1-2x)}$

$\dfrac{3}{(1-2x)^2} = 3(1-2x)^{-2}$

8.1

$\qquad = 3[1 - 2(-2x) +$
$\qquad\qquad 3(-2x)^2 - 4(-2x)^3 + \ldots]$
$\qquad = 3 + 12x + 36x^2 + 96x^3 + \ldots$

$\dfrac{1}{(1+x)} = (1+x)^{-1}$

$\qquad = 1 - x + x^2 - x^3 + \ldots$

$\dfrac{2}{(1-2x)} = 2(1-2x)^{-1}$

$\qquad = 2[1 + 2x + (2x)^2 + (2x)^3 + \ldots]$
$\qquad = 2 + 4x + 8x^2 + 16x^3 + \ldots$

Hence

$\dfrac{9x}{(1+x)(1-2x)^2} = (3 + 12x + 36x^2 + 96x^3)$

$\qquad\qquad - (1 - x + x^2 - x^3)$
$\qquad\qquad - (2 + 4x + 8x^2 + 16x^3) + \ldots$
$\qquad\qquad = 9x + 27x^2 + 81x^3 + \ldots$

12 Prove that for all real x,

$$0 < \dfrac{1}{x^2 + 10x + 27} \leqslant \dfrac{1}{2}.$$

Sketch the curve $y = \dfrac{1}{x^2 + 10x + 27}$.

• The implications of $0 < \dfrac{1}{x^2 + 10x + 27} \leqslant \dfrac{1}{2}$

are

(i) $x^2 + 10x + 27$ is always positive, and
(ii) $x^2 + 10x + 27 \geqslant 2$ for all real x.
These conditions are both satisfied if
$x^2 + 10x + 27 \geqslant 2$ for all real x.
Let $f(x) = x^2 + 10x + 27$
$\qquad = (x + 5)^2 + 2$.
Since $(x + 5)^2 \geqslant 0$ for all real x, $f(x) \geqslant 2$ for all real x.

Thus $0 < \dfrac{1}{x^2 + 10x + 27} \leqslant \dfrac{1}{2}$ for all real x.

When $x = -5$, $f(x) = 2$, $\Rightarrow y = \frac{1}{2}$ (maximum point).

7.1

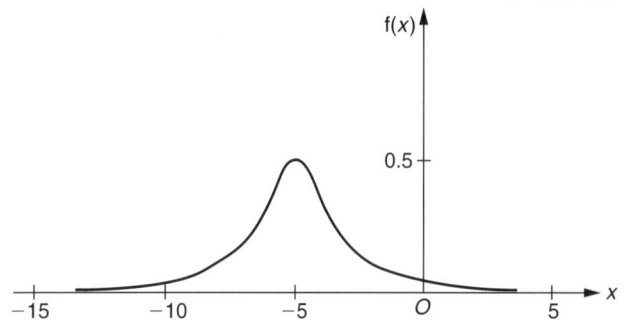

13 Find the solution set of the inequality

$$\dfrac{8}{x} < x + 2 \quad (x \in \mathbb{R}, x \neq 0)$$

• *Do not multiply by x, since x may be negative*

Multiply by x^2
$\quad 8x < x^2(x + 2)$
$\quad\; 0 < x^3 + 2x^2 - 8x$
$\quad\; 0 < x(x + 4)(x - 2)$

3.1

From the sketch it can be seen that

$$x > 2 \text{ or } -4 < x < 0$$

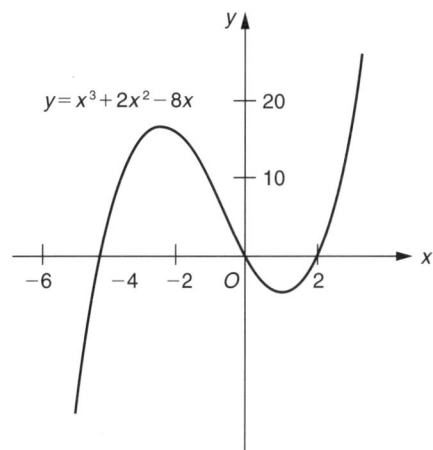

14 Find the set of real values of x for which

$$\frac{x(x-3)}{x-2} > 2.$$

• If $\dfrac{x(x-3)}{x-2} > 2$ then

$$\frac{x(x-3)}{x-2} - 2 > 0$$

$$\frac{x(x-3) - 2(x-2)}{x-2} > 0$$

$$\frac{x^2 - 3x - 2x + 4}{x-2} > 0$$

$$\frac{(x-4)(x-1)}{x-2} > 0 \qquad \boxed{3.1}$$

For a fraction to be positive either all the factors must be positive or an even number of factors negative for given values of x

		0	1	2	3	4	
$x-4$		$-$	$-$		$-$		$+$
$x-1$		$-$	$+$		$+$		$+$
$x-2$		$-$	$-$		$+$		$+$
$f(x)$		$-$	$+$		$-$		$+$

True if $x > 4$ or if $1 < x < 2$.

The inequality must not be multiplied by $(x - 2)$ since for some values of x, $x - 2$ is negative
An alternative method is to multiply by $(x - 2)^2$, which is never negative for real values of x

$$x(x-3)(x-2) > 2(x-2)^2$$
$$\Rightarrow x(x-3)(x-2) - 2(x-2)^2 > 0$$
$$(x-2)(x^2 - 3x - 2x + 4) > 0$$
$$(x-2)(x^2 - 5x + 4) > 0$$
$$(x-2)(x-1)(x-4) > 0$$

Use the same number line as for the first method.

15 The triangular region R has vertices A, B and C and is defined by

$$x \geq 0 \qquad y + 2 \geq 0 \qquad 2y + x - 2 \leq 0$$

(a) By drawing suitable straight lines on a sketch, show and label the region R.
(b) Find the coordinates of A, B and C.
   The point P lies in R and is such that
   $PA = PB = PC$.
(c) Determine the coordinates of P.
(d) Hence find the radius of the circle with centre P which passes through A, B and C.

(L Specimen Paper, 1996)

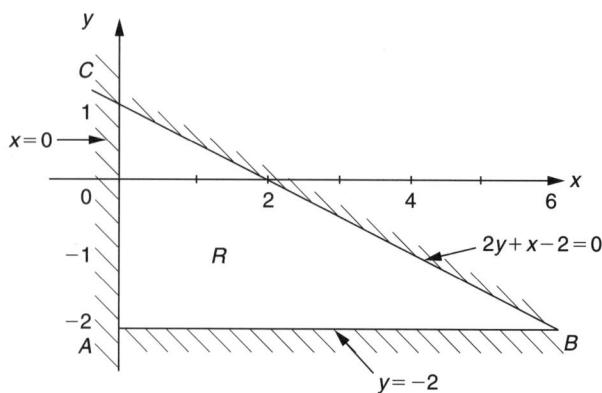

To decide which side of a line R should be, try any suitable point in the inequality

For $2y + x - 2 \leq 0$ (bounded by $2y + x - 2 = 0$) try $(0, 0)$: $-2 \leq 0$ is valid
∴ R lies on the same side of the line as the origin
Shade the unwanted regions

$$\boxed{3.1}$$

At A: $x = 0, y = -2 \Rightarrow A(0, -2)$
At B: $y = -2$ and $2y + x - 2 = 0$,
         $\therefore x = 6 \Rightarrow B(6, -2)$ $\boxed{2.2a}$
At C: $x = 0$ and $2y + x - 2 = 0$,
         $\therefore y = 1 \Rightarrow C(0, 1)$

Let P have coordinates (x, y).
Then $(PA)^2 = (x - 0)^2 + (y + 2)^2$
              $= x^2 + y^2 + 4y + 4$
      $(PB)^2 = (x - 6)^2 + (y + 2)^2$
              $= x^2 + y^2 - 12x + 4y + 40$
      $(PC)^2 = (x - 0)^2 + (y - 1)^2$
              $= x^2 + y^2 - 2y + 1$

Since $(PA)^2$ and $(PC)^2$ do not contain a term in x they are easier to solve first

$(PA)^2 = (PC)^2 \Rightarrow$
   $x^2 + y^2 + 4y + 4 = x^2 + y^2 - 2y + 1$
   $\Rightarrow 6y = -3, y = -\frac{1}{2}$
$(PB)^2 = (PC)^2 \Rightarrow$
   $x^2 + y^2 - 12x + 4y + 40 = x^2 + y^2 - 2y + 1$
   $\Rightarrow -12x + 6y + 39 = 0$
But $y = -\frac{1}{2} \therefore 12x = 36, x = 3$
Hence P is the point $(3, -\frac{1}{2})$.
Radius of the circle is PA (or PB or PC)

$$|PA| = \sqrt{3^2 + \tfrac{1}{2}^2 - 2 + 4} = \sqrt{\frac{45}{4}}$$

$$= \frac{3}{2}\sqrt{5}$$

This could be simplified using the theory of a circle. The angle in a semicircle is a right-angle. $\angle CAB$ is a right angle $\Rightarrow$ CB is a diameter of the circle through A, B and C

Mid-point of CB = $(\dfrac{6+0}{2}, \dfrac{-2+1}{2})$, i.e. $(3, -\frac{1}{2})$

Length CB = $\sqrt{(6-0)^2 + (-2-1)^2} = \sqrt{45}$ or $3\sqrt{5}$

$\therefore$ Radius = $\dfrac{3}{2}\sqrt{5}$

16 (a) On the same diagram, sketch the graphs of

$$y = \frac{1}{x-a} \text{ and } y = 4\,|\,x-a\,|$$

where a is a positive constant. Show clearly the coordinates of any points of intersection with the coordinate axes.

(b) Hence, or otherwise, find the set of values of x for which

$$\frac{1}{x-a} < 4\,|\,x-a\,|$$

(L Specimen Paper, 1996)

• (a)

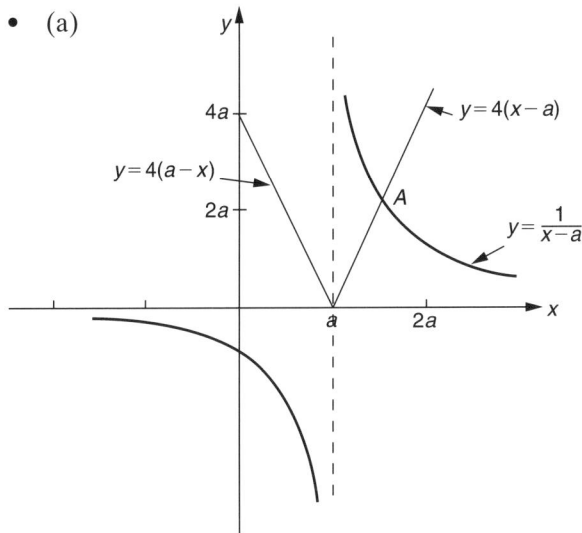

At A, $\quad 4x - 4a = \dfrac{1}{x-a}$

$(x-a)^2 = \frac{1}{4}$

$x - a = \pm\frac{1}{2}$

By inspection, $x = a + \frac{1}{2}$

(b) $\dfrac{1}{x-a} < 4\,|\,x-a\,|$ for $x < a$ and $x > a + \frac{1}{2}$

Exercises

1 The roots of the equation $x^2 - 4x + 2 = 0$ are denoted by α, β. Find the quadratic equation

whose roots are $\dfrac{\alpha+1}{\beta}$ and $\dfrac{\beta+1}{\alpha}$. (WJEC, 1994)

2 On the same diagram, sketch the graphs of

$$y = x \text{ and } y = |\,2x - 1\,|$$

(a) Find the coordinates of the points of intersection of the two graphs.
(b) Hence, or otherwise, find the set of values of x for which $|\,2x - 1\,| > x$. (L, 1992)

3 A householder wants to build a shed with a rectangular floor base in a triangular plot of land at the rear of the house, as shown. Naturally, the greatest floor area is desirable. Determine the appropriate dimensions for the floor.

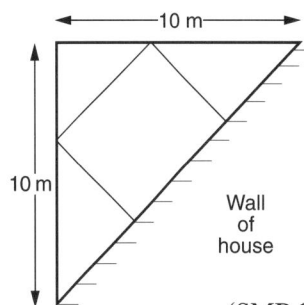

(SMP 16–19, pre-1993)

4 Solve the equation $\sqrt{(3x-2)} - \sqrt{(10-x)} = 2$ (when $\sqrt{}$ denotes positive root only).

5 Solve for a and b the simultaneous equations

$$a^2 + b^2 = \frac{13}{4}$$

$$ab = \frac{-3}{2}$$

6 The graphs of $y = (x+2)(x-3)$ and $y = 6 - 2x$ are sketched.

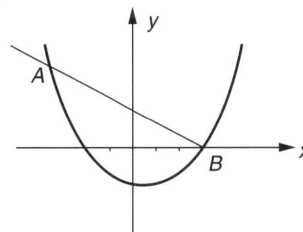

(a) Find the coordinates of the points A and B.
(b) Hence state the range of values for x for which $(x+2)(x-3) < 6 - 2x$.

(SMP Specimen Paper, 1991/92)

7 By treating the following equation as a quadratic in e^x, find the two values of x satisfying
$e^{2x} - 5e^x + 6 = 0$. (WJEC, 1994)

8 Given that f$(x) \equiv \dfrac{7}{(3x-1)(x+2)}$, express f$(x)$

in partial fractions. Sketch the curve $y = f(x)$ showing the asymptotes and the points of intersection of the curve with the axes.

9 One root of the equation $2x^3 - 3x^2 + px + 30 = 0$ is -3. Find the value of p and the other roots.

(SEB, 1992)

10 Solve the simultaneous equations

$$\log_2 xy = 7$$

$$\log_2 \frac{x^2}{y} = 5 \qquad \text{(AEB, 1991)}$$

11 Given that $g(x) \equiv \dfrac{3}{(1 + x)^2(1 + 2x)}$, express

g(x) in partial fractions and hence find the first three terms in the series when $g(x)$ is expanded in a series of ascending powers of $\dfrac{1}{x}$, stating the set

of values for which the series is valid.

12 Express $\dfrac{36 - 2x}{(2x + 1)(9 + x^2)}$ in partial fractions.

13 $f(x) \equiv \dfrac{9 - 3x - 12x^2}{(1 - x)(1 + 2x)}$

Given that $f(x) \equiv A + \dfrac{B}{1 - x} + \dfrac{C}{1 + 2x}$:

(a) Find the values of constants A, B and C.
(b) Given that $|x| < \frac{1}{2}$, expand $f(x)$ in ascending powers of x up to and including the term x^3, simplifying each coefficient.
(c) Hence, or otherwise, find the value of $f'(0)$.

(L, 1993)

14 Find the sets of values of x for which

(a) $|3x - 5| < 6$
(b) $(3x - 1)(x + 3) > 0$
(c) $|x^2 - 9| < 8$

15 Prove that, for all real x,

$$0 < \frac{1}{x^2 - 5x + 9} \leqslant \frac{4}{11}.$$

Sketch the curve $y = \dfrac{1}{x^2 - 5x + 9}$.

16 (*You must not use your calculator in this question.*)
If $\alpha = 3 + \sqrt{5}$ and $\beta = 3 - \sqrt{5}$, obtain $\alpha + \beta$ and $\alpha\beta$. Hence write down a quadratic equation, with integer coefficients, having α and β as its roots.

Show that $\alpha^3 + \beta^3 \equiv (\alpha + \beta)(\alpha^2 - \alpha\beta + \beta^2)$. Express $\alpha^2 + \beta^2$ in terms of $\alpha + \beta$ and $\alpha\beta$. Hence find the exact numerical value of $\alpha^3 + \beta^3$.

Hence, or otherwise, obtain a quadratic equation, with integer coefficients, having α^3 and β^3 as its two roots.

(O&C, 1993)

17 The graph of the curve with equation $y = 2x^3 + x^2 - 13x + a$ crosses the x-axis at the point $(2, 0)$.

(a) Find the value of a and hence write down the coordinates of the point at which the curve crosses the y-axis.
(b) Find, algebraically, the coordinates of the other points at which the curve crosses the x-axis.

(SEB, 1994)

18 (a) Sketch the graphs of $f(x) = |x + 2|$ and $g(x) = 3|x - 2|$ on the same set of axes.
(b) Find the set of real values of x for which $|x + 2| > 3|x - 2|$.

Brief Solutions

1 $\alpha + \beta = 4$, $\alpha\beta = 2$

Sum of roots $\dfrac{\alpha + 1}{\beta} + \dfrac{\beta + 1}{\alpha} \equiv \dfrac{\alpha^2 + \beta^2 + \alpha + \beta}{\alpha\beta}$

$$= \frac{(\alpha + \beta)^2 - 2\alpha\beta + (\alpha + \beta)}{\alpha\beta} = 8$$

Product of roots $= \dfrac{\alpha\beta + (\alpha + \beta) + 1}{\alpha\beta} = \frac{7}{2}$

2.3

Required equation is
$x^2 - 8x + \frac{7}{2} = 0 \Rightarrow 2x^2 - 16x + 7 = 0$

2 At A: $x = 1 - 2x \Rightarrow x = \frac{1}{3}, y = \frac{1}{3}$
At B: $x = 2x - 1 \Rightarrow x = 1, y = 1$
$|2x - 1| > x$ when $x < \frac{1}{3}$ or $x > 1$.

3.1

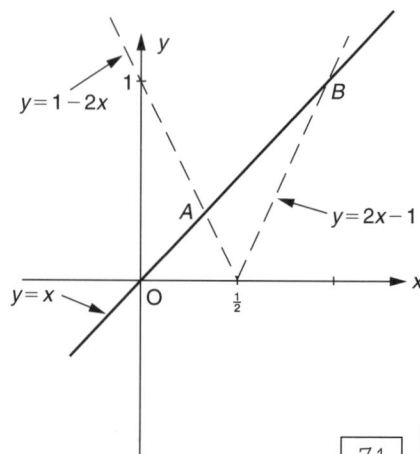

7.1

3 Angles are 45°. Dimensions as shown.
$AB = \sqrt{2}x \quad BC = 10 - \sqrt{2}x \quad BG = x$
$y = 10\sqrt{2} - 2BG = 10\sqrt{2} - 2x$
Area $a = xy = 10\sqrt{2}x - 2x^2$
$\qquad a = -2(x^2 - 5\sqrt{2}x)$

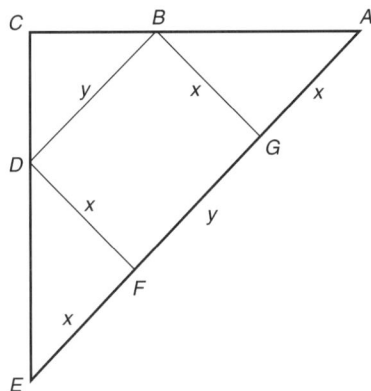

Find the maximum value by

$\quad$(i)$\quad \dfrac{\mathrm{d}a}{\mathrm{d}x} = -2(2x - 5\sqrt{2})$

$\qquad\qquad = 0$ when $x = \dfrac{5}{\sqrt{2}}$

$\quad$or (ii) Complete the square:

$\qquad a = -2\left(\left(x - \dfrac{5\sqrt{2}}{2}\right)^2 - \dfrac{25}{2}\right)$

From (ii) the maximum value of a is 25 when

$x = \dfrac{5\sqrt{2}}{2}, y = 5\sqrt{2}.$ $\boxed{\text{5.3 and 1.2}}$

4 $\sqrt{(3x - 2)} = 2 + \sqrt{(10 - x)}.$
Square both sides:
$3x - 2 = 4 + (10 - x) + 4\sqrt{(10 - x)}$
$\qquad\quad 4x - 16 = 4\sqrt{(10 - x)}$
Squaring again gives $x^2 - 7x + 6 = 0 \Rightarrow x = 6$ or 1.

$\boxed{\text{2.1a and 2.2a}}$

CHECK both of these in the original equation
$\quad x = 6:\qquad$ LHS $= \sqrt{16} = 4$
$\qquad\qquad\qquad$ RHS $= 2 + \sqrt{4} = 4$
$\quad x = 1:\qquad$ LHS $= \sqrt{1} = 1$
$\qquad\qquad\qquad$ RHS $= 2 + \sqrt{9} \neq 1$
$\therefore x = 6.$

5 $a^2 + 2ab + b^2 = \tfrac{1}{4}$, so $a + b = \pm\tfrac{1}{2}$.
Also, $a^2 - 2ab + b^2 = \tfrac{25}{4}$, so $a - b = \pm\tfrac{5}{2}$.

$\boxed{\text{2.1b and 7.2}}$

Having rearranged to
$\qquad a = \pm \sqrt{\tfrac{13}{4} - b^2}$

$\qquad a = -\dfrac{3}{2b}$

$\quad a + b = \tfrac{1}{2}, a - b = \tfrac{5}{2} \Rightarrow a = \tfrac{3}{2}, b = -1$
$\quad a + b = -\tfrac{1}{2}, a - b = \tfrac{5}{2} \Rightarrow a = 1, b = -\tfrac{3}{2}$
$\quad a + b = \tfrac{1}{2}, a - b = -\tfrac{5}{2} \Rightarrow a = -1, b = \tfrac{3}{2}$
$\quad a + b = -\tfrac{1}{2}, a - b = -\tfrac{5}{2} \Rightarrow a = -\tfrac{3}{2}, b = 1$

6 At A and B,
$(x + 2)(x - 3) = 6 - 2x \Rightarrow x^2 + x - 12 = 0$
$\Rightarrow x = -4$ or $3, y = 14$ or 0.
(a) $A(-4, 14), B(3, 0)$
(b) $(x + 2)(x - 3) < 6 - 2x$ when $-4 < x < 3$ $\boxed{\text{7.1}}$

7 $(\mathrm{e}^x)^2 - 5\mathrm{e}^x + 6 = 0 \quad \therefore \mathrm{e}^x = 2$ or $3 \Rightarrow x = \ln 2$ or $\ln 3$

$\boxed{\text{2.1a and 2.2a; 1.2}}$

8 $\dfrac{7}{(3x - 1)(x + 2)} \equiv \dfrac{3}{(3x - 1)} - \dfrac{1}{(x + 2)}.$ $\boxed{\text{4.3}}$

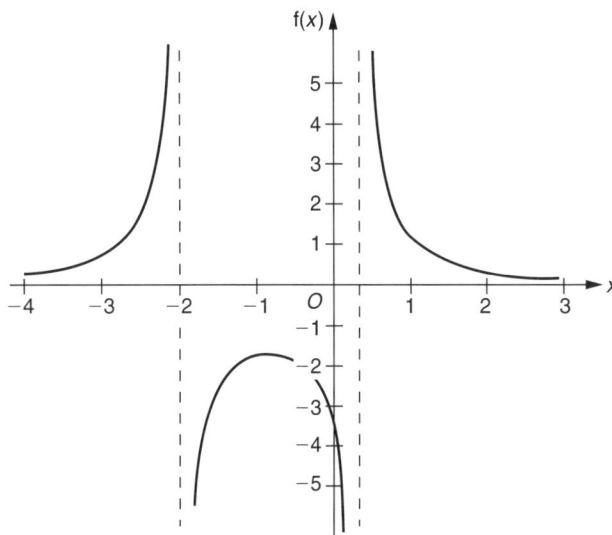

9 Substitute $x = -3$ to obtain $p = -17$.
$(x + 3)(2x^2 - 9x + 10) = 0 \Rightarrow x = -3, 2\tfrac{1}{2}$ or 2

$\boxed{\text{2.1a and 2.2a}}$

10 $\log_2 x + \log_2 y = 7$ $\qquad\qquad$ (1)
$2\log_2 x - \log_2 y = 5$ $\qquad\qquad$ (2)
Equations 1 and 2 give $\log_2 x = 4 \Rightarrow x = 2^4 = 16$,
$\log_2 y = 3 \Rightarrow y = 2^3 = 8$ ($x = 16, y = 8$)

11 $g(x) = \dfrac{-6}{1 + x} - \dfrac{3}{(1 + x)^2} + \dfrac{12}{1 + 2x}$ $\boxed{\text{4.3}}$

To obtain series in ascending powers of $\dfrac{1}{x}$, express

$\dfrac{1}{1 + x}$ in the form $\dfrac{1}{x}\left(1 + \dfrac{1}{x}\right)^{-1}$ and then use the

binomial

Valid for $\left|\dfrac{1}{x}\right| < 1 \Rightarrow |x| > 1 \Rightarrow x < -1$ or $x > 1.$

Valid | Not valid | Valid

$$\xleftarrow{\qquad} \underset{-1}{\bigg|} \qquad \underset{1}{\bigg|} \xrightarrow{\qquad}$$

$\boxed{8.1}$

$$\frac{-6}{1+x} = \frac{-6}{x}\left(1 + \frac{1}{x}\right)^{-1}$$

$$= \frac{-6}{x}\left(1 - \frac{1}{x} + \frac{1}{x^2} - \frac{1}{x^3} + \frac{1}{x^4} + \ldots\right)$$

$$\left(\text{for } \left|\frac{1}{x}\right| < 1\right)$$

$$\frac{-3}{(1+x)^2} = \frac{-3}{x^2}\left(1 + \frac{1}{x}\right)^{-2}$$

$$= \frac{-3}{x^2}\left(1 - \frac{2}{x} + \frac{3}{x^2} - \frac{4}{x^3} + \ldots\right)$$

$$\left(\text{for } \left|\frac{1}{x}\right| < 1\right)$$

$$\frac{12}{1+2x} = \frac{12}{2x}\left(1 + \frac{1}{2x}\right)^{-1}$$

$$= \frac{6}{x}\left(1 - \frac{1}{2x} + \frac{1}{4x^2} - \frac{1}{8x^3} + \frac{1}{16x^4} \ldots\right)$$

$$\left(\text{for } \left|\frac{1}{2x}\right| < 1\right)$$

Hence $g(x) = \dfrac{3}{2x^3} - \dfrac{15}{4x^4} + \dfrac{51}{8x^5} + \ldots$

$$\left(\text{valid for } \left|\frac{1}{x}\right| < 1\right).$$

12 $\dfrac{36 - 2x}{(2x + 1)(9 + x^2)} = \dfrac{A}{2x + 1} + \dfrac{Bx + C}{9 + x^2}$

$$= \frac{4}{2x + 1} - \frac{2x}{9 + x^2} \qquad \boxed{4.3}$$

13 $A + \dfrac{B}{(1 - x)} + \dfrac{C}{(1 + 2x)}$

$$= \frac{A(1 + x - 2x^2) + B(1 + 2x) + C(1 - x)}{(1 - x)(1 + 2x)}$$

$$\equiv \frac{9 - 3x - 12x^2}{(1 - x)(1 + 2x)}$$

Compare numerators:
$x = 1: 3B = -6 \Rightarrow B = -2$
$x = -\frac{1}{2}: \frac{3}{2}C = 7\frac{1}{2} \Rightarrow C = 5$
Constant term $A + B + C = 9 \Rightarrow A = 6$
$f(x) = 6 - 2(1 - x)^{-1} + 5(1 + 2x)^{-1}$
$\qquad = 9 - 12x + 18x^2 - 42x^3$
See particular series in Chapter 8
$f'(x) = -12 + \text{terms in } x: f'(0) = -12$

14 (a) $|3x - 5| < 6$
$\qquad \Rightarrow 3x - 5 < 6 \qquad$ or $\qquad 5 - 3x < 6$
$\qquad \Rightarrow x < \frac{11}{3} \qquad$ or $\qquad x > -\frac{1}{3}.$
$\qquad$ Thus $-\frac{1}{3} < x < \frac{11}{3}.$

(b) $(3x - 1)(x + 3) > 0.$
$\qquad$ From the sketch, $f(x) = (3x - 1)(x + 3),$
$\qquad f(x) > 0 \qquad$ when $x < -3 \qquad$ or $\qquad x > \frac{1}{3}.$

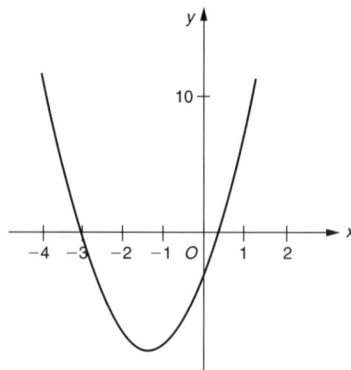

(c) $|x^2 - 9| < 8 \Rightarrow x^2 - 9 < 8 \qquad$ or $\qquad 9 - x^2 < 8.$
$\qquad$ These give respectively $x^2 < 17 \qquad$ or $\qquad x^2 > 1.$
$\qquad$ Thus $-\sqrt{17} < x < -1 \qquad$ or $\qquad 1 < x < \sqrt{17}.$

15 Let $y = \dfrac{1}{x^2 - 5x + 9}$

$$\Rightarrow yx^2 - 5xy + 9y - 1 = 0. \text{ (1)}$$

x is real, so $25y^2 \geqslant 4y(9y - 1) \Rightarrow 0 \geqslant y(11y - 4).$
Thus $0 \leqslant y \leqslant \frac{4}{11}.$
If $y = 0$, equation 1 gives $-1 = 0$, which is
unacceptable, hence $y > 0.$
If $y = \frac{4}{11}$, equation 1 gives $(2x - 5)^2 = 0$
$\Rightarrow x = \frac{5}{2}.$
Hence $0 < \dfrac{1}{x^2 - 5x + 9} \leqslant \dfrac{4}{11}.$

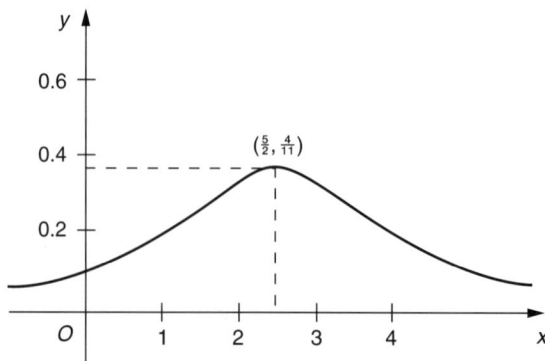

16 $\alpha + \beta = 6, \alpha\beta = 4 \Rightarrow x^2 - 6x + 4 = 0$ has roots
$3 \pm \sqrt{5}$
Without your calculator – see the question
Multiply $(\alpha + \beta)(\alpha^2 - \alpha\beta + \beta^2)$ and obtain $\alpha^3 + \beta^3$

$$\alpha^2 + \beta^2 = (\alpha + \beta)^2 - 2\alpha\beta$$

Substitute
$(\alpha^3 + \beta^3) = (\alpha + \beta)((\alpha + \beta)^2 - 3\alpha\beta) = 144$
$\Rightarrow \alpha^3\beta^3 = 64.$
Required equation is $x^2 - 144x + 64 = 0.$

2.3 for checking your answers

17 (a) $0 = 2(2)^3 + (2)^2 - 26 + a \Rightarrow a = 6.$ Curve
crosses the y-axis at $(0, 6).$
(b) $y = (x - 2)(2x^2 + 5x - 3)$
$= (x - 2)(x + 3)(2x - 1).$
Curve crosses the x-axis at $(-3, 0)$ and $(\frac{1}{2}, 0).$

4.1

18 (a)

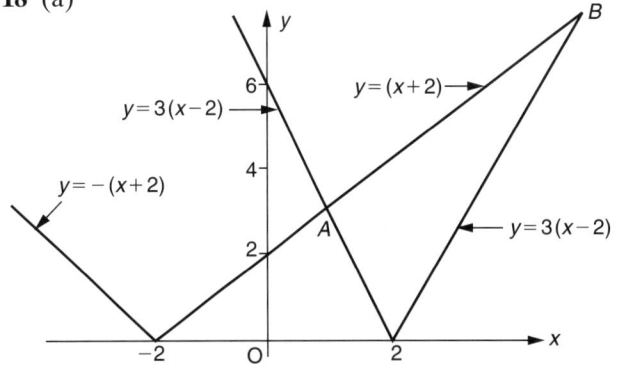

(b) At A: $6 - 3x = x + 2 \Rightarrow x = 1, y = 3$
At B: $3x - 6 = x + 2 \Rightarrow x = 4, y = 6$
$|x + 2| > 3|x - 2|$ in the interval $1 < x < 4.$

5

Graphs and coordinate geometry

AS Level			A Level						
CORE	MODULAR	TRADITIONAL	CORE	MODULAR	TRADITIONAL	Topic	Date attempted	Date completed	Self-assessment
✓	✓	✓	✓	✓	✓	**Lines**			
	✓	✓		✓	✓	**Circles**			
	✓	✓		✓	✓	**Ellipses and other conics**			
	✓	✓	✓	✓	✓	**Graphs of common functions**			
			✓	✓	✓	**Graphs of parametric equations**			

Fact Sheet

Line

- Equations

  (i) $y = mx + c$ m = gradient, c = y-intercept
  (ii) $y - y_1 = m(x - x_1)$ line passes through the point (x_1, y_1)
  (iii) $ax + by + c = 0$ general equation

- Distance of the point (h, k) from $ax + by + c = 0$ is

$$\left| \frac{ah + bk + c}{\sqrt{a^2 + b^2}} \right|$$

- The angle between two lines with gradients m_1 and m_2 is α where

$$\tan \alpha = \frac{m_1 - m_2}{1 + m_1 m_2}$$

  If $m_1 = m_2$, $\tan \alpha = 0$ $\Rightarrow$ lines are parallel.
  If $m_1 m_2 = -1$, $\tan \alpha = \infty$ $\Rightarrow$ lines are perpendicular.

Circle

(i) $(x - h)^2 + (y - k)^2 = r^2$ centre (h, k), radius r
    Parametric form: $x = h + r \cos t$, $y = k + r \sin t$

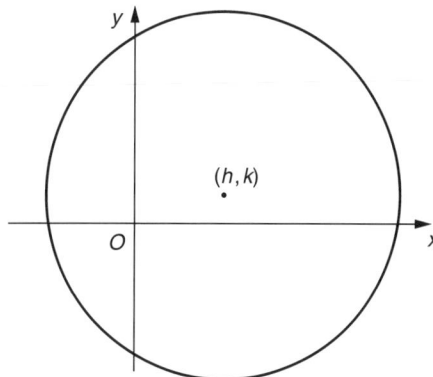

(ii) $x^2 + y^2 + 2gx + 2fy + c = 0$ centre $(-g, -f)$, radius $\sqrt{g^2 + f^2 - c}$

Ellipse

$$\frac{x^2}{a^2} + \frac{y^2}{b^2} = 1$$

Parametric form: $x = a \cos t$, $y = b \sin t$

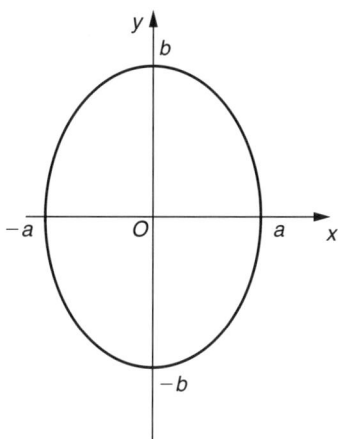

Parabola

$$y^2 = 4ax$$

Parametric form: $x = at^2$, $y = 2at$

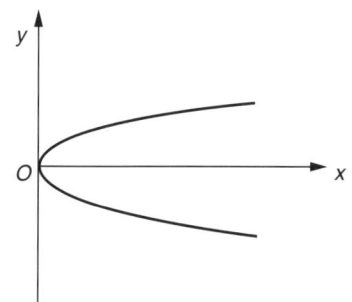

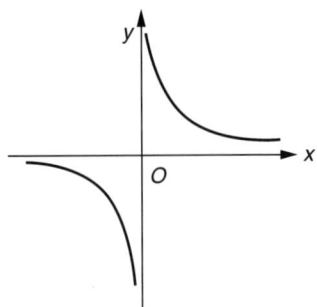

Rectangular Hyperbola

$$xy = c^2$$

Parametric form: $x = ct$, $y = \dfrac{c}{t}$

General Quadratic

$$y = ax^2 + bx + c$$

$a > 0$ $a < 0$

General Cubic

$$y = ax^3 + bx^2 + cx + d$$

$a > 0$ or   or

$a < 0$ or   or

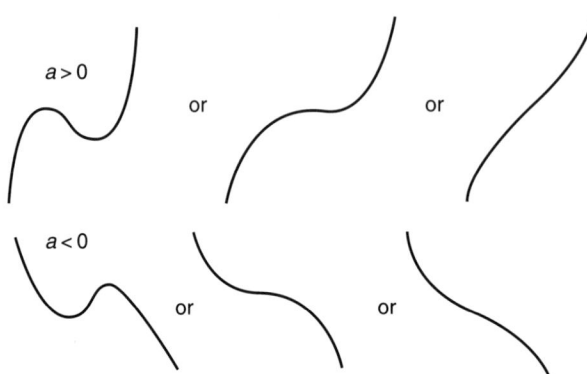

kx^n for Other Values of n

$n = \dfrac{1}{q}$, where q is a positive odd integer

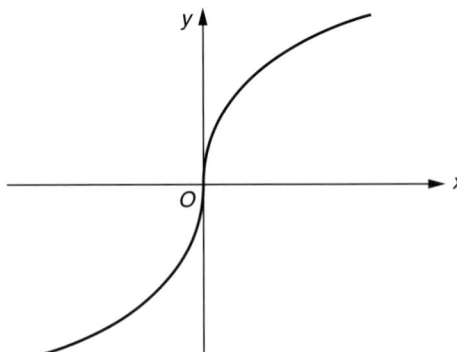

n is a negative even integer

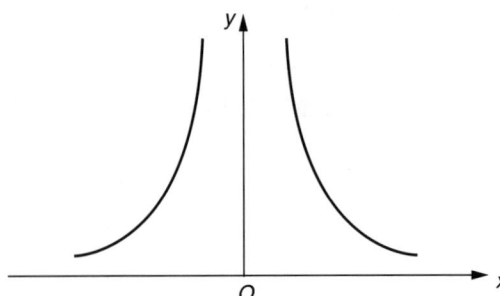

n is a negative odd integer

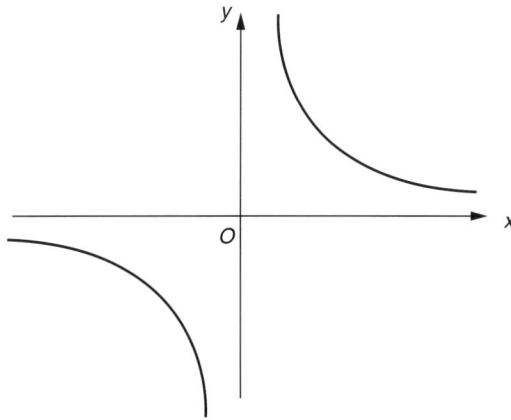

Worked Examples

1 The circle with equation
$x^2 + y^2 + 8x + 6y - 56 = 0$ has radius

A 5   **B** 9   **C** $\sqrt{56}$ **D** $\sqrt{103}$ **E** $\sqrt{156}$

- 'Complete the square' in x and y:

$$x^2 + 8x + y^2 + 6y = 56$$

Add $(4)^2$ and $(3)^2$ to each side:

$$(x^2 + 8x + 16) + (y^2 + 6y + 9) = 56 + 16 + 9$$
$$= 81$$
$$(x + 4)^2 + (y + 3)^2 = 81$$

centre $(-4, -3)$, radius 9   <u>Answer **B**</u>

2 If the points $P(h, k)$, $Q(1, 3)$ and $R(-2, 7)$ are
collinear, then the relationship connecting h and k
could be

A $3h + 4k = 5$ **B** $3k - 4h = 5$
C $3k + 4h = 13$ **D** $3h - 4k = 13$
E $3k + 4h = -13$

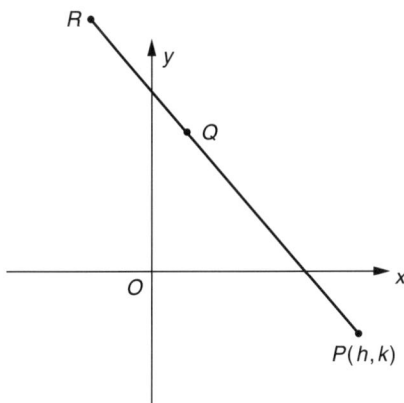

- PQ must have the same gradient as QR.

$$m = \frac{y_1 - y_2}{x_1 - x_2}$$

$$\frac{k - 3}{h - 1} = \frac{3 - 7}{1 - (-2)} \quad \left(= \frac{-4}{3} \right)$$

$\Rightarrow 3k - 9 = -4h + 4$ or $4h + 3k = 13$ <u>Answer **C**</u>

3 The line $2y = 3x + 6$ meets the y-axis at C. The
gradient of the line joining C to $A(4, -3)$ is

A $\frac{9}{4}$ **B** $\frac{2}{3}$ **C** $-\frac{2}{3}$ **D** $-\frac{3}{2}$ **E** $-\frac{9}{4}$ (SEB, 1993)

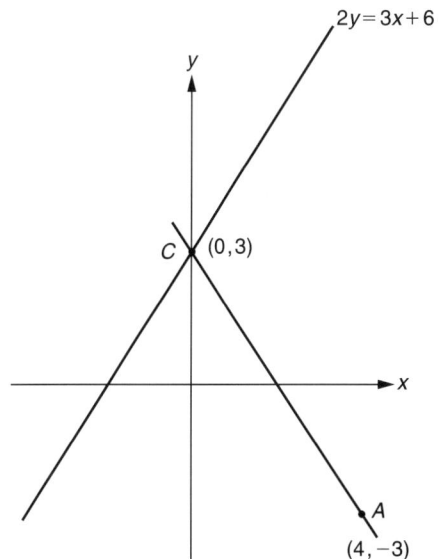

- At $C, x = 0, y = 3$

$$C(0, 3), A(4, -3) \quad \Rightarrow \quad \text{gradient} = \frac{3 - (-3)}{0 - 4}$$
$$= -\frac{3}{2}$$
<u>Answer **D**</u>

4 Sketch the curve $9x^2 + y^2 = 36$ and give the
coordinates of any intersections with the axes.
(UCLES, 1988)

- This is the equation of an ellipse   7.2
When $x = 0, y^2 = 36 \Rightarrow y = \pm 6$
$y = 0, x^2 = 4 \Rightarrow x = \pm 2$

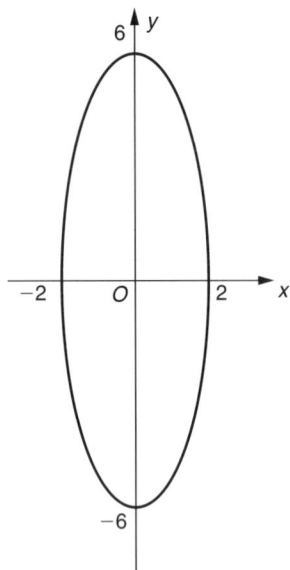

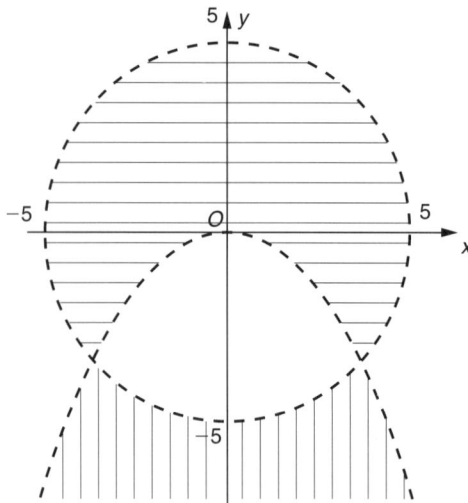

$(x^2 + y^2 - 25)(x^2 + 4y) < 0$ will be the shaded area.
Boundaries not included − draw them with a broken line

5 The centre of the circle
$2x^2 + 2y^2 + 12x - 4y + 1 = 0$ is the point
A $(-12, 4)$ **B** $(-6, 2)$ **C** $(6, -2)$ **D** $(3, -1)$
E $(-3, 1)$ (SEB, 1993)

• *Divide the equation by 2, and complete the square*
$x^2 + y^2 + 6x - 2y + \frac{1}{2} = 0$
$(x^2 + 6x + (3)^2) + (y^2 - 2y + 1^2) = -\frac{1}{2} + 3^2 + 1^2$
$(x + 3)^2 + (y - 1)^2 = 9\frac{1}{2}$
Centre $(-3, 1)$.   Answer **E**

6 Sketch on the same diagram the curves whose
equations are $x^2 + y^2 = 25$ and $x^2 + 4y = 0$. (Do not
calculate the points of intersection.) Shade in your
diagram the regions of the plane for which

$(x^2 + y^2 - 25)(x^2 + 4y) < 0$

• $x^2 + y^2 = 25$ is a circle centre $(0, 0)$, radius 5.
$x^2 + y^2 - 25 < 0$ for all points inside the circle
(shaded horizontally).
$x^2 + 4y = 0$ is a parabola $y = -\frac{1}{4}x^2$.
$x^2 + 4y < 0$ for points below the parabola (shaded
vertically).

7 Sketch the graph of $y = (x - 2)^2 - 4$, showing
clearly on your graph the coordinates of any
stationary points and of the intersections with the
axes.

Find the coordinates of the stationary points on the
graph of

$y = (x - 2)^3 - 12(x - 2)$

and sketch the graph, giving the exact coordinates
(in surd form where appropriate) of the
intersections with the axes.

Find the set of values of x for which

$(x - 2)^3 > 12(x - 2)$

Find the set of values of k for which the equation

$(x - 2)^3 - 12(x - 2) + k = 0$

has exactly one real root.   (UCLES, 1988)

• $y = (x - 2)^2 - 4$ is a parabola, line of symmetry
$x = 2$, minimum point $(2, -4)$.
$y = 0$ when $x = 0$ or 4. $\boxed{7.1}$

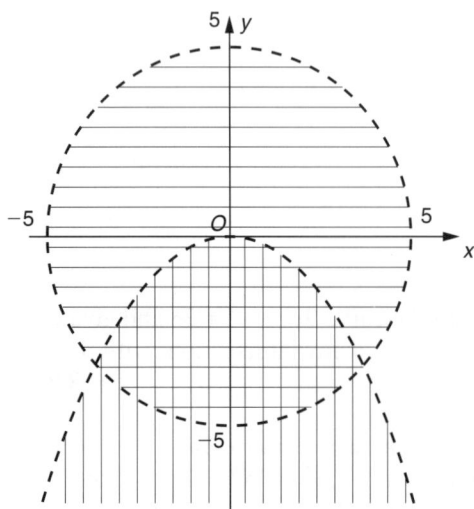

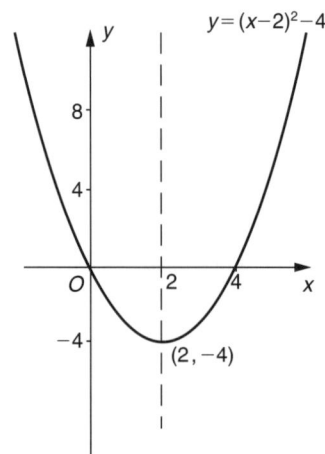

$y = (x - 2)^3 - 12(x - 2) = (x - 2)((x - 2)^2 - 12)$

$y = 0$ when $x = 2$, or $x - 2 = \pm\sqrt{12}$

$\Rightarrow \quad x = 2 \pm 2\sqrt{3}$

$\dfrac{dy}{dx} = 3(x - 2)^2 - 12 = 3((x - 2)^2 - 4)$

$\dfrac{dy}{dx} = 0$ when $x = 0$ or 4 (from above)

$\Rightarrow \quad y = 16$ or -16

Stationary points at $(0, 16)$ and $(4, -16)$. $\boxed{7.1}$

When $(x - 2)^3 > 12(x - 2)$ then $y > 0$

$\Rightarrow \quad -2\sqrt{3} + 2 < x < 2 \quad$ or $\quad x > 2\sqrt{3} + 2$

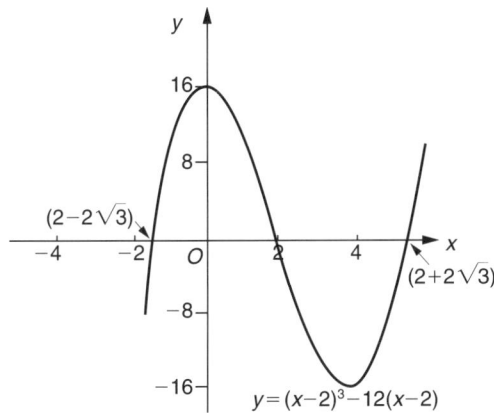

$y = (x-2)^3 - 12(x-2)$

If $(x - 2)^3 - 12(x - 2) + k = 0$ has exactly one real root then

$\quad k > 16$, i.e. graph is translated by $\begin{pmatrix} 0 \\ 16^+ \end{pmatrix}$

or $\quad k < -16$, i.e. graph is translated by $\begin{pmatrix} 0 \\ -(16^+) \end{pmatrix}$.

8 The figure shows a sketch of the curve with equation $y = f(x)$.

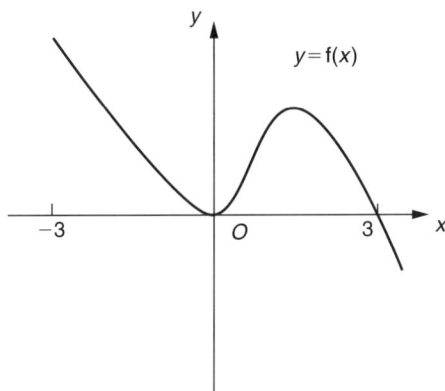

$y = f(x)$

In separate diagrams show, for $-3 \leqslant x \leqslant 3$, sketches of the curves with equations

(a) $y = f(-x)$
(b) $y = -f(x)$
(c) $y = f(|x|)$ $\boxed{7.4}$

Mark on each sketch the x-coordinate of any point, or points, where the curve touches or crosses the x-axis. (L, 1993)

- (a) $y = f(-x)$ is a reflection of $f(x)$ in the y-axis.
  (b) $y = -f(x)$ is a reflection of $f(x)$ in the x-axis.
  (c) $y = f(|x|)$ is a reflection of the curve $f(x)$ ($x \geqslant 0$) in the y-axis.

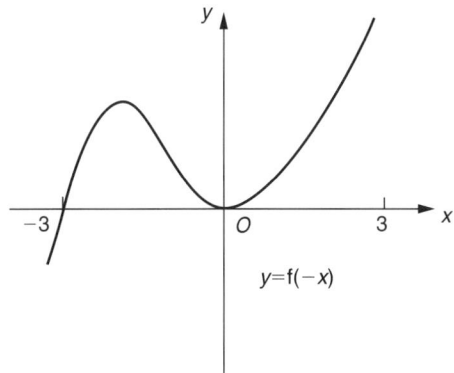

$y = f(-x)$

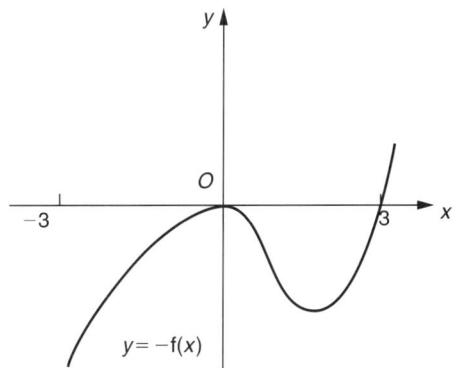

$y = -f(x)$

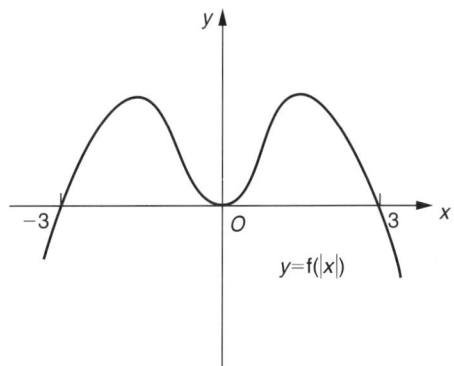

$y = f(|x|)$

Do not confuse $f(|x|)$ with $|f(x)|$
$y = |f(x)|$ is a reflection in the x-axis of the graph
$y = f(x)$ where y is negative

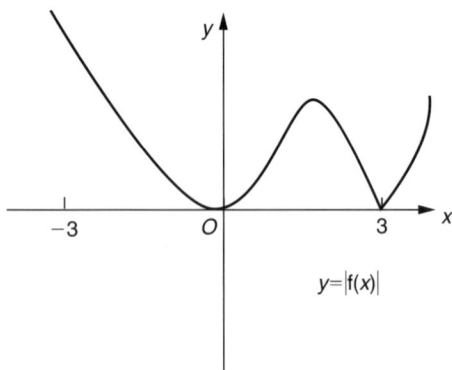

$y=|f(x)|$

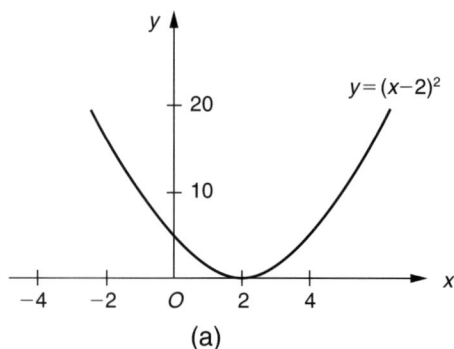

(a)

9 Express $5x^2 - 20x + 16$ in the form $a(x - b)^2 + c$. Show how the graph of $y = 5x^2 - 20x + 16$ may be obtained from the graph of $y = x^2$ by appropriate translations and one-way stretches. List these transformations clearly in the order of application.

• $5x^2 - 20x + 16 = 5(x^2 - 4x + \frac{16}{5})$

Completing the square with $x^2 - 4x$,

$x^2 - 4x + 4 = (x - 2)^2$

Thus

$x^2 - 4x + \frac{16}{5} = x^2 - 4x + 4 - \frac{4}{5} = (x - 2)^2 - \frac{4}{5}$

Hence

$5x^2 - 20x + 16 = 5[(x - 2)^2 - \frac{4}{5}]$
$\qquad\qquad\qquad = 5(x - 2)^2 - 4$

$y = 5x^2 - 20x + 16$ may be obtained from $y = x^2$ by:

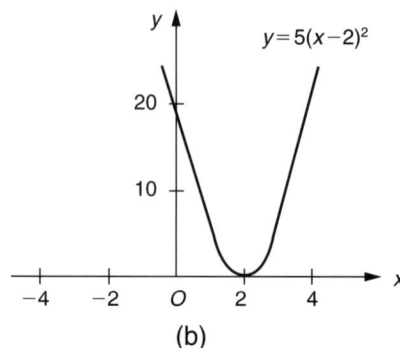

(b)

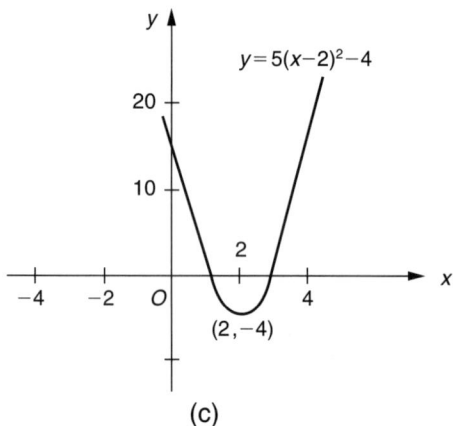

(c)

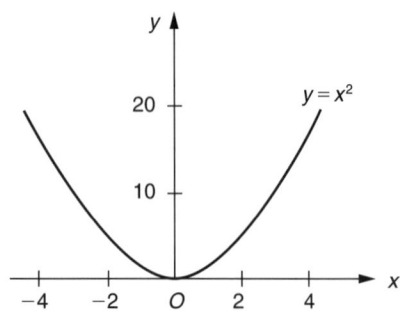

(a) Translation $\begin{pmatrix} 2 \\ 0 \end{pmatrix}$, i.e. 2 units in the direction of the x-axis.

(b) One-way stretch, or scaling, of factor 5 parallel to the y-axis.

(c) Translation $\begin{pmatrix} 0 \\ -4 \end{pmatrix}$, i.e. -4 units in the direction of the y-axis.

10 A function $g(x)$ of period 2π is defined by

$$g(x) = x^2 \quad \text{for } 0 \leqslant x \leqslant \frac{\pi}{2}$$

$$g(x) = \frac{\pi^2}{4} \quad \text{for } \frac{\pi}{2} < x \leqslant \pi$$

Given also that $g(x) = g(-x)$ for all x, sketch the graph of $g(x)$ for $-2\pi \leqslant x \leqslant 2\pi$. (L)

• $g(x) = x^2 \quad$ for $0 \leqslant x \leqslant \frac{\pi}{2}$ gives figure (a) (p. 61).

$g(x) = \frac{\pi^2}{4} \quad$ for $\frac{\pi}{2} < x \leqslant \pi$ gives figure (b) (p. 61).

Combining these two gives figure (c) (p. 61). $g(x) = g(-x)$ extends the graph to figure (d) (p. 61).

$g(x)$ has a period of 2π, so this part of the curve is one period. Extending it to $-2\pi \leqslant x \leqslant 2\pi$ gives figure (e) (p. 61).

(a)   (b)   (c)

(d)

(e)

11 (i)   Show that $(x - y)$ is a factor of

$$2x^3 - 3x^2y + y^3$$

and factorize the expression completely.

(ii)   Derive the equation of the tangent to the curve $y = x^{2/3}$ at the point $P(t^3, t^2)$.

(iii) If the tangent at P cuts the curve again at the point $Q(s^3, s^2)$ show that

$$2s + t = 0$$

(iv) For which values of t is the line PQ perpendicular to the tangent at Q? (NICCEA)

• (i) $2x^3 - 3x^2y + y^3 = (x - y)(2x^2 - xy - y^2)$
$$= (x - y)(2x + y)(x - y)$$
$$= (x - y)^2(2x + y)$$

(ii) $y = x^{2/3}$ $x = t^3$ $\dfrac{dx}{dt} = 3t^2$ $y = t^2$ $\dfrac{dy}{dt} = 2t$

$$\therefore \dfrac{dy}{dx} = \dfrac{2}{3t}$$

Equation of the tangent is $y - t^2 = \dfrac{2}{3t}(x - t^3)$

$$\Rightarrow \quad 3ty = 2x + t^3$$

(iii) If $Q(s^3, s^2)$ lies on the tangent at P then
$$3ts^2 = 2s^3 + t^3$$
$$\Rightarrow \quad 2s^3 - 3ts^2 + t^3 = 0$$

$$\Rightarrow \quad s = t \text{ (twice) or } s = -\dfrac{t}{2} \text{ (from (i))}$$

$$\Rightarrow \quad 2s + t = 0$$

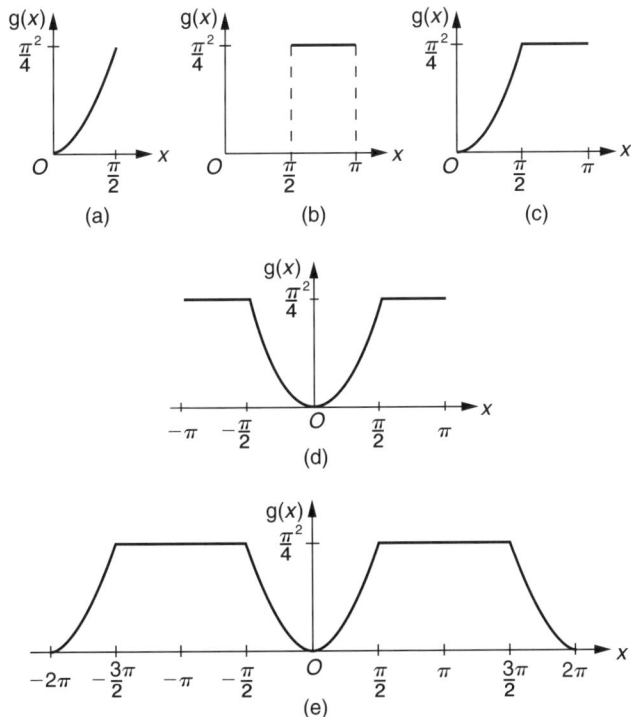

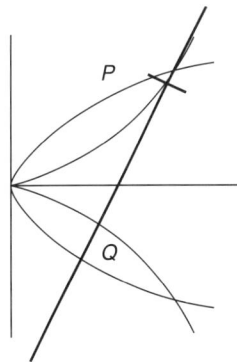

(iv) Gradient of the tangent at Q is $\dfrac{2}{3s} = -\dfrac{4}{3t}$.

Gradient of PQ is $\dfrac{2}{3t}$.

These are perpendicular if $-\dfrac{8}{9t^2} = -1$

$$\Rightarrow \quad t^2 = \dfrac{8}{9}$$

$$\Rightarrow \quad t = \pm\dfrac{2\sqrt{2}}{3}$$

Exercises

1   The equation of the line through the points $(1, -2)$ and $(-5, 6)$ is
A $4x - 3y = -1$ **B** $4x + 3y = -2$
C $3x + 4y = -5$ **D** $3x - 4y = 11$
E none of these

2   The locus of the points equidistant from the centres of the circles whose equations are
$x^2 + y^2 - 2x + 4y - 5 = 0$ and
$x^2 + y^2 - 10x - 8y + 2 = 0$ has equation
A $8x + 12y - 7 = 0$ **B** $6x + 2y + 3 = 0$
C $4x + 6y - 5 = 0$ **D** $2x + 3y - 9 = 0$
E none of these

3   The centre of the circle
$3x^2 + 3y^2 - 15x - 6y + 2 = 0$ is the point
A $(15, 6)$ **B** $(7.5, 3)$ **C** $(-5, -2)$ **D** $(-15, -6)$
E $(2.5, 1)$

4 A and B are the points $(2, 4)$ and $(4, 10)$ respectively. Find
(a) the equation of AB
(b) the equation of the perpendicular bisector of AB
(c) the equations of circles through A which touch both of the coordinate axes
(d) the perpendicular distance of the centres of the circles from the perpendicular bisector of AB
(e) the point other than A at which the line AB intersects the smaller circle.

5 The voltage V displayed on a screen is a function of the time t seconds. The graph of V is periodic with period 3 seconds and

$$V(t) = \begin{cases} 3 + t & \text{for } 0 \leq t \leq 2 \\ 9 - 2t & \text{for } 2 \leq t \leq 3 \end{cases}$$

Sketch the graph of $V(t)$ for $0 \leq t \leq 9$.

6 (i) Sketch the graph of the function f where

f: $x \to |3x - 1|$ for all $x \in \mathbb{R}$

(ii) Find the *two* values of x such that $|3x - 1| = x$.

(iii) Hence find the range of values of x for which $|3x - 1| > x$. (NICCEA, 1994)

7 Give two reasons why this graph cannot be the

graph of $\dfrac{1}{1 + x^2}$. (SMP, 1993)

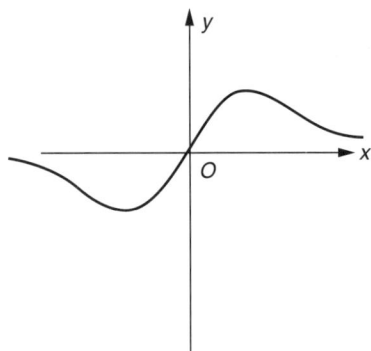

8 (a) Sketch, on the same diagram, the graphs of

$$y = \frac{1}{x} \text{ and } y = x - \frac{3}{2}.$$

Find the solution set of the inequality

$$x - \frac{3}{2} > \frac{1}{x}.$$

(b) Sketch, on separate diagrams, the graphs of

$$y = |x| \quad y = |x - 3| \quad y = |x - 3| + |x + 3|$$

Find the solution set of the equation

$$|x - 3| + |x + 3| = 6. \quad \text{(UCLES, 1989)}$$

9 A curve is defined parametrically by

$$x = t^2, y = \frac{2}{t}, t \neq 0.$$

(a) Sketch the curve.

(b) Find $\dfrac{dy}{dx}$ in terms of t.

Hence find the equation of the normal to the curve at the point where $t = 2$. (OLE, 1993)

10 Find the equation of the circle having the line

joining the points $(2, 1)$ and $(4, 7)$ as diameter. Find the equations of the tangents from the point $(3, 0)$ to this circle.

Brief Solutions

1 $(1, -2)$ does not satisfy **A**, but does satisfy **B**, **C** and **D**.

$(-5, 6)$ satisfies **B**, but not **C** or **D**. <u>Answer **B**</u>

2 Centres of circles are $C_1(1, -2)$ and $C_2(5, 4)$. $P(x, y)$ is equidistant from C_1 and C_2 if $(PC_1)^2 = (PC_2)^2$.

$\Rightarrow \quad (x - 1)^2 + (y + 2)^2 = (x - 5)^2 + (y - 4)^2$

$\Rightarrow \quad 8x + 12y = 36 \quad \Rightarrow \quad 2x + 3y = 9 \quad$ <u>Answer **D**</u>

3 $x^2 + y^2 - 5x - 2y + \frac{2}{3} = 0$ has centre $(\frac{5}{2}, \frac{2}{2})$.

<u>Answer **E**</u>

4 (a) $\dfrac{y - 4}{4 - 10} = \dfrac{x - 2}{2 - 4} \quad \Rightarrow \quad y = 3x - 2$

(b) Midpoint of AB is $(3, 7)$, perpendicular bisector has gradient $-\frac{1}{3}$, so equation is

$$y - 7 = -\tfrac{1}{3}(x - 3) \quad \Rightarrow \quad 3y = -x + 24$$

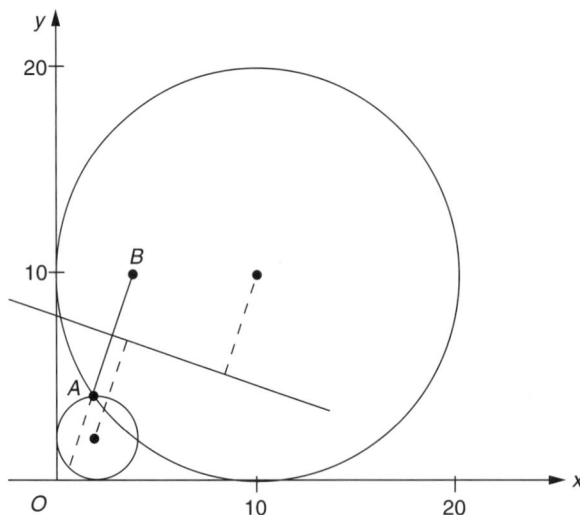

(c) If the circle has centre (h, k), in order to touch both axes $h = \pm k$ and radius $= |h|$, the equation must have the form $(x - h)^2 + (y \pm h)^2 = h^2$.

To go through $(2, 4)$, the circle lies in first quadrant, so $h = +k$ and

$(2 - h)^2 + (4 - h)^2 = h^2$

$\Rightarrow \quad h = 10 \quad \text{or} \quad h = 2$

Circles are $\quad (x - 10)^2 + (y - 10)^2 = 100$

and $\quad (x - 2)^2 + (y - 2)^2 = 4.$

(d) Centres are $(10, 10)$ and $(2, 2)$.

Perpendicular distance of $(10, 10)$ from $x + 3y - 24 = 0$ is

$$\left| \frac{10 + (3)(10) - 24}{\sqrt{(1^2 + 3^2)}} \right| = \frac{16}{\sqrt{10}}$$

Perpendicular distance of $(2, 2)$ from
$x + 3y - 24 = 0$ is

$$\left| \frac{2 + (3)(2) - 24}{\sqrt{(1^2 + 3^2)}} \right| = \frac{16}{\sqrt{10}}.$$

(e) AB has equation $y = 3x - 2$, and the smaller circle has equation $(x - 2)^2 + (y - 2)^2 = 4$. Substituting gives

$$(x - 2)^2 + (3x - 4)^2 = 4 \Rightarrow x = 2 \text{ or } x = \tfrac{4}{5}$$

Required point is $(\tfrac{4}{5}, \tfrac{2}{5})$.

5 This graph repeats itself every 3 seconds.
For $0 \leqslant t \leqslant 3$

t	0	1	2	3
$V(t)$	3	4	5	3

Join the points with straight lines.

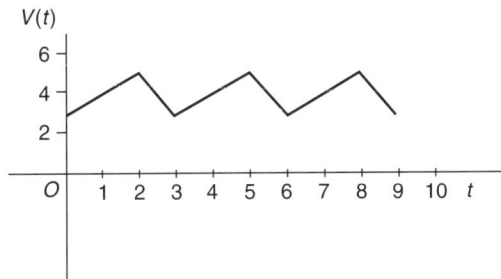

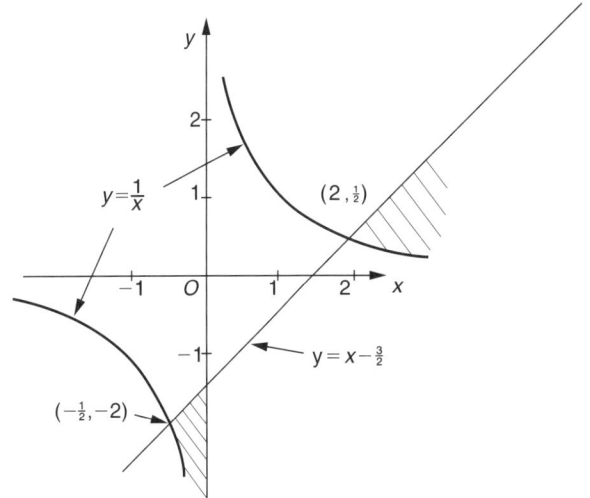

(b) The sketches are as shown.   7.1

6 At A: $1 - 3x = x \Rightarrow x = \tfrac{1}{4}, y = \tfrac{1}{4}$ 7.1
At B: $3x - 1 = x \Rightarrow x = \tfrac{1}{2}, y = \tfrac{1}{2}$
$|3x - 1| > x$ when $x < \tfrac{1}{4}$ or $x > \tfrac{1}{2}$ 3.1

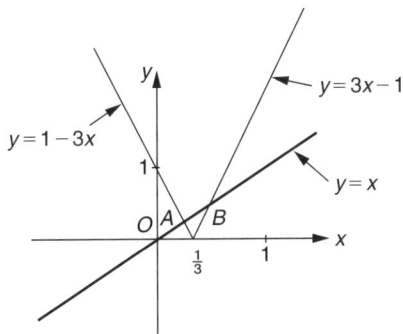

7 (i) $\dfrac{1}{1 + x^2}$ is positive for all x.

(ii) When $x = 0$, $y = 1$ not 0.

8 (a) At points of intersection $\dfrac{1}{x} = x - \dfrac{3}{2}$

$\Rightarrow x = -\tfrac{1}{2}$ or 2.   7.1

$x - \dfrac{3}{2} > \dfrac{1}{x}$ when $-\tfrac{1}{2} < x < 0$ or $x > 2$.   3.1

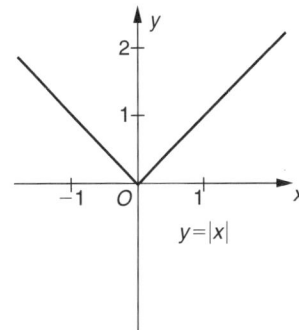

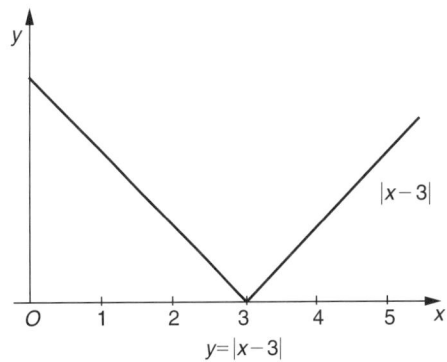

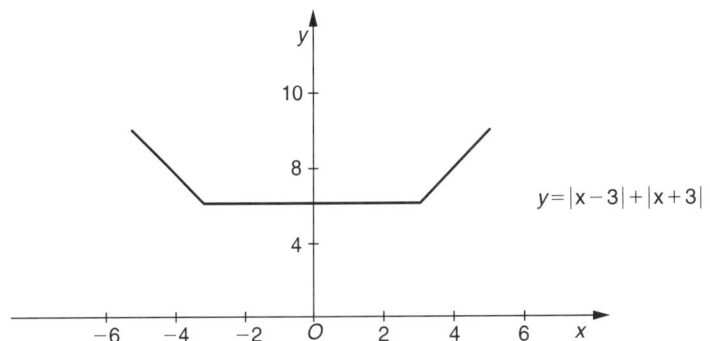

$|x - 3| + |x + 3| = 6$ for $-3 \leqslant x \leqslant 3$.   2.1a and 2.2a

9 (a)

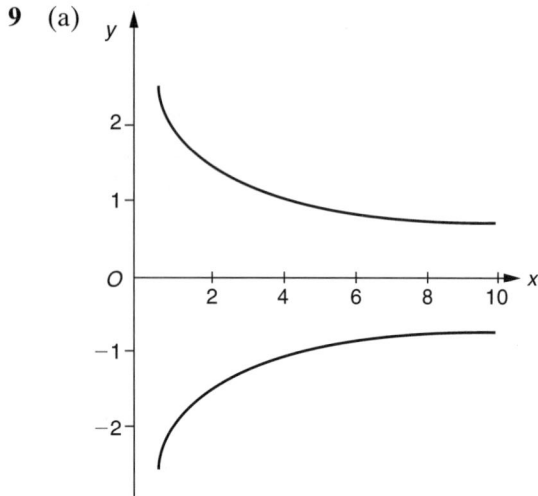

Since $x = t^2$, $x > 0$ for all t

t	±1	±2	±3
x	1	4	9
y	±2	±1	$\pm\frac{2}{3}$

(b) As $t \to 0$, $y \to \pm\infty$, $x \to 0^+$

$$\frac{dx}{dt} = 2t, \frac{dy}{dt} = \frac{-2}{t^2} \Rightarrow \frac{dy}{dx} = -\frac{1}{t^3}$$

When $t = 2$, $\dfrac{dy}{dx} = \dfrac{-1}{8}$

$\Rightarrow$ gradient of the normal = 8.

Equation of the normal is $y - 1 = 8(x - 4)$
$$\Rightarrow \quad y = 8x - 31$$

10 Centre of circle is $(3, 4)$.

Radius $= \sqrt{(1 + 9)} = \sqrt{10}$
$\Rightarrow \quad (x - 3)^2 + (y - 4)^2 = 10.$
Line through $(3, 0)$ is $y = m(x - 3)$.
To be a tangent, perpendicular distance from $(3, 4) = \sqrt{10}$.

$$\frac{m(3) - 4 - 3m}{\sqrt{(m^2 + 1^2)}} = \pm\sqrt{10} \quad \Rightarrow \quad 10m^2 = 6$$

$$\Rightarrow \quad m = \pm\sqrt{\tfrac{3}{5}}$$

Equations of tangents are $\quad y = \pm\sqrt{\tfrac{3}{5}}(x - 3)$.

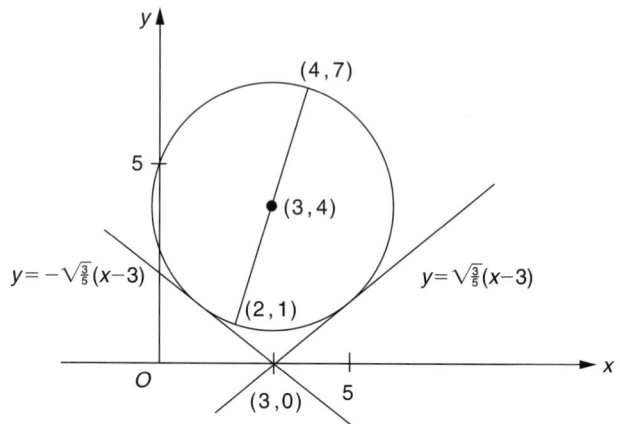

6

Plane trigonometry

AS Level			A Level			Topic	Date attempted	Date completed	Self-assessment
CORE	MODULAR	TRADITIONAL	CORE	MODULAR	TRADITIONAL				
✓	✓	✓		✓	✓	Sine and cosine formulae			
✓	✓	✓		✓	✓	Trigonometric functions			
	✓	✓	✓	✓	✓	Compound angle identities			
	✓	✓		✓	✓	(r, α) formula			

Fact Sheet

Angle Measure – Degrees and Radians

- 1 radian is defined as the angle subtended at the centre of a circle by an arc equal in length to the radius.

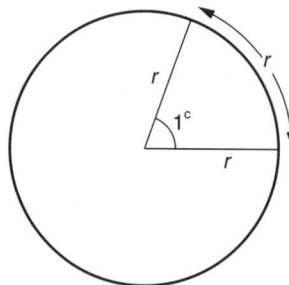

- 1 radian $= 1^c = \dfrac{180}{\pi}$ degrees; 1 degree $= 1° = \dfrac{\pi}{180}$ radians.

- Common conversions: since $180° = \pi$ radians, simple fractions of $180°$ can easily be expressed in radians, for example

$$30° = \frac{180°}{6} = \frac{\pi}{6} \text{ radians} \qquad 240° = \frac{4}{3} \times 180° = \frac{4\pi}{3} \text{ radians}$$

Circular Measure

- Length of an arc $= r\theta$
  Area of a sector $= \frac{1}{2}r^2\theta$ (θ measured in radians)

- Length of an arc $= \dfrac{\pi r\theta}{180}$
  (θ measured in degrees)
  Area of a sector $= \dfrac{\pi r^2\theta}{360}$

Trigonometry of the Right-Angled Triangle

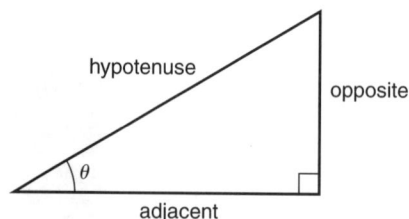

- $\sin\theta = \dfrac{\text{opposite}}{\text{hypotenuse}}$

- $\cos\theta = \dfrac{\text{adjacent}}{\text{hypotenuse}}$

- $\tan\theta = \dfrac{\text{opposite}}{\text{adjacent}} \left(= \dfrac{\sin\theta}{\cos\theta} \right)$

- Exact values of $\sin\theta$, $\cos\theta$ and $\tan\theta$ for some common angles follow from these right-angled triangles (values for $0°$ and $90°$ have also been included):

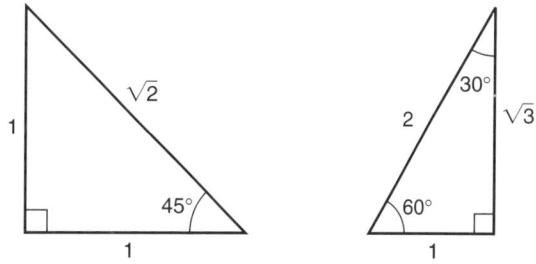

	0°	30°	45°	60°	90°
sin θ	0	$\dfrac{1}{2}$	$\dfrac{1}{\sqrt{2}}$	$\dfrac{\sqrt{3}}{2}$	1
cos θ	1	$\dfrac{\sqrt{3}}{2}$	$\dfrac{1}{\sqrt{2}}$	$\dfrac{1}{2}$	0
tan θ	0	$\dfrac{1}{\sqrt{3}}$	1	$\sqrt{3}$	undefined

Sine and Cosine Rules

(Although these can be applied to right-angled triangles, it is simpler in such cases to use the definitions given above.)

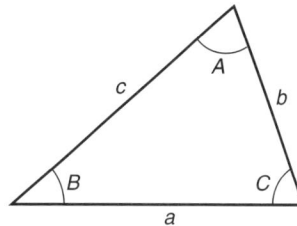

- Sine rule

$$\frac{a}{\sin A} = \frac{b}{\sin B} = \frac{c}{\sin C} = 2R \quad \text{where } R \text{ is the radius of the circumcircle}$$

- Cosine rule

$$a^2 = b^2 + c^2 - 2bc \cos A \quad \text{or} \quad \cos A = \frac{b^2 + c^2 - a^2}{2bc}$$

3D Lines and Planes

- Lines which are not parallel but do not intersect are called **skew lines**.
  Examples:
  (i) Edges AB and $A'D'$ of the cube shown.

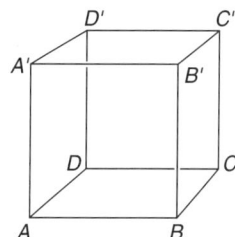

(ii) Edges VA and CB of the pyramid shown.

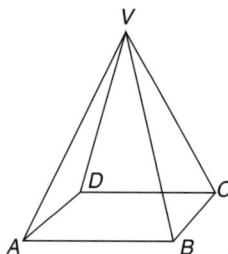

- Angle between a line and a plane

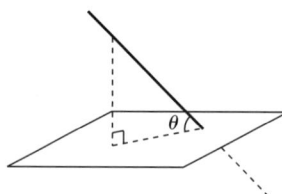

- Angle between two planes

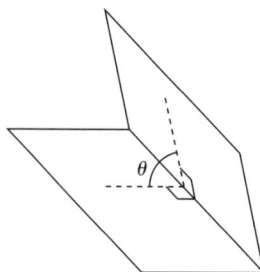

Sine, Cosine and Tangent of Any Angle

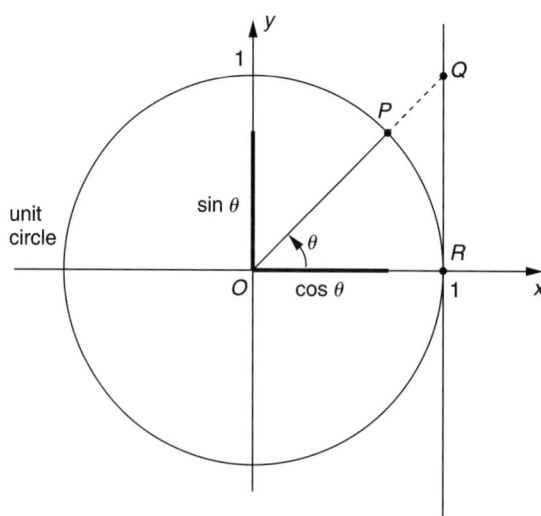

- $\sin \theta$ = projection of OP on the y-axis (= y-coordinate of P)
- $\cos \theta$ = projection of OP on the x-axis (= x-coordinate of P)

- $\tan \theta = QR = \dfrac{\sin \theta}{\cos \theta}$

Relating Angles in Other Quadrants to an Angle in the First Quadrant

- All the following relationships can easily be deduced from the symmetry of the graphs of the functions.

θ in 2nd quadrant	θ in 3rd quadrant	θ in 4th quadrant
$\sin \theta = \sin (180° - \theta)$	$\sin \theta = -\sin (\theta - 180°)$	$\sin \theta = -\sin (360° - \theta)$
$\cos \theta = -\cos (180° - \theta)$	$\cos \theta = -\cos (\theta - 180°)$	$\cos \theta = \cos (360° - \theta)$
$\tan \theta = -\tan (180° - \theta)$	$\tan \theta = \tan (\theta - 180°)$	$\tan \theta = -\tan (360° - \theta)$

Also, for negative angles,
$$\sin (-\theta) = -\sin \theta$$
$$\cos (-\theta) = \cos \theta$$
$$\tan (-\theta) = -\tan \theta$$

- The signs of $\sin \theta$, $\cos \theta$ and $\tan \theta$ in each of the quadrants can be summarized as follows – the diagram shows where each is *positive*:

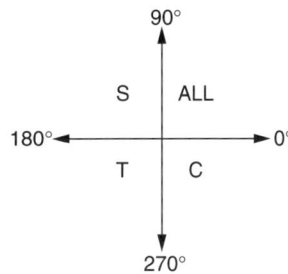

Cosecant, Secant and Cotangent

- $\operatorname{cosec} \theta = \dfrac{1}{\sin \theta}$

- $\sec \theta = \dfrac{1}{\cos \theta}$

- $\cot \theta = \dfrac{1}{\tan \theta}$

Graphs of the Three Trigonometric Functions

- $\sin \theta$ Period 360° or 2π radians
  $\sin \theta = \sin [180n° + (-1)^n \theta]$
  Amplitude $= 1$

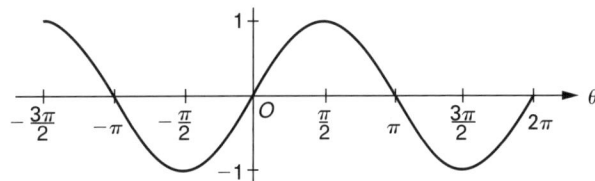

- $\cos \theta$ Period 360° or 2π radians
  $\cos \theta = \cos (360n° \pm \theta)$
  Amplitude $= 1$

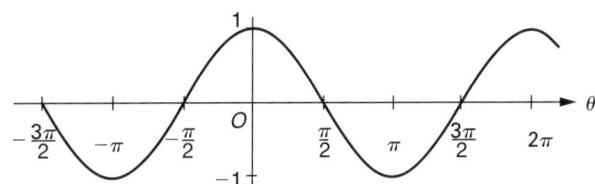

- tan θ Period 180° or π radians

  $\tan \theta = \tan (180n° + \theta)$

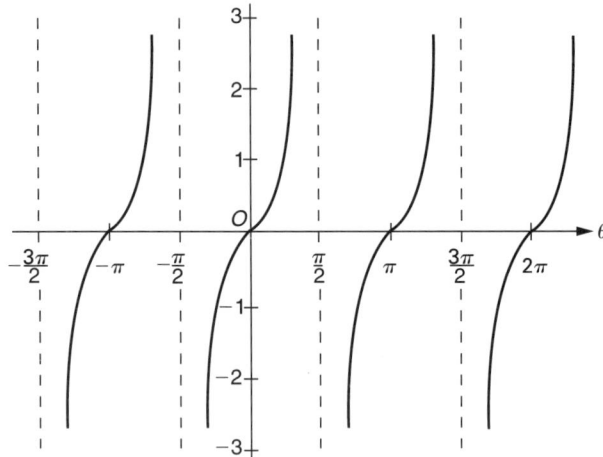

Transformations of the Graphs of sin x and cos x

- $a \sin bx$ is a stretch parallel to the y-axis with factor a, and a stretch parallel to the x-axis with factor $\dfrac{1}{b}$. It therefore has **amplitude** a,

  and **period** $\dfrac{360°}{b}\left(\text{or } \dfrac{2\pi}{b} \text{ radians}\right)$.

- $a \sin (bx + c)$ is a translation of $a \sin bx$ by $\begin{bmatrix} -c/b \\ 0 \end{bmatrix}$. It therefore has the same amplitude and period as $a \sin bx$,

  but it also has a **phase shift** of $\dfrac{-c}{b}$ from $a \sin bx$.

- $a \sin (bx + c) + d$ is a translation of $a \sin (bx + c)$ by $\begin{bmatrix} 0 \\ d \end{bmatrix}$.

All of these apply in the same way to $\cos x$.

Modelling Oscillations

- $A \sin (\omega t + \phi)$ represents an oscillation of amplitude A, time period

  $T = \dfrac{2\pi}{\omega}$ with a phase shift of $\dfrac{-\phi}{\omega}$ relative to $A \sin \omega t$.

Trigonometric Identities

Pythagorean Identities

- $\sin^2 A + \cos^2 A \equiv 1$
- $\tan^2 A + 1 \equiv \sec^2 A$
- $\cot^2 A + 1 \equiv \csc^2 A$

Compound Angle Identities

- Addition

  $\sin (A \pm B) \equiv \sin A \cos B \pm \cos A \sin B$

  $\cos (A \pm B) \equiv \cos A \cos B \mp \sin A \sin B$

  $\tan (A \pm B) \equiv \dfrac{\tan A \pm \tan B}{1 \mp \tan A \tan B}$

- Factor

$$\sin A + \sin B \equiv 2 \sin \left(\frac{A + B}{2} \right) \cos \left(\frac{A - B}{2} \right)$$

$$\sin A - \sin B \equiv 2 \cos \left(\frac{A + B}{2} \right) \sin \left(\frac{A - B}{2} \right)$$

$$\cos A + \cos B \equiv 2 \cos \left(\frac{A + B}{2} \right) \cos \left(\frac{A - B}{2} \right)$$

$$\cos A - \cos B \equiv -2 \sin \left(\frac{A + B}{2} \right) \sin \left(\frac{A - B}{2} \right)$$

- Product
  $\sin A \cos B \equiv [\sin (A + B) + \sin (A - B)]/2$
  $\cos A \sin B \equiv [\sin (A + B) - \sin (A - B)]/2$
  $\cos A \cos B \equiv [\cos (A + B) + \cos (A - B)]/2$
  $\sin A \sin B \equiv [\cos (A - B) - \cos (A + B)]/2$

Multiple Angle

- $\sin 2A \equiv 2 \sin A \cos A$
- $\cos 2A \equiv \cos^2 A - \sin^2 A \equiv 2 \cos^2 A - 1 \equiv 1 - 2 \sin^2 A$

- $\tan 2A \equiv \dfrac{2 \tan A}{1 - \tan^2 A}$

- $\cos^2 A \equiv \dfrac{1 + \cos 2A}{2}$

- $\sin^2 A \equiv \dfrac{1 - \cos 2A}{2}$

- $\sin 3A \equiv 3 \sin A - 4 \sin^3 A$
- $\cos 3A \equiv 4 \cos^3 A - 3 \cos A$

(r, α) Formula

- If a and b are positive,

$$a \cos \theta \pm b \sin \theta \equiv r \cos (\theta \mp \alpha_1)$$
$$b \sin \theta \pm a \cos \theta \equiv r \sin (\theta \pm \alpha_2)$$

- $r^2 = a^2 + b^2$, $\tan \alpha_1 = \dfrac{b}{a}$, $\tan \alpha_2 = \dfrac{a}{b}$

  (r positive, α_1 and α_2 acute)

't' Formulae

- If $t = \tan \dfrac{x}{2}$, then

$$\sin x \equiv \frac{2t}{1 + t^2}, \cos x \equiv \frac{1 - t^2}{1 + t^2}, \tan x \equiv \frac{2t}{1 - t^2}$$

Hints for Solutions

(i) When in doubt, or if more than two trigonometric functions are present, change all of the functions into sines and cosines.

(ii) Compare left-hand side and right-hand side. If the angles are all the same, e.g. all θ, use the basic identities and the Pythagorean identities. If the angles are different, use the compound angle identities.

(iii) In identity questions, start with the more complicated side, and work on one side at a time.

Solving Trigonometric Equations

* $\sin \theta = k$

1st solution:	$\theta_1 = \arcsin k$	(= principal value)
2nd solution:	$\theta_2 = 180° - \theta_1$	
Other solutions:	First two solutions + multiples of 360°	

* $\cos \theta = k$

1st solution:	$\theta_1 = \arccos k$	(= principal value)
2nd solution:	$\theta_2 = 360° - \theta_1$	
Other solutions:	First two solutions + multiples of 360°	

* $\tan \theta = k$

1st solution:	$\theta_1 = \arctan k$	(= principal value)
Other solutions:	First solution + multiples of 180°	

Note: If the angles are in radians then proceed as above except substitute π for 180° and 2π for 360°.

*   General Solutions

    $\sin \theta = k \Rightarrow \theta = 180n + (-1)^n \, \text{PV}$
    (PV = principal value = $\arcsin k$ or $\sin^{-1} k$)
    $\cos \theta = k \Rightarrow \theta = 360n \pm \text{PV}$
    (PV = principal value = $\arccos k$ or $\cos^{-1} k$)
    $\tan \theta = k \Rightarrow \theta = 180n + \text{PV}$
    (PV = principal value = $\arctan k$ or $\tan^{-1} k$)

    $\sin \theta = \sin \phi \Rightarrow \theta = 180n + (-1)^n \, \phi$
    $\cos \theta = \cos \phi \Rightarrow \theta = 360n \pm \phi$
    $\tan \theta = \tan \phi \Rightarrow \theta = 180n + \phi$

Note: If the angles are in radians then proceed as above except substitute π for 180° and 2π for 360°.

Worked Examples

1

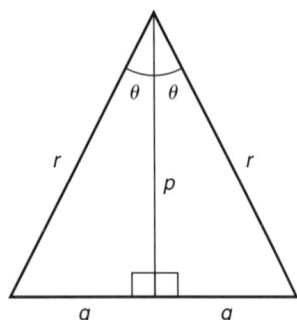

In the triangle above, $\sin 2\theta$ equals

A $\dfrac{pq}{r}$ **B** $\dfrac{pq}{r^2}$ **C** $\dfrac{2q}{r}$ **D** $\dfrac{2pq}{r}$ **E** $\dfrac{2pq}{r^2}$ (SEB, 1993)

* $\sin 2\theta = 2 \sin \theta \cos \theta$ (double angle formula)

$$= 2\left(\frac{q}{r}\right)\left(\frac{p}{r}\right)$$

$$= \frac{2pq}{r^2}$$

Answer **E**

2

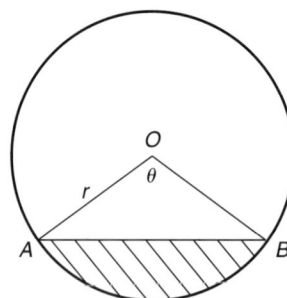

The diagram shows a circle with centre O and radius r, and a chord AB which subtends an angle θ radians at O. Express the area of the shaded segment bounded by the chord AB in terms of r and θ.

Given that the area of this segment is one-third of the area of triangle OAB, show that $3\theta - 4 \sin \theta = 0$.

Find the positive value of θ satisfying $3\theta - 4 \sin \theta = 0$ to within 0.1 radians, by tabulating values of $3\theta - 4 \sin \theta$ and searching for a sign change, or otherwise.

(UCLES Specimen Paper, 1994)

- Area of sector $OAB = \frac{1}{2}r^2\theta$
  Area of $\triangle OAB = \frac{1}{2}r^2 \sin \theta$
  $\therefore$ Area of shaded segment $= \frac{1}{2}r^2(\theta - \sin \theta)$

  Given $\qquad \frac{1}{2}r^2(\theta - \sin \theta) = \frac{1}{3}(\frac{1}{2}r^2 \sin \theta)$
  it follows that $\qquad \theta - \sin \theta = \frac{1}{3}\sin \theta$
  $\qquad \qquad \therefore 3\theta - 3\sin \theta = \sin \theta$
  $\qquad \qquad \quad 3\theta - 4\sin \theta = 0$

θ	$f(\theta) = 3\theta - 4\sin \theta$
1	-0.3659
2	2.3628
1.5	0.5102
1.25	-0.0459
1.35	0.1471

  7.5

  $\therefore 1.25 < \theta < 1.35 \quad \Rightarrow \quad \theta = 1.3$ to 1 d.p.
  This is basically a bisection process with the final value 1.35 chosen instead of 1.375 in order to establish the answer correct to 1 decimal place, without the need for another step

3 A pyramid has a horizontal square base $ABCD$ of side 10 cm and the vertex V is 6 cm vertically above the centre of the base.

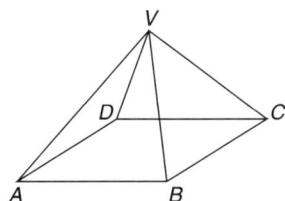

(a) Calculate the angle between VA and the horizontal, giving your answer to the nearest $0.1°$.
(b) Calculate the angle between the planes VAB and VCD, giving your answer to the nearest $0.1°$.
(c) Calculate the perpendicular distance from C to the edge AV (extended if necessary), giving your answer to the nearest 0.01 cm. (OLE, 1993)

●

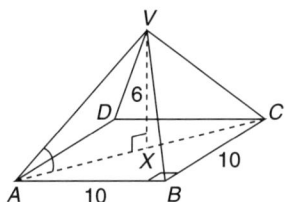

(a) The base is square so V lies directly above X, the mid-point of AC.
  $\triangle ABC$: $\qquad AC^2 = AB^2 + BC^2$
  $\qquad \qquad \quad AC = 10\sqrt{2}\left(=\sqrt{200}\right)$
  $\qquad \qquad \therefore AX = 5\sqrt{2}$
  $\qquad \therefore \angle VAC = \tan^{-1}\left(\dfrac{6}{5\sqrt{2}}\right)$
  $\qquad \qquad \qquad = 40.3°$

(b)

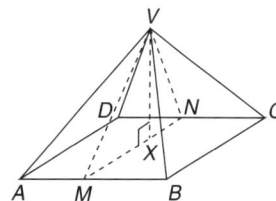

The required angle is $\angle MVN$, where M and N are the mid-points of AB and DC respectively.

$\triangle MVX$: $\qquad \angle MVX = \tan^{-1}\left(\dfrac{5}{6}\right)$

$\qquad \therefore \angle MVN = 2\tan^{-1}\left(\dfrac{5}{6}\right)$

$\qquad \qquad \qquad = 79.6°$

(c)

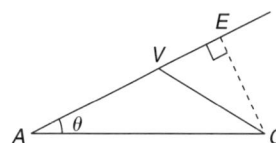

$CE = AC \sin \theta \; (\theta \text{ found in part (a)})$
$\qquad = 10\sqrt{2} \sin 40.3°$
$\qquad = 9.15$ cm

1.1

4 With the usual notation in a triangle XYZ, $x = 7$, $y = 4$, $z = 5$. Without using tables, find $\cos X$ as a fraction in its lowest terms and prove that

$$\sin Y = \frac{8\sqrt{6}}{35}.$$

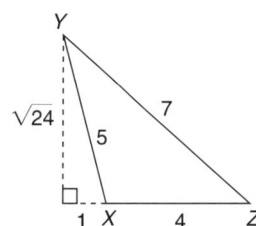

- Using the cosine rule for triangle XYZ,

$$\cos X = \frac{y^2 + z^2 - x^2}{2yz} = \frac{16 + 25 - 49}{(2)(4)(5)} = -\frac{1}{5}$$

If $\cos X = -\frac{1}{5}$ and $\sin^2 X + \cos^2 X = 1$ then

$$\sin^2 X = 1 - \left(-\frac{1}{5}\right)^2 = \frac{24}{25}$$

$$\Rightarrow \quad \sin X = \frac{\sqrt{24}}{5} = \frac{2\sqrt{6}}{5}$$

(positive value only since $0° < X < 180°$).
Using the sine rule for triangle XYZ,

$$\frac{\sin Y}{y} = \frac{\sin X}{x} \quad \Rightarrow$$

$$\sin Y = \frac{y \sin X}{x} = \frac{(4)(2\sqrt{6})}{(7)(5)} = \frac{8\sqrt{6}}{35}$$

5 In the diagram below ABC is an arc of a circle of radius 3 which is centred at O. *In terms of* π write down
(i) the length of the arc AB and
(ii) the area of the sector OAB.
(iii) Given that the area of the sector OAC is $6\pi/5$, obtain the value of the angle θ (in degrees).

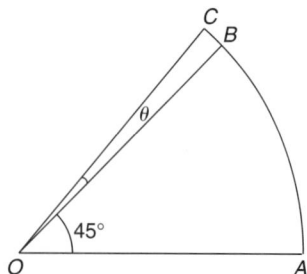

(NICCEA, 1993)

- (i) Arc $AB = 3 \times \dfrac{\pi}{4} = \dfrac{3\pi}{4}$

  (ii) Area of sector $OAB = \frac{1}{2} \cdot 3^2 \cdot \dfrac{\pi}{4} = \dfrac{9\pi}{8}$

  (iii) Area of sector $OAC = \frac{1}{2} \cdot 3^2 \cdot \left(\dfrac{\pi}{4} + \theta\right) = \dfrac{6\pi}{5}$

  $$\Rightarrow \quad \frac{9\pi}{4} + 9\theta = \frac{12\pi}{5}$$

  $$9\theta = \frac{3\pi}{20}$$

  $$\theta = \frac{\pi}{60} \text{ radians}$$

  $$= \frac{\pi}{60} \times \frac{180°}{\pi}$$

  $$= 3°$$

6

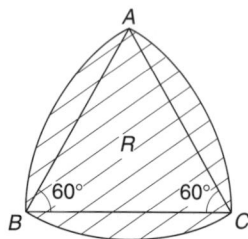

The triangle ABC is equilateral with each side of length 6 cm. With centre A and radius 6 cm, a circular arc is drawn joining B to C. Similar arcs are drawn with centres B and C and with radii 6 cm joining C to A and A to B respectively, as shown in

the figure. The shaded region R is bounded by the 3 arcs AB, BC and CA. Calculate, giving your answer in cm^2 to 3 significant figures,
(a) the area of triangle ABC,
(b) the area of R. (L, 1993)

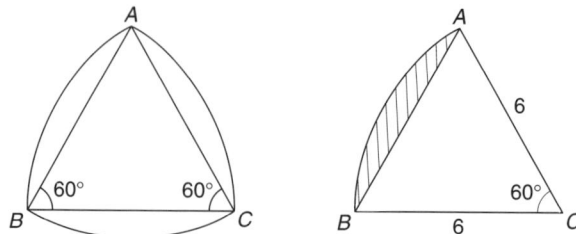

(a) Area of sector $ABC = \dfrac{60}{360}\pi 6^2 = 6\pi$ cm^2

Area of $\triangle ABC = \frac{1}{2}(6)(6)\sin 60°$
$= 9\sqrt{3}$ cm^2 ($= 15.6$ cm^2)

(b) Area of segment (shaded) $= 6\pi - 18\sin 60°$
$= 6\pi - 9\sqrt{3}$ cm^2

$\therefore$ Area of $R = 9\sqrt{3} + 3(6\pi - 9\sqrt{3})$
$= 18(\pi - \sqrt{3})$ cm^2 ($= 25.4$ cm^2)

7 $ABCD$ is one face of a cube and AA', BB', CC' and DD' are edges. The point E divides AA' internally in the ratio 2:3. Find
(a) the angle between CE and $C'D'$
(b) the angle between the planes BCE and ABC.

- (a) Lines CE and $C'D'$ are skew. The angle between them is the same as that between CE and any line parallel to $C'D'$, such as CD.

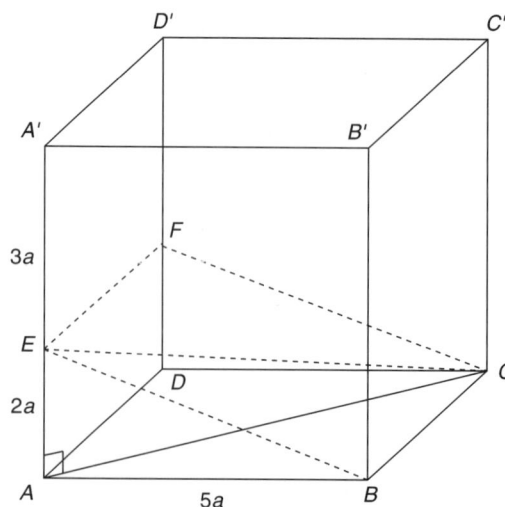

Let the length of each edge be $5a$ (since AA' has to be divided internally in the ratio 2:3).

The length of the diagonal of a face is $5\sqrt{2}a$.

In triangle EAC,
$EA = 2a$, $AC = 5\sqrt{2}a$, $\angle EAC = 90°$.

Hence $EC^2 = (2a)^2 + (5\sqrt{2}a)^2 = 54a^2$
$$\Rightarrow \quad EC = \sqrt{54}a$$

In triangle ECD,

$$\angle EDC = 90° \quad \Rightarrow \quad \cos C = \frac{5a}{\sqrt{54}a}$$

$$\angle C = 47.1°$$

Therefore the angle between CE and $C'D'$ is $47.1°$.

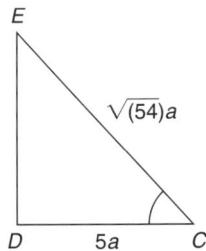

(b) The angle between planes BCE and ABC is $\angle EBA$.

$$\tan EBA = \frac{EA}{AB} = \frac{2a}{5a} \quad \Rightarrow \quad \angle EBA = 21.8°$$

Therefore the angle between the planes BCE and ABC is $21.8°$.

8 The diagram below shows the graph of $y = 2 \sin 2x + 1$ for $0 \leq x \leq \pi$.

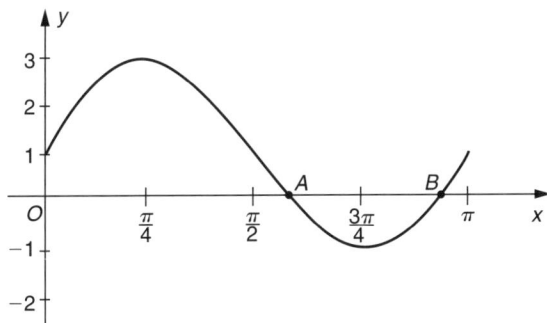

(a) Find the coordinates of A and B (as shown in the diagram) by solving an appropriate equation algebraically.
(b) The points $(0, 2)$ and $(\pi, 0)$ are joined by a straight line l. In how many points does l intersect the given graph?
(c) C is the point on the given graph with an x-coordinate of $\frac{\pi}{2}$. Explain whether C is above, below or on the line l.   (SEB, 1993)

• (a) At A and B, $y = 0$:

$$2 \sin 2x + 1 = 0$$
$$\sin 2x = -\tfrac{1}{2}$$
$$2x = \frac{7\pi}{6}, \frac{11\pi}{6}$$

$$x = \frac{7\pi}{12}, \frac{11\pi}{12} \qquad \boxed{9.1}$$

So the coordinates of A are $\left(\frac{7\pi}{12}, 0\right)$

and of B are $\left(\frac{11\pi}{12}, 0\right)$.

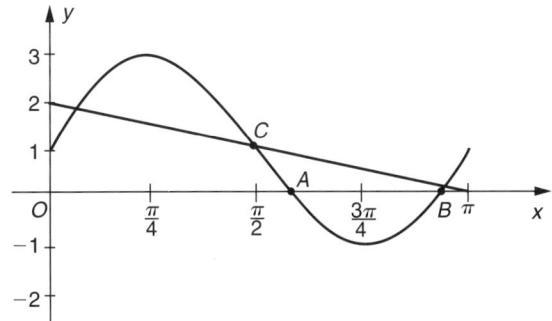

(b) 3

(c) On the line l, when $x = \frac{\pi}{2}$, $y = 1$.

On the curve, when $x = \frac{\pi}{2}$, $y = 2 \sin \pi + 1 = 1$.

Therefore the point C is on the line l.

9 (a) Write down the greatest and least values of the expression

$$3 - \cos 2x$$

as x varies.
Sketch the graph of the curve with equation $y = 3 - \cos 2x$ for $0 \leq x \leq 2\pi$, stating the values of x for which the curve has maximum and minimum values.
(b) Solve the equation

$$3 - \cos 2x = 7 \cos x$$

for values of x in the interval $0 \leq x \leq 2\pi$.
        (OLE, 1993)

• (a) $-1 \leq \cos 2x \leq 1$
$\therefore \quad 2 \leq 3 - \cos 2x \leq 4$
$\therefore$ Greatest value = 4, least value = 2

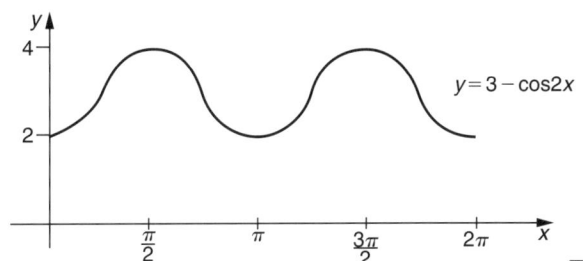

$y = 3 - \cos 2x$

$$\boxed{7.1}$$

Maximum values are when $x = \frac{\pi}{2}$ and $\frac{3\pi}{2}$.

Minimum values are when $x = 0$, π and 2π.

$3 - \cos 2x$ can be thought of as $-\cos 2x + 3$, so it is obtained from $\cos x$ by a stretch in the x direction with scale factor $\frac{1}{2}$ ($\cos 2x$) followed by a reflection in the x-axis ($-\cos 2x$) followed by a translation of $+3$ in the y direction

(b)
$$3 - \cos 2x = 7 \cos x$$
$$3 - (2 \cos^2 x - 1) = 7 \cos x \qquad \boxed{9.1}$$
$$2 \cos^2 x + 7 \cos x - 4 = 0$$
$$(2 \cos x - 1)(\cos x + 4) = 0$$
$$\cos x = \tfrac{1}{2} \text{ or } \cos x = -4 \text{ (no solutions)}$$

$$x = \frac{\pi}{3}, \frac{5\pi}{3}$$

10 A mass is suspended from the end of a spring as shown. The spring is stretched and then released so that it oscillates. The distance, d cm, of the mass from the support t seconds after its release is given by

$$d = 10 + 3 \cos 4t$$

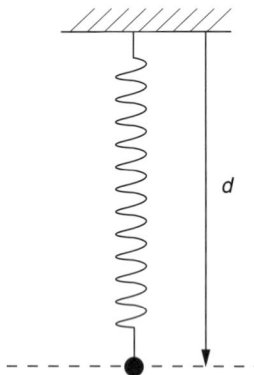

(a) State the distance of the mass from the support when the spring is released.
Find the period of oscillation of the mass.
(b) Calculate the time when d is first 9 cm.
(c) Calculate the speed of the mass 2 seconds after its release and state whether it is moving up or down. (NEAB, 1994)

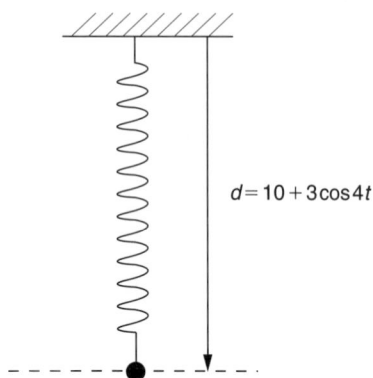

(a) When $t = 0$, $d = 10 + 3 \cos 0 = 13$ cm

$$\text{Period} = \frac{2\pi}{4} = \frac{\pi}{2} \text{ seconds}$$

(b) $9 = 10 + 3 \cos 4t$
$$\Rightarrow \quad \cos 4t = -\tfrac{1}{3}$$

We want the first positive solution of this equation:
$$4t = 1.9106 \qquad \text{Calculator must be in radians}$$
$$t = 0.48 \text{ seconds}$$

(c) velocity = rate of change of displacement

$$v = \frac{\mathrm{d}}{\mathrm{d}t}(10 + 3 \cos 4t)$$

$$= -12 \sin 4t$$

When $t = 2$, $v = -12 \sin 8 = -11.87$.
The negative sign indicates that the velocity is in the opposite direction to the (positive) displacement
So its speed is 11.87 cm s^{-1} and it is moving upwards.

11 (a) Find the general solution, in degrees, of the equation

$$\sin x + \sin 2x + \sin 3x = \sin x \sin 2x$$

(b) Find the general solution, in radians, of the equation

$$\sin 2\theta + \cos 2\theta = \sin \theta - \cos \theta + 1$$

• (a) By the factor formula,

$$\sin x + \sin 3x = 2 \sin 2x \cos x$$

Equation becomes

$$2 \sin 2x \cos x + \sin 2x - \sin x \sin 2x = 0$$
$$\Rightarrow \quad \sin 2x(2 \cos x - \sin x + 1) = 0$$

Either $\sin 2x = 0$, or $2 \cos x - \sin x + 1 = 0$.
When $\sin 2x = 0$, $2x = 180n°$, so $x = 90n°$.
When $2 \cos x - \sin x + 1 = 0$, then use either '$t$-substitutions' or compound angles.
$2 \cos x - \sin x = r \cos (x + \alpha)$, where
$r = \sqrt{(2^2 + 1^2)}$, and $\tan \alpha = 0.5$, so $\alpha = 26.6°$.

$$\sqrt{5} \cos (x + 26.6°) = -1$$

$$\Rightarrow \quad \cos (x + 26.6°) = -\frac{1}{\sqrt{5}}$$

$$\Rightarrow \quad x + 26.6° = 360n° \pm 116.6°$$

$x = 360n° + 90°$ or $360n° - 143.1°$.
$x = 360n° + 90°$ is included in the earlier solution $x = 90n°$.
Hence general solution:
$x = 90°n$ or $360n° - 143.1°$ (n is any integer).

(b) $\sin 2\theta + \cos 2\theta = \sin \theta - \cos \theta + 1$

It would be convenient to get rid of the $+1$ on the RHS
Either change $\cos 2\theta$ to $1 - 2 \sin^2 \theta$
or change $\cos 2\theta$ to $\cos^2 \theta - \sin^2 \theta$ and 1 to $\cos^2 \theta + \sin^2 \theta$

So
$2 \sin \theta \cos \theta + 1 - 2 \sin^2 \theta = \sin \theta - \cos \theta + 1$
(or $2 \sin \theta \cos \theta + \cos^2 \theta - \sin^2 \theta$

$= \sin\theta - \cos\theta + \cos^2\theta + \sin^2\theta)$, giving

$2\sin\theta\cos\theta - 2\sin^2\theta + \cos\theta - \sin\theta = 0$

$\Rightarrow \quad 2\sin\theta(\cos\theta - \sin\theta) + (\cos\theta - \sin\theta) = 0$

$\Rightarrow \quad (\cos\theta - \sin\theta)(2\sin\theta + 1) = 0.$

Either

$2\sin\theta + 1 = 0 \quad \Rightarrow \sin\theta = -0.5$

$$\Rightarrow \theta = n\pi + (-1)^n\left(-\frac{\pi}{6}\right)$$

or

$\cos\theta - \sin\theta = 0 \quad \Rightarrow \quad \tan\theta = 1$

$$\Rightarrow \quad \theta = n\pi + \frac{\pi}{4}$$

General solution:

$$\theta = n\pi + (-1)^{n+1}\left(\frac{\pi}{6}\right) \text{ or } (4n+1)\frac{\pi}{4}.$$

12 The figure shows a rectangle $PQSU$ containing a triangle PTR, right-angled at T.

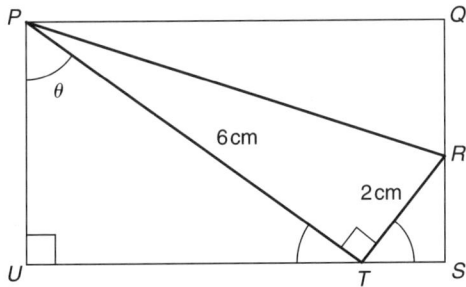

Given that PT has length 6 cm, TR is of length 2 cm and angle UPT is denoted by θ, show that

$$US = (6\sin\theta + 2\cos\theta)\text{ cm}$$

and express RQ in a similar form.
(a) In the special case when the rectangle $PQSU$ has perimeter 19 cm, show that

$$4\cos\theta + 3\sin\theta = 4.75$$

and hence find the two possible values of θ, to the nearest $0.1°$.
(b) Show that the triangle PQR has area A cm² where A can be written in the form

$$A = 8\sin 2\theta + 6\cos 2\theta$$

Hence determine the greatest value of A and the value of θ at which this occurs. (OLE, 1993)

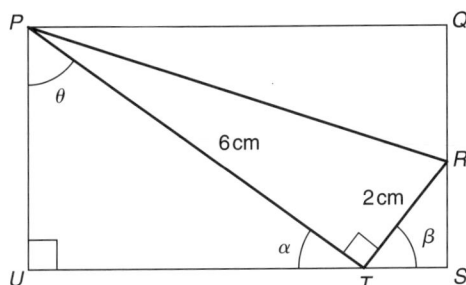

$$\alpha = 90 - \theta$$
$$\beta = 90 - \alpha = \theta$$
$$US = UT + TS$$
$$= (6\sin\theta + 2\cos\theta)\text{ cm}$$
$$RQ = SQ - SR$$
$$= UP - SR$$
$$= (6\cos\theta - 2\sin\theta)\text{ cm}$$

(a) $PQSU = 2(US + UP)$
$\qquad = 2(6\sin\theta + 2\cos\theta + 6\cos\theta) = 19$
$\qquad \Rightarrow \quad 6\sin\theta + 8\cos\theta = 9.5$
$\qquad \Rightarrow \quad 4\cos\theta + 3\sin\theta = 4.75$

Let
$4\cos\theta + 3\sin\theta = R\cos(\theta - \alpha)$
$\qquad\qquad = R\cos\theta\cos\alpha +$
$\qquad\qquad\quad R\sin\theta\sin\alpha$

Comparing coefficients of $\cos\theta$ and of $\sin\theta$:

$$4 = R\cos\alpha \quad (1) \qquad \boxed{9.2}$$
$$3 = R\sin\alpha \quad (2)$$

$(2) \div (1) \quad \Rightarrow \tan\alpha = \frac{3}{4} \Rightarrow \quad \alpha = 36.87°$
$(1)^2 + (2)^2 \quad \Rightarrow \quad 25 = R^2 \quad \Rightarrow \quad R = 5$

$\therefore 5\cos(\theta - 36.87°) = 4.75$
$\qquad \cos(\theta - 36.87°) = 0.95 \qquad \boxed{9.1}$
$\qquad\qquad \theta - 36.87° = 18.19°, -18.19°$
$\qquad\qquad\qquad \theta = 18.7°, 55.1°$

(b)　Area $\triangle PQR = A = \frac{1}{2}(PQ)(RQ)$
$\qquad\qquad = \frac{1}{2}(US)(RQ)$
$\qquad\qquad = \frac{1}{2}(6\sin\theta + 2\cos\theta)(6\cos\theta - 2\sin\theta)$
$\qquad\qquad = \frac{1}{2}(36\sin\theta\cos\theta - 12\sin^2\theta + 12\cos^2\theta$
$\qquad\qquad\qquad - 4\cos\theta\sin\theta)$
$\qquad\qquad = 6(\cos^2\theta - \sin^2\theta) + 16\sin\theta\cos\theta$
$\qquad\qquad = 6\cos 2\theta + 8\sin 2\theta$

Let $\quad 6\cos 2\theta + 8\sin 2\theta = R\cos(2\theta - \alpha)$
$\qquad\qquad = R\cos 2\theta\cos\alpha + R\sin 2\theta\sin\alpha$

Compare coefficients of $\cos 2\theta$ and of $\sin 2\theta$:

$$6 = R\cos\alpha \qquad (1)$$
$$8 = R\sin\alpha \qquad (2)$$

$(2) \div (1) \quad \Rightarrow \tan\alpha = \frac{4}{3} \Rightarrow \alpha = 53.13°$
$(1)^2 + (2)^2 \quad \Rightarrow \quad 100 = R^2 \Rightarrow \quad R = 10$
$\qquad\qquad\qquad\qquad\qquad\qquad \boxed{9.2}$

$\therefore A = 10\cos(2\theta - 53.13°)$

This has a maximum value of 10 when
$2\theta - 53.13° = 0$, i.e. when $\theta = 26.6°$
$(= \frac{1}{2}\tan^{-1}(\frac{4}{3}))$

The obvious alternative method for finding the maximum value of A would be to use calculus

$$\left(\frac{dA}{d\theta} = 0 \text{ at the maximum point}\right)$$

13 (a) (i) Show that

$$\tan 3\theta = \frac{3 \tan \theta - \tan^3 \theta}{1 - 3 \tan^2 \theta}$$

(ii) If $\theta = \tan^{-1}\left(\frac{1}{2}\right)$ and $\alpha = \tan^{-1}\left(\frac{9}{13}\right)$ show, **without using a calculator**, that $\tan(3\theta - \alpha) = 1$.

(b) (i) Express

$$4 \sin \theta - 3 \cos \theta$$

in the form $R \sin(\theta - \alpha)$, where R is positive and α is an acute angle.

(ii) Hence, or otherwise, find the greatest and least values of the expression

$$\frac{1}{10 - 3 \cos \theta + 4 \sin \theta} \qquad \text{(NICCEA, 1994)}$$

• (a) (i) $\tan 3\theta = \tan(2\theta + \theta)$

This is an identity so we could replace $=$ with $\equiv$

$$\Rightarrow \quad \tan 3\theta = \frac{\tan 2\theta + \tan \theta}{1 - \tan 2\theta \tan \theta}$$

(addition formula)

$$= \frac{\dfrac{2 \tan \theta}{1 - \tan^2 \theta} + \tan \theta}{1 - \left(\dfrac{2 \tan \theta}{1 - \tan^2 \theta}\right) \tan \theta}$$

(double angle formula)

$$\frac{\dfrac{2 \tan \theta + \tan \theta (1 - \tan^2 \theta)}{1 - \tan^2 \theta}}{\dfrac{1 - \tan^2 \theta - 2 \tan^2 \theta}{1 - \tan^2 \theta}}$$

$$= \frac{3 \tan \theta - \tan^3 \theta}{1 - 3 \tan^2 \theta}$$

(ii) $\tan(3\theta - \alpha) = \dfrac{\tan 3\theta - \tan \alpha}{1 + \tan 3\theta \tan \alpha}$ \qquad (1)

(addition formula)

But $\theta = \tan^{-1}\left(\frac{1}{2}\right) \Rightarrow \tan \theta = \frac{1}{2}$

$$\Rightarrow \quad \tan 3\theta = \frac{3\left(\frac{1}{2}\right) - \left(\frac{1}{2}\right)^3}{1 - 3\left(\frac{1}{2}\right)^2}$$

(using result from part (i))

$$\Rightarrow \quad \tan 3\theta = \frac{\frac{11}{8}}{\frac{1}{4}}$$

$$= \frac{11}{2}$$

and $\alpha = \tan^{-1}\left(\frac{9}{13}\right) \Rightarrow \tan \alpha = \frac{9}{13}$

Using (1) $\Rightarrow \tan(3\theta - \alpha) = \dfrac{\frac{11}{2} - \frac{9}{13}}{1 + \left(\frac{11}{2}\right)\left(\frac{9}{13}\right)}$

$$= \frac{\dfrac{143 - 18}{26}}{\dfrac{26 + 99}{26}}$$

$$= \frac{125}{125}$$

$$= 1$$

(b) (i) $4 \sin \theta - 3 \cos \theta \equiv R \sin(\theta - \alpha)$
$\equiv R \sin \theta \cos \alpha - R \cos \theta \sin \alpha$

(addition formula)

Comparing coefficients of $\sin \theta$ and of $\cos \theta$:

$$4 = R \cos \alpha \qquad (1)$$
$$3 = R \sin \alpha \qquad (2)$$

$(2) \div (1) \Rightarrow \tan \alpha = \frac{3}{4}, \; \alpha = \tan^{-1}\left(\frac{3}{4}\right)$
$$= 36.87°$$

`9.2`

$(1)^2 + (2)^2 \quad \Rightarrow \quad 25 = R^2$
$$\Rightarrow \quad R = 5$$

$\therefore 4 \sin \theta - 3 \cos \theta = 5 \sin(\theta - 36.87°)$

(ii) $\dfrac{1}{10 - 3 \cos \theta + 4 \sin \theta}$

$$= \frac{1}{10 + 5 \sin(\theta - 36.87°)}$$

Greatest value $= \dfrac{1}{10 + 5(-1)} = \dfrac{1}{5}$.

`7.1`

Least value $= \dfrac{1}{10 + 5(1)} = \dfrac{1}{15}$.

14 Given that

$$\cos \theta + \sqrt{3} \sin \theta \equiv R \cos(\theta - \alpha)$$

where $R > 0$ and $0 \leqslant \alpha \leqslant \dfrac{\pi}{2}$, calculate the exact values of R and α.

(a) Solve each of the following equations in the interval $0 \leqslant \theta \leqslant 2\pi$, giving your answers in terms of π:

(i) $\cos \theta + \sqrt{3} \sin \theta = \sqrt{3}$

(ii) $\cos 2\theta + \sqrt{3} \sin 2\theta = 2 \cos \theta$

(b) Find the coordinates of
(i) the maximum point and

(ii) the minimum point of the curve with the equation

$$y = \frac{1}{\cos x + \sqrt{3} \sin x + 3}, \quad -\pi \leqslant x \leqslant \pi$$

(AEB, 1994)

• $\cos \theta + \sqrt{3} \sin \theta \equiv R \cos (\theta - \alpha)$
$\equiv R \cos \theta \cos \alpha + R \sin \theta \sin \alpha$

Compare coefficients of $\sin \theta$ and of $\cos \theta$:

$$\sqrt{3} = R \sin \alpha \qquad (1)$$
$$1 = R \cos \alpha \qquad (2)$$

$(1) \div (2) \quad \tan \alpha = \sqrt{3} \Rightarrow \quad \alpha = \frac{\pi}{3} \qquad \boxed{9.2}$

$(1)^2 + (2)^2 \qquad 4 = R^2 \Rightarrow \quad R = 2$

(a) (i) $\cos \theta + \sqrt{3} \sin \theta = \sqrt{3}$

$$\therefore 2 \cos \left(\theta - \frac{\pi}{3} \right) = \sqrt{3}$$

$$\cos \left(\theta - \frac{\pi}{3} \right) = \frac{\sqrt{3}}{2}$$

$$\theta - \frac{\pi}{3} = -\frac{\pi}{6}, \frac{\pi}{6}, \ldots$$

$$\theta = \frac{\pi}{6}, \frac{\pi}{2} \qquad \boxed{9.1}$$

(ii) $\cos 2\theta + \sqrt{3} \sin 2\theta = 2 \cos \theta$

$$2 \cos \left(2\theta - \frac{\pi}{3} \right) = 2 \cos \theta$$

$$\cos \left(2\theta - \frac{\pi}{3} \right) = \cos \theta$$

$$\Rightarrow \quad 2\theta - \frac{\pi}{3} = \pm\theta, 2\pi \pm \theta, 4\pi \pm \theta, \ldots$$

Taking the different possibilities in turn:

$$2\theta - \frac{\pi}{3} = -\theta + 2\pi n \Rightarrow 3\theta = \frac{\pi}{3} + 2\pi n$$

$$\Rightarrow \theta = \frac{\pi}{9} + \frac{2}{3}\pi n$$

$$2\theta - \frac{\pi}{3} = \theta + 2\pi n \Rightarrow \theta = \frac{\pi}{3} + 2\pi n$$

Solutions are $\theta = \dfrac{\pi}{9}, \dfrac{\pi}{3}, \dfrac{7\pi}{9}, \dfrac{13\pi}{9} \qquad \boxed{9.1}$

(b) $\qquad y = \dfrac{1}{\cos x + \sqrt{3} \sin x + 3}$

$$= \frac{1}{2 \cos \left(x - \dfrac{\pi}{3} \right) + 3}$$

The function will be maximum when the denominator is minimum and vice versa.

The maximum value of $\cos \theta$ is $+1$ when $\theta = 0$ and the minimum value of $\cos \theta$ is -1 when $\theta = -\pi, \pi, \ldots$

Therefore the maximum value of $\cos \left(x - \dfrac{\pi}{3} \right)$

is 1 when $x - \dfrac{\pi}{3} = 0$, i.e. $x = \dfrac{\pi}{3}$, and the

minimum value of $\cos \left(x - \dfrac{\pi}{3} \right)$ is -1 when

$$x - \frac{\pi}{3} = -\pi \text{ (or } +\pi), \text{ i.e. } x = -\frac{2\pi}{3}$$

$$\left(\text{or } \frac{4\pi}{3}, \text{ which is outside the range} \right).$$

It follows that the minimum value of y is

$$\frac{1}{2(1) + 3} = \frac{1}{5} \text{ when } x = \frac{\pi}{3}$$

$\Rightarrow$ minimum point is $\left(\dfrac{\pi}{3}, \dfrac{1}{5} \right)$

and the maximum value of y is

$$\frac{1}{2(-1) + 3} = 1 \text{ when } x = -\frac{2\pi}{3}$$

$\Rightarrow$ maximum point is $\left(-\dfrac{2\pi}{3}, 1 \right). \qquad \boxed{7.1}$

15 Prove the identity

$$(\cos A + \cos B)^2 + (\sin A + \sin B)^2 \equiv 2 + 2 \cos (A - B)$$

(a) Given that
$f(x) = (\cos 7x + \cos x)^2 + (\sin 7x + \sin x)^2$,
  (i) show that $f(x) = 4 \cos^2 3x$
  (ii) write down the greatest and least values of $f(x)$
  (iii) evaluate $\displaystyle\int_0^{\pi/12} \frac{1}{f(x)} \, dx.$

(b) Solve the equation
$$(\cos 4\theta + \cos \theta)^2 + (\sin 4\theta + \sin \theta)^2 = 2\sqrt{3} \sin 3\theta$$

giving the general solution in degrees.

(AEB, 1993)

- LHS $=\cos^2 A + 2\cos A\cos B + \cos^2 B$
  $\qquad + \sin^2 A + 2\sin A\sin B + \sin^2 B$
  $= (\cos^2 A + \sin^2 A) + (\cos^2 B + \sin^2 B)$
  $\qquad + 2(\cos A\cos B + \sin A\sin B)$
  $= 2 + 2\cos (A - B)$

  $(\cos A + \cos B)^2 + (\sin A + \sin B)^2$
  $= 2 + 2\cos (A - B)$

  (a) (i) Use the proven identity with $A = 7x$ and $B = x$ to give

  $\quad f(x) = 2 + 2\cos (7x - x)$
  $\qquad = 2 + 2\cos 6x$
  $\qquad = 2 + 2(2\cos^2 3x - 1)$
  $\qquad = 4\cos^2 3x$

  (ii) Greatest value = 4, least value = 0.

  (iii) $\displaystyle\int_0^{\pi/12} \frac{1}{4\cos^2 3x}\,dx = \tfrac14\int_0^{\pi/12} \sec^2 3x\,dx$

  $\qquad\qquad = \tfrac{1}{12}[\tan 3x]_0^{\pi/12}$

  $\qquad\qquad = \tfrac{1}{12}\left(\tan \frac{\pi}{4} - \tan 0\right)$

  $\qquad\qquad = \tfrac{1}{12}$ $\boxed{6.2 \text{ and } 1.2}$

  (b) $(\cos 4\theta + \cos \theta)^2 + (\sin 4\theta + \sin \theta)^2$
  $\qquad\qquad = 2\sqrt3 \sin 3\theta$
  $\therefore 2 + 2\cos (4\theta - \theta) = 2\sqrt3 \sin 3\theta$
  $\qquad 2 + 2\cos 3\theta = 2\sqrt3 \sin 3\theta$
  $\qquad \sqrt3 \sin 3\theta - \cos 3\theta = 1$
  Let $\sqrt3 \sin 3\theta - \cos 3\theta = R \sin (3\theta - \alpha)$
  where $R^2 (\sqrt3)^2 + (1)^2 \Rightarrow R = 4$

  and $\tan \alpha = \dfrac{1}{\sqrt3}$ (see Fact Sheet)

  $\therefore \alpha = 30°$

  $\therefore 2\sin (3\theta - 30) = 1$
  $\quad \sin (3\theta - 30) = \tfrac12$
  $\quad 3\theta - 30 = 180n + (-1)^n(30)$
  $\qquad \theta = 60n + (-1)^n(10) + 10$
  $\qquad = 10(6n + 1 + (-1)^n)$

16 Solve the equation

$\quad 9\cos^2 x - 6\cos x - 0.21 = 0, 0° \leqslant x \leqslant 360°$

giving each answer in degrees to 1 decimal place.
$\qquad\qquad$ (L, 1992)

- $9\cos^2 x - 6\cos x - 0.21 = 0$

This is a quadratic equation in $\cos x$, so using the quadratic formula,

$\cos x = \dfrac{-(-6) \pm \sqrt{(-6)^2 - 4(9)(-0.21)}}{2(9)}$

$\qquad = 0.7$ or $-0.0\dot3$

$\cos x = 0.7 \quad \Rightarrow \quad x = 45.6°, 314.4°$
$\cos x = -0.0\dot3 \quad \Rightarrow \quad x = 91.9°, 268.1°$

$\therefore x = 45.6°, 91.9°, 268.1°, 314.4°$ $\boxed{9.2}$

17 (a) Show that $7\cos x - 4\sin x$ may be expressed in the form $R\cos (x + \alpha)$, where R is $\sqrt{65}$ and

$\tan \alpha = \dfrac{4}{7}.$

(b) Find, in radians to 2 decimal places, the smallest positive value of x for which $7\cos x - 4\sin x$ takes its maximum value.

(c) Find, in radians to 2 decimal places, the two smallest positive values of x for which

$7\cos x - 4\sin x = 4.88$

The curve C has equation

$y = (7\cos x - 4\sin x + 4)^{1/2}$

(d) Take corresponding values of y at $x = 0, \dfrac{\pi}{6}, \dfrac{\pi}{3}$

and $\dfrac{\pi}{2}$ for the curve C and use the trapezium

rule to find an estimate for the area of the finite region bounded by the curve C, the y-axis and

the x-axis for $0 \leqslant x \leqslant \dfrac{\pi}{2}$ giving your answer to

1 decimal place. $\qquad$ (L, 1993)

- (a) $7\cos x - 4\sin x = R\cos (x + \alpha)$
  $\qquad\qquad = R\cos x\cos \alpha - R\sin x\sin \alpha$

  Compare coefficients of $\cos x$ and $\sin x$:

  $7 = R\cos \alpha \qquad\qquad (1)$
  $4 = R\sin \alpha \qquad\qquad (2)$

  $(2) \div (1) \quad \Rightarrow \quad \tan \alpha = \dfrac{4}{7}$ as required

  $(1)^2 + (2)^2 \quad \Rightarrow \quad 7^2 + 4^2 = R^2(\cos^2 \alpha + \sin^2 \alpha)$
  $\qquad\qquad\qquad 65 = R^2(1)$
  $\qquad\qquad\qquad R = \sqrt{65}$ as required

  (b) $\sqrt{65} \cos (x + \alpha) = \sqrt{65}$
  $\qquad$ when $x + \alpha = 2\pi n$
  $\qquad\qquad x = 2\pi n - \alpha$

  Minimum positive value of x is $2\pi - \tan^{-1}\dfrac{4}{7}$ which is 5.76 to 2 d.p. $\boxed{7.1}$

(c) $7 \cos x - 4 \sin x = 4.88$
$\sqrt{65} \cos (x + \alpha) = 4.88$

$$\cos (x + \alpha) = \frac{4.88}{\sqrt{65}}$$

$$x + \alpha = 0.9206\ldots, 2\pi - 0.9206$$

$$x = 0.40, 4.84 \text{ radians} \quad \boxed{9.1}$$

(d) $y = (7 \cos x - 4 \sin x + 4)^{1/2} \quad \boxed{6.3}$

By the trapezium rule:

$$\text{Area} = \frac{\pi}{12}\left(y_0 + y\left(\frac{\pi}{2}\right) + 2\left(y\left(\frac{\pi}{6}\right) + y\left(\frac{\pi}{3}\right)\right)\right)$$

$$= \frac{\pi}{12}(3.3166 + 0 + 2(2.8394 + 2.009))$$

$$= 3.4 \text{ (to 1 d.p.)}$$

18 Find all the solutions in the interval $0° < x < 360°$ of the equation

$$3 \tan x + 2 \cos x = 0 \quad \text{(AEB, 1993)}$$

• $$3 \tan x + 2 \cos x = 0$$

$$3\left(\frac{\sin x}{\cos x}\right) + 2 \cos x = 0$$

$$3 \sin x + 2 \cos^2 x = 0$$
$$3 \sin x + 2(1 - \sin^2 x) = 0$$
$$2 \sin^2 x - 3 \sin x - 2 = 0$$
$$(2 \sin x + 1)(\sin x - 2) = 0$$
$$\sin x = -\tfrac{1}{2} \text{ or } \sin x = 2 \text{ (no solutions)}$$
$$x = 210°, 330° \quad \boxed{9.1}$$

Exercises

1 Which two of the following have the same least period?

(1) $\sin x$ (2) $\cos 2x$ (3) $1 + \sin 2x$ (4) $2 \cos \dfrac{x}{2}$

(SEB, 1993)

A (1) and (2) **B** (2) and (3) **C** (3) and (4)
D (1) and (4) **E** (1) and (3)

2 Two parts A and B of a wooden jigsaw fit together as shown in the diagram below.

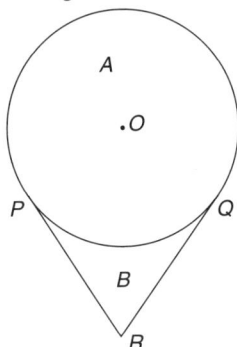

Part A is a circle, centre O, of radius 6 cm. The chord PQ has length 10 cm. Part B is bounded by the tangents to the circle at P and Q, which meet at R, and the minor arc PQ of the circle.
(i) Find the area of the minor sector OPQ.
(ii) Find the area of Part B. (NICCEA, 1994)

3 The tetrahedron $PQRS$ has a horizontal triangular base PQR and $PQ = PR = 8$ cm, $QR = 12$ cm. The vertex S is above the level of PQR and $SQ = SR = 10$ cm, $SP = 8$ cm.

(a) Show that the perpendicular distance from S to the plane PQR is $\sqrt{57}$ cm.
Find, to the nearest degree, the acute angle between
(b) the edge SQ and the horizontal
(c) the plane SQR and the horizontal. (L, 1993)

4 ABC is a plane triangle with $AB = 4$ cm, $AC = 5$ cm and angle $ACB = 30°$. Find, by calculation, the two possible lengths of BC, each correct to 3 significant figures, and the corresponding values of angle ABC, correct to the nearest tenth of a degree.

5 With the usual notation for triangle ABC prove that

$$\cos A = \frac{b^2 + c^2 - a^2}{2bc}$$

(assume that the triangle is acute-angled).
If two circles of radii 5 cm and 7 cm, centres C_1 and C_2 respectively, intersect at A and B, and $C_1C_2 = 10$ cm, calculate
(a) $\angle AC_1C_2$
(b) the area of triangle AC_1C_2
(c) the area common to both circles.
Give all answers to one decimal place.

6

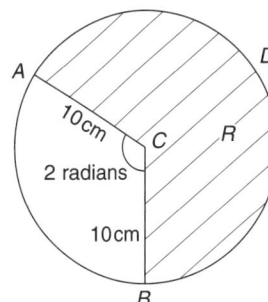

The figure represents a circle of radius 10 cm and centre C. Points A and B are taken on the circumference of the circle so that $\angle ACB = 2$ radians. The shaded region R is bounded by the radii CA and CB and the major arc ADB, as shown.
Calculate
(a) the perimeter of R

(b) the area of R
(c) the area, in cm², of $\triangle CAB$, giving your answer to 1 decimal place. (L, 1993)

7

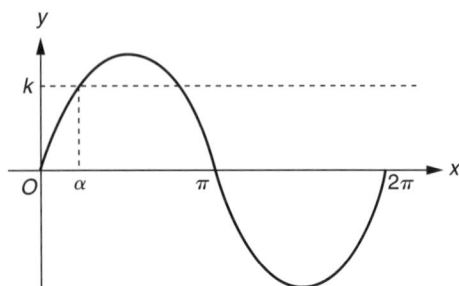

The diagram shows part of the graph of $y = \sin x$, where x is measured in radians, and values α on the x-axis and k on the y-axis such that $\sin \alpha = k$. Write down, in terms of α,
(i) a value of x between $\frac{1}{2}\pi$ and π such that $\sin x = k$
(ii) two values of x between 3π and 4π such that $\sin x = -k$. (UCLES Specimen Paper, 1994)

8 (i) Sketch on the same axes the graphs of

$y = \cos 2x$ and $y = 3 \sin x - 1$ for $0 \leqslant x \leqslant 2\pi$

(ii) Show that these curves meet at points whose x-coordinates are solutions of the equation

$2 \sin^2 x + 3 \sin x - 2 = 0$

(iii) Solve this equation to find the values of x in terms of π for $0 \leqslant x \leqslant 2\pi$. (MEI, 1993)

9 (i) The lengths of the sides of a triangle are 4 cm, 5 cm and 6 cm. The size of the largest angle of the triangle is θ.
(a) Calculate the value of $\cos \theta$.

(b) Hence, or otherwise, show that

$\sin \theta = \dfrac{a\sqrt{7}}{b}$, where a and b are integers.

(ii) Given that $0 \leqslant x \leqslant \pi$, find the values of x for which
(a) $\sin 3x = 0.5$

(b) $\cot \left(x + \dfrac{\pi}{2}\right) = 1$. (L Specimen Paper, 1996)

10 Find expressions for $\dfrac{\sin 2\theta}{\sin \theta}$, $\dfrac{\sin 3\theta}{\sin \theta}$ and $\dfrac{\sin 4\theta}{\sin \theta}$ in terms of $\cos \theta$.

11 (a) (i) Given that

$a \cos \theta + b \sin \theta \equiv r \sin (\theta + \alpha)$

(where $r > 0$) show that
$r = \sqrt{a^2 + b^2}$

and obtain an expression for $\tan \alpha$ in terms of a and b.
(ii) Determine all the values of θ between 0° and 360° satisfying the equation

$4 \sin \theta + 2 \cos (\theta + 30°) = 1$

(b) Prove the identity

$\tan A + \cot A \equiv 2 \operatorname{cosec} 2A$

The angle θ lies between 0 and $\dfrac{\pi}{4}$ and

$\sin 2\theta = \dfrac{4}{5}$. Without using a calculator, find the

numerical value of $\tan \theta$. (WJEC, 1993)

12 By squaring both sides of the identity $\sin^2 \theta + \cos^2 \theta \equiv 1$, prove that

$4(\sin^4 \theta + \cos^4 \theta) \equiv 3 + \cos 4\theta$

(a) Evaluate

$$\int_0^{\pi/8} \frac{\sin 4\theta}{\sin^4 \theta + \cos^4 \theta}\, d\theta$$

(b) Find the general solution in radians of the equation

$4(\sin^4 \theta + \cos^4 \theta) = 2 - \cos 2\theta$ (AEB, 1994)

13 Show that

$\cos (A - B) - \sin (A + B)$
$\qquad = (\cos A - \sin A)(\cos B - \sin B)$
for all A and B.

Find, in degrees, all solutions of the equation

$\cos x - \sin 3x = 0$

for which $-90° \leqslant x \leqslant 90°$. (AEB, 1984)

14 Express $3 \cos x - 4 \sin x$ in the form $A \cos (x + \alpha)$, where $A > 0$ and α is acute, stating the value of α to the nearest 0.1°.

(a) Given that $f(x) = \dfrac{24}{3 \cos x - 4 \sin x + 7}$:

(i) write down the greatest and least values of $f(x)$ and the values of x to the nearest 0.1° in the interval $-180° < x < 180°$ at which these occur

(ii) find the general solution, in degrees, of the equation

$f(x) = \dfrac{16}{3}$.

(b) Solve the equation $3 \cos x - 4 \sin x = 5 \cos 3x$, giving your answers to the nearest 0.1° in the interval $0 < x < 180°$. (OLE, 1992)

15 Prove the identity $\dfrac{1}{\sin 2x} + \dfrac{1}{\tan 2x} \equiv \cot x$.

Hence
(a) show that $\cot 15° = 2 + \sqrt{3}$

(b) solve the equation $\dfrac{1}{\sin 3\theta} + \dfrac{1}{\tan 3\theta} = 2$

giving your answers in the interval $0 < \theta < 180°$, to the nearest $0.1°$. (OLE, 1992)

16 Find, in radians, the general solution of the equation

$$\cos 2\theta + \cos \theta = 0 \qquad \text{(L, 1994)}$$

17 Find the positive constant R and the acute angle A for which

$$\cos x + \sin x \equiv R \cos (x - A)$$

(a) Find the general solution of the equation

$$\cos x + \sin x = 1$$

(b) Deduce the greatest value of $\sin x + \cos x$.
(L Specimen Paper, 1996)

18 Prove that $\sin 3A = 3 \sin A - 4 \sin^3 A$. Given that $\sin 3x = \sin^2 x$, find three possible values for $\sin x$.

Hence find all of the solutions of the equation $\sin 3x = \sin^2 x$ for $90° \leqslant x \leqslant 270°$. Using the same axes, sketch the graphs of the functions $\sin 3x$ and $\sin^2 x$ for $90° \leqslant x \leqslant 270°$. Find the subset of values of x for which $\sin 3x < \sin^2 x$.

(ii) Area of B = Area of $POQR$ − Area of minor sector OPQ
$$= 2 \times \tfrac{1}{2} \times 6 \times 6 \tan \theta - 35.46$$
$$= 18.81 \text{ cm}^2$$

3

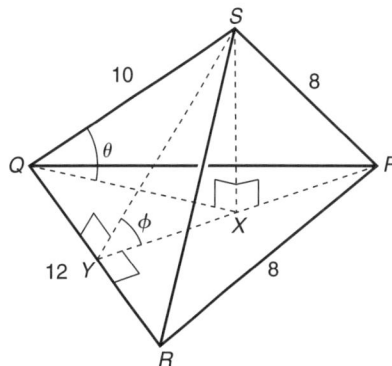

(a) $PY = \sqrt{64 - 36} = 2\sqrt{7}$
$SY = \sqrt{100 - 36} = 8$

$\triangle SYP$ is iscoceles $\Rightarrow SX = \sqrt{57}$

(b) In $\triangle SQX$, we seek $\sin^{-1}\left(\dfrac{SX}{SQ}\right) = 49°$

(c) In $\triangle SYX$, $\sin^{-1}\left(\dfrac{SX}{SY}\right) = 71°$

4

Brief Solutions

1 Periods (1) 2π, (2) π, (3) π, (4) 4π
Answer **B** $\boxed{7.1}$

2

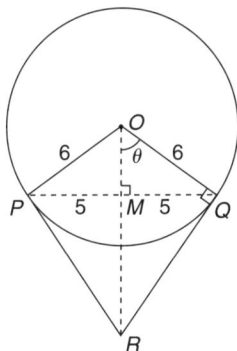

(i) $\theta = 56.4°$

Area of sector $OPQ = \pi\left(\dfrac{56.4}{180}\right)6^2 = 35.46 \text{ cm}^2$

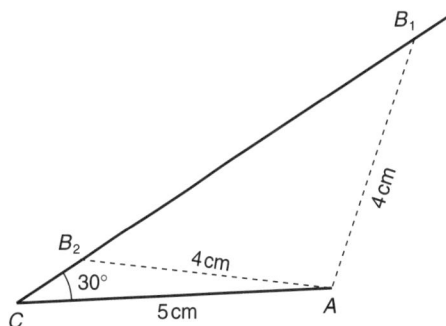

$\triangle ABC$ cosine rule:
$16 = BC^2 + 25 - 2(BC)(5) \cos 30°$
$\Rightarrow BC = 7.45$ or 1.21 cm

$\triangle ABC$ sine rule:

$\sin \angle ABC = \dfrac{5 \sin 30°}{4} = 0.625$

$\Rightarrow \angle ABC = 38.7°$ or $141.3°$

When $BC = 7.45$ cm, $\angle ABC = 38.7°$;
when $BC = 1.21$ cm, $\angle ABC = 141.3°$.

5

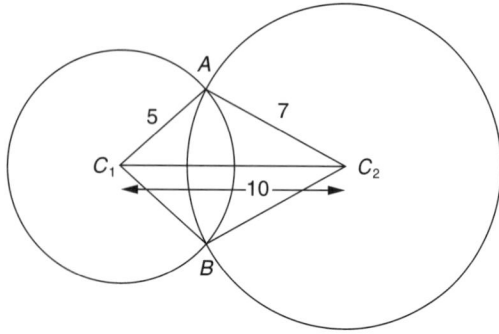

(a) $\triangle AC_1C_2$ cosine rule:

$$\cos \angle AC_1C_2 = \frac{25 + 100 - 49}{100}$$

$$\Rightarrow \quad \angle AC_1C_2 = 40.5°$$

(b) Area of $\triangle AC_1C_2 = \frac{1}{2}(5)(10) \sin (40.5°)$
$= 16.2$ cm^2
(c) Sine rule: $\angle AC_2C_1 = 27.7°$
Area of sector $AC_1B = 17.67$ cm^2,
area of sector $AC_2B = 23.69$ cm^2
Area common to both circles
= sum of sectors − area AC_1BC_2
= 8.9 cm^2.

6 (a) Arc $ADB = 20(\pi - 1)$,
perimeter of $R = 20\pi$ cm = 62.8 cm.
(b) Area of $R = 100(\pi - 1)$ cm^2 = 214 cm^2.
(c) Area of $\triangle CAB = 45.5$ cm^2.

7

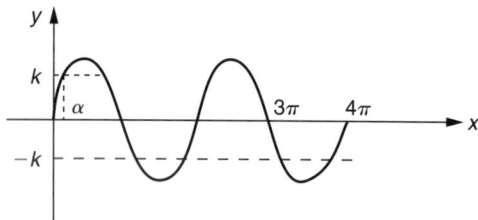

(i) $\pi - \alpha$
(ii) $3\pi + \alpha, 4\pi - \alpha$

Answers follow directly from the symmetry of the sine curve

8

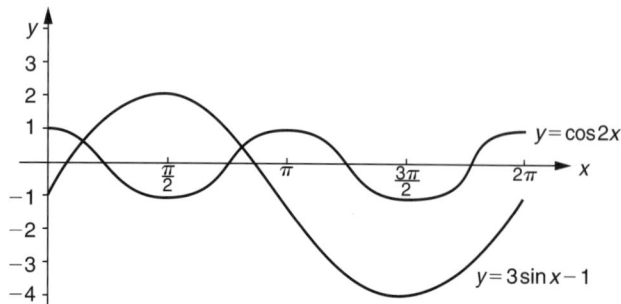

(i) The curves are sketched.
(ii) The curves meet when
$\cos 2x = 1 - 2 \sin^2 x = 3 \sin x - 1$ 　7.1

(iii)
$$2 \sin^2 x + 3 \sin x - 2 = 0$$
$$(2 \sin x - 1)(\sin x + 2) = 0 \quad \boxed{9.1}$$
$$x = \frac{\pi}{6} \text{ or } \frac{5\pi}{6}$$

$\sin x = -2$ has no solutions

9 (i) (a) Cosine rule: $\cos \theta = \frac{1}{8}$

(b) $\sin \theta = \sqrt{\left(1 - \frac{1}{64}\right)} = \frac{3\sqrt{7}}{8}; a = 3, b = 8$ 　1.2

(ii) (a) $3x = \frac{\pi}{6}, \frac{5\pi}{6}, \frac{13\pi}{6}, \frac{17\pi}{6}$, etc.

$x = \frac{\pi}{18}, \frac{5\pi}{18}, \frac{13\pi}{18}, \frac{17\pi}{18}$ in range $0 \leq x \leq \pi$ 　9.1

Find all the relevant values of $3x$ first; then divide each by 3

(b) $\cot \left(x + \frac{\pi}{2}\right) = 1 \Rightarrow \tan \left(x + \frac{\pi}{2}\right) = 1$,

$x + \frac{\pi}{2} = \frac{\pi}{4}, \frac{5\pi}{4}, \frac{9\pi}{4}$, etc.,

$x = \frac{3\pi}{4}$ in range $0 \leq x \leq \pi$ 　9.1

10 $\sin 2\theta = 2 \sin \theta \cos \theta$,
$\sin 3\theta = 3 \sin \theta - 4 \sin^3 \theta = \sin \theta (4 \cos^2 \theta - 1)$,
$\sin 4\theta = 4 \sin \theta \cos \theta (2 \cos^2 \theta - 1)$

Hence $\frac{\sin 2\theta}{\sin \theta} = 2 \cos \theta$, $\frac{\sin 3\theta}{\sin \theta} = 4 \cos^2 \theta - 1$,

$\frac{\sin 4\theta}{\sin \theta} = 4 \cos \theta (2 \cos^2 \theta - 1)$

11 (a) (i) Bookwork
(ii) $4 \sin \theta + 2 \cos (\theta + 30°) = 1$
$\Rightarrow 3 \sin \theta + \sqrt{3} \cos \theta = 1$
$\Rightarrow \sqrt{12} \sin (\theta + 30°) = 1$
$\theta + 30° = 16.78°, 180 - 16.78°$,
$360 + 16.78°$ etc.,
$\theta = 133.22°, 346.78°$ for $0° < \theta < 360°$.

(b) LHS $= \frac{\sin A}{\cos A} + \frac{\cos A}{\sin A} \equiv \frac{1}{\cos A \sin A}$

$= \frac{2}{\sin 2A} = 2 \operatorname{cosec} 2A = $ RHS 　9.1

$$\csc 2\theta = \frac{5}{4}; \text{ from above}$$

$$\tan \theta + \cot \theta = 2\left(\frac{5}{4}\right)$$

$$\Rightarrow \quad 2\tan^2 \theta - 5\tan \theta + 2 = 0$$

$$\Rightarrow \quad \tan \theta = 2 \text{ or } \tfrac{1}{2};$$

$$\text{for } 0 < \theta < \frac{\pi}{4}, \ \tan \theta = \tfrac{1}{2}$$

12 $\sin^4 \theta + 2\sin^2 \theta \cos^2 \theta + \cos^4 \theta = 1$

$\Rightarrow \quad 2(\sin^4 \theta + \cos^4 \theta) + \sin^2 2\theta = 2$

$\Rightarrow \quad 4(\sin^4 \theta + \cos^4 \theta) + 1 - \cos 4\theta = 4$

$\Rightarrow \quad 4(\sin^4 \theta + \cos^4 \theta) = 3 + \cos 4\theta$

(a) $I = \displaystyle\int_0^{\pi/8} \frac{4\sin 4\theta}{3 + \cos 4\theta}\, d\theta$

$\text{Let } u = 3 + \cos 4\theta, \dfrac{du}{d\theta} = -4\sin 4\theta$

$$I = \int -\frac{1}{u}\, du = \left[-\ln(3 + \cos 4\theta)\right]_0^{\pi/8}$$

$$= \ln\left(\frac{4}{3}\right) \qquad \boxed{6.2 \text{ and } 1.2}$$

(b) $3 + \cos 4\theta = 2 - \cos 2\theta$

$\Rightarrow \quad 3 + 2\cos^2 2\theta - 1 = 2 - \cos 2\theta$

$\Rightarrow \quad \cos 2\theta (2\cos 2\theta + 1) = 0$

$$\cos 2\theta = 0 \quad \Rightarrow \quad 2\theta = (2n+1)\frac{\pi}{2}$$

$$\Rightarrow \quad \theta = (2n+1)\frac{\pi}{4}$$

$$\text{or } \cos 2\theta = -\tfrac{1}{2} \quad \Rightarrow \quad 2\theta = 2n\pi \pm \frac{2\pi}{3}$$

$$\Rightarrow \quad \theta = n\pi \pm \frac{\pi}{3}$$

13 RHS $= \cos A \cos B + \sin A \sin B$
$\quad\quad\quad - (\cos A \sin B + \sin A \cos B)$
$\quad\quad = \cos(A - B) - \sin(A + B)$
$\quad\quad = $ LHS

$\cos x - \sin 3x = 0$

$\Rightarrow \quad (\cos 2x - \sin 2x)(\cos x - \sin x) = 0$

$\Rightarrow \quad \tan 2x = 1$

$\Rightarrow \quad x = -67.5°, 22.5°$

or $\tan x = 1, x = 45°$ in the range $-90° \leqslant x \leqslant 90°$

$\boxed{9.1}$

14 $5\cos(x + 53.1°)$ $\qquad\qquad \boxed{9.2}$

(a) (i) $\text{f}(x) = \dfrac{24}{5\cos(x + 53.1°) + 7}$

$$\text{Max f}(x) = \frac{24}{5(-1) + 7} = 12$$

$$\text{when } x = 126.9°$$
$$\text{Min f}(x) = 2 \text{ when } x = -53.1° \quad \boxed{7.1}$$

(ii) $\dfrac{24}{5\cos(x + 53.1°) + 7} = \dfrac{16}{3}$

$\Rightarrow \quad \cos(x + 53.1°) = -\tfrac{1}{2}$

$\Rightarrow \quad x = 360n + 66.9°$ or
$\quad\quad\quad x = 360n - 173.1°$

(b) $\cos(x + 53.1°) = \cos 3x$

$\Rightarrow \quad x + 53.1° = 360n \pm 3x$

$\Rightarrow \quad 4x = 360n - 53.1°$

$\Rightarrow \quad x = 76.7°, 166.7°$

or $-2x = 360n - 53.1°$

$\Rightarrow \quad x = 26.6°$ in the range $0 < x < 180°$

$\boxed{9.1}$

15 LHS $= \dfrac{1 + \cos 2x}{\sin 2x} = \dfrac{2\cos^2 x}{2\sin x \cos x} = \cot x = $ RHS

(a) $x = 15°$ gives $\cot 15° = \dfrac{1}{\sin 30°} + \dfrac{1}{\tan 30°}$

$$= 2 + \sqrt{3}$$

(b) $\cot \dfrac{3\theta}{2} = 2 \quad \Rightarrow \quad \tan \dfrac{3\theta}{2} = \dfrac{1}{2}$

$\Rightarrow \theta = 17.7°, 137.7°$ in the interval $0 < \theta < 180°$

$\boxed{9.1}$

16 $2\cos^2 \theta - 1 + \cos \theta = 0$

$\Rightarrow \quad (2\cos \theta - 1)(\cos \theta + 1) = 0$

$\Rightarrow \quad \cos \theta = \tfrac{1}{2}, \theta = 2n\pi \pm \dfrac{\pi}{3},$

or $\cos \theta = -1, \theta = (2n+1)\pi$

17 (a) $\sqrt{2}\cos\left(x - \dfrac{\pi}{4}\right) = 1 \quad \Rightarrow \quad x - \dfrac{\pi}{4} = 2n\pi \pm \dfrac{\pi}{4},$

$$x = 2n\pi \text{ or } x = 2n\pi + \dfrac{\pi}{2} \qquad \boxed{9.2}$$

(b) $\sqrt{2}\cos\left(x - \dfrac{\pi}{4}\right)$, greatest value $\sqrt{2}$

18 $\sin 3A = \sin 2A \cos A + \cos 2A \sin A$
$\quad\quad\quad = 2\sin A \cos^2 A + \sin A - 2\sin^3 A$
$\quad\quad\quad = 3\sin A - 4\sin^3 A$

$3 \sin x - 4 \sin^3 x = \sin^2 x$

$\Rightarrow \quad \sin x \,(4 \sin^2 x + \sin x - 3) = 0$

$\Rightarrow \quad \sin x = 0$, or 0.75 or -1

$x = 180n°$, $x = 48.6 + 360n°$ or $131.4 + 360n°$,
$x = 270 + 360n°$

In range $90° \leqslant x \leqslant 270°$, $x = 131.4°, 180°, 270°$

$\boxed{9.1}$

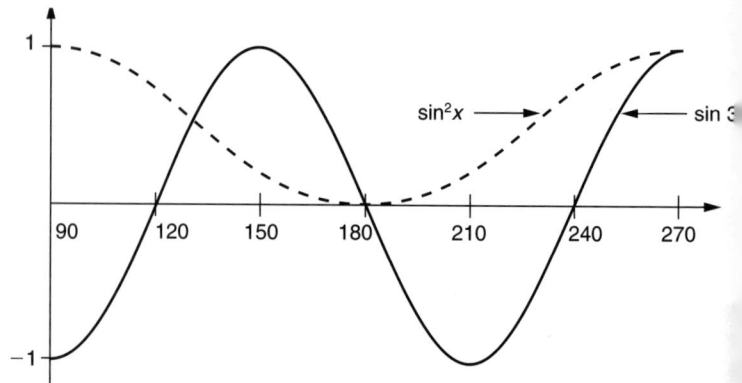

$\sin 3x < \sin^2 x$ when $90° < x < 131.4°$ or
$180° < x < 270°$

$\boxed{7.1}$

7

Vectors

AS Level			A Level			Topic	Date attempted	Date completed	Self-assessment
CORE	MODULAR	TRADITIONAL	CORE	MODULAR	TRADITIONAL				
✓	✓	✓	✓	✓	✓	Vector algebra			
✓	✓	✓	✓	✓	✓	Scalar products			
✓		✓	✓	✓	✓	Lines and planes			
					✓	Vector products			

Fact Sheet

Algebra

- A vector **a** has magnitude or modulus |**a**| = a and direction **â** (where **â** is a unit vector): $\mathbf{a} = a\hat{\mathbf{a}}$ or $\hat{\mathbf{a}} = \dfrac{\mathbf{a}}{a}$.

- If λ is any positive scalar then $\lambda\mathbf{a}$ has magnitude λa and direction **â**. If λ is negative then $\lambda\mathbf{a}$ has magnitude $|\lambda|a$ and direction $-\hat{\mathbf{a}}$, i.e. in the opposite direction.

Position Vectors

If O is an origin then $\overline{OA}$ is the position vector of A relative to O. The usual notation is $\overline{OA} = \mathbf{a}$, $\overline{OB} = \mathbf{b}$, etc.

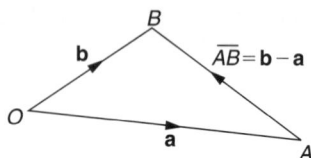

Displacement Vectors

If $\overline{OA} = \mathbf{a}$ and $\overline{OB} = \mathbf{b}$ then $\overline{AB}$ is the displacement vector from A to B, and is given by $\overline{AB} = \mathbf{b} - \mathbf{a}$.

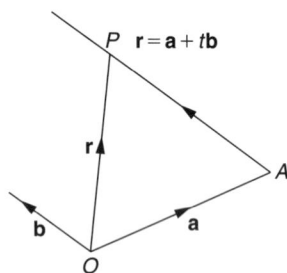

Equations of Lines

If a line is drawn through a point A, parallel to the vector **b**, the position vector of any point P on the line is given by $\mathbf{r} = \mathbf{a} + t\mathbf{b}$, where t is a parameter.

Cartesian Components

i, **j** and **k** are unit vectors in the positive directions of the x, y and z axes, and are called base vectors. A point P with coordinates (x, y, z) has a position vector

$$\mathbf{r} = x\mathbf{i} + y\mathbf{j} + z\mathbf{k} \text{ or } \mathbf{r} = \begin{pmatrix} x \\ y \\ z \end{pmatrix} \text{ or occasionally } (x, y, z).$$

The magnitude or modulus of the vector $\mathbf{r} = x\mathbf{i} + y\mathbf{j} + z\mathbf{k}$ is $\sqrt{(x^2 + y^2 + z^2)}$.
The direction ratios of the vector are (x, y, z).
The direction cosines of the vector are

$$\frac{x}{\sqrt{x^2 + y^2 + z^2}}, \frac{y}{\sqrt{x^2 + y^2 + z^2}}, \frac{z}{\sqrt{x^2 + y^2 + z^2}}.$$

Scalar Product

The scalar product of the vectors **a** and **b** is written $\mathbf{a} \cdot \mathbf{b}$ and defined $\mathbf{a} \cdot \mathbf{b} = |\mathbf{a}|\,|\mathbf{b}| \cos\theta = ab \cos\theta$ where θ is the angle between **a** and **b**.

(i) If **a** and **b** are parallel then $\mathbf{a} \cdot \mathbf{b} = ab$.
(ii) If $\mathbf{a} = \mathbf{b}$ then $\mathbf{a} \cdot \mathbf{b} = \mathbf{a} \cdot \mathbf{a} = a^2$.
(iii) If **a** and **b** are perpendicular then $\mathbf{a} \cdot \mathbf{b} = 0$.
(iv) If **a** and **b** are given in Cartesian form, $\mathbf{a} = a_1\mathbf{i} + a_2\mathbf{j} + a_3\mathbf{k}$, $\mathbf{b} = b_1\mathbf{i} + b_2\mathbf{j} + b_3\mathbf{k}$, then $\mathbf{a} \cdot \mathbf{b} = a_1b_1 + a_2b_2 + a_3b_3$.
(v) If the angle between the vectors **a** and **b** is θ then

$$\cos\theta = \frac{a_1b_1 + a_2b_2 + a_3b_3}{ab} = \frac{\mathbf{a} \cdot \mathbf{b}}{ab}$$

Equations of Planes

(i) The equation of the plane containing points A, B and C is given by
$\mathbf{r} = \mathbf{a} + \lambda(\overline{AB}) + \mu(\overline{AC})$ where λ and μ are parameters.

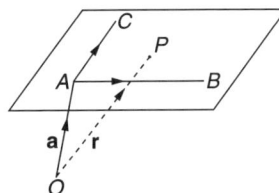

(ii) The equation of the plane containing point A and with a normal vector $\mathbf{n}$ is
$\mathbf{r} \cdot \mathbf{n} = \text{constant} = \mathbf{a} \cdot \mathbf{n}$.

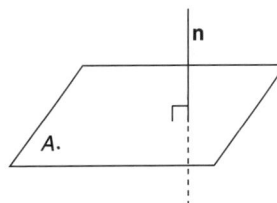

Vector Product

The vector product between $\mathbf{a}$ and $\mathbf{b}$ is written as $\mathbf{a} \times \mathbf{b}$.
If $\mathbf{a} = a_1\mathbf{i} + a_2\mathbf{j} + a_3\mathbf{k}$ and $\mathbf{b} = b_1\mathbf{i} + b_2\mathbf{j} + b_3\mathbf{k}$ then

$$\mathbf{a} \times \mathbf{b} = \begin{vmatrix} \mathbf{i} & \mathbf{j} & \mathbf{k} \\ a_1 & a_2 & a_3 \\ b_1 & b_2 & b_3 \end{vmatrix}$$

$$= (a_2b_3 - a_3b_2)\mathbf{i} - (a_1b_3 - a_3b_1)\mathbf{j} + (a_1b_2 - a_2b_1)\mathbf{k}$$

The resultant **vector** is perpendicular to $\mathbf{a}$ and $\mathbf{b}$ and has a magnitude of $ab \sin \theta$ where θ is the angle between $\mathbf{a}$ and $\mathbf{b}$.
$ab \sin \theta = 2 \times$ area of triangle ABO.

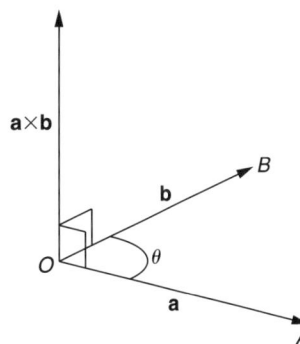

Vector Product by the Column Method

To find the $\mathbf{i}$, $\mathbf{j}$ and $\mathbf{k}$ components use the two rows below the letter $\mathbf{i}$, $\mathbf{j}$ or $\mathbf{k}$.

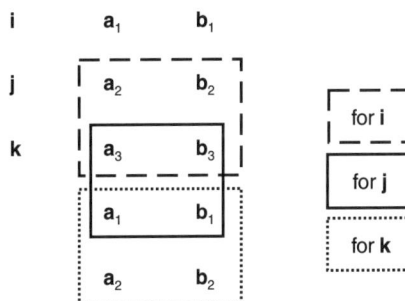

Worked Examples

1 $\mathbf{a} = \begin{pmatrix} \frac{2}{11} \\ p \\ -\frac{9}{11} \end{pmatrix}$ is a unit vector. Which of the following could be the value of p?

(i) $-\dfrac{4}{11}$, (ii) $\dfrac{6}{11}$ (iii) $\dfrac{18}{11}$ (iv) $-\dfrac{6}{11}$

A (i) only **B** (i) and (iv) only **C** (iii) only
D (ii) and (iv) only **E** none of these

• $|\mathbf{a}| = \sqrt{\dfrac{4}{121} + p^2 + \dfrac{81}{121}} = \sqrt{\dfrac{85}{121} + p^2}$

But $\mathbf{a}$ is a unit vector with magnitude 1

$\therefore p^2 = \dfrac{36}{121} \;\Rightarrow\; p = \pm\dfrac{6}{11}$ Answer **D**

2 In triangle ABC, E lies on BC with $BE:EC = 2:3$; F lies on CA with $CF:FA = 3:4$ and G lies on AB produced, with $GB:GA = 1:2$. The position vectors of A, B and C relative to an origin O are $\mathbf{a}$, $\mathbf{b}$ and $\mathbf{c}$. Deduce that E, F and G lie on a straight line

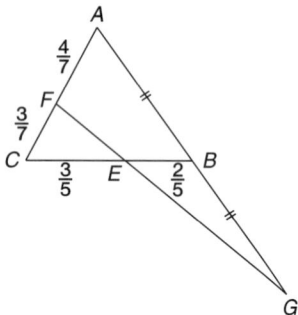

•

If $BE:EC = 2:3$ then $BE = \frac{2}{5}BC$
$$EC = \tfrac{3}{5}BC$$
Position vector of E is $\overline{OB} + \overline{BE}$
$$= \mathbf{b} + \tfrac{2}{5}(\mathbf{c} - \mathbf{b})$$
$$\mathbf{e} = \tfrac{3}{5}\mathbf{b} + \tfrac{2}{5}\mathbf{c}$$
Similarly for F $\mathbf{f} = \tfrac{4}{7}\mathbf{c} + \tfrac{3}{7}\mathbf{a}$
For G, $\overline{OG} = \overline{OA} + 2\overline{AB} = \mathbf{a} + 2(\mathbf{b} - \mathbf{a})$
$$\mathbf{g} = 2\mathbf{b} - \mathbf{a}$$

$\overline{GE} = \mathbf{e} - \mathbf{g} = \tfrac{3}{5}\mathbf{b} + \tfrac{2}{5}\mathbf{c} - 2\mathbf{b} + \mathbf{a}$
$$= \mathbf{a} - \tfrac{7}{5}\mathbf{b} + \tfrac{2}{5}\mathbf{c}$$
$\overline{EF} = \mathbf{f} - \mathbf{e} = \tfrac{4}{7}\mathbf{c} + \tfrac{3}{7}\mathbf{a} - \tfrac{3}{5}\mathbf{b} - \tfrac{2}{5}\mathbf{c}$
$$= \tfrac{3}{7}(\mathbf{a} - \tfrac{7}{5}\mathbf{b} + \tfrac{2}{5}\mathbf{c})$$
$\therefore \overline{EF} = \tfrac{3}{7}\overline{GE}$

Hence E, F, G lie on a straight line.

3 AB is the diameter of a circle centre O and P is any point on the circumference. Use vectors to show that $\angle APB = 90°$.

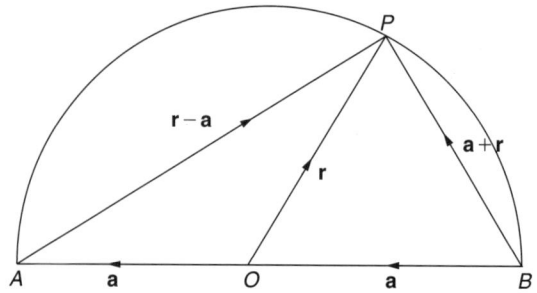

•

$\overline{AP} = \mathbf{r} - \mathbf{a} \quad \overline{BP} = \mathbf{a} + \mathbf{r}$
$\overline{AP} \cdot \overline{BP} = (\mathbf{r} - \mathbf{a}) \cdot (\mathbf{a} + \mathbf{r})$
$$= \mathbf{r} \cdot \mathbf{a} + \mathbf{r} \cdot \mathbf{r} - \mathbf{a} \cdot \mathbf{a} - \mathbf{a} \cdot \mathbf{r}$$
$$= r^2 - a^2$$
But $|\mathbf{r}| = |\mathbf{a}| = $ radius of the circle
$\Rightarrow \overline{AP} \cdot \overline{BP} = 0 \;\Rightarrow\; \angle APB = 90°$

4 With respect to an origin O, the position vectors of the points L and M are $2\mathbf{i} - 3\mathbf{j} + 3\mathbf{k}$ and $5\mathbf{i} + \mathbf{j} + c\mathbf{k}$ respectively, where c is a constant.

The point N is such that $OLMN$ is a rectangle.

(a) Find the value of c.
(b) Write down the position vector of N.
(c) Find, in the form $\mathbf{r} = \mathbf{p} + t\mathbf{q}$, an equation of the line MN.   (L Specimen Paper, 1996)

• Draw a diagram for $OLMN$. Do not attempt to plot the points. Use the diagram to find the coordinates of the unknown points

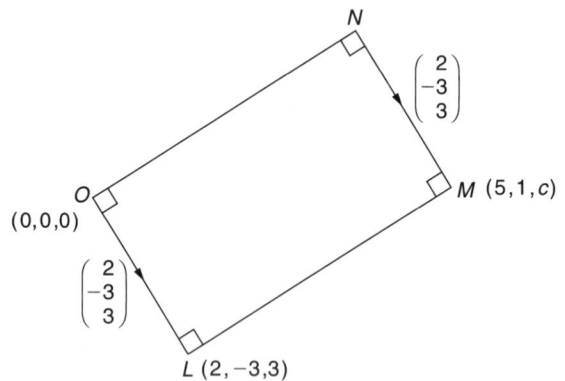

Coordinates of L and M are $(2, -3, 3)$ and $(5, 1, c)$ respectively.
OL is perpendicular to $LM \;\Rightarrow \overline{OL} \cdot \overline{LM} = 0$

(a) $\overline{LM} = \overline{OM} - \overline{OL} = \begin{pmatrix} 3 \\ 4 \\ c - 3 \end{pmatrix}$

$\overline{OL} \cdot \overline{LM} = 0$

$\Rightarrow \begin{pmatrix} 2 \\ -3 \\ 3 \end{pmatrix} \cdot \begin{pmatrix} 3 \\ 4 \\ c - 3 \end{pmatrix} = 6 - 12 + 3c - 9 = 0$

$\Rightarrow c = 5$

(b) $\overline{LM} = \begin{pmatrix} 3 \\ 4 \\ 2 \end{pmatrix} = \overline{ON}$

∴ position vector $\mathbf{n} = 3\mathbf{i} + 4\mathbf{j} + 2\mathbf{k}$

(c) Line NM has equation
$\mathbf{r} = \mathbf{n} + \lambda\,(\overline{NM})$

$$\mathbf{r} = \begin{pmatrix} 3 \\ 4 \\ 2 \end{pmatrix} + \lambda \begin{pmatrix} 2 \\ -3 \\ 3 \end{pmatrix}$$

$\left[\text{ or use } \mathbf{r} = \overline{OM} + \mu\overline{MN} \text{ to give}\right.$

$$\mathbf{r} = \begin{pmatrix} 5 \\ 1 \\ 5 \end{pmatrix} + \mu \begin{pmatrix} -2 \\ 3 \\ -3 \end{pmatrix}\Bigg].$$

5 The vectors $\mathbf{a}$, $\mathbf{b}$, $\mathbf{c}$ are given by

$\mathbf{a} = \mathbf{i} + 2\mathbf{j} + \mathbf{k}$
$\mathbf{b} = \mathbf{i} + \mathbf{j} + 2\mathbf{k}$
$\mathbf{c} = \mathbf{j} + \mathbf{k}$

A fourth vector is given by $\mathbf{d} = 3\mathbf{i} + 6\mathbf{j} + 5\mathbf{k}$.
Find the values of α, β, γ such that
$\mathbf{d} = \alpha\mathbf{a} + \beta\mathbf{b} + \gamma\mathbf{c}$. (WJEC, 1994)

• $\mathbf{d} = (\alpha + \beta)\mathbf{i} + (2\alpha + \beta + \gamma)\mathbf{j} + (\alpha + 2\beta + \gamma)\mathbf{k}$ ▣

2.2b

$\Rightarrow$
$\quad \alpha + \beta = 3$ (1)
$\quad 2\alpha + \beta + \gamma = 6$ (2)
$\quad \alpha + 2\beta + \gamma = 5$ (3)
From (2) − (3), $\alpha - \beta = 1$ (4)
From (1) and (4), $\alpha = 2$, $\beta = 1$.
Substitute into (2): $4 + 1 + \gamma = 6 \Rightarrow \gamma = 1$.

6 With respect to a fixed origin O, the points A and B have position vectors $2\mathbf{i} + 3\mathbf{j} + 6\mathbf{k}$ and $2\mathbf{i} + 4\mathbf{j} + 4\mathbf{k}$.

(a) Calculate $|\overline{OA}\,|$, $|\overline{OB}\,|$ and, by using the scalar product $\overline{OA} \cdot \overline{OB}$, calculate the value of the cosine of angle AOB.

(b) The point C has position vector $5\mathbf{i} + 12\mathbf{j} + 6\mathbf{k}$. Show that OC and AB are perpendicular. Show also that the line through O and C intersects the line through A and B and find the position vector of the point E where they intersect.

(c) Given that $\overline{AE} = \lambda\overline{EB}$, find the value of λ and explain briefly why λ is negative. (AEB, 1990)

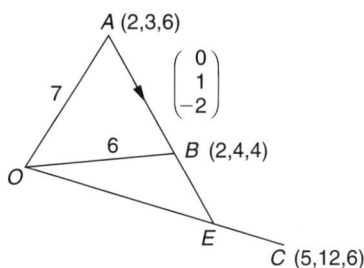
A (2,3,6)
7
$\begin{pmatrix} 0 \\ 1 \\ -2 \end{pmatrix}$
6
B (2,4,4)
O
E
C (5,12,6)

(a) $|\overline{OA}\,| = \sqrt{2^2 + 3^3 + 6^2} = 7$

$|\overline{OB}\,| = \sqrt{2^2 + 4^2 + 4^2} = 6$

$\overline{OA} \cdot \overline{OB} = |\overline{OA}\,||\overline{OB}\,|\cos A\hat{O}B$

$$\begin{pmatrix} 2 \\ 3 \\ 6 \end{pmatrix} \cdot \begin{pmatrix} 2 \\ 4 \\ 4 \end{pmatrix} = 7 \times 6 \times \cos A\hat{O}B$$

$\Rightarrow 4 + 12 + 24 = 42\cos A\hat{O}B$

$\Rightarrow \cos A\hat{O}B = \dfrac{40}{42} = \dfrac{20}{21}$

(b) OC and AB are perpendicular if $\overline{OC} \cdot \overline{AB} = 0$

$$\overline{AB} = \begin{pmatrix} 0 \\ 1 \\ -2 \end{pmatrix}$$

$$\overline{OC} \cdot \overline{AB} = \begin{pmatrix} 5 \\ 12 \\ 6 \end{pmatrix} \cdot \begin{pmatrix} 0 \\ 1 \\ -2 \end{pmatrix} = 0 + 12 - 12 = 0$$

∴ $\overline{OC}$ is perpendicular to $\overline{AB}$.

Equation of the line OC is $\mathbf{r}_1 = s\begin{pmatrix} 5 \\ 12 \\ 6 \end{pmatrix}$.

Equation of the line AB is $\mathbf{r}_2 = \begin{pmatrix} 2 \\ 3 \\ 6 \end{pmatrix} + t\begin{pmatrix} 0 \\ 1 \\ -2 \end{pmatrix}$.

These lines intersect if $5s = 2$ (1)
$\qquad\qquad\qquad 12s = 3 + t$ (2)
and $\qquad 6s = 6 - 2t$ (3)

From (1), $s = \frac{2}{5}$.
In (2), $\frac{24}{5} = 3 + t \Rightarrow t = \frac{9}{5}$.
The solutions for s and t obtained from two of the equations must be checked in the third
In (3), LHS $= \frac{12}{5}$
$\qquad\qquad$ RHS $= 6 - \frac{18}{5} = \frac{12}{5}$
Hence the lines intersect at the point E with position vector $2\mathbf{i} + 4.8\mathbf{j} + 2.4\mathbf{k}$

(c) If $\overline{AE} = \lambda\overline{EB}$

$$\overline{AE} = \begin{pmatrix} 0 \\ 1.8 \\ -3.6 \end{pmatrix} \qquad \overline{EB} = \begin{pmatrix} 0 \\ -0.8 \\ 1.6 \end{pmatrix} = -\tfrac{4}{9}\overline{AE}$$

Therefore $\overline{AE} = -\tfrac{9}{4}\overline{EB}$.
E divides the line AB externally (on AB produced) in the ratio $9:-4$.

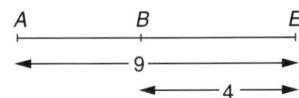
A
B
E
9
4

7 The parallelogram $ABCD$ forms the base of a pyramid, which has the origin O as its vertex. The coordinates of A, B, C are $(11, 3, -8)$, $(8, 9, -2)$, $(4, 11, -6)$ respectively.

(i) Find the vectors $\overline{AB}$ and $\overline{BC}$. Find the coordinates of D. Show that $ABCD$ is a rectangle and calculate the lengths of AB and BC.

(ii) The vector

$$\mathbf{p} = \begin{pmatrix} l \\ m \\ 1 \end{pmatrix}$$

is normal to the plane $ABCD$. Use the fact that $\mathbf{p} \cdot \overline{AB} = 0$ to obtain an equation connecting l and m.

Find the vector equation of the plane $ABCD$ in the form $\mathbf{r} \cdot \mathbf{n} = h$, giving the values of $\mathbf{n}$ and h.

(iii) Write down the vector equation of the line through O perpendicular to the plane $ABCD$. Use your answer to (ii) to calculate the coordinates of the point of intersection of this line and the plane $ABCD$.

Calculate the volume of the pyramid $OABCD$.

(iv) Calculate the acute angle between the line OA and the plane $ABCD$. (AEB, 1994)

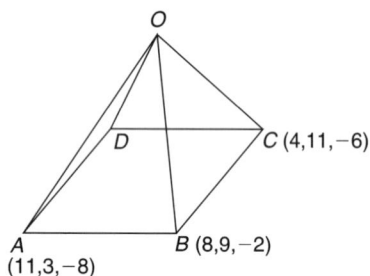

O, D, $C (4,11,-6)$, $A (11,3,-8)$, $B (8,9,-2)$

$ABCD$ is a parallelogram $\Rightarrow \overline{AB} = \overline{DC}$

(i) $\overline{AB} = -3\mathbf{i} + 6\mathbf{j} + 6\mathbf{k} = \overline{DC}$

$\overline{BC} = -4\mathbf{i} + 2\mathbf{j} - 4\mathbf{k} = \overline{AD}$

Coordinates of D are $(7, 5, -12)$

$\overline{AB} \cdot \overline{BC} = 12 + 12 - 24 = 0$

Hence $\overline{AB}$ is perpendicular to $\overline{BC}$.

But $ABCD$ is a parallelogram with one angle $90° \Rightarrow$ a rectangle.

$|AB| = \sqrt{9 + 36 + 36} = 9$
$|BC| = \sqrt{16 + 4 + 16} = 6$

(ii) $\mathbf{p} \cdot \overline{AB} = \begin{pmatrix} l \\ m \\ 1 \end{pmatrix} \cdot \begin{pmatrix} -3 \\ 6 \\ 6 \end{pmatrix} = -3l + 6m + 6$

Since $\mathbf{p}$ is perpendicular to $\overline{AB}$,

$-3l + 6m + 6 = 0$ or $-l + 2m + 2 = 0$ (1)

Similarly, $\mathbf{p} \cdot \overline{BC} = 0 \Rightarrow$

$-4l + 2m - 4 = 0$ (2)

(1) $-$ (2) $\Rightarrow 3l + 6 = 0 \Rightarrow l = -2$, $m = -2$. Hence

$$\mathbf{p} = \begin{pmatrix} -2 \\ -2 \\ 1 \end{pmatrix}$$

The vector equation of $ABCD$ is

$$\mathbf{r} \cdot \begin{pmatrix} -2 \\ -2 \\ 1 \end{pmatrix} = \text{constant}$$

Substituting $r = 11\mathbf{i} + 3\mathbf{j} - 8\mathbf{k}$ gives

$$\mathbf{r} \cdot \begin{pmatrix} -2 \\ -2 \\ 1 \end{pmatrix} = -36$$

$$n = \begin{pmatrix} -2 \\ -2 \\ 1 \end{pmatrix} \quad h = -36$$

(iii) The line through O perpendicular to $ABCD$ is $\mathbf{r} = \lambda \begin{pmatrix} -2 \\ -2 \\ 1 \end{pmatrix}$. This meets $ABCD$ when

$$\lambda \begin{pmatrix} -2 \\ -2 \\ 1 \end{pmatrix} \cdot \begin{pmatrix} -2 \\ -2 \\ 1 \end{pmatrix} = -36 \Rightarrow \lambda = -4.$$

Hence the point of intersection is $(8, 8, -4)$.
The distance from O is 12.
Volume of the pyramid is
$\frac{1}{6} \times 9 \times 6 \times 12 = 108$.

(iv) Direction $\overline{OA} = \begin{pmatrix} 11 \\ 3 \\ -8 \end{pmatrix}$.

Direction of the normal to $ABCD$ is $\begin{pmatrix} -2 \\ -2 \\ 1 \end{pmatrix}$.

If the angle between the line OA and $ABCD$ is θ then

$$\sin \theta = \frac{\left| \begin{pmatrix} 11 \\ 3 \\ -8 \end{pmatrix} \cdot \begin{pmatrix} -2 \\ -2 \\ 1 \end{pmatrix} \right|}{\sqrt{121 + 9 + 64} \sqrt{4 + 4 + 1}}$$

$$= \frac{36}{\sqrt{194} \cdot \sqrt{9}} \Rightarrow \theta = 59.5°$$

8 Find a unit vector which is perpendicular to the vector $4\mathbf{i} + 4\mathbf{j} - 7\mathbf{k}$ and the vector $3\mathbf{i} - 2\mathbf{j} + \mathbf{k}$.

• *Method 1 – Use of scalar products*
If **a** and **b** are perpendicular then **a** · **b** = 0
Let the required vector be $p\mathbf{i} + q\mathbf{j} + r\mathbf{k}$. Then

$$\begin{pmatrix} 4 \\ 4 \\ -7 \end{pmatrix} \cdot \begin{pmatrix} p \\ q \\ r \end{pmatrix} = 4p + 4q - 7r = 0 \qquad (1)$$

$$\begin{pmatrix} 3 \\ -2 \\ 1 \end{pmatrix} \cdot \begin{pmatrix} p \\ q \\ r \end{pmatrix} = 3p - 2q + r = 0 \qquad (2)$$

(1) + 2 × (2): $10p - 5r = 0 \implies r = 2p$
From (2): $q = \frac{5}{2}p$
Choosing a numerical value for p (such as 2), we have $(2\mathbf{i} + 5\mathbf{j} + 4\mathbf{k})$ is a vector perpendicular to $3\mathbf{i} - 2\mathbf{j} + \mathbf{k}$ and to $4\mathbf{i} + 4\mathbf{j} - 7\mathbf{k}$.
The unit vector is

$$\frac{(2\mathbf{i} + 5\mathbf{j} + 4\mathbf{k})}{\sqrt{4 + 25 + 16}} = \frac{1}{\sqrt{45}}(2\mathbf{i} + 5\mathbf{j} + 4\mathbf{k})$$

Method 2 – Use of vector products
A vector perpendicular to $4\mathbf{i} + 4\mathbf{j} - 7\mathbf{k}$ and $3\mathbf{i} - 2\mathbf{j} + \mathbf{k}$ is

$$\begin{vmatrix} \mathbf{i} & \mathbf{j} & \mathbf{k} \\ 4 & 4 & -7 \\ 3 & -2 & 1 \end{vmatrix}$$

That is $(4 - 14)\mathbf{i} - (4 + 21)\mathbf{j} + (-8 - 12)\mathbf{k}$
or $-10\mathbf{i} - 25\mathbf{j} - 20\mathbf{k}$

Unit vector is $\dfrac{1}{\sqrt{45}}(2\mathbf{i} + 5\mathbf{j} + 4\mathbf{k})$

In each method the vector could also be

$$-\frac{1}{\sqrt{45}}(2\mathbf{i} + 5\mathbf{j} + 4\mathbf{k})$$

9 The point A has position vector $\mathbf{i} + 3\mathbf{j} - 4\mathbf{k}$ relative to a fixed origin O. The vector equation of the line l_1 passing through A is

$$\mathbf{r} = \mathbf{i} + 3\mathbf{j} - 4\mathbf{k} + s(2\mathbf{i} + \mathbf{j} + 3\mathbf{k})$$

where s is a scalar parameter.

(a) The point $B(a, b, 5)$ lies on l_1. Determine the values of the constants a and b.
(b) Determine the acute angle between vector $\overline{OA}$ and the line l_1, giving your answer to the nearest 0.1°.
(c) Write down, in terms of a parameter t, the vector equation of the line l_2 which passes through $C(1, 2, -1)$ and $D(4, 2, 8)$. Show that the lines l_1 and l_2 intersect and find the position vector of the point of intersection.
(d) Prove that the shortest distance from the point with position vector $3\mathbf{i} + 6\mathbf{j} + 3\mathbf{k}$ to the line l_1 is $\sqrt{6}$.

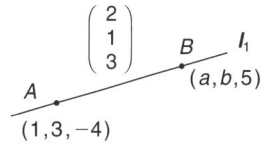

• (a) A general point on line l_1 is
$(1 + 2s, 3 + s, -4 + 3s)$.
At B, $-4 + 3s = 5 \implies s = 3$
so B is the point $(7, 6, 5)$.
(b) Direction vector of $\overline{OA}$ is $\mathbf{i} + 3\mathbf{j} - 4\mathbf{k}$.
Direction vector of l_1 is $2\mathbf{i} + \mathbf{j} + 3\mathbf{k}$.

If the angle between $\overline{OA}$ and l_1 is θ then

$$\cos\theta = \frac{|(\mathbf{i} + 3\mathbf{j} - 4\mathbf{k}) \cdot (2\mathbf{i} + \mathbf{j} + 3\mathbf{k})|}{\sqrt{1^2 + 3^2 + 4^2}\sqrt{2^2 + 1^2 + 3^2}} = \frac{7}{\sqrt{364}}$$

$\theta = 68.5°$

(c) l_2 has direction vector $3\mathbf{i} + 9\mathbf{k}$.
Equation of l_2 is $\mathbf{r} = (\mathbf{i} + 2\mathbf{j} - \mathbf{k}) + t(3\mathbf{i} + 9\mathbf{k})$.
If l_1 and l_2 intersect $\quad 1 + 3t = 1 + 2s \qquad (1)$
$\qquad\qquad\qquad 2 = 3 + s \qquad (2)$
$\qquad\qquad -1 + 9t = -4 + 3s \qquad (3)$
From equation 2, $s = -1 \implies$ in equation 1, $t = -\frac{2}{3}$.
Check in equation 3: LHS = -7 = RHS.
Hence l_1 and l_2 intersect at $(-1, 2, -7)$ with position vector $-\mathbf{i} + 2\mathbf{j} - 7\mathbf{k}$.
(d) Distance between $(3, 6, 3)$ and a general point $(1 + 2s, 3 + s, -4 + 3s)$ on l_1 is

$$h = \sqrt{(2 - 2s)^2 + (3 - s)^2 + (7 - 3s)^2}.$$

$h^2 = 14s^2 - 56s + 62$
Either complete the square $14(s - 2)^2 + 6$

The least value of h^2 is 6 when $s = 2$.

Or differentiate $\dfrac{\mathrm{d}}{\mathrm{d}s}(h^2) = 28s - 56$
$\qquad\qquad\qquad\qquad\qquad = 0$ when $s = 2$
$\dfrac{\mathrm{d}^2(h^2)}{\mathrm{d}s^2} = 28$

This is positive, hence there is a minimum value at $s = 2$.
Hence minimum value of h^2 is $56 - 112 + 62 = 6$

Therefore the shortest distance is $\sqrt{6}$ when $s = 2$.

10 Three cables exert forces that act in a horizontal plane on the top of a telegraph pole.
(a) Find the resultant of these three forces, in terms of **i** and **j**.
(b) A fourth cable is attached to the top of the telegraph pole to keep the pole in equilibrium. Find the force exerted by this fourth cable, in terms of **i** and **j**.
(c) Show that the magnitude of the fourth force is 192 N, correct to 3 s.f.

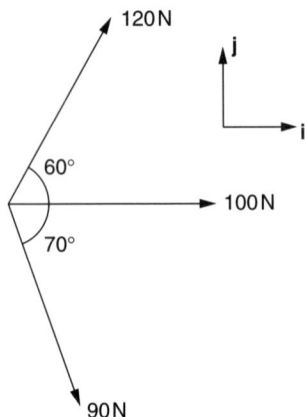

(d) On a diagram show clearly the direction in which the fourth force acts.

(e) The fourth cable does not lie in the same horizontal plane as the other three cables and the tension in the cable is in fact 200 N. Find the angle between this cable and the horizontal plane. (AEB Specimen Paper, 1994)

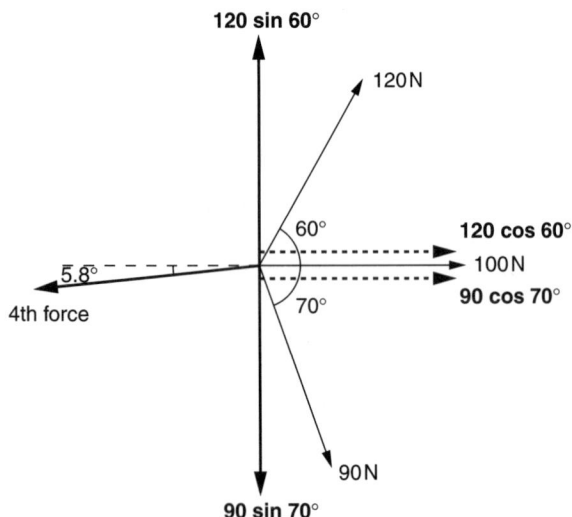

(a) Sum of forces

$$= \frac{\mathbf{i}}{\mathbf{j}}\binom{100}{0} + \binom{120\cos 60°}{120\sin 60°} + \binom{90\cos 70°}{-90\sin 70°}$$

$$= 190.8\mathbf{i} + 19.4\mathbf{j}$$

(b) For equilibrium the extra force is

$$-190.8\mathbf{i} - 19.4\mathbf{j}$$

(c) The magnitude of this force
$$= \sqrt{(190.8)^2 + (19.4)^2} = 192 \text{ N to 3 s.f.}$$

(d) See diagram.

(e) If the cable makes an angle θ with the horizontal then $200\cos\theta = 192 \Rightarrow \theta = 16.3°$

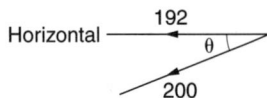

Exercises

1 P, Q, R and S are 4 points such that $\overline{PQ}$ represents
$$\begin{pmatrix} 1 \\ 2 \\ 1 \end{pmatrix}, \overline{QR} \text{ represents } \begin{pmatrix} 2 \\ 1 \\ 2 \end{pmatrix} \text{ and } \overline{RS} = 2\overline{PQ}. \text{ Given}$$
that P is the point $(1, 1, 1)$, then S is the point

A $(6, 8, 6)$ **B** $(4, 4, 4)$ **C** $(5, 6, 5)$
D $(4, 6, 4)$ **E** none of these   (SEB, 1993)

2 With respect to a fixed origin O, the points A, B and C have position vectors $\mathbf{i} + \mathbf{j} + 8\mathbf{k}, \mathbf{i} + 2\mathbf{j} + 6\mathbf{k}$ and $3\mathbf{i} + 12\mathbf{j} + 6\mathbf{k}$ respectively.

(i) Show that $\overline{OC}$ and $\overline{AB}$ are perpendicular.
(ii) Show that the line through O and C intersects the line through A and B, and find the position vector of the point where they intersect.
    (NICCEA, 1992)

3 If $\mathbf{i}, \mathbf{j}$ and $\mathbf{k}$ are mutually perpendicular vectors, and
$\mathbf{A} = \mathbf{i} + 2\mathbf{j} + 3\mathbf{k}$ $\mathbf{B} = 3\mathbf{i} + \mathbf{k}$ $\mathbf{C} = -2\mathbf{i} + 3\mathbf{j}$
calculate $|\mathbf{A} + 2\mathbf{B} - \mathbf{C}|$ and $\mathbf{B} \cdot (\mathbf{A} - \mathbf{C})$. Find a vector of length $2\sqrt{26}$ perpendicular to $\mathbf{A}$ and $\mathbf{B}$.

4 Given that A and B have position vectors
$\mathbf{a} = \mathbf{i} + 2\mathbf{j} + 3\mathbf{k}$ and $\mathbf{b} = -2\mathbf{i} + 5\mathbf{j} - \mathbf{k}$ respectively, determine $\overline{AB}, |\overline{AB}|$, the direction ratios of the line AB and the direction cosines of $\overline{AB}$.

5 The lines l_1 and l_2 are given by
$$\mathbf{r}_1 = \begin{pmatrix} 3 \\ 1 \\ 4 \end{pmatrix} + s\begin{pmatrix} 5 \\ -2 \\ 3 \end{pmatrix} \qquad \mathbf{r}_2 = \begin{pmatrix} -3 \\ 5 \\ 2 \end{pmatrix} + t\begin{pmatrix} 1 \\ 0 \\ 2 \end{pmatrix}$$
respectively.

Show that the lines l_1 and l_2 are skew and find the acute angle between them.

6 The lines l_1 and l_2 are given by
$$\mathbf{r}_1 = \begin{pmatrix} 3 \\ 2 \\ 1 \end{pmatrix} + \lambda\begin{pmatrix} 1 \\ 2 \\ 2 \end{pmatrix} \text{ and } \mathbf{r}_2 = \begin{pmatrix} 2 \\ 3 \\ 2 \end{pmatrix} + \mu\begin{pmatrix} 2 \\ 1 \\ -2 \end{pmatrix}.$$

Show that l_1 and l_2 are perpendicular. Find the values of λ and μ if $\mathbf{r}_2 - \mathbf{r}_1$ is perpendicular to l_1 and l_2.
Hence find the length of the common perpendicular of l_1 and l_2.

7 $PQRS$ is a quadrilateral. The mid-points of PQ, QR, RS and SP are A, B, C and D respectively. Prove that $ABCD$ is a parallelogram (by use of vectors).

8 The point P has position vector $(1 + \mu)\mathbf{i} + (3 - 2\mu)\mathbf{j} + (4 + 2\mu)\mathbf{k}$ where μ is a variable parameter. The point Q has position vector $4\mathbf{i} + 2\mathbf{j} + 3\mathbf{k}$.

    (a) The points P_0 and P_1 are the positions of P when $\mu = 0$ and $\mu = 1$ respectively. Calculate the size of the angle P_0QP_1, giving your answer to the nearest degree.

    (b) Show that $PQ^2 = (3\mu - 1)^2 + 10$ and hence, or otherwise, find the position vector of P when it is closest to Q.   (AEB, 1991)

9 Three of the corners of a crystal are at the points $A(2, 1, 3)$, $B(1, 3, 6)$ and $C(3, 4, 8)$.

    (a) Find, in column vector form, the equations of the lines AB and AC.

    (b) Calculate the angle between the lines AB and AC.

    (c) Write down, in column vector form, the vector equation of the plane ABC.   (SMP, 1993)

Brief Solutions

1

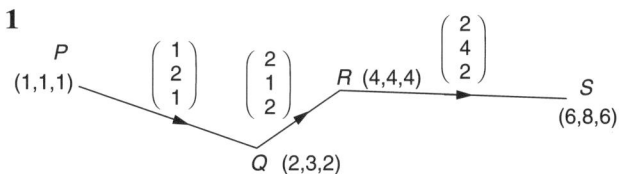

    <u>Answer **A**</u>

2

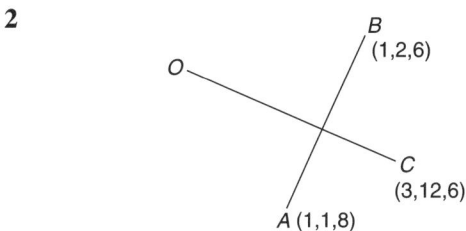

  (i) $\overline{OC} = \begin{pmatrix} 3 \\ 12 \\ 6 \end{pmatrix}$ $\overline{AB} = \begin{pmatrix} 0 \\ 1 \\ -2 \end{pmatrix}$ $\overline{OC} \cdot \overline{AB} = 0$

      $\therefore OC \perp AB$

  (ii)   General points on $\overline{OC}$ and $\overline{AB}$ are

      $\begin{pmatrix} 3\lambda \\ 12\lambda \\ 6\lambda \end{pmatrix}$ and $\begin{pmatrix} 1 \\ 1 + \mu \\ 8 - 2\mu \end{pmatrix}$

      If lines intersect $3\lambda = 1$ $\Rightarrow$ $\lambda = \frac{1}{3}$
                            $12\lambda = 1 + \mu$ $\Rightarrow$ $\mu = 3$
      check-line $6\lambda = 8 - 2\mu$ $\Rightarrow$ $2 = 2$
      Lines intersect at $(1, 4, 2)$.

3 $\mathbf{A} + 2\mathbf{B} - \mathbf{C} = 9\mathbf{i} - \mathbf{j} + 5\mathbf{k}$
    $|\mathbf{A} + 2\mathbf{B} - \mathbf{C}| = \sqrt{81 + 1 + 25} = \sqrt{107}$

  $\mathbf{B} \cdot (\mathbf{A} - \mathbf{C}) = \begin{pmatrix} 3 \\ 0 \\ 1 \end{pmatrix} \cdot \begin{pmatrix} 3 \\ -1 \\ 3 \end{pmatrix} = 12$

  Vector perpendicular to A and B is

  $\begin{vmatrix} \mathbf{i} & \mathbf{j} & \mathbf{k} \\ 1 & 2 & 3 \\ 3 & 0 & 1 \end{vmatrix} = 2\mathbf{i} + 8\mathbf{j} - 6\mathbf{k}$, magnitude $2\sqrt{26}$.

4 $\overline{AB} = \begin{pmatrix} -3 \\ 3 \\ -4 \end{pmatrix}$ $|\overline{AB}| = \sqrt{34}$

  Direction ratios $-3:3:-4$

  Direction cosines $\dfrac{-3}{\sqrt{34}}:\dfrac{3}{\sqrt{34}}:\dfrac{-4}{\sqrt{34}}$

5   General points on l_1 and l_2 are
  $(3 + 5s, 1 - 2s, 4 + 3s)$ and $(-3 + t, 5, 2 + 2t)$.
  To intersect: $3 + 5s = -3 + t$ (1)
                  $1 - 2s = 5$ (2)
                  $4 + 3s = 2 + 2t$ (3)
  From equation 2, $s = -2$; from equation 1, $t = -4$; from equation 3, $t = -2$. Contradiction $\therefore$ the lines are skew.

  Direction vectors $\begin{pmatrix} 5 \\ -2 \\ 3 \end{pmatrix}$ and $\begin{pmatrix} 1 \\ 0 \\ 2 \end{pmatrix}$

  $\cos \theta = \dfrac{5 + 0 + 6}{\sqrt{38}\,\sqrt{5}}$ $\Rightarrow$ $\theta = 37.1°$

6   Direction vectors $\begin{pmatrix} 1 \\ 2 \\ 2 \end{pmatrix}$ and $\begin{pmatrix} 2 \\ 1 \\ -2 \end{pmatrix}$

  Scalar product $= 0$ $\therefore$ the lines are perpendicular

  $\mathbf{r}_2 - \mathbf{r}_1 = \begin{pmatrix} -1 - \lambda + 2\mu \\ 1 - 2\lambda + \mu \\ 1 - 2\lambda - 2\mu \end{pmatrix}$

  $(\mathbf{r}_2 - \mathbf{r}_1) \cdot \begin{pmatrix} 1 \\ 2 \\ 2 \end{pmatrix} = 0$ $\Rightarrow$ $3 - 9\lambda = 0$ $\lambda = \frac{1}{3}$

  $(\mathbf{r}_2 - \mathbf{r}_1) \cdot \begin{pmatrix} 2 \\ 1 \\ -2 \end{pmatrix} = 0$ $\Rightarrow$ $-3 + 9\mu = 0$ $\mu = \frac{1}{3}$

  $\mathbf{r}_2 - \mathbf{r}_1 = \begin{pmatrix} -\frac{2}{3} \\ \frac{2}{3} \\ -\frac{1}{3} \end{pmatrix}$

  Shortest distance $= \frac{1}{3}\sqrt{4 + 4 + 1} = 1$.

  Alternatively use the formula $|(\mathbf{a}_1 - \mathbf{a}_2) \cdot \hat{\mathbf{n}}|$ where $\mathbf{a}_1$ and $\mathbf{a}_2$ are the position vectors of points A_1 and

A_2 on the lines l_1 and l_2 and $\hat{\mathbf{n}}$ is the unit vector perpendicular to lines l_1 and l_2.

The vector perpendicular to $\begin{pmatrix} 1 \\ 2 \\ 2 \end{pmatrix}$ and $\begin{pmatrix} 2 \\ 1 \\ -2 \end{pmatrix}$

is $\begin{vmatrix} \mathbf{i} & \mathbf{j} & \mathbf{k} \\ 1 & 2 & 2 \\ 2 & 1 & -2 \end{vmatrix} = -6\mathbf{i} + 6\mathbf{j} - 3\mathbf{k}$

Unit vector $\hat{\mathbf{n}} = \dfrac{1}{3}\begin{pmatrix} -2 \\ 2 \\ -1 \end{pmatrix}$

$\mathbf{a}_1 - \mathbf{a}_2 = \begin{pmatrix} 1 \\ -1 \\ -1 \end{pmatrix}$

Shortest distance $= \dfrac{1}{3}\left| \begin{pmatrix} 1 \\ -1 \\ -1 \end{pmatrix} \cdot \begin{pmatrix} -2 \\ 2 \\ -1 \end{pmatrix} \right|$

$= | \tfrac{1}{3}(-3) |$

$= 1$

7 Taking O as the origin

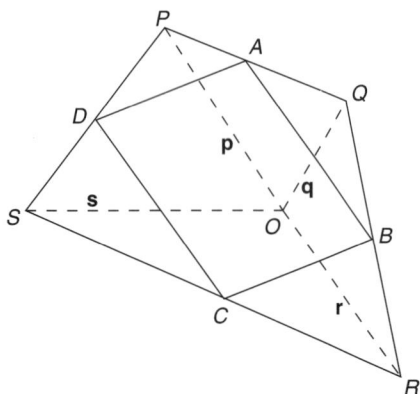

Position vectors of A and B are $\frac{1}{2}(\mathbf{p} + \mathbf{q})$ and $\frac{1}{2}(\mathbf{q} + \mathbf{r})$

Position vectors of D and C are $\frac{1}{2}(\mathbf{p} + \mathbf{s})$ and $\frac{1}{2}(\mathbf{s} + \mathbf{r})$

$\overline{AB} = \mathbf{b} - \mathbf{a} = \frac{1}{2}(\mathbf{r} - \mathbf{p})$ $\overline{DC} = \mathbf{c} - \mathbf{d} = \frac{1}{2}(\mathbf{r} - \mathbf{p})$

$\therefore AB$ is equal and parallel to DC

$\Rightarrow$ $ABCD$ is a parallelogram.

8 (a) $P_0(1, 3, 4)$ $P_1(2, 1, 6)$ $Q(4, 2, 3)$

$\overline{P_0Q} = \begin{pmatrix} 3 \\ -1 \\ -1 \end{pmatrix}$ $\overline{P_1Q} = \begin{pmatrix} 2 \\ 1 \\ -3 \end{pmatrix}$

$\cos P_0QP_1 = \dfrac{\begin{pmatrix} 3 \\ -1 \\ -1 \end{pmatrix} \cdot \begin{pmatrix} 2 \\ 1 \\ -3 \end{pmatrix}}{\sqrt{11} \cdot \sqrt{14}}$

$\angle P_0QP_1 = 50°$

(b) $PQ^2 = (\mu - 3)^2 + (1 - 2\mu)^2 + (1 + 2\mu)^2$
$= 9\mu^2 - 6\mu + 11 = (3\mu - 1)^2 + 10$
PQ^2 has a least value of 10 when $\mu = \frac{1}{3}$.
Position vector of P is $\frac{1}{3}(4\mathbf{i} + 7\mathbf{j} + 14\mathbf{k})$.

9 (a) Line AB: $\mathbf{r} = \begin{pmatrix} 2 \\ 1 \\ 3 \end{pmatrix} + \lambda \begin{pmatrix} -1 \\ 2 \\ 3 \end{pmatrix}$

Line AC: $\mathbf{r} = \begin{pmatrix} 2 \\ 1 \\ 3 \end{pmatrix} + \mu \begin{pmatrix} 1 \\ 3 \\ 5 \end{pmatrix}$

(b) $\overline{AB} \cdot \overline{AC} = |AB| \, |AC| \cos B\hat{A}C$

$\Rightarrow$ $\cos B\hat{A}C = \dfrac{20}{\sqrt{14}\sqrt{35}}$ $\Rightarrow$ $B\hat{A}C = 25.4°$

(c) Vector equation of ABC is

$\mathbf{r} = \begin{pmatrix} 2 \\ 1 \\ 3 \end{pmatrix} + s \begin{pmatrix} -1 \\ 2 \\ 3 \end{pmatrix} + t \begin{pmatrix} 1 \\ 3 \\ 5 \end{pmatrix}.$

In (b), remember that only the direction vectors are used to find the angle between two lines

8

Sequences and series

AS Level			A Level			Topic	Date attempted	Date completed	Self-assessment
CORE	MODULAR	TRADITIONAL	CORE	MODULAR	TRADITIONAL				
✓	✓	✓	✓	✓	✓	**Arithmetic and geometric series**			
	✓	✓	✓	✓	✓	**Binomial series**			
	✓	✓	✓	✓	✓	**Standard series**			

Fact Sheet

Definitions

A sequence is a set of terms which follow a definite mathematical rule.

$$2, 4, 6, 8 \tag{1}$$
$$\text{and} \quad 1, 4, 9, 16 \tag{2}$$

are sequences.

The sum of these sequences is called a series or progression.
Individual terms are usually written $u_1, u_2, u_3, \ldots, u_n$.

$u_n = 2n$ refers to the sequence (1)
$u_n = n^2$ refers to the sequence (2)
$S_n = $ sum of the first n terms

Specific Series or Progressions

Arithmetic Progression or A.P.

$$u_n = a + (n - 1)d$$

First term $= a$, common difference $= d$.

$$S_n = \text{sum of the first } n \text{ terms} = \frac{n}{2}(2a + (n - 1)d) \text{ or } \frac{n}{2}(\text{first term} + \text{last term})$$

Condition for p, q and r to be in arithmetic progression is $p + r = 2q$.

Geometric Progression or G.P.

$$u_n = ar^{n-1}$$

where a is the first term, r is the common ratio.

Series of n terms is $a + ar + ar^2 + ar^3 + \ldots + ar^{n-1}$

$$S_n = \frac{a(1 - r^n)}{1 - r} \text{ or } \frac{a(r^n - 1)}{r - 1}$$

Sum to infinity exists if $|r| < 1$ (or $-1 < r < 1$)

$$S_\infty \text{ (or just } S) = \frac{a}{1 - r}$$

Condition for x, y, z to form a geometric progression is $xz = y^2$.

Σ Notation for Standard Series

- $$\sum_{r=1}^{n} r^\alpha = 1^\alpha + 2^\alpha + 3^\alpha + \ldots + (n - 1)^\alpha + n^\alpha$$

where n is a positive integer.

- $$\sum_{r=1}^{n} ar^2 = a + a(2)^2 + a(3)^2 + \ldots + a(n)^2 = a \sum_{r=1}^{n} r^2$$

- $$\sum_{r=1}^{n} (ar^2 + br + c) = a \sum_{r=1}^{n} r^2 + b \sum_{r=1}^{n} r + cn \quad \left(\sum_{r=1}^{n} 1 = n\right)$$

Useful Results

- $$\sum_{r=1}^{n} r = \frac{n(n + 1)}{2}$$

- $$\sum_{r=1}^{n} r^2 = \frac{n(n + 1)(2n + 1)}{6}$$

- $$\sum_{r=1}^{n} r^3 = \frac{n^2(n + 1)^2}{4}$$

Change of Limits

- $$\sum_{r=m}^{n} f(r) = \sum_{r=1}^{n} f(r) - \sum_{1}^{m-1} f(r)$$

Binomial Series $(1 + x)^n$

(a) If n is a positive integer this is a **finite** series with $n + 1$ terms.

(i) $$(1 + x)^n = 1 + nx + \frac{n(n - 1)}{1.2}x^2 + \frac{n(n - 1)(n - 2)}{1.2.3}x^2 + \ldots$$
$$+ nx^{n-1} + x^n \quad (n + 1 \text{ terms})$$

This can be expressed as

$$1 + {}^nC_1 x + {}^nC_2 x^2 + {}^nC_3 x^3 + \ldots$$

or $$1 + \binom{n}{1}x + \binom{n}{2}x^2 + \binom{n}{3}x^3 + \ldots$$

where $${}^nC_r = \binom{n}{r} = \frac{n!}{r!(n - r)!}$$

These coefficients can be calculated quickly using the ${}^nC_r$ function key on your calculator. For example:

$${}^7C_4 \rightarrow 7 \boxed{{}^nC_r} 4 = 35$$

If your calculator does not have a ${}^nC_r$ function key but does have $x!$, then:

$${}^7C_4 \rightarrow \boxed{7} \boxed{x!} \boxed{\div} \boxed{4} \boxed{x!} \boxed{=} \boxed{\div} \boxed{3} \boxed{x!} \boxed{=} \boxed{35}$$

(ii) For small values of n the binomial coefficients may be quickly determined by Pascal's triangle

```
              1
            1  1
          1  2  1
        1  3  3  1
      1  4  6  4  1
    1  5  10  10  5  1
  1  6  15  20  15  6  1   etc.
```

(iii) $$(a + bx)^n = a^n \left(1 + \frac{bx}{a}\right)^n$$

$$= a^n \left(1 + n\left(\frac{bx}{a}\right) + \frac{n(n - 1)}{1.2}\left(\frac{bx}{a}\right)^2 + \ldots + n\left(\frac{bx}{a}\right)^{n-1} + \left(\frac{bx}{a}\right)^n\right)$$

$$= \sum_{r=0}^{n} \binom{n}{r} a^{n-r} b^r$$

(b) If n is not a positive integer:

$$(1 + x)^n = 1 + nx + \frac{n(n-1)}{1.2}x^2 + \frac{n(n-1)(n-2)}{1.2.3}x^3 + \ldots \text{ if } |x| < 1$$

This is an **infinite** series which can only be used when $|x| < 1$.

Similarly $(a + bx)^n = a^n\left(1 + \frac{bx}{a}\right)^n$

$$= a^n\left(1 + n\frac{bx}{a} + \frac{n(n-1)}{1.2}\left(\frac{bx}{a}\right)^2 + \ldots\right) \text{ when } \left|\frac{bx}{a}\right| < 1$$

(c) Particular series:

$$(1 + x)^{-1} = 1 - x + x^2 - x^3 + \ldots + (-1)^n x^n + \ldots \qquad |x| < 1$$
$$(1 - x)^{-1} = 1 + x + x^2 + x^3 + \ldots + x^n + \ldots \qquad |x| < 1$$
$$(1 + x)^{-2} = 1 - 2x + 3x^2 - 4x^3 + \ldots + (-1)^n(n+1)x^n + \ldots \qquad |x| < 1$$
$$(1 - x)^{-2} = 1 + 2x + 3x^2 + 4x^3 + \ldots + (n+1)x^n + \ldots \qquad |x| < 1$$

Standard Series

- $\sin x = x - \dfrac{x^3}{3!} + \dfrac{x^5}{5!} + \ldots + \dfrac{(-1)^{n-1}x^{2n-1}}{(2n-1)!} + \ldots$

- $\cos x = 1 - \dfrac{x^2}{2!} + \dfrac{x^4}{4!} + \ldots + \dfrac{(-1)^{n-1}x^{2(n-1)}}{(2n-2)!} + \ldots$

These are valid for all x, expressed in *radians*.

- $e^x = 1 + x + \dfrac{x^2}{2!} + \dfrac{x^3}{3!} + \ldots + \dfrac{x^n}{n!} + \ldots$ valid for all x

- $\ln(1 + x) = x - \dfrac{x^2}{2} + \dfrac{x^3}{3} + \ldots + (-1)^{n-1}\dfrac{x^n}{n} + \ldots$ valid for $-1 < x \leqslant 1$

- $\ln(1 - x) = -\left(x + \dfrac{x^2}{2} + \dfrac{x^3}{3} + \ldots + \dfrac{x^n}{n} + \ldots\right)$ valid for $-1 \leqslant x < 1$

- $\ln\left(\dfrac{1+x}{1-x}\right) = 2\left(x + \dfrac{x^3}{3} + \dfrac{x^5}{5} + \ldots + \dfrac{x^{2n-1}}{2n-1} + \ldots\right)$ valid for $-1 < x < 1$

Worked Examples

1 Given that $(1 + x)^{10} = 1 + px + qx^2 + \ldots + x^{10}$, find the values of p and q.

(L, 1993)

- Using the binomial expansion

$$(1 + x)^{10} = 1 + 10x + \frac{(10)(9)}{(1)(2)}x^2 + \ldots$$

$$\Rightarrow \quad p = 10, q = 45$$

2 On 1 January 1962, a new town was created from a collection of villages which at that time had a combined population of 10 000. Given that the new town population increased by 10% per year:

(a) State the growth factor for the population.
(b) State the population after t years.
(c) During which year will the population reach 25 000?   (SMP)

- (a) The growth factor is 110% or 1.1.
  This is a geometric progression with ratio 1.1.
  (b) At the beginning of year 1 the population was 10 000.
  At the *end* of year 1 and the beginning of year 2 the population was $10\ 000 \times (1.1)^1 = 11\ 000$
  After t years the population is $10\ 000 \times (1.1)^t$.
  (c) When $10\ 000 \times (1.1)^t = 25\ 000$
  $$(1.1)^t = 2.5$$
  Take logs: $t \log(1.1) = \log(2.5)$

2.1a and 2.2a

$$t = \frac{\log (2.5)}{\log (1.1)} = 9.61, \text{ i.e. during the}$$
10th year

Hence the town's population reaches 25 000 during 1971.

3 In an arithmetic progression, the 8th term is twice the 3rd term and the 20th term is 110.
(a) Find the common difference.
(b) Determine the sum of the first 100 terms.
(AEB, 1992)

• (a) $u_3 = a + 2d$, $u_8 = a + 7d$, $u_{20} = a + 19d$
  (i) $a + 7d = 2(a + 2d) \Rightarrow a = 3d$ (1)
  (ii) $a + 19d = 110$ (2)

2.1b and 2.2b

  Replace a by $3d$: $22d = 110$
            $d = 5$, $a = 15$ (from equation 1)
  Hence the common difference is 5.
  (b) $S_{100} = \frac{100}{2}(30 + 99(5)) = 26\ 250$
  Sum of the first 100 terms is 26 250.

8.3

4 By using an infinite geometric progression show that $0.6\dot{5}\dot{1}$ is equal to $\frac{43}{66}$.

• $0.6\dot{5}\dot{1} = 0.6 + 0.051 + 0.000\ 51 + \ldots$
        $= 0.6 + \text{a G.P.}$
  The G.P. has $a = 0.051$ and $r = 0.01$

  $$\Rightarrow S_\infty = \frac{0.051}{1 - \frac{1}{100}} \quad \text{using } S_\infty = \frac{a}{1 - r}$$

  $$= \frac{51}{990}$$

  Hence $0.6\dot{5}\dot{1} = \frac{6}{10} + \frac{51}{990}$
  $$= \frac{645}{990}$$
  $$= \frac{43}{66}$$

5 Given that x is sufficiently small, use the approximations $\sin x \approx x$ and $\cos x \approx 1 - \frac{1}{2}x^2$ to show that

$$\frac{\cos x}{1 + \sin x} \approx 1 - x + \frac{1}{2}x^2$$

A student estimates the value of $\dfrac{\cos x}{1 + \sin x}$ when $x = 0.1$ by evaluating the approximation $1 - x + \frac{1}{2}x^2$ when $x = 0.1$.
Find, to 3 d.p., the percentage error made by the student. (L, 1993)

• $$\frac{\cos x}{1 + \sin x} \approx \frac{1 - \frac{1}{2}x^2}{1 + x}$$

Rewrite $\dfrac{1}{1 + x}$ as

$(1 + x)^{-1}$ and expand by the binomial series

$$\approx (1 - \tfrac{1}{2}x^2)(1 + x)^{-1}$$
$$\approx (1 - \tfrac{1}{2}x^2)(1 - x + x^2 + \ldots)$$
$$\approx 1 - x - \tfrac{1}{2}x^2 + x^2 + \text{higher powers of } x$$
$$\approx 1 - x + \tfrac{1}{2}x^2$$

When $x = 0.1$, $1 - x + \frac{1}{2}x^2 = 0.905$.

But $\dfrac{\cos x}{1 + \sin x} = 0.904\ 686$ when $x = 0.1$ radians

Error $= 0.000\ 313\ 75$

% error $= \dfrac{\text{error}}{\text{correct value}} \times 100$

      $= 0.035\%$ to 3 d.p.

6 Use the binomial expansion to find a quadratic approximation for

$$\frac{1}{(1 + 2x)^{1/3}} - \frac{1}{(9 - 4x)^{3/2}}$$

where x is small enough for terms in x^3 and higher powers to be negligible.

• $$\frac{1}{(1 + 2x)^{1/3}} = (1 + 2x)^{-1/3}$$

$$= 1 + (-\tfrac{1}{3})(2x) + \frac{(-\tfrac{1}{3})(-\tfrac{4}{3})}{(1)(2)}(2x)^2$$
$$+ \text{ higher powers of } x$$

$$= 1 - \frac{2x}{3} + \frac{8x^2}{9} + \text{ higher powers of } x$$
$$|x| < \tfrac{1}{2}$$

$$\frac{1}{(9 - 4x)^{3/2}} = \frac{1}{9^{3/2}\left(1 - \dfrac{4x}{9}\right)^{3/2}}$$ Remember to take out the factor of 9 before putting the $(\text{bracket})^{-3/2}$

$$= \frac{1}{27}\left(1 - \frac{4x}{9}\right)^{-3/2}$$

$$= \frac{1}{27}\left(1 + \left(-\frac{3}{2}\right)\left(-\frac{4x}{9}\right)\right.$$

$$\left. + \frac{(-\tfrac{3}{2})(-\tfrac{5}{2})}{(1)(2)}\left(-\frac{4x}{9}\right)^2 \right) \quad |x| < \tfrac{9}{4}$$

$$= \frac{1}{27}\left[1 + \frac{2x}{3} + \frac{10x^2}{27} + \ldots\right]$$

Hence

$$\frac{1}{(1 + 2x)^{1/3}} - \frac{1}{(9 - 4x)^{3/2}}$$

$$\approx 1 - \frac{2x}{3} + \frac{8x^2}{9} - \frac{1}{27} - \frac{2x}{81} - \frac{10x^2}{729} \ldots$$

$$= \frac{26}{27} - \frac{56x}{81} + \frac{638x^2}{729} \ldots \quad |x| < \tfrac{1}{2}$$

8.1

7 Write down the first 4 terms in the expansion of e^x. Obtain the first 4 terms in the expansion of $(x^2 + x)e^x$. Deduce that $(x^2 + x)(e^x - 1)$ is always positive when x is positive.

• $$e^x = 1 + x + \frac{x^2}{2} + \frac{x^3}{3!} + \text{positive terms.}$$

$$(x^2 + x)e^x = (x^2 + x)\left(1 + x + \frac{x^2}{2} + \frac{x^3}{3!} + \ldots\right)$$

$$= x^2 + x^3 + \frac{x^4}{2} + \frac{x^5}{3!} + x + x^2$$
$$+ \frac{x^3}{2} + \frac{x^4}{3!} + \ldots$$

$$= x + 2x^2 + \frac{3x^3}{2} + \frac{2x^4}{3} + \text{positive terms}$$

$$(x^2 + x)(e^x - 1) = (x^2 + x)e^x - (x^2 + x)$$

$$= x^2 + \frac{3x^3}{2} + \frac{2x^4}{3} + \text{positive terms}$$

$\boxed{8.1}$

All the terms are positive.
Therefore $(x^2 + x)(e^x - 1)$ is always positive, provided x is positive.

8 Find the first three non-zero terms in the expansion of $\ln\left[(1 + 3x)(1 - 2x)\right]$ in a series of ascending powers of x.

• $\ln\left[(1 + 3x)(1 - 2x)\right] = \ln(1 + 3x) + \ln(1 - 2x)$

$$\ln(1 + 3x) = 3x - \frac{(3x)^2}{2} + \frac{(3x)^3}{3} - \ldots$$
$$\qquad\qquad -\tfrac{1}{3} \leqslant x < \tfrac{1}{3}$$

$$= 3x - \frac{9x^2}{2} + 9x^3 - \ldots \qquad (1)$$

$$\ln(1 - 2x) = -2x - \frac{(2x)^2}{2} - \frac{(2x)^3}{3} - \ldots$$
$$\qquad\qquad -\tfrac{1}{2} < x \leqslant \tfrac{1}{2}$$

$$= -\left(2x + 2x^2 + \frac{8x^3}{3} \ldots\right) \qquad (2)$$

Adding expansions (1) and (2):

$\ln\left[(1 + 3x)(1 - 2x)\right]$

$$= (3x - 2x) + \left(-\frac{9x^2}{2} - 2x^2\right) + \left(9x^3 - \frac{8x^3}{3}\right) \ldots$$

$$= x - \frac{13x^2}{2} + \frac{19x^3}{3} \ldots \quad -\tfrac{1}{3} \leqslant x < \tfrac{1}{3}$$

$\boxed{8.1}$

9 Express $\dfrac{9x - 9}{9 + 3x - 2x^2}$ as a series in ascending

powers of x as far as and including the term in x^3. Give the range of values of x for which this is valid.

• Changing this into 2 fractions by partial fractions will simplify the expansion

$$\frac{9x - 9}{9 + 3x - 2x^2} = \frac{9x - 9}{(3 - x)(3 + 2x)}$$

$$\equiv \frac{A}{3 - x} + \frac{B}{3 + 2x}$$

Since the denominators are linear, use the 'cover over' method

$\Rightarrow$ putting $x = 3$ into $\dfrac{9x - 9}{(3 + 2x)}$ (i.e. covering over the

$(3 - x)$ bracket) gives $A = 2$ etc.

Putting $x = -\tfrac{3}{2}$ into $\dfrac{9x - 9}{3 - x}$ gives $B = -5$.

Hence $\dfrac{9x - 9}{9 + 3x - 2x^2} = \dfrac{2}{3 - x} - \dfrac{5}{3 + 2x}$ $\boxed{4.3}$

$$= \frac{2}{3\left(1 - \dfrac{x}{3}\right)} - \frac{5}{3\left(1 + \dfrac{2x}{3}\right)}$$

$$= \frac{2}{3}\left(1 - \frac{x}{3}\right)^{-1} - \frac{5}{3}\left(1 + \frac{2x}{3}\right)^{-1}$$

$$= \frac{2}{3}\left(1 + \frac{x}{3} + \frac{x^2}{9} + \frac{x^3}{27} + \ldots\right)$$
$$- \frac{5}{3}\left(1 - \frac{2x}{3} + \frac{4x^2}{9} - \frac{8x^3}{27} + \ldots\right)$$

$$= -1 + \frac{4x}{3} - \frac{2x^2}{3} + \frac{14x^3}{27} \ldots$$

$\boxed{8.1}$

The series for $\left(1 - \dfrac{x}{3}\right)^{-1}$ is valid for $|x| < 3$

$\left(1 + \dfrac{2x}{3}\right)^{-1}$ is valid for $|x| < \tfrac{3}{2}$

Hence the combined series is valid for $|x| < \tfrac{3}{2}$ or $-\tfrac{3}{2} < x < \tfrac{3}{2}$.

10 (a) The sum of the first n terms of a series is $1 - (\tfrac{1}{4})^n$. Obtain the values of the first three terms of this series. What is the sum to infinity of this series?

(b) An arithmetic progression is such that the sum of the first n terms is $2n^2$ for all positive integral values of n. Find, by substituting two values of n or otherwise, the first term and the common difference.

• (a) If $S_n = 1 - (\frac{1}{4})^n$ then

$S_1 = u_1 = \frac{3}{4}$

$S_2 = u_1 + u_2 = \frac{15}{16} \Rightarrow u_2 = \frac{3}{16}$

$S_3 = S_2 + u_3 = \frac{63}{64} \Rightarrow u_3 = \frac{3}{64}$

Hence the first 3 terms of the series are $\frac{3}{4}, \frac{3}{16}, \frac{3}{64}$.
This is a G.P. with first term $\frac{3}{4}$, ratio $\frac{1}{4}$.

$S_\infty = \dfrac{\frac{3}{4}}{1 - \frac{1}{4}} = 1$

This could have been deduced from $1 - (\frac{1}{4})^n$ as $n \to \infty$

(b) $S_n = 2n^2$

$S_1 = 2 = u_1$

$S_2 = 8 = u_1 + u_2 \Rightarrow u_2 = 6$

Hence the first term is 2,
common difference = 4.

11 Prove that, for $|x| < 1$,

$$\sqrt{\frac{1+x}{1-x}} \approx 1 + x + \frac{x^2}{2} + \frac{x^3}{2} + \dots$$

By a suitable choice of a value for x, prove that

$$\sqrt{5} \approx \frac{1630}{729}$$

• $\sqrt{\dfrac{1+x}{1-x}} = (1+x)^{1/2}(1-x)^{-1/2}$ ‖ 8.1 ‖

$(1+x)^{1/2} = 1 + \frac{1}{2}x + \dfrac{(\frac{1}{2})(-\frac{1}{2})}{1.2}x^2 + \dfrac{(\frac{1}{2})(-\frac{1}{2})(-\frac{3}{2})}{1.2.3}x^3 + \dots$

$= 1 + \dfrac{x}{2} - \dfrac{x^2}{8} + \dfrac{x^3}{16} + \dots \quad |x| < 1$

$(1-x)^{-1/2} = 1 + \frac{1}{2}x + \dfrac{(-\frac{1}{2})(-\frac{3}{2})}{1.2}(-x)^2$

$+ \dfrac{(-\frac{1}{2})(-\frac{3}{2})(-\frac{5}{2})}{1.2.3}(-x)^3 + \dots$

$= 1 + \dfrac{x}{2} + \dfrac{3x^2}{8} + \dfrac{5x^3}{16} + \dots \quad |x| < 1$ ‖ 8.1 ‖

Hence $\sqrt{\dfrac{1+x}{1-x}} = \left(1 + \dfrac{x}{2} - \dfrac{x^2}{8} + \dfrac{x^3}{16} + \dots\right)$

$\times \left(1 + \dfrac{x}{2} + \dfrac{3x^2}{8} + \dfrac{5x^3}{16} + \dots\right)$

$= 1 + x + \dfrac{x^2}{2} + \dfrac{x^3}{2} \dots \quad |x| < 1$

To find a suitable value of x, look at the answer given. A denominator of 729 from terms up to x^3 hints at a

denominator of $(729)^{1/3}$.
Try $x = \frac{1}{9}$

$\sqrt{\dfrac{1+x}{1-x}} = \sqrt{\frac{10}{8}} = \sqrt{\frac{5}{4}} = \frac{1}{2}\sqrt{5}$

If $x = \frac{1}{9}, \frac{1}{2}\sqrt{5} = 1 + \frac{1}{9} + \frac{1}{162} + \frac{1}{1458}$

$= \frac{815}{729}$

Hence $\sqrt{5} = \frac{1630}{729}$

12 Find the values of (i) $\ln(1.1)$, (ii) $\ln(1.2)$ to 5 d.p.

• $\ln(1 + x) = x - \dfrac{x^2}{2} + \dfrac{x^3}{3} - \dfrac{x^4}{4} + \dfrac{x^5}{5} \dots$

(i) $\ln(1.1) = 0.1 - \dfrac{(0.1)^2}{2} + \dfrac{(0.1)^3}{3} - \dfrac{(0.1)^4}{4} + \dfrac{(0.1)^5}{5}$

$= 0.095\,31$ to 5 d.p.

(ii) $\ln(1.2) = 0.2 - \dfrac{(0.2)^2}{2} + \dfrac{(0.2)^3}{3} - \dfrac{(0.2)^4}{4}$

$+ \dfrac{(0.2)^5}{5} - \dfrac{(0.2)^6}{6} + \dfrac{(0.2)^7}{7}$

$- \dfrac{(0.2)^8}{8} \dots$

$= 0.182\,32$ to 5 d.p.

Using $\ln\left(\dfrac{1+x}{1-x}\right)$ could shorten the working.

If $\dfrac{1+x}{1-x} = 1.2$ then $x = \frac{1}{11}$

$\ln\left(\dfrac{1 + \frac{1}{11}}{1 - \frac{1}{11}}\right) = \ln(1 + \frac{1}{11}) - \ln(1 - \frac{1}{11})$

$= 2\left(\frac{1}{11} + \dfrac{(\frac{1}{11})^3}{3} + \dfrac{(\frac{1}{11})^5}{5} \dots\right)$

$= 0.182\,32$ to 5 d.p.

Exercises

1 Show that $\displaystyle\sum_{r=1}^{n} r(3r+1) = n(n+1)^2$.

Hence evaluate $\displaystyle\sum_{r=31}^{60} r(3r+1)$. (L, 1992)

2 The numbers $\dfrac{1}{t}, \dfrac{1}{t-1}, \dfrac{1}{t+2}$ are the first, second and third terms respectively of a geometric series. Find

(a) the value of t

(b) the 19th term of the series giving your answers to 3 s.f.

(c) the sum to infinity of the series.

(L Specimen Paper, 1996)

3 The sum of the first twenty terms of an arithmetic progression is 50, and the sum of the next twenty terms is −50. Find the sum of the first hundred terms of the progression.

4 A sequence is defined by $u_{n+2} = 3u_{n+1} − 2u_n$ and $u_1 = 2$, $u_2 = 5$.

(a) Find u_3 and u_4.

(b) Write down a definition for the sequence in the form

$$u_{n+1} = au_n + b.$$ (SMP)

5 (a) Use the binomial theorem to obtain the expansion of $(1 + x)^{-1}$ up to and including the term in x^3.

(b) Show that $1 − 2x + 2x^2$ is the quadratic approximation to the function $\dfrac{1 − x}{1 + x}$ near the origin.

(c) Calculate the percentage error when using the approximation to work out the value of

$\dfrac{1 − x}{1 + x}$ when $x = 0.25$. (O&C SMP)

6 (a) A savings account offers 4% interest every six months. What would the annual rate of interest be for a saver who adds the interest to the original amount and reinvests it for a further six months?

(b) What half-yearly interest rate (to 1 d.p.) would be equivalent to a 10% annual rate?

(O&C SMP, 1994)

7 The first, second and third terms of a geometric series are p, p^2 and q respectively when $p < 0$. The first, second and third terms of an arithmetic series are p, q and p^2 respectively.

(a) Show that $p = −\frac{1}{2}$ and find the value of q.

(b) Find the sum to infinity of the geometric series.

(AEB, 1991)

8 Use the binomial expansion to show that

$$\frac{x^2}{\sqrt{4 − x^2}} = \frac{1}{2}x^2 + \frac{1}{16}x^4 + kx^6 + \dots \quad (|x| < 2)$$

for some constant k, and state its value.

Hence show, by integrating the first three terms of the series, that the value of the integral

$$I = \int_0^1 \frac{x^2}{\sqrt{4 − x^2}} \, dx \text{ is approximately 0.1808.}$$

(AEB, 1991)

9 The fourth term of an arithmetic progression is 8 and the sum of the first nine terms is 48. Find the first term and the sum of the first sixteen terms.

(AEB Specimen Paper, 1994)

10 Use the expansions of e^x and $\sin x$ to write down the expansions, in ascending powers of x as far as the term in x^3, of e^{-x} and $\sin 3x$. Show that the expansion, in ascending powers of x as far as the term in x^3, of $e^{-x} \sin 3x$ is $3x − 3x^2 − 3x^3$.

Hence, or otherwise, find the expansion, in ascending powers of x as far as the term in x^3, of $e^{-2x} \sin^2 3x$. (O&C, 1993)

11 (a) Given that $b > 0$, S_n is defined by

$$S_n = \ln 32 + \ln 32b + \ln 32b^2 + \dots$$
$$+ \ln 32(b^{n-1})$$

Explain why this is an arithmetic series. Write down the value of S_n in terms of b and n. If $S_{11} = 0$, find the value of b in a form which does not involve logarithms.

(b) T_n is defined by

$$T_n = e + e^{1+x} + e^{1+2x} + e^{1+3x} + \dots + e^{1+(n-1)x}$$

Explain why this is a geometric series. Write down the value of T_n in terms of x and n. For what values of x does the sum to infinity exist? Write down the value of this sum when it does exist.

(c) Write down the expansion of $(x + 1)^6$ in descending powers of x. Hence, or otherwise, prove that if x is positive then

$$(x + 1)^6 > x^6 + 6x^5$$

Prove by induction, or otherwise, that

$$\sum_{r=1}^{n-1} r^5 < \frac{n^6}{6}, \text{ for any positive integer } n \text{ greater}$$

than 1. (O&C, 1993)

12 A geometric series has first term 1 and common ratio r. Given that the sum to infinity of the series is 5, find the value of r.

Find the least value of n for which the sum of the first n terms of the series exceeds 4.9. (UCLES, 1991)

13 (a) Find the sum of the arithmetic progression 1, 4, 7, 10, …, 1000. Every third term is removed (i.e. 7, 16, etc.). Find the sum of the remaining terms.

(b) The rth term, u_r, of a series is given by

$$u_r = (\tfrac{1}{3})^{3r-2} + (\tfrac{1}{3})^{3r-1}$$

Express $\displaystyle\sum_{r=1}^{n} u_r$ in the form $A\left(1 − \dfrac{B}{27^n}\right)$, where A and B are constants. Find the sum to infinity of the series. (UCLES, 1991)

14 Evaluate, correct to the nearest whole number,

$$\sum_{r=1}^{100} (0.99)^r. \qquad \text{(UCLES, 1990)}$$

15 The first three terms of a geometric series are $1, r$ and s where r is negative. Given also that $1, s$ and r are the first three terms of an arithmetic series, show that $2r^2 - r - 1 = 0$. Hence find the value of r.

Find the sum to infinity of the geometric series.
(AEB, 1992)

Brief Solutions

1 $\displaystyle\sum_{r=1}^{n} r(3r + 1) = \sum_{r=1}^{n} 3r^2 + \sum_{r=1}^{n} r$

$$= \frac{3n(n + 1)(2n + 1)}{6} + \frac{n(n + 1)}{2}$$

$$= \frac{n(n + 1)}{2}(2n + 1 + 1) = n(n + 1)^2$$

$$\sum_{r=31}^{60} r(3r + 1) = \sum_{r=1}^{r=60} r(3r + 1) - \sum_{r=1}^{r=30} r(3r + 1)$$

$$= 60(61)^2 - 30(31)^2$$
$$= 194\,430$$

2 The condition for a, b and c to be consecutive terms in a G.P. is $ac = b^2$

(a) $\dfrac{1}{t}\dfrac{1}{t + 2} = \dfrac{1}{(t - 1)^2} \Rightarrow t^2 + 2t = t^2 - 2t + 1$

$$\Rightarrow t = \tfrac{1}{4}$$

Terms are $4, -\tfrac{4}{3}, \tfrac{4}{9}, \ldots \quad a = 4, r = -\tfrac{1}{3}$

(b) 19th term $= ar^{18} = \dfrac{4}{3^{18}} = 1.03 \times 10^{-8}$ to 3 s.f.

(c) Sum to infinity $= \dfrac{a}{1 - r} = 3$

3 A.P.: $S_{20} = 10(2a + 19d)$

$$\Rightarrow \quad 20a + 190d = 50 \qquad (1)$$

$$S_{40} = 20(2a + 39d) = 0 \Rightarrow d = -\frac{2a}{39} \qquad (2)$$

From equations 1 and 2, $a = \tfrac{39}{8}, d = -\tfrac{1}{4}$ $\boxed{8.3}$

$$S_{100} = 50(\tfrac{39}{4} + (-\tfrac{99}{4})) = -750$$

4 (a) $u_{n+2} = 3u_{n+1} - 2u_n$
$u_1 = 2, u_2 = 5, u_3 = 11, u_4 = 23$ $\boxed{8.3}$
(b) By inspection $u_{n+1} = 2u_n + 1$

5 (a) $(1 + x)^{-1} = 1 - x + x^2 - x^3 \ldots$ for $|x| < 1$
(b) $(1 - x)(1 + x)^{-1} = 1 - 2x + 2x^2 +$ higher powers of x

(c) When $x = 0.25$

$$\frac{1 - x}{1 + x} \approx 1 - 0.5 + 0.125 = 0.625$$

Exact value is $0.6 \Rightarrow \%$ error $= \dfrac{0.025}{0.6} \times 100$

$$= 4.17\%$$

6 (a) $£A \xrightarrow{6\text{ months}} 1.04A \xrightarrow{6\text{ months}} (1.04)^2 A = 1.0816A$
Annual rate 8.16%
(b) If the annual rate is 10%, ratio $= 1.1$.

For six months, ratio $= \sqrt{1.1} = 1.049$, i.e. 4.9% every six months.

7 (a) G.P.: $p, p^2, q \Rightarrow q = p^3$
A.P.: $p, q, p^2 \Rightarrow p + p^2 = 2q$
$\Rightarrow 2p^3 - p^2 - p = 0$
$\Rightarrow p = 0, -\tfrac{1}{2} \text{ or } 1$
$p < 0 \Rightarrow p = -\tfrac{1}{2}, q = -\tfrac{1}{8}$

(b) S_∞ of G.P. $= \dfrac{-\tfrac{1}{2}}{1 + \tfrac{1}{2}} = -\tfrac{1}{3}$

8 $\dfrac{x^2}{2}\left(1 - \dfrac{x^2}{4}\right)^{-1/2}$

$$= \frac{x^2}{2}\left(1 + \frac{x^2}{8} + \frac{(-\tfrac{1}{2})(-\tfrac{3}{2})}{(1)(2)}\left(-\frac{x^2}{4}\right)^2 + \ldots\right)$$

$$= \frac{x^2}{2} + \frac{x^4}{16} + \frac{3x^6}{256} + \ldots \qquad \boxed{8.1}$$

$$k = \frac{3}{256}$$

$$\int_0^1 \frac{x^2}{\sqrt{4 - x^2}}\,dx = \left[\frac{x^3}{6} + \frac{x^5}{80} + \frac{3x^7}{1792} + \ldots\right]_0^1$$

$$= 0.1808 \text{ to 4 d.p.} \qquad \boxed{6.2}$$

9 $a + 3d = 8 \quad \tfrac{9}{2}(2a + 8d) = 48 \Rightarrow 9a + 36d = 48$
$\Rightarrow d = -\tfrac{8}{3}, a = 16$
$S_{16} = 8(32 + 15(-\tfrac{8}{3})) = -64$

10 $e^{-x} = 1 - x + \dfrac{x^2}{2} - \dfrac{x^3}{6} \ldots \sin 3x = 3x - \dfrac{9x^3}{2} \ldots$

$$e^{-x}\sin 3x = 3x - 3x^2 + \frac{3x^3}{2} - \frac{9x^3}{2} \ldots$$

$$= 3x - 3x^2 - 3x^3 \ldots$$

$$e^{-2x}\sin^2 3x = (e^{-x}\sin 3x)^2 = 9(x^2 - 2x^3 \ldots)$$

$\boxed{8.1}$

11 (a) $\ln(32b) = \ln 32 + \ln b$,
$\ln(32b^2) = \ln 32 + 2\ln b$
This is an A.P. with common difference $= \ln b$

$$S_n = \frac{n}{2}(2\ln 32 + (n-1)\ln b)$$
$$= \ln(32^n b^{n(n-1)/2})$$

$S_{11} = \ln 32^{11}b^{55} = 0 \quad \Rightarrow \quad 32^{11}b^{55} = 1 \quad \Rightarrow \quad b = \frac{1}{2}$
b must be positive

(b) $T_n = e + e.e^x + e.(e^x)^2 + \ldots$
i.e. G.P. with common ratio e^x

$$T_n = \frac{e(e^{nx} - 1)}{e^x - 1} \quad S_\infty \text{ exists if } x < 0$$

$$S_\infty = \frac{e}{1 - e^x}$$

(c) $(x+1)^6$
$= x^6 + 6x^5 + 15x^4 + 20x^3 + 15x^2 + 6x + 1$
$(x+1)^6 - x^6 - 6x^5$
$= 15x^4 + 20x^3 + 15x^2 + 6x + 1$
which is > 0 when $n > 0$
$\therefore (x+1)^6 > x^6 + 6x^5$

$$\Rightarrow \quad x^5 < \frac{(x+1)^6 - x^6}{6}$$

When $x = 1, 1^5 < \dfrac{2^6 - 1}{6}$.

$$x = 2, 2^5 < \frac{3^6 - 2^6}{6} \text{ etc.}$$

$$\sum_1^{n-1} r^5 < \frac{(n)^6 - 1}{6} = \frac{n^6}{6} - \frac{1}{6}$$

i.e. $\displaystyle\sum_1^{n-1} r^5 < \frac{n^6}{6}$ for $n > 1$

12 $1 + r + r^2 + \ldots: \quad S_\infty = \dfrac{1}{1-r} = 5 \quad \Rightarrow \quad r = 0.8$

$$S_n = \frac{1 - (0.8)^n}{0.2} = 5(1 - (0.8)^n) = 5 - 5(0.8)^n$$

$\boxed{3.1}$

$S_n > 4.9$ then $\dfrac{0.1}{5} > (0.8)^n$

Take logs: $\ln(0.02) > n \ln(0.8)$
Since $\ln(0.8)$ is negative, dividing by $\ln(0.8)$ reverses the inequality

$$n > \frac{\ln(0.02)}{\ln(0.8)} = 17.53$$

Least value of n is 18.

13 (a) $1, 4, 7, \ldots, 1000$ has 334 terms:
$S_{334} = \frac{334}{2}(1 + 1000) = 167\ 167$
$7, 16, 25, \ldots, 997$ has 111 terms:
$S_{111} = \frac{111}{2}(7 + 997) = 55\ 722$
Sum of remaining terms $= 111\ 445$

(b) $u_r = (\frac{1}{3})^{3r-2} + (\frac{1}{3})^{3r-1} = 9(\frac{1}{3})^{3r} + 3(\frac{1}{3})^{3r}$
$= 12(\frac{1}{3})^{3r}$ or $12(\frac{1}{27})^r$
This is a term of a G.P. with common ratio $\frac{1}{27}$, $a = \frac{12}{27}$.

$$\text{Sum of G.P.} = \frac{\frac{12}{27}(1 - (\frac{1}{27})^n)}{1 - \frac{1}{27}} = \frac{6}{13}\left(1 - \frac{1}{27^n}\right)$$

$$\Rightarrow \quad A = \frac{6}{13}, B = 1$$

$$S_\infty = \frac{6}{13}$$

14 $\displaystyle\sum_{r=1}^{100}(0.99)^r = $ G.P. with $a = 0.99$, ratio $= 0.99$

$$S_{100} = \frac{0.99(1 - 0.99^{100})}{0.01} = 63 \text{ to the nearest integer}$$

$\boxed{8.3}$

15 $1, r, s$ in a G.P. $\Rightarrow s = r^2$ $\qquad$ (1)
$1, s, r$ in an A.P. $\Rightarrow 1 + r = 2s$ $\qquad$ (2)
From equations 1 and 2, $2r^2 - r - 1 = 0$
$\Rightarrow (2r+1)(r-1) = 0 \Rightarrow r = -\frac{1}{2}$

$\boxed{\text{2.1a and 2.2a}}$

G.P. $1, -\frac{1}{2}, \frac{1}{4}, \ldots \Rightarrow S_\infty = \dfrac{1}{1 + \frac{1}{2}} = \frac{2}{3}$

9

Functions

AS Level			A Level			Topic	Date attempted	Date completed	Self-assessment
CORE	MODULAR	TRADITIONAL	CORE	MODULAR	TRADITIONAL				
✓	✓	✓	✓	✓	✓	**Definitions and notation**			
✓	✓	✓	✓	✓	✓	**f(x) → af(x) → f(a+x) → f(ax)**			
	✓	✓		✓	✓	**Modulus function**			
✓	✓	✓	✓	✓	✓	**Inverse functions**			
✓	✓	✓	✓	✓	✓	**Graphical representation**			
			✓	✓	✓	**Growth and decay $a^x = b$**			

Fact Sheet

Notation

(i)   f:$x \rightarrow y$ is the function which maps x onto y (also $x \rightarrow y$).
(ii)   f(x) is the image of x under function f $\Rightarrow$ f(x) = y.
(iii) f^{-1} is the inverse function of f.
(iv)   f $\cdot$ g is the function mapping x onto f(x) $\cdot$ g(x).
(v)   fg or f $\circ$ g is the composite function f(g(x)) or f operating on the result of g.
(vi)   f $\circ$ f is often written f^2.
(vii) f + g is the function mapping x onto f(x) + g(x).
(viii) f' is the derived function of f.

Definitions

- *Domain*. The values of x for which the function is defined. The most common of these is $\mathbb{R}$, the set of real numbers $\{x: x \in \mathbb{R}\}$.
- *Range*. The values of f(x) arising from the domain.

Limits of a Domain or Range

- If a function has discontinuities or is not defined for some values of x then this can be shown by inequalities.
  Example 1: f:$x \rightarrow \log x$ has a domain $\{x: x > 0\}$.
  Example 2: If f(x) = x $\{x: 0 \leqslant x \leqslant 2\}$
                              = $2x$ $\{x: 2 < x \leqslant 4\}$
  then the first part of the domain includes $x = 2$, but the second does not. If $x = 2$ is included in both parts the function will have two different values at that point. Since functions can only be single-valued this is not acceptable. The same function can be written

  f:$x \rightarrow x$ [0, 2]
  f:$x \rightarrow 2x$ (2, 4].

  [ ] imply that the limit points are included.
  ( ) imply that the limit points are not included.

  Graphically, included limit points are represented by •, excluded limit points by ○.

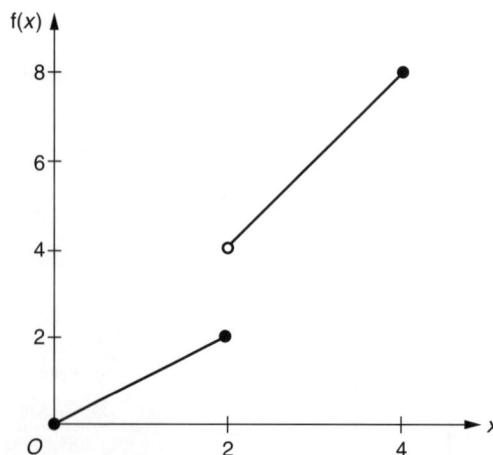

- A function and its inverse can be recognized graphically since the reflection of f(x) in the line $y = x$ gives $f^{-1}(x)$.
- If the inverse is to be a function its domain may have to be restricted. For example the inverse function of $\cos x$ is $\cos^{-1} x$ with domain $\{x: -1 \leqslant x \leqslant 1\}$.

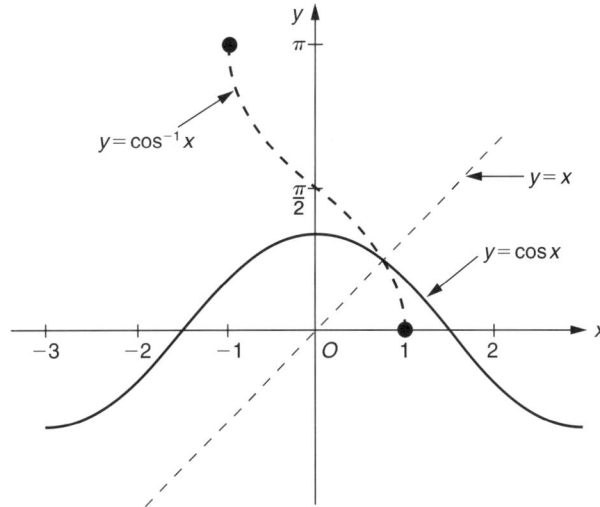

Graphical Solutions of Equations

If an equation $y = f(x)$ is complicated, split $f(x)$ into $g(x) + h(x)$ where g and h are easily sketched. Then the solutions of $f(x) = 0$ are given by the points of intersection of the graphs $y = g(x)$ and $y = -h(x)$.

For example, if $y = 2 \sin x - x$ then $y = 0$ when $y = 2 \sin x$ and $y = +x$ intersect.

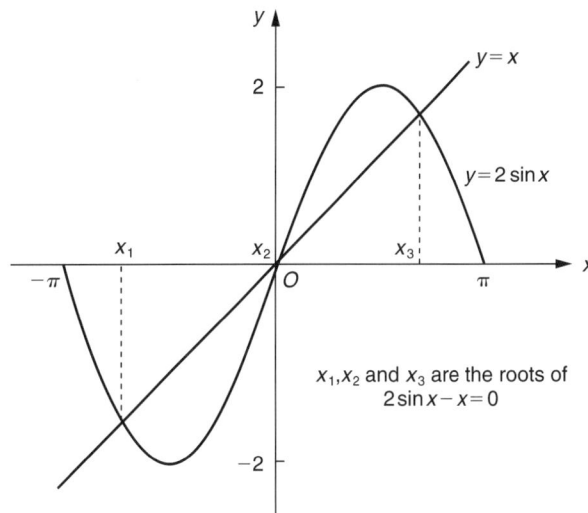

x_1, x_2 and x_3 are the roots of $2 \sin x - x = 0$

Odd and Even Functions

- *Definition:*
  If $f(-x) = f(x)$ then $f(x)$ is an even function.
  If $f(-x) = -f(x)$ then $f(x)$ is an odd function.
- *Properties:*
  Even functions are symmetrical about the line $x = 0$.
  Odd functions have half-turn rotational symmetry about O.
  A function containing only even powers of x (including x^0) is an even function, but one containing only odd functions is not necessarily odd.

- *Examples:*
  Even: Odd:

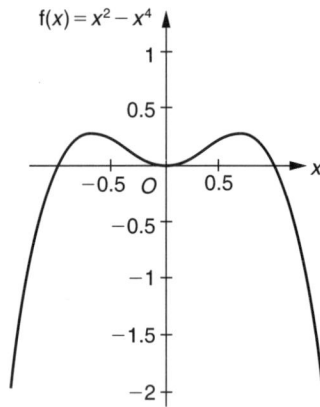

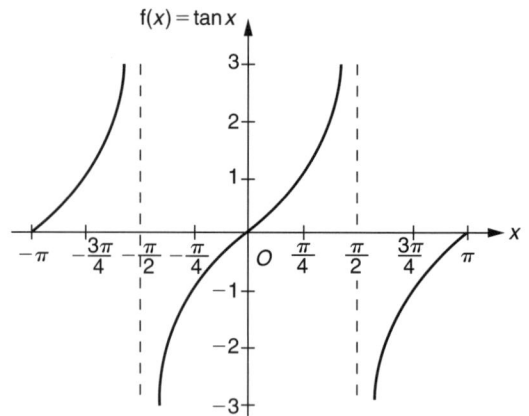

Neither even nor odd:

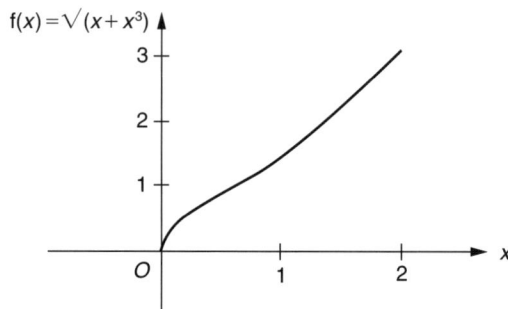

Gradient function f'(x)

A sketch of $y = f'(x)$ is obtained from the graph of $f(x)$. At stationary points on $y = f(x)$, $f'(x) = 0$, i.e. points on the x-axis. Non-stationary points of inflexion on $y = f(x)$ give stationary points on $y = f'(x)$.

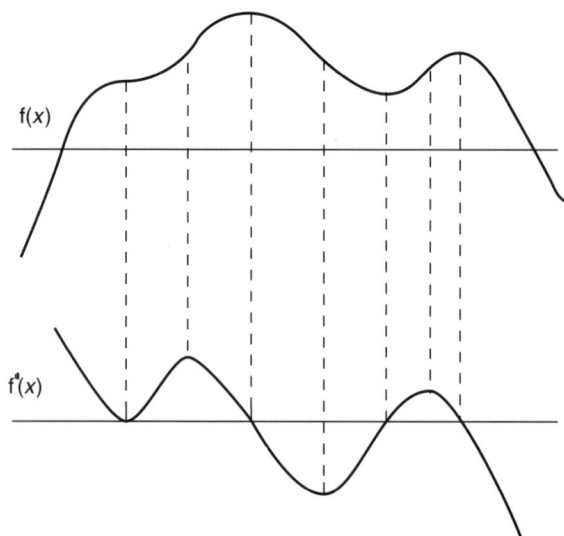

Worked Examples

1 The function f(x) is defined for all values of x except $x = 0$ and is an odd function, i.e. f($-x$) = $-$f(x).

(a) Part of the graph of y = f(x) is given below. Copy and complete the sketch.

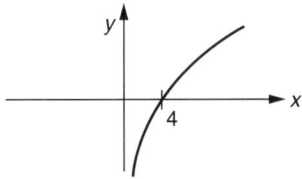

(b) Draw a separate sketch to illustrate the graph of

y = f($x + 3$)

showing clearly where the graph will intercept the x-axis.

(NEAB SMP, 1993)

• (a) If f($-x$) = $-$f(x) then f(-4) $-$ $-$f(4) = 0
If f(2) = $-\alpha$ then f(-2) = $+\alpha$
$(x, y) \rightarrow (-x, -y)$
i.e. the part of the graph shown is rotated 180° about the origin

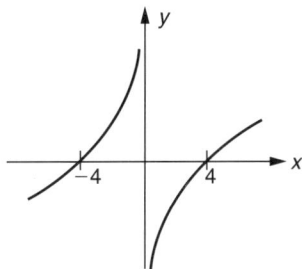

(b) y = f($x + 3$) is a translation of f(x) of $(-3, 0)$ or $\begin{pmatrix} -3 \\ 0 \end{pmatrix}$.

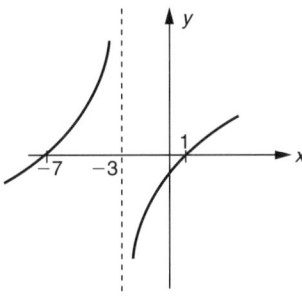

2 The functions f and g are defined by

$\text{f}:x \rightarrow e^{2x} \qquad x \in \mathbb{R}$
$\text{g}:x \rightarrow x^2 \qquad x \in \mathbb{R}$

Find $\text{f}^{-1}(x)$ and gf(x).
Sketch the graph of y = $\text{f}^{-1}(x - 2)$. (UCLES, 1991)

• Flow chart for f(x) is $x \xrightarrow{\times 2} 2x \xrightarrow{exp} e^{2x}$

Flow chart for $\text{f}^{-1}(x)$ is $x \xrightarrow{\ln} \ln x \xrightarrow{\div 2} \tfrac{1}{2} \ln x$

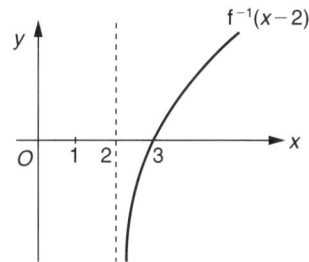

$\text{f}:x \rightarrow e^{2x}$
$\text{f}^{-1}:x \rightarrow \tfrac{1}{2} \ln x$
$\text{gf}(x) = (e^{2x})^2 = e^{4x}$
$\text{f}^{-1}(x - 2) = \tfrac{1}{2} \ln (x - 2)$
(a translation of $(2, 0)$ of $\text{f}^{-1}(x)$)

3 In four separate diagrams, illustrate the following four sets of points:

(i) $S = \{(x, y): y \leqslant 6$, where $x, y \in \mathbb{R}\}$
(ii) $T = \{(x, y): y = x^2 - 5x + 6$, where $x, y \in \mathbb{R}\}$
(iii) $U = \{(x, y): y > |x^2 - 5x + 6|$, where $x, y \in \mathbb{R}\}$
(iv) $S \cap U$

(O&C, 1992)

• (i)

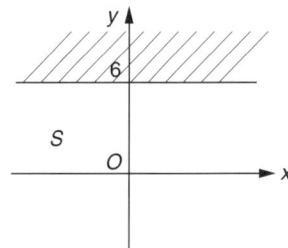

Line y = 6 is included in the sets

(ii)

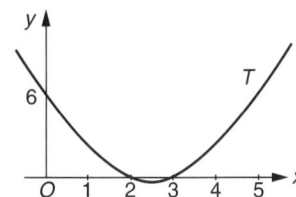

$x^2 - 5x + 6 = (x - 2)(x - 3)$

T is the set of points on the curve

(iii)

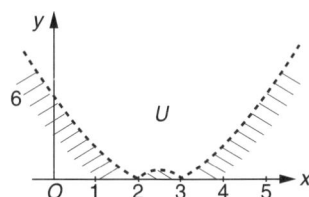

The points on the curve are not included in U

(iv)

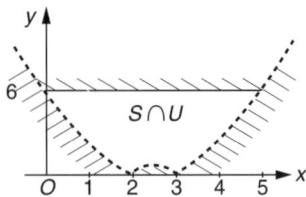

Line $y = 6$ is included
Curve $|x^2 - 5x + 6|$ is not included

4 The function f with domain $\{x: x \geqslant 0\}$ is defined by

$$f(x) = 4x^2 - 1$$

(a) State the range of f and sketch the graph of f.
(b) Explain why the inverse function f^{-1} exists and find $f^{-1}(x)$.
(c) Given that g has domain $\{x: x \geqslant 0\}$ and is defined by

$$g(x) = \sqrt{(x + 6)}$$

solve the equation $fg(x) = 2f(x)$
giving your answer to 3 significant figures.

(OLE, 1993)

• (a) Range of $f(x)$ is $f(x) \geqslant -1$

7.1

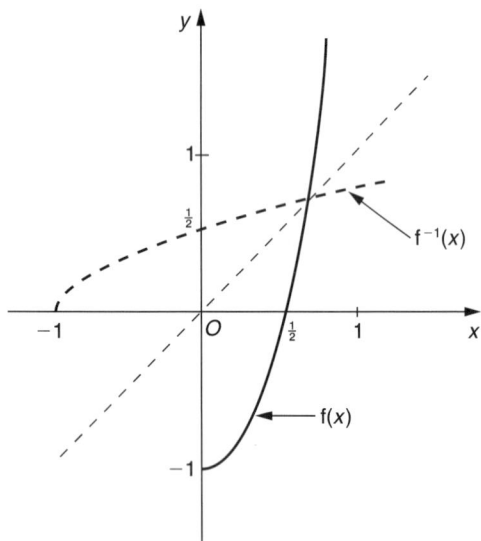

(b) $f(x)$ is one-one in the given domain $\therefore f^{-1}(x)$ exists.

$$f(x) = x \xrightarrow{\text{square}} x^2 \xrightarrow{\times 4} 4x^2 \xrightarrow{-1} 4x^2 - 1$$

$$f^{-1}(x) = x \xrightarrow{+1} x + 1 \xrightarrow{\div 4} \frac{x+1}{4} \xrightarrow{\text{sq. root}}$$

$$\pm \sqrt{\frac{x+1}{4}}$$

$f^{-1}(x)$ could be $+\sqrt{\dfrac{x+1}{4}}$ or $-\sqrt{\dfrac{x+1}{4}}$

depending on the domain of $f(x)$.

$f^{-1}(x)$ is the reflection of f(x) in the line $x = y \Rightarrow$ in this example

$$f^{-1}(x) = \sqrt{\frac{x+1}{4}} \quad \{x: x \geqslant -1\}$$

(c) $g(x) = \sqrt{(x + 6)}$ $\{x: x \geqslant 0\}$
$fg(x) = 4(\sqrt{x + 6})^2 - 1 = 4x + 23$ $\quad x \geqslant 0$
$fg(x) = 2f(x) \Rightarrow 4x + 23 = 8x^2 - 2$
$\Rightarrow 8x^2 - 4x - 25 = 0$
$\Rightarrow x = 2.04$ or -1.54

But $x \geqslant 0 \therefore x = 2.04$

2.5

5

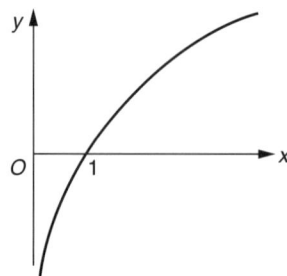

The diagram shows the graph of $y = \ln x$. Sketch the graph of $y = \ln(x + a)$ where a is a constant such that $a > 1$, and state the coordinates of the points of intersection of the graph with the axes.

(UCLES, 1993)

• $y = \ln(x + a)$ is a translation of $(-a, 0)$.
The graph cuts the x-axis at $(1 - a, 0)$ and the y-axis at $(0, \ln a)$.

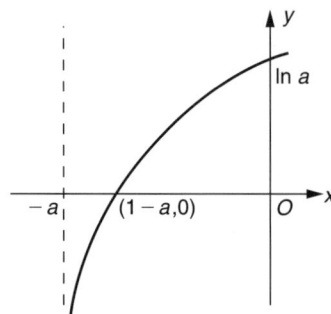

6 (a) The function f is defined for $x > 1$ by

$$f(x) = 3x^2 - 6x + 2$$

(i) By considering the derivative of $f(x)$, show that the inverse function f^{-1} exists.
(ii) Obtain an expression for $f^{-1}(x)$ and state the domain of f^{-1}.
(iii) Sketch f and f^{-1} on the same axes.
(iv) State the geometrical relationship between the graphs of f, f^{-1} and the line $y = x$. Use this to find the coordinates of the point of intersection of the graphs of f and f^{-1}.

(b) A further function g is defined for $x > 5$ by

$$g(x) = \frac{1}{x - 5}$$

(i)   Find an expression for $g \circ f(x)$.

(ii) Given that the domain of $g \circ f$ is $x > k$, find the smallest possible value of k.

(WJEC, 1994)

• (a) (i) $f(x) = 3x^2 - 6x + 2$ $f'(x) = 6x - 6$
When $x > 1, f'(x) > 0$ $\Rightarrow$ no turning points $\Rightarrow$ $f^{-1}(x)$ is single-valued.

(ii) To find the inverse function complete the square so that x only occurs once in f(x)

$$f(x) = 3(x^2 - 2x + \tfrac{2}{3})$$
$$= 3((x-1)^2 - \tfrac{1}{3})$$
$$= 3(x-1)^2 - 1 \qquad \boxed{4.2}$$

Flow chart for f(x) is

$$x \xrightarrow{-1} (x-1) \xrightarrow{\text{square}} (x-1)^2$$

$$\xrightarrow{\times 3} 3(x-1)^2 \xrightarrow{-1} 3(x-1)^2 - 1$$

Flow chart for $f^{-1}(x)$ is

$$x \xrightarrow{+1} (x+1) \xrightarrow{\div 3} \frac{x+1}{3} \xrightarrow{\sqrt{}}$$

$$\pm\sqrt{\frac{x+1}{3}} \xrightarrow{+1} \pm\sqrt{\frac{x+1}{3}} + 1$$

$f^{-1}(x)$ could be $\sqrt{\dfrac{x+1}{3}} + 1$ or

$-\sqrt{\dfrac{x+1}{3}} + 1$ depending on the domain of f(x).

Look at the graph of f(x) to decide which it is

$$f^{-1}(x) = 1 + \sqrt{\frac{x+1}{3}} \quad x \geqslant -1$$

(iii) The sketches are as shown.
$\boxed{7.1}$

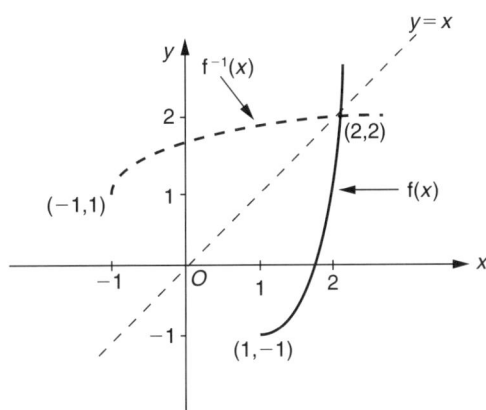

(iv) $f(x)$ and $f^{-1}(x)$ intersect when $f(x) = x$ since $f^{-1}(x)$ is the reflection of $f(x)$ in the line $y = x$

$\boxed{\text{2.1a and 2.2a}}$

$f(x) = x$ when $3x^2 - 6x + 2 = x$
$$\Rightarrow (3x - 1)(x - 2) = 0$$
$$x = \tfrac{1}{3} \text{ or } 2$$

Since $x > 1$, $f(x)$ and $f^{-1}(x)$ intersect at $(2, 2)$.

(b) $g(x) = \dfrac{1}{x-5}$ $(x > 5)$

(i) $g \circ f(x) = \dfrac{1}{3x^2 - 6x - 3}$ for $f(x) > 5$

(ii) The least value for k is the value of x when $f(x) = 5$
$$\Rightarrow 3x^2 - 6x + 2 = 5$$
$$\Rightarrow 3x^2 - 6x - 3 = 0$$
$$\Rightarrow x^2 - 2x - 1 = 0$$

$$x = \frac{2 \pm \sqrt{4+4}}{2} = 1 + \sqrt{2}$$

$(1 - \sqrt{2} < 1 \Rightarrow$ exclude)

Hence the smallest possible value of k is $1 + \sqrt{2}$ $(= 2.414)$.

7 Pair up two of the functions with their gradient functions.

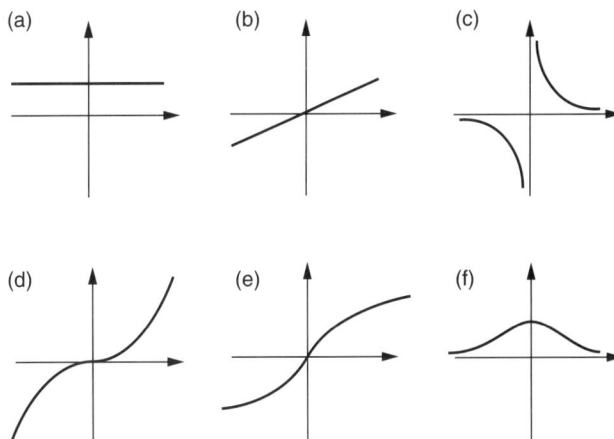

(a)

(b)

(c)

(d)

(e)

(f)

For the two functions you have **not** paired up, explain why each one cannot be the gradient function of the other.   (O&C SMP)

• (a) is the gradient function of a straight line with a positive gradient $\Rightarrow$ (a) is the gradient function of (b).
(e) has positive gradient for all x, maximum gradient at $x = 0$ and the gradient function is symmetrical about the y-axis $\Rightarrow$ (f) is the gradient function of (e).

(i) (c) has infinite negative gradient at $x = 0$; (d) is zero at $x = 0$.

(ii) (d) has zero gradient at $x = 0$; (c) $y \to \infty$ at $x = 0$.

Hence from (i), (d) cannot be the gradient function of (c); and from (ii), (c) cannot be the gradient function of (d).

8 The function f is defined for the domain $x \geq 0$ by $f(x) = 4 - x^2$.

(a) Sketch the graph of f and state the range of f.
(b) Describe a single transformation whereby the graph of $y = f(x)$ may be obtained from the graph of $y = x^2$ for $x \geq 0$.
(c) The inverse of f is denoted by f^{-1}. Find an expression for $f^{-1}(x)$ and state the domain of f^{-1}.
(d) Show, by reference to a sketch, or otherwise, that the solution to the equation $f(x) = f^{-1}(x)$ can be obtained from the quadratic equation $x^2 + x - 4 = 0$. Determine the solution of $f(x) = f^{-1}(x)$, giving your value to two decimal places. (OLE, 1992)

• (a)

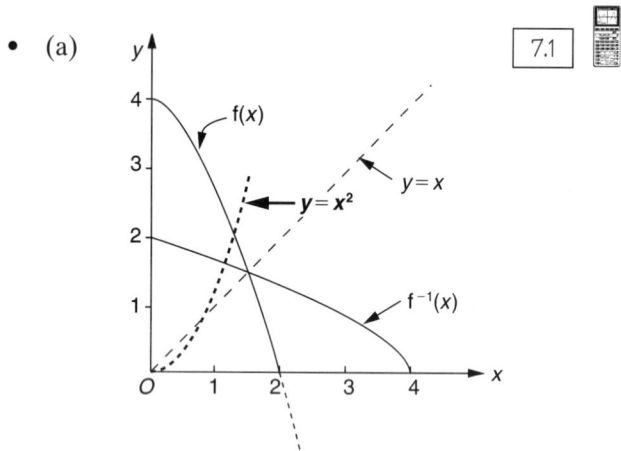

7.1

Range of f is $f(x) \leq 4$.
(b) $y = f(x)$ is obtained from $y = x^2$ by a reflection in the line $y = 2$. *See diagram*
(c) $f^{-1}(x) = \sqrt{4 - x}$
domain $x \leq 4$. *Draw a flow chart*
(d) $f^{-1}(x)$ is a reflection of $f(x)$ in the line $y = x$. Hence the point of intersection lies on the line $y = x$.

2.1a and 2.2a

$4 - x^2 = x \Rightarrow x^2 + x - 4 = 0$
$\Rightarrow x = 1.56$ (take the root which satisfies $x > 0$)

9 The figure shows the shape of the graph of the function f, where

$f: x \to xe^x \quad x \in \mathbb{R}$

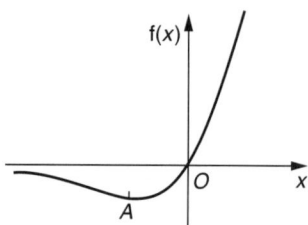

(a) Determine the coordinates of the stationary point A and hence write down the range of f.
(b) Sketch the graph of the curve $y = (x - 2)e^{x-2}$ and mark on the sketch the coordinates of any points where the curve crosses the coordinate axes. (AEB, 1987)

• (a) $f(x) = xe^x \Rightarrow f'(x) = xe^x + e^x = e^x(x + 1)$
At A, $f'(x) = 0 \therefore x = -1 \Rightarrow f(x) = -e^{-1}$

A is the point $(-1, -e^{-1})$
Range of $f(x)$ is $f(x) \geq -e^{-1}$
(b) $y = (x - 2)e^{x-2}$ is a translation of $f(x)$ by $(2, 0)$.

7.1

7.1

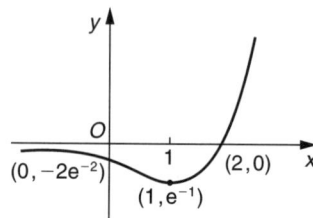

Exercises

1 The functions f and g are defined by $f(x) = \cos x$, $g(x) = \sqrt{(1 - 9x^2)}$. Find the composite function fg and state its domain and range. Can gf be formed? Give reasons for your answer.

2 The functions f and g are defined by

$f: x \to x^2 - 3 \quad x \in \mathbb{R}$
$g: x \to 2x + 5 \quad x \in \mathbb{R}$

Find in similar form the composite function $f \circ g$. Sketch on separate axes the graphs of f and $f \circ g$. Hence, or otherwise, show that the range of f corresponding to the domain $-4 \leq x \leq 4$ is $-3 \leq f(x) \leq 13$, and find the range of $f \circ g$ corresponding to this domain. (AEB, 1986)

3 (i) The function f is defined by

$$f(x) = \frac{1}{1 - x} \quad (x \neq 0, x \neq 1)$$

Given that $h = f \circ f$, obtain expressions for
(a) $h(x)$
and (b) $h \circ f(x)$.
Hence, or otherwise, obtain an expression for $f^{-1}(x)$.
(ii) The function g is defined by

$g(x) = |x - 2| \quad (-\infty < x < \infty)$

Sketch the graphs of f and g on the same axes. Find the x-coordinate of the single point of intersection of the two graphs.
(iii) Solve the equation

$g \circ g(x) = 0$ (WJEC, 1992)

4 (a) The function f is defined, for real values of x, by $f(x) = 1 + e^{-x}$. Express $f^{-1}(x)$ in terms of x, where $x > 1$.

(b)

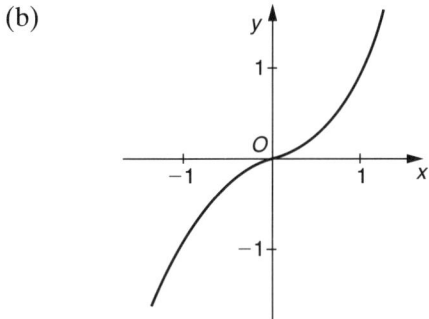

The diagram shows the graph of $y = g(x)$.
Sketch the graph of $y = g^{-1}(x)$. (UCLES, 1993)

5 Which of the following statements is true for the functions defined by

$$f(x) = x - \frac{1}{x}, \quad x > 1$$

$$g(x) = \frac{2}{1 - x}, \quad x > 1?$$

A f is increasing, g is decreasing
B f is increasing, g is increasing
C f is decreasing, g is decreasing
D f is decreasing, g is increasing
E none of these

6 Sketch on separate axes the graphs of

(i) $f(x) \to |2x + 3|$
(ii) $g(x) \to 2|x| + 3$
(iii) $h(x) \to |x^2 - 4|$
(iv) $k(x) \to |x|^2 - 2|x| - 3$

7 The graph of a function f is illustrated below.

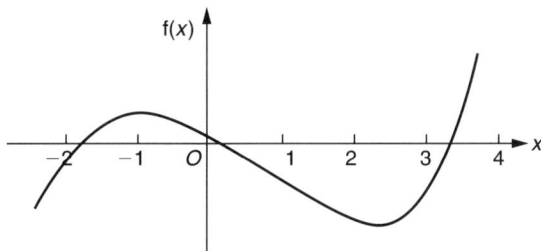

(a) From the graph, write down values, to the nearest integer, for the roots of

(i) $f(x) = 0$
(ii) $f'(x) = 0$

(b) Sketch the graph of $y = f'(x)$, clearly indicating the scale on the x-axis. (NEAB SMP, 1994)

8 The function f is defined by

$$f: x \to \frac{1}{2 - x} + 3 \quad x \in \mathbb{R}, x \neq 2$$

(a) Calculate $f(5)$ and $ff(5)$.
State the value of k $(k \neq 2)$ for which $ff(k)$ is not defined.
(b) The inverse of f is f^{-1}. Find an expression for $f^{-1}(x)$ and state the domain of f^{-1}.
(AEB, 1994)

9 (a) Make x the subject of the formula $y = \dfrac{3 - x}{x + 2}$.
Hence, given that $f(x) = \dfrac{3 - x}{x + 2}$, find the inverse function f^{-1}.

(b) Given that $g(x) = \dfrac{1}{x}$, find $fg(x)$. Hence show that $fg(x)$ can be put in the form

$$fg(x) = \frac{3x + a}{bx + 1}$$ giving the values of a and b.
(O&C SMP)

10 State whether each of the following statements is true or false. If you decide that a statement is false explain why this is the case.

(i) $x^2 - 6x + 8 = 0 \implies x = 2$
(ii) $f: x \to x^2$ and $h: x \to x^4 \implies fh \to x^6$
(iii) $|x + 2| < 2 \implies x < 0$

11

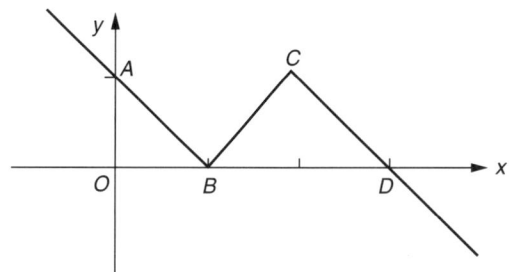

The graph of $y = f(x)$ is shown above. The points A, B, C and D have coordinates $(0, 1)$, $(1, 0)$, $(2, 1)$ and $(3, 0)$ respectively. Sketch, separately, the graphs of

(i) $y = f(2x)$
(ii) $f(x + 3)$

stating in each case the coordinates of the points corresponding to A, B, C and D. (UCLES, 1990)

Brief Solutions

1 $f(x) = \cos x$, $g(x) = \sqrt{(1 - 9x^2)}$
$\implies fg = \cos[\sqrt{(1 - 9x^2)}]$.
Domain of $f(x) \in \mathbb{R}$, domain of $g(x)$ is $-\frac{1}{3} \leq x \leq \frac{1}{3}$
$\implies$ domain of $fg(x)$ is $-\frac{1}{3} \leq x \leq \frac{1}{3}$.
Range of $fg(x)$ is $\cos(1) \leq fg \leq 1$.
$gf(x) = \sqrt{(1 - 9\cos^2 x)}$ defined for $|\cos x| \leq \frac{1}{3}$ thus $gf(x)$ exists only if the domain of $f(x)$ is restricted to

$$\cos^{-1}(\tfrac{1}{3}) \leq x \leq \cos^{-1}(-\tfrac{1}{3}) \text{ such as } 1.23 < x < 1.91$$

2 $f \circ g = (2x + 5)^2 - 3 = 4x^2 + 20x + 22 \quad x \in \mathbb{R}$

$f(x)$

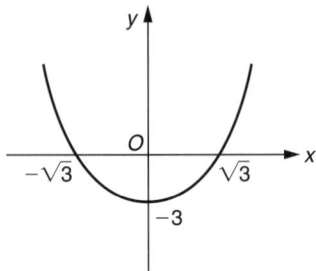

$f \circ g(x)$

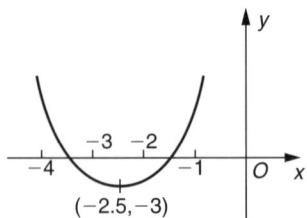

7.1 and 7.6

$f(-4) = f(4) = 13$ Range of $f(x)$ is $-3 \leqslant f(x) \leqslant 13$
$f \circ g$ has minimum value of -3
$f \circ g(4) = 166$ Range of $f \circ g$ is $-3 \leqslant f \circ g \leqslant 166$

3 (i) $h(x) = \dfrac{1}{1 - \dfrac{1}{1-x}} = 1 - \dfrac{1}{x}$

$h \circ f = 1 - (1 - x) = x$

$f^{-1}(x) = h = 1 - \dfrac{1}{x}$

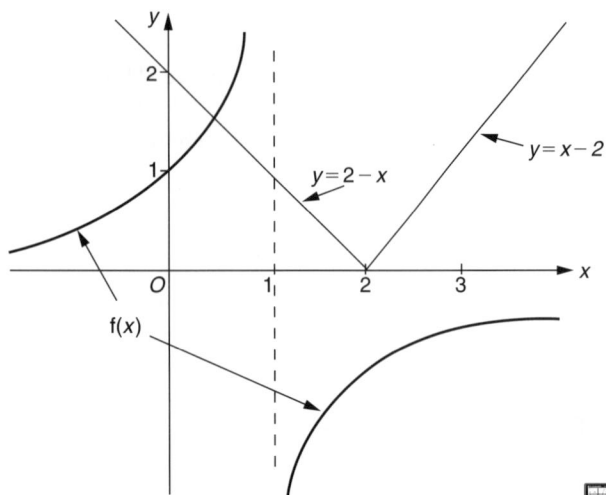

7.1

(ii) $f(x)$ and $g(x)$ intersect when $\dfrac{1}{1-x} = 2 - x$
and $x < 2$

$\Rightarrow \quad x = 0.382$

(iii) $g \circ g = ||x - 2| - 2|$
$\qquad = |x - 4| \text{ or } |-x|$
$g \circ g = 0$ when $x = 4$ or 0

2.5

4 (a) $f(x) = 1 + e^{-x} \quad f^{-1}(x) = -\ln(x - 1)$
(b)

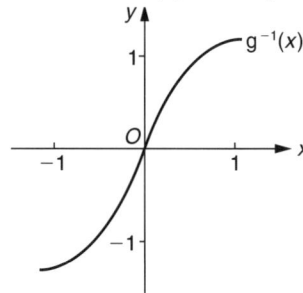

5 $f(x) = x - \dfrac{1}{x}, x > 1; \quad f'(x) = 1 + \dfrac{1}{x^2}$

7.1

$g(x) = \dfrac{2}{1 - x}, x > 1; \quad g'(x) = \dfrac{2}{(1 - x)^2}$

$f'(x)$ and $g'(x)$ are both positive for $x > 1$. **Answer B**

6 (i)

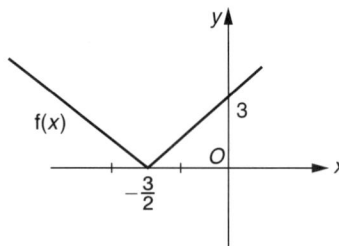

(ii)

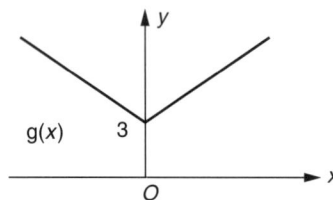

(iii)

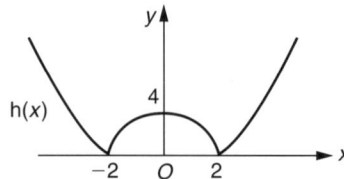

(iv)

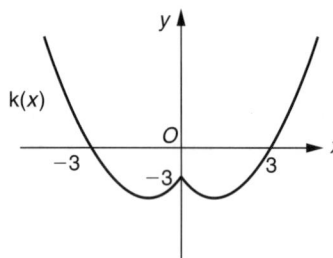

7.1

7 Roots of $f(x) = 0$ are $3, 0, -2$ to the nearest integer.
Roots of $f'(x) = 0$ are $2, -1$ to the nearest integer.

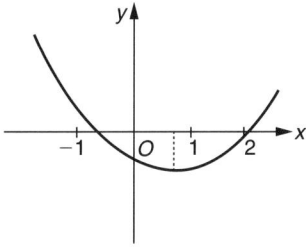

8 (a) $f(5) = -\frac{1}{3} + 3 = 2\frac{2}{3}$ $f(2\frac{2}{3}) = \frac{3}{2} = ff(5)$ $\boxed{7.6}$

ff(k) is not defined when $f(k) = 2$, i.e. $k = 3$.

(b) Flow chart for $f(x)$:

$$x \xrightarrow{\text{subt from 2}} 2 - x \xrightarrow{\text{divide into 1}}$$

$$\frac{1}{2 - x} \xrightarrow{\text{add 3}}$$

$f^{-1}(x) = 2 - \dfrac{1}{x - 3}$ (reversing the flow chart for $f(x)$)

Domain of $f^{-1}(x)$: $x \in \mathbb{R}, x \neq 3$.

9 (a) $y = \dfrac{3 - x}{x + 2}$ $\Rightarrow$ $yx + 2y = 3 - x$
$\Rightarrow$ $x(y + 1) = 3 - 2y$

$\Rightarrow$ $x = \dfrac{3 - 2y}{y + 1}$

$f^{-1}(x) = \dfrac{3 - 2x}{x + 1}$

(b) $fg(x) = \dfrac{3 - \dfrac{1}{x}}{\dfrac{1}{x} + 2} = \dfrac{3x - 1}{1 + 2x}$ $\Rightarrow$ $a = -1, b = 2$

10 (i) false $x = 2 \ or \ 4$
(ii) false fh $\rightarrow (x^4)^2 = x^8$
(iii) $|x + 2| < 2, x < 0$ $-4 \leqslant x \leqslant 0$ True
(a necessary but not sufficient condition)

11 A: $f(0) = 1$ B: $f(1) = 0$ C: $f(2) = 1$ D: $f(3) = 0$
$f(2x)$:
$A(0, 1)$ $B(\frac{1}{2}, 0)$ $C(1, 1)$ $D(1\frac{1}{2}, 0)$
Stretch factor $\frac{1}{2}$

$f(x + 3)$:
$A(-3, 1)$ $B(-2, 0)$ $C(-1, 1)$ $D(0, 0)$

Translation $\begin{pmatrix} -3 \\ 0 \end{pmatrix}$

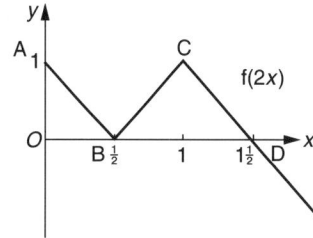

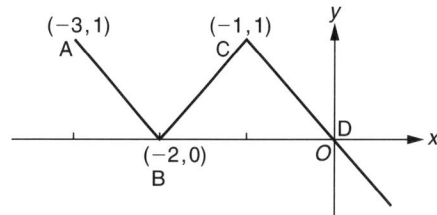

10

$$\int \frac{x^2}{\sqrt{1-x^2}}\,dx$$

Differentiation

AS Level			A Level			Topic	Date attempted	Date completed	Self-assessment
CORE	MODULAR	TRADITIONAL	CORE	MODULAR	TRADITIONAL				
✓	✓	✓	✓	✓	✓	Gradients			
	✓	✓	✓	✓	✓	Standard results/formulae			
✓	✓	✓	✓	✓	✓	Stationary points, tangents, normals			
	✓	✓		✓	✓	Implicit			
	✓	✓		✓	✓	Parametric			
			✓	✓		Maclaurin's series			

Fact Sheet

If $y = f(x)$ then

$$\frac{dy}{dx} = f'(x) = \lim_{\delta x \to 0} \frac{f(x + \delta x) - f(x)}{\delta x}$$

or $\quad f'(a) = \lim_{b \to a} \dfrac{f(a) - f(b)}{a - b}$

Product

$$\frac{d}{dx}(uv) = u\frac{dv}{dx} + v\frac{du}{dx}$$

Quotient

$$\frac{d}{dx}\left(\frac{u}{v}\right) = \frac{v\dfrac{du}{dx} - u\dfrac{dv}{dx}}{v^2}$$

Composite Function (Function of a Function)

If $y = f(u)$ and $u = g(x)$ then $\dfrac{dy}{dx} = \left(\dfrac{dy}{du}\right)\left(\dfrac{du}{dx}\right).$

This is known as the chain rule.

Standard Results

$y = f(x)$	$\dfrac{dy}{dx} = f'(x)$
x^n	nx^{n-1}
e^x	e^x
$a^x\ (a > 0)$	$a^x \ln a$
$\ln x$	$\dfrac{1}{x}$
$\log_a x$	$\dfrac{1}{x \ln a}$
$\sin x$	$\cos x$
$\cos x$	$-\sin x$
$\tan x$	$\sec^2 x$
$\operatorname{cosec} x$	$-\operatorname{cosec} x \cot x$
$\sec x$	$\sec x \tan x$
$\cot x$	$-\operatorname{cosec}^2 x$
$\sin^{-1} x$	$\dfrac{1}{\sqrt{(1 - x^2)}}$
$\cos^{-1} x$	$\dfrac{-1}{\sqrt{(1 - x^2)}}$
$\tan^{-1} x$	$\dfrac{1}{1 + x^2}$

Implicit Functions

$$\frac{d}{dx}(f(y)) = (f'(y))\left(\frac{dy}{dx}\right) \quad \text{where } f'(y) = \frac{d}{dy}(f(y)).$$

Parametric Functions

If $x = x(t)$ and $y = y(t)$ then $\dfrac{dy}{dx} = \dfrac{\frac{dy}{dt}}{\frac{dx}{dt}} = \dfrac{\dot{y}}{\dot{x}}$

where $\dot{x} = \dfrac{dx}{dt}$ and $\dot{y} = \dfrac{dy}{dt}$.

Tangents and Normals at (x_0, y_0)

If $\dfrac{dy}{dx} = m$ when evaluated at (x_0, y_0) then:

(i)   the tangent at (x_0, y_0) may be written as

$$y - y_0 = m(x - x_0)$$

(ii) the normal at (x_0, y_0) may be written as

$$y - y_0 = \frac{-1}{m}(x - x_0)$$

Maxima and Minima

Stationary points occur when $\dfrac{dy}{dx} = 0$.

At such points, if

(i) $\dfrac{d^2y}{dx^2} < 0$ the point is a maximum

(ii) $\dfrac{d^2y}{dx^2} > 0$ the point is a minimum

(iii) $\dfrac{d^2y}{dx^2} = 0$ the point can be maximum, minimum or a point of inflexion.

Points of Inflexion

If $\dfrac{d^2y}{dx^2} = 0$ at $x = x_0$, and there is a change of sign of $\dfrac{d^2y}{dx^2}$ as x passes through x_0,

then it is a point of inflexion. This is independent of the value of $\dfrac{dy}{dx}$ at $x = x_0$.

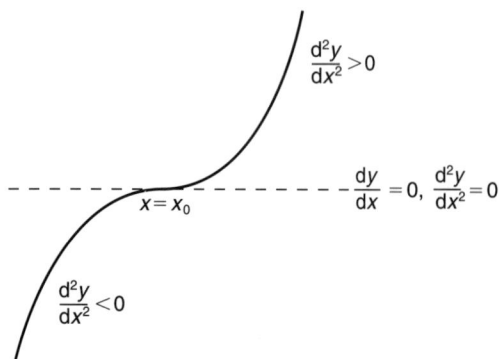

$\dfrac{d^2y}{dx^2} > 0$

$\dfrac{dy}{dx} = 0, \dfrac{d^2y}{dx^2} = 0$

$x = x_0$

$\dfrac{d^2y}{dx^2} < 0$

$\dfrac{d^2y}{dx^2} < 0$

$x = x_0, \dfrac{dy}{dx} \neq 0, \dfrac{d^2y}{dx^2} = 0$

$\dfrac{d^2y}{dx^2} > 0$

Inflexion and stationary point

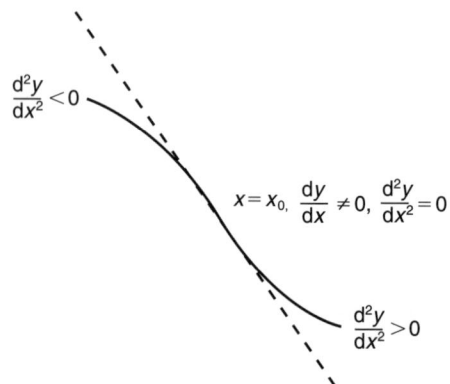

Inflexion but not a stationary point

Rates of Change

$f'(x)$ or $\dfrac{dy}{dx}$ represents the rate of increase of a function $f(x)$ or y with respect to x.

The rate of decrease of $f(x)$ or y is $-f'(x)$ or $-\dfrac{dy}{dx}$.

Connected Rates of Change

If $y = y(x)$ and the rate of change of x with respect to t is $\dfrac{dx}{dt}$ then the rate of change of y with respect to t is $\dfrac{dy}{dt}$ given by $\dfrac{dy}{dt} = \left(\dfrac{dy}{dx}\right)\left(\dfrac{dx}{dt}\right)$.

Small Increments

If $y = f(x)$ then $\dfrac{\delta y}{\delta x} \approx f'(x) \quad \Rightarrow \quad \delta y \approx f'(x)\delta x$.

Maclaurin's Series

$$f(x) = f(0) + \frac{x}{1}f'(0) + \frac{x^2}{2!}f''(0) + \ldots + \frac{x^n}{n!}f^n(0) +$$

Worked Examples

1 Differentiate the following functions with respect to x:

(i) $x \ln x$

(ii) $\sqrt{1 + x^2}$ (WJEC, 1994)

• (i) $x \ln x$ is a product $\Rightarrow$ use $\dfrac{d}{dx}(uv) = u\dfrac{dv}{dx} + v\dfrac{du}{dx}$

$$\frac{d}{dx}(x \ln x) = x\frac{1}{x} + 1 \ln x = 1 + \ln x \quad \boxed{5.1}$$

(ii) $\sqrt{1 + x^2}$ is a composite function $(g(x))^{1/2}$ – use the chain rule

$\sqrt{1 + x^2} = (1 + x^2)^{1/2}$
$g(x) = 1 + x^2 (=u) \quad f(u) = u^{1/2}$

$$\frac{d}{dx}(\sqrt{1 + x^2}) = 2x \cdot \tfrac{1}{2}(1 + x^2)^{-1/2}$$

$$= \frac{x}{\sqrt{1 + x^2}} \quad \boxed{5.1}$$

2 (a) Differentiate with respect to x
  (i) $(1 + 3x^2)^{1/2}$
  (ii) $e^{2x} \sec 3x$
  (iii) $(1 + x^3)/(1 + x)$

(b) If $x = (1 + t)^{1/2}$, $y = t(1 + t)^{1/2}$ determine $\dfrac{dy}{dx}$

and $\dfrac{d^2y}{dx^2}$ in terms of the parameter t. (O&C, 1991)

• (a) (i) Let $f(x) = (1 + 3x^2)^{1/2}$
  Chain rule: $u = 1 + 3x^2$, $f(x) = u^{1/2}$

Then $f'(x) = 6x \cdot \tfrac{1}{2}u^{-1/2} = \dfrac{3x}{\sqrt{1 + 3x^2}}$

$$\frac{d}{dx}(1 + 3x^2)^{1/2} = \frac{3x}{\sqrt{1 + 3x^2}} \quad \boxed{5.1}$$

(ii) $g(x) = e^{2x} \sec 3x$ Product rule

$$g'(x) = \frac{d}{dx}(e^{2x} \sec 3x)$$

$$= e^{2x} \cdot 3 \sec 3x \tan 3x + 2e^{2x} \sec 3x \quad \boxed{5.1}$$

$$= e^{2x} \sec 3x \,(3 \tan 3x + 2)$$

(iii) Let $h(x) = \dfrac{(1 + x^3)}{1 + x}$ Quotient rule

$$h'(x) = \frac{d}{dx}\left(\frac{1 + x^3}{1 + x}\right)$$

$$= \frac{(1 + x)(3x^2) - (1 + x^3)(1)}{(1 + x)^2} \quad \boxed{5.1}$$

$$= \frac{3x^2 + 3x^3 - 1 - x^3}{(1 + x)^2}$$

$$= \frac{2x^3 + 3x^2 - 1}{(1 + x)^2}$$

(b) $x = (1 + t)^{1/2} \qquad \dfrac{dx}{dt} = \dfrac{1}{2(1 + t)^{1/2}}$

$y = t(1 + t)^{1/2} \qquad \dfrac{dy}{dt} = (1 + t)^{1/2} + \dfrac{t}{2(1 + t)^{1/2}}$

$$= \frac{2 + 3t}{2(1 + t)^{1/2}}$$

$$\frac{dy}{dx} = \frac{dy}{dt} \cdot \frac{dt}{dx} = 2 + 3t$$

$$\frac{d^2y}{dx^2} = \frac{d}{dt}\left(\frac{dy}{dx}\right) \cdot \frac{dt}{dx}$$

$$= 3 \cdot 2(1 + t)^{1/2}$$

$$= 6(1 + t)^{1/2}$$

3 The equation of a curve is $y = \dfrac{4x}{(x-1)^2}$. State the

equations of the asymptotes of the curve, and use differentiation to find the coordinates of the turning point on the curve.

Sketch, on separate diagrams, the graphs of

$y = \dfrac{4x}{(x-1)^2}$ and $y^2 = \dfrac{4x}{(x-1)^2}$. (UCLES, 1993)

• $y = \dfrac{4x}{(x-1)^2}$

As $x \to +\infty$, $y \to 0^+$ ∴ $y = 0$ is an asymptote
As $x \to 1$, $y \to \infty$ ∴ $x = 1$ is an asymptote (double)

$$\frac{dy}{dx} = \frac{(x-1)^2 \cdot 4 - 4x \cdot 2(x-1)}{(x-1)^4}$$

$$= \frac{4(x-1) - 8x}{(x-1)^3}$$

$\dfrac{dy}{dx} = 0$ when $x - 1 = 2x$, i.e. $x = -1$ $\Rightarrow$ $y = -1$

The turning point is $(-1, -1)$. $\boxed{7.1}$

$y = \dfrac{4x}{(x-1)^2}$

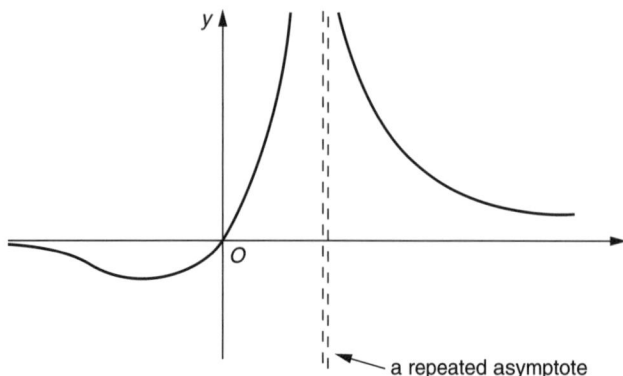

$y^2 = \dfrac{4x}{(x-1)^2}$ $\boxed{7.2}$

← a repeated asymptote

For $y^2 = \dfrac{4x}{(x-1)^2}$, $\dfrac{4x}{(x-1)^2} \geqslant 0$ when $x \geqslant 0$ (from 1st

graph)

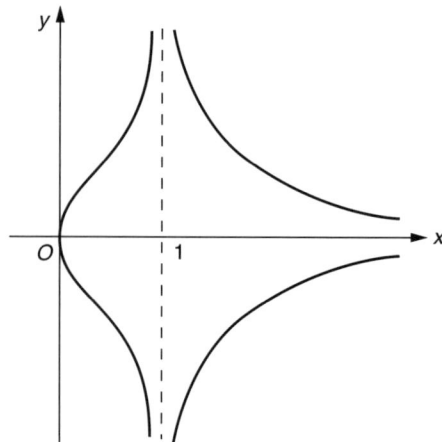

As $x \to 0$, $y \to \pm 2\sqrt{x}$ i.e. infinite gradient (as the curve $y^2 = 4x$).
As $x \to \infty$, $y^2 \to 0$ ∴ $y \to 0^+$ and 0^-

4 A camera is placed at C, 50 metres from a straight section of a racing track. The camera rotates to film a racing car travelling at 70 m s^{-1} along the straight. t seconds after passing the camera, the car has travelled x metres and the camera has turned through θ radians.

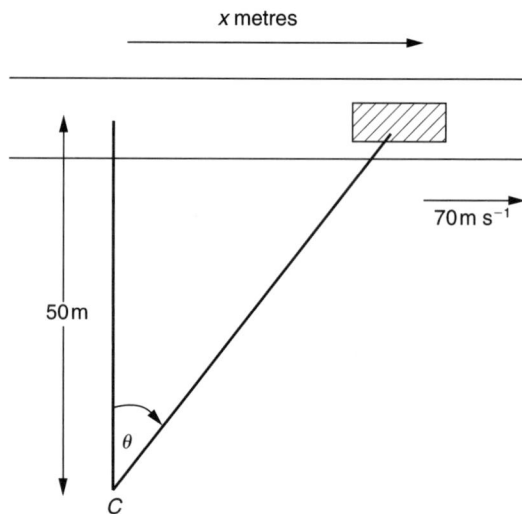

In this question you may need the result

$$\frac{d}{d\theta}(\tan \theta) = \frac{1}{\cos^2 \theta}$$

(a) Express x in terms of θ. Hence find $\dfrac{dx}{d\theta}$

(b) Explain what $\dfrac{dx}{dt}$ and $\dfrac{d\theta}{dt}$ represent.

(c) Show that $\dfrac{dx}{dt} = \dfrac{50}{\cos^2 \theta} \cdot \dfrac{d\theta}{dt}$. Hence obtain an

expression for $\dfrac{d\theta}{dt}$.

(d) Describe the rate of rotation of the camera as the car first comes towards and then passes the camera. (O&C SMP, 1993)

• (a) $\dfrac{x}{50} = \tan\theta \therefore x = 50\tan\theta$

$\dfrac{\mathrm{d}x}{\mathrm{d}\theta} = \dfrac{50}{\cos^2\theta}$

(b) $\dfrac{\mathrm{d}x}{\mathrm{d}t}$ is the rate at which x is increasing,
i.e. $70\ \mathrm{m\ s^{-1}}$ (linear velocity of the car)

$\dfrac{\mathrm{d}\theta}{\mathrm{d}t}$ is the rate at which θ is increasing,
i.e. angular velocity (rate of rotation of the camera)

(c) $\dfrac{\mathrm{d}x}{\mathrm{d}t} = \dfrac{\mathrm{d}x}{\mathrm{d}\theta}\,\dfrac{\mathrm{d}\theta}{\mathrm{d}t}$

$= \dfrac{50}{\cos^2\theta}\cdot\dfrac{\mathrm{d}\theta}{\mathrm{d}t}$

Therefore $\dfrac{\mathrm{d}\theta}{\mathrm{d}t} = \dfrac{\cos^2\theta}{50}\cdot\dfrac{\mathrm{d}x}{\mathrm{d}t}$

$= \dfrac{7\cos^2\theta}{5}$

(d) As the car approaches the point on the road which is nearest to C, $\dfrac{\mathrm{d}\theta}{\mathrm{d}t} = \tfrac{7}{5}$ rad/s (maximum value).

$\dfrac{\mathrm{d}\theta}{\mathrm{d}t}$ then decreases as x increases, i.e. the rate of rotation of the camera increases approaching C and then decreases after the car passes C.

5 The function f defined by

$$f(x) = ax^2 + 8x + 2 \quad \text{if } x \in (-\infty, 1)$$
$$\quad\ \ = b(2x - 3)^2 \quad \text{if } x \in [1, \infty)$$

is continuous and has a continuous derivative for all values of x. Find the values of the constants a and b. Find the stationary points and sketch the graphs of f and f' in the neighbourhood of $x = 1$.

• $f(x) = ax^2 + 8x + 2 \qquad f'(x) = 2ax + 8$
$x \in (-\infty, 1)$
$f(x) = b(2x - 3)^2 \qquad f'(x) = 4b(2x - 3)$
$x \in [1, \infty)$
Since the curve and the gradient are continuous, $f(x)$ is single-valued at $x = 1$ and $f'(x)$ is single-valued at $x = 1$.

$f(1) = a + 10 = b$ (1)
$f'(1) = 2a + 8 = -4b$ (2)

Solving equations 1 and 2: $a = -8, b = 2$ [7.1]
For $x \in (-\infty, 1)$:
$f(x) = -8x^2 + 8x + 2$
$f'(x) = -16x + 8$

Stationary point when $x = \tfrac{1}{2}$, $f(x) = 4$ [5.3]
For $x \in [1, \infty)$:
$f(x) = 2(2x - 3)^2$
$f'(x) = 8(2x - 3)$ [7.1]

Stationary point when $x = \tfrac{3}{2}$, $f(x) = 0$ [5.3]

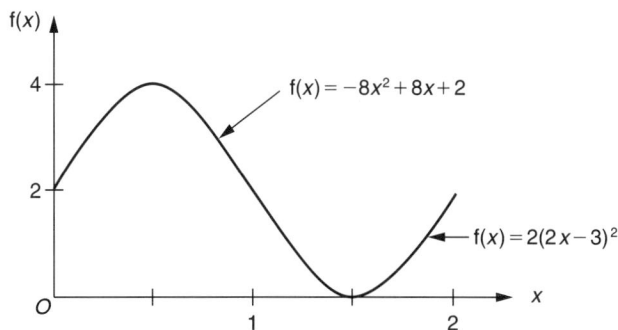

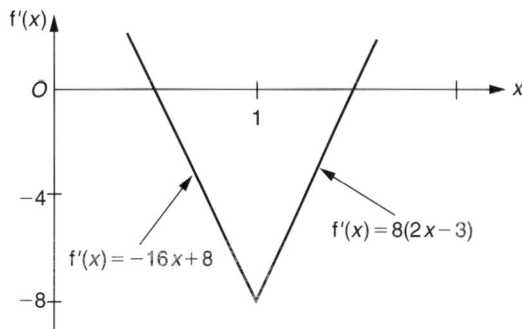

6 An oil production platform, $9\sqrt{3}$ km offshore, is to be connected by a pipeline to a refinery on shore, 100 km down the coast from the platform as shown in the diagram.

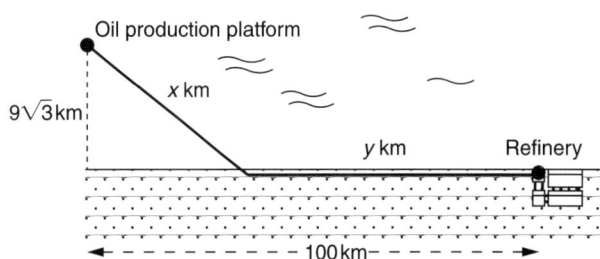

The length of underwater pipeline is x km and the length of pipeline on land is y km. It costs £2 million to lay each kilometre of pipeline underwater and £1 million to lay each kilometre of pipeline on land.

(a) Show that the total cost of this pipeline is £$C(x)$ million where

$$C(x) = 2x + 100 - (x^2 - 243)^{1/2}$$

(b) Show that $x = 18$ gives a minimum cost for this pipeline.
Find this minimum cost and the corresponding total length of the pipeline. (SEB, 1993)

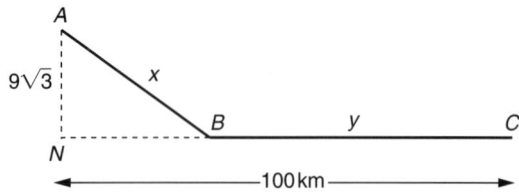

Cost = $(2x + y)$ million pounds.
From triangle ABN, $(NB)^2 = x^2 - 243$
$\Rightarrow \quad NB = (x^2 - 243)^{1/2}$.
Hence $y = 100 - (x^2 - 243)^{1/2}$.
Total cost is $2x + 100 - (x^2 - 243)^{1/2}$ million pounds.

$$C(x) = 2x + 100 - (x^2 - 243)^{1/2}$$

$$\frac{dC}{dx} = 2 - \frac{x}{(x^2 - 243)^{1/2}}$$

$$\frac{dC}{dx} = 0 \text{ when } x = 2(x^2 - 243)^{1/2}$$

$$x^2 = 4x^2 - 972$$

$$x = \sqrt{324} = 18$$

Check that this gives a minimum value

When $x = 17$, $\dfrac{dC}{dx}$ is negative;

when $x = 19$, $\dfrac{dC}{dx}$ is positive

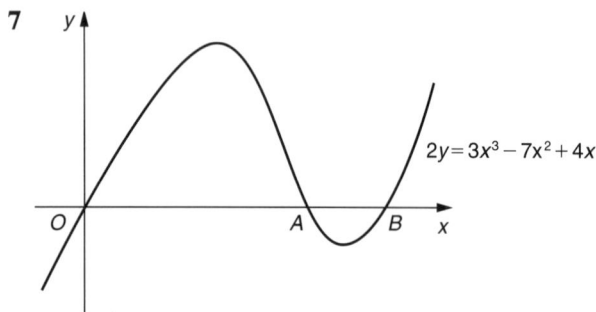

∴ Minimum cost occurs when $x = 18$.
Minimum cost $C = £127$ million.
Total length of the pipeline $(x + y) = (18 + 91)$ km
$\qquad\qquad\qquad\qquad\qquad\qquad = 109$ km

7

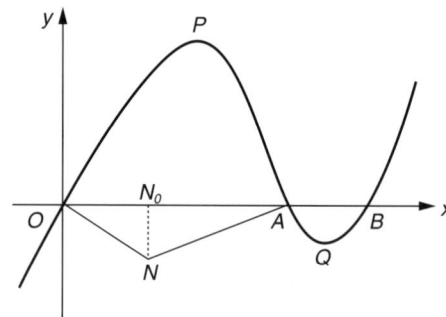

$2y = 3x^3 - 7x^2 + 4x$

The figure shows a sketch of part of the curve C with equation

$$2y = 3x^3 - 7x^2 + 4x$$

which meets the x-axis at the origin O, the point A $(1, 0)$ and the point B.

(a) Find the coordinates of B.
The normals to the curve C at the points O and A meet at the point N.
(b) Find the coordinates of N.
(c) Calculate the area of $\triangle OAN$.

At the points P and Q, on the curve C, $\dfrac{dy}{dx} = 0$.

Given that the x-coordinates of P and Q are x_1 and x_2,
(d) find the values of
  (i) $x_1 + x_2$
  (ii) $x_1 x_2$ (L, 1993)

•

$2y = 3x^3 - 7x^2 + 4x$
$\quad = x(3x^2 - 7x + 4)$
$\quad = x(3x - 4)(x - 1)$

Hence points on the x-axis are $(0, 0)$, $(1, 0)$ and $(\frac{4}{3}, 0)$.

(a) Coordinates of B are $(\frac{4}{3}, 0)$.
(b) The gradient of the curve C is

$$\frac{dy}{dx} = \tfrac{9}{2}x^2 - 7x + 2.$$

At $(0, 0)$, $\dfrac{dy}{dx} = 2 \Rightarrow$ the gradient of the normal is $-\frac{1}{2}$.

At $(1, 0)$, $\dfrac{dy}{dx} = -\frac{1}{2} \Rightarrow$ the gradient of the normal is 2.

Hence the equations of the normals at O and A are $2y + x = 0$ and $y = 2x - 2$.
At N, $x = -2y \Rightarrow y = -4y - 2$
$\Rightarrow \quad y = -\frac{2}{5}, x = \frac{4}{5}$, i.e. $N(\frac{4}{5}, -\frac{2}{5})$.

2.1b and 2.2b

(c) Area of $\triangle OAN$ is $\frac{1}{2}(OA)(NN_0) = \frac{1}{2}(1)(\frac{2}{5}) = \frac{1}{5}$.

(d) At P and Q, $\dfrac{dy}{dx} = 0 \Rightarrow \frac{9}{2}x^2 - 7x + 2 = 0$

  or $\quad 9x^2 - 14x + 4 = 0$

If the roots are x_1 and x_2 then

$$x_1 + x_2 = \frac{14}{9} \qquad \left(\frac{-b}{a}\right)$$

$$x_1 x_2 = \frac{4}{9} \qquad \left(\frac{c}{a}\right)$$

8 Given that

$$y = (\tan x)^x$$

use logarithmic differentiation to obtain an expression for $\dfrac{dy}{dx}$ in terms of x alone. (WJEC, 1994)

- $y = (\tan x)^x$

  Take logs of both sides of the equation:

  $$\ln y = \ln (\tan x)^x = x \ln (\tan x)$$

  Differentiate implicitly:

  $$\frac{1}{y}\frac{dy}{dx} = \ln (\tan x) + x\frac{\sec^2 x}{\tan x}$$

  $$\sec^2 x = \frac{1}{\cos^2 x}, \ \tan x = \frac{\sin x}{\cos x}$$

  $$\therefore \frac{dy}{dx} = (\tan x)^x \left[\ln \tan x + \frac{x}{\sin x \cos x} \right] \quad \boxed{5.1}$$

9 *In this question, you will need the following formulae:*

Volume of cylinder $= \pi r^2 h$;
Curved surface area of cylinder $= 2\pi r h$;
Volume of sphere $= \frac{4}{3}\pi r^3$;
Surface area of sphere $= 4\pi r^2$.

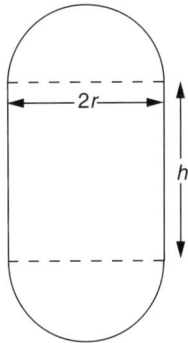

The figure shows a cross-section of a water container in the shape of a circular cylinder of height h and radius r with a hemisphere of radius r at each end.

Write down expressions for the volume V and external surface area S of the container in terms of h and r.

Given that $V = 36\pi$, show that

$$S = \frac{72\pi}{r} + \frac{4\pi r^2}{3}$$

(a) Find the minimum value of S.
(b) It is decided to take $r = 4$. Show that any subsequent small increase δr in r results in an increase δS in S given by

$$\delta S \approx \frac{37\pi}{6}\delta r$$

(c) A different radius results in the value of S

being 125. Show that one possible value of r lies between 4.07 and 4.08.

Another possible value of r lies between 2 and 3. Find its value correct to 1 decimal place.
(WJEC, 1994)

- Volume = volume of a sphere radius r + volume of a cylinder radius r and height h

  $$= \frac{4}{3}\pi r^3 + \pi r^2 h$$

  $$V = \pi r^2\left(\frac{4}{3}r + h\right)$$

  Surface area = surface area of a sphere
  + curved surface area of a cylinder
  $$= 4\pi r^2 + 2\pi r h$$
  $$= 2\pi r(2r + h)$$

  If $V = 36\pi$ then $r^2\left(\dfrac{4r}{3} + h\right) = 36$

  $$\frac{4r}{3} + h = \frac{36}{r^2}$$

  $$h = \frac{36}{r^2} - \frac{4r}{3}$$

  $$S = 2\pi r\left(2r + \frac{36}{r^2} - \frac{4r}{3}\right)$$

  $$= 2\pi r\left(\frac{2r}{3} + \frac{36}{r^2}\right)$$

  $$= \frac{72\pi}{r} + \frac{4\pi r^2}{3}$$

(a) S is a minimum when $\dfrac{dS}{dr} = 0$

  $$\Rightarrow -\frac{72\pi}{r^2} + \frac{8\pi r}{3} = 0 \quad \boxed{5.1}$$

  $$r^3 = 27 \quad \Rightarrow \quad r = 3 \qquad \frac{d^2 S}{dr^2} = \frac{144\pi}{r^3} + \frac{8\pi}{3} > 0$$

  Minimum value of S is $24\pi + 12\pi = 36\pi$

(b) If $r = 4$, $\dfrac{dS}{dr} = -4.5\pi + \dfrac{32\pi}{3} = \dfrac{\pi(-27 + 64)}{6}$

  $$= \frac{37\pi}{6}$$

  $$\therefore \delta S \approx \frac{37\pi}{6}\delta r$$

(c) When $r = 4$, $S = 123.57$.

  For an increase in S of 1.43, $\delta r \approx \dfrac{1.43 \times 6}{37\pi}$

  $$= 0.074$$

i.e. r lies between 4.07 and 4.08
If $r = 2$, $S = 129.8$ and $\delta S = -4.8$

$$\frac{\mathrm{d}S}{\mathrm{d}r} = \left(\frac{-72}{r^2} + \frac{8r}{3}\right)\pi = -12.67\pi$$

$$\therefore \delta r \approx \frac{-4.8}{-12.67\pi} = 0.1$$

$$\therefore r = 2.1$$

If $r = 2.1$, $S = 126.2$ and $\delta S = -1.2$

$$\frac{\mathrm{d}S}{\mathrm{d}r} = -33.7 \qquad \delta r = \frac{-1.2}{-33.7} = 0.03$$

$$\therefore r = 2.1 \text{ to 1 d.p.}$$

Alternatively, solve $125 - \dfrac{72\pi}{r} + \dfrac{4\pi r^2}{3} = 0$ on a GC

2.1a and 2.2a

10 Let $-\dfrac{\pi}{2} < x < \dfrac{\pi}{2}$.

(i) Show that the Maclaurin series for $\sec^2 x$ is given by

$$1 + x^2 + \frac{2x^4}{3} + \dots$$

(ii) Deduce the Maclaurin series for $\tan^2 x$ up to and including the term in x^4.
(iii) Use Maclaurin's theorem, or otherwise, to show that

(a) $\tan x = x + \dfrac{x^3}{3} + \dfrac{2x^5}{15} + \dots$

(b) $\log_e \sec x = \dfrac{x^2}{2} + \dfrac{x^4}{12} + \dfrac{x^6}{45} + \dots$

(NICCEA, 1994)

• (i) Let $f(x) = \sec^2 x$. Then

$f'(x) = 2\sec^2 x \tan x$
$\quad = 2f(x)\tan x$
$f''(x) = 2f'(x)\tan x + 2f(x)\sec^2 x$
$\quad = 2f'(x)\tan x + 2[f(x)]^2$
$f'''(x) = 2f''(x)\tan x + 2f'(x)\sec^2 x$
$\quad + 4f(x)f'(x)$
$\quad = 2f''(x)\tan x + 6f(x)f'(x)$
$f''''(x) = 2f'''(x)\tan x + 2f''(x)\sec^2 x$
$\quad + 6(f'(x))^2 + 6f(x)f''(x)$
$\quad = 2f'''(x)\tan x + 8f(x)f''(x) + 6(f'(x))^2$

When $x = 0$: $f(x) = 1$, $f'(x) = 0$, $f''(x) = 2$,
$f'''(x) = 0$, $f''''(x) = 16$.

Hence $\sec^2 x = 1 + \dfrac{2x^2}{2!} + \dfrac{16x^4}{4!} + \dots$

$$= 1 + x^2 + \frac{2x^4}{3} + \dots$$

(ii) Since $\sec^2 x = \tan^2 x + 1$

$$\tan^2 x = x^2 + \frac{2x^4}{3} + \dots$$

(iii) (a) *Either*

$$\tan x = \int \sec^2 x \, \mathrm{d}x$$

$$= a + x + \frac{x^3}{3} + \frac{2x^5}{15} + \dots$$

$\tan x = 0$ when $x = 0 \quad \Rightarrow \quad a = 0$

$$\therefore \tan x = x + \frac{x^3}{3} + \frac{2x^5}{15} + \dots$$

or

$g(x) = \tan x$, $g'(x) = \sec^2 x = f(x)$,
$g''(x) = f'(x)$, etc.

$$\therefore \tan x = g(0) + x + \frac{2x^3}{3!} + \frac{16x^5}{5!} + \dots$$

$$= x + \frac{x^3}{3} + \frac{2x^5}{15} + \dots$$

(b) $h(x) = \log_e \sec x$,

$$h'(x) = \frac{\sec x \tan x}{\sec x} = g(x), \text{ etc.}$$

$$\therefore \log_e \sec x = 0 + 0(x) + \frac{1x^2}{2!} + \frac{2x^4}{4!}$$

$$+ \frac{16x^6}{6!} + \dots$$

$$= \frac{x^2}{2} + \frac{x^4}{12} + \frac{x^6}{45} + \dots$$

Alternative Solution

• Let $y = \sec^2 x$, then if $\dfrac{\mathrm{d}y}{\mathrm{d}x} = y_1$, $\dfrac{\mathrm{d}^2 y}{\mathrm{d}x^2} = y_2$, etc.,

$y_1 = 2\sec^2 x \tan x$
$y_2 = 2\sec^4 x + 4\sec^2 x \tan^2 x$
$\quad = 6\sec^4 x - 4\sec^2 x = 6y^2 - 4y$
$y_3 = 12yy_1 - 4y_1$
$y_4 = 12yy_2 + 12y_1^2 - 4y_2$
etc.

When $x = 0$, $y = 1$
$y_1 = 0$
$y_2 = 2$
$y_3 = 0$
$y_4 = 16$
etc.

Then, $y = 1 + \dfrac{2x^2}{2!} + \dfrac{16x^4}{4!} + \dots$

or $\sec^2 x = 1 + x^2 + \dfrac{2x^4}{3} + \dots$ (1)

11 Given that $xy^2 = 3x^2 + 2y^3$, find $\dfrac{dy}{dx}$ giving your answer in terms of x and y.

- This question combines the product rule and implicit differentiation

$$\frac{d}{dx}(xy^2) = 1 \cdot y^2 + x\left(2y \cdot \frac{dy}{dx}\right)$$

Remember the $\dfrac{dy}{dx}$ when differentiating f(y)

Differentiate the equation **with respect to x**.

$$y^2 + 2xy\frac{dy}{dx} = 6x + 6y^2 \cdot \frac{dy}{dx}$$

$$\frac{dy}{dx}(2xy - 6y^2) = 6x - y^2$$

$$\frac{dy}{dx} = \frac{6x - y^2}{2xy - 6y^2}$$

Exercises

1 Given that

$$y = e^{-x^2}$$

find expressions for (i) $\dfrac{dy}{dx}$ and (ii) $\dfrac{d^2y}{dx^2}$.

Hence find the x-coordinates of the two points on the graph of y for which $\dfrac{d^2y}{dx^2}$ is equal to zero.

Show that these are both points of inflexion.
(WJEC, 1992)

2 (a) Differentiate with respect to x
    (i) $\sec(2x^2)$
    (ii) $x \arctan x$

  (b) Given that $x = \ln(1 + t^2)$ and $y = \dfrac{t}{1 + t^2}$,
    obtain $\dfrac{dy}{dx}$ in terms of t. (O&C, 1993)

3 Differentiate with respect to t
  (i) $e^{2t} \cos t$
  (ii) $\sin(t^2 + 4)$ (UCLES, 1988)

4 The function f, whose incomplete graph is shown in the diagram, is defined by $f(x) = x^4 - 2x^3 + 2x - 1$. Find the coordinates of the stationary points and justify their nature. (SEB, 1993)

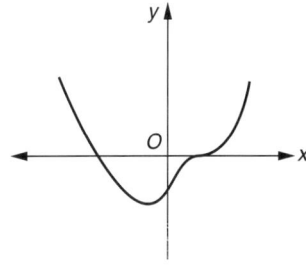

5 (i) Express $\log_e(x^2 e^{-1})$ in terms of $\log_e x$ only.
  (ii) Use the product rule to obtain the derivative of $x \log_e x$.
  (iii) If $y = (x^2 e^{-1})^x$, use logarithms to show that

$$\frac{dy}{dx} = y(1 + 2\log_e x)$$

  (iv) Derive the equation of the tangent to the curve $y = (x^2 e^{-1})^x$ at the point $(1, e^{-1})$. (NICCEA, 1993)

6 A curve is defined parametrically by

$$x = \frac{2t}{1 + t}, y = \frac{t^2}{1 + t}.$$

Prove that the normal to the curve at the point $(1, \tfrac{1}{2})$ has equation

$$6y + 4x = 7$$

Determine the coordinates of the other point of intersection of this normal with the curve.
(AEB, 1990)

7 (a) Find by differentiation the x-coordinate of the stationary point of the curve

$$y = x^2 - k^2 \ln\left(\frac{x}{a}\right)$$

    where k and a are positive constants, and determine the nature of the stationary point.
  (b) The tangent to the curve $y = \ln x$ at the point A, with coordinates $(a, \ln a)$, passes through the origin. Find the value of a.

    The normal to the curve $y = \ln x$ at the point B, with coordinates $(b, \ln b)$, passes through the origin. Find an equation satisfied by b and deduce that B lies on the curve $y = -x^2$.
(UCLES, 1993)

8 (i) Divide $(2x - x^2)$ by $(1 + x)$.
  (ii) Use Maclaurin's theorem to obtain the series expansion of

$$f(x) = 2\log_e(1 + x) + \frac{2x - x^2}{1 + x}$$

    up to and including the term involving x^3.
  (iii) Use the substitution $x = e^{2y} - 1$ in the equation $f(x) = 0$ to obtain the equation $e^{2y} + 3e^{-2y} = 4(y + 1)$.

(iv) Taking $y = 1$ as an approximate solution to this equation, use the Newton–Raphson method once to determine the solution to three s.f.
(NICCEA, 1992)

9 A straight line having a positive gradient m passes through the point $(-1, 2)$ and cuts the coordinate axes at X and Y. Find, in terms of m, the coordinates of X and Y.

(i) Show that the area A of the triangle OXY, where O is the origin, is given by

$$A = \frac{(m + 2)^2}{2m}$$

(ii) Show that A has a minimum value of 4.
(WJEC, 1994)

10 (a) Use Maclaurin's theorem to verify that

$$\frac{1}{1 + e^x} = \frac{1}{2} - \frac{x}{4} + \frac{x^3}{48} + \dots$$

Deduce, or obtain, the corresponding expansion (as far as the term in x^3)

$$\frac{e^x}{1 + e^x}$$

(b) Show that the equation

$$x^3 - 18x + 2 = 0$$

has a root lying between 4 and 5.

Use the Newton–Raphson method of successive approximations to locate the root correct to one decimal place. (NICCEA, 1993)

11 The equation of a curve is $2x^3 + 4xy + 9y^2 = -16$.

Find the values of $\dfrac{dy}{dx}$ and $\dfrac{d^2y}{dx^2}$ at the point

where the curve meets the negative x-axis.

Brief Solutions

1 $y = e^{-x^2}: \dfrac{dy}{dx} = -2xe^{-x^2}, \dfrac{d^2y}{dx^2} = e^{-x^2}(-2 + 4x^2)$ [5.1]

$\dfrac{d^2y}{dx^2} = 0$ when $x = \pm\dfrac{1}{\sqrt{2}}$ $\dfrac{dy}{dx} = \mp\sqrt{2}\,e^{-1/2}$

i.e. non-zero $\Rightarrow$ Points of inflection

2 (a) (i) Chain rule: $f'(x) = 4x \sec (2x^2) \tan (2x^2)$

(ii) Product rule: $g'(x) = \arctan x + \dfrac{x}{1 + x^2}$ [5.1]

(b) Parametric: $\dfrac{dy}{dt} = \dfrac{1 - t^2}{(1 + t^2)^2}$ $\dfrac{dx}{dt} = \dfrac{2t}{(1 + t^2)}$

$$\frac{dy}{dx} = \frac{(1 - t^2)}{2t(1 + t^2)}$$

3 (i) Product: $\dfrac{d}{dt}(e^{2t} \cos t) = e^{2t}(2 \cos t - \sin t)$ [5.1]

(ii) Chain rule: $\dfrac{d}{dt}(\sin(t^3 + 4)) = 3t^2 \cos (t^3 + 4)$

4 $f'(x) = 2(2x^3 - 3x^2 + 1) = 2(x - 1)^2(2x + 1)$ [5.1]

Repeated root at $x = 1$
$\Rightarrow$ inflexion at $x = 1, y = 0$.
$x = -\frac{1}{2}$: $f''(x)$ is positive
$\Rightarrow$ minimum point at $(-\frac{1}{2}, -1\frac{11}{16})$

5 (i) $\log_e(x^2 e^{-1}) = 2 \log_e x - 1$ [5.1]

(ii) $\dfrac{d}{dx}(x \log_e x) = 1 + \log_e x$

(iii) $y = (x^2 e^{-1})^x \Rightarrow \log_e y = x \log_e(x^2 e^{-1})$

$\dfrac{1}{y}\dfrac{dy}{dx} = 2 \log_e x - 1 + 2$ from (i) and (ii)

$\dfrac{dy}{dx} = y(2 \log_e x + 1)$

(iv) Gradient at $(1, e^{-1}) = e^{-1}$. Equation of tangent
$y - e^{-1} = e^{-1}(x - 1) \Rightarrow x = ey$

6 $\dfrac{dx}{dt} = \dfrac{2}{(1 + t)^2}$ $\dfrac{dy}{dt} = \dfrac{t(2 + t)}{(1 + t)^2}$ $\dfrac{dy}{dx} = \dfrac{t}{2}(2 + t)$

$x = 1, y = \frac{1}{2} \Rightarrow t = 1 \Rightarrow \dfrac{dy}{dx} = \dfrac{3}{2}$

Normal: $y = -\dfrac{2}{3}x + \dfrac{7}{6} \Rightarrow 6y + 4x = 7$

Substitute parametric form for x and y into $6y + 4x = 7$ and solve for t:

$t = 1$ or $-\frac{7}{6}$
$t = -\frac{7}{6} \Rightarrow x = 14, y = -\frac{49}{6}$

7 (a) $\dfrac{dy}{dx} = 2x - \dfrac{k^2}{x}$ Stationary point $x = \dfrac{k}{\sqrt{2}}$

(must be positive since $\dfrac{x}{a}$ must be positive)

$\dfrac{d^2y}{dx^2} = 2 + \dfrac{k^2}{x^2} > 0$ ∴ minimum point

(b) $\dfrac{dy}{dx} = \dfrac{1}{x}$ Equation of tangent at $(a, \ln a)$ is

$y = \dfrac{1}{a}(x - a) + \ln a$

Substitute $(0, 0)$: $\ln a = 1 \Rightarrow a = e$
Equation of normal at $(b, \ln b)$ is
$y = -bx + b^2 + \ln b$
Substitute $(0, 0)$: $-b^2 = +\ln b \Rightarrow -x^2 = y$

8 (i) $\dfrac{2x - x^2}{1 + x} = -x + 3 - \dfrac{3}{(x + 1)}$

(ii) $f'(x) = \dfrac{2}{1 + x} - 1 + \dfrac{3}{(1 + x)^2}$

$f''(x) = -\dfrac{2}{(1 + x)^2} - \dfrac{6}{(1 + x)^3}$

$f'''(x) = \dfrac{4}{(1 + x)^3} + \dfrac{18}{(1 + x)^4}$

By Maclaurin's theorem

$f(x) = 4x - 4x^2 + \dfrac{11x^3}{3} \cdots$ $\boxed{8.1}$

(iii) $0 = 2\ln(e^{2y}) + \dfrac{2e^{2y} - 2 - (e^{2y} - 1)^2}{e^{2y}}$

$0 = 4y + 2 - 2e^{-2y} - e^{2y} + 2 - e^{-2y}$
$\Rightarrow e^{-2y} + 3e^{-2y} = 4(y + 1)$

(iv) $g(y) = e^{2y} + 3e^{-2y} - 4(y + 1)$
$g'(y) = 2e^{2y} - 6e^{-2y} - 4$ $\boxed{5.1}$

$g(1) = -0.2049 \quad g'(1) = 9.9661$

$y_1 = 1 - \dfrac{-0.2049}{9.9661} = 1.02$ to three significant figures

9 Equation of XY is $y = mx + m + 2$:

$X\left(\dfrac{-(m + 2)}{m}, 0\right), Y(0, m + 2)$

(i) Area $A = \dfrac{(m + 2)^2}{2m}$ $\dfrac{dA}{dm} = \dfrac{m^2 - 4}{2m^2}$ $\dfrac{d^2A}{dm^2} = \dfrac{4}{m^3}$

(ii) $\dfrac{dA}{dm} = 0 \Rightarrow m = 2 \Rightarrow \dfrac{d^2A}{dm^2} = \dfrac{1}{2}$ (positive)

$\Rightarrow$ minimum area of 4 when $m = 2$

10 (a) $y = \dfrac{1}{1 + e^x}$

Using notation $\dfrac{dy}{dx} = y_1, \dfrac{d^2y}{dx^2} = y_2$, etc.:

$y_1 = \dfrac{-e^x}{(1 + e^x)^2} = \dfrac{-(1 + e^x) + 1}{(1 + e^x)^2} = -y_0 + (y_0)^2$

$y_2 = -y_1 + 2y_0 y_1$
$y_3 = -y_2 + 2y_1^2 + 2y_0 y_2$
When $x = 0$: $y_0 = \frac{1}{2} \quad y_1 = -\frac{1}{4} \quad y_2 = 0 \quad y_3 = \frac{1}{8}$
Hence the series for y is $\frac{1}{2} - \frac{1}{4}x + \frac{1}{48}x^3 + \ldots$

Since $\dfrac{e^x}{1 + e^x} = 1 - y$ the series for $\dfrac{e^x}{1 + e^x}$ is

$\frac{1}{2} + \frac{1}{4}x - \frac{1}{48}x^3 + \ldots$

(b) $f(x) = x^3 - 18x + 2$
$f(4) = -6, f(5) = 37 \therefore f(x) = 0$ between 4 and 5

(continuous function) $\boxed{2.4b}$
Newton–Raphson approximation:
(See page 159)
$x_0 = 4, x_1 = 4.2, x_2 = 4.186$
Root between 4 and 5 is 4.2 (1 d.p.)

11 *This is an implicit equation*

$2x^3 + 4xy + 9y^2 = -16$ (1)

$\Rightarrow 6x^2 + 4y + 4x\dfrac{dy}{dx} + 18y\dfrac{dy}{dx} = 0$ (2)

$\Rightarrow 12x + 4\dfrac{dy}{dx} + 4\dfrac{dy}{dx} + 4x\dfrac{d^2y}{dx^2}$

$+ 18\left(\dfrac{dy}{dx}\right)^2 + 18y\dfrac{d^2y}{dx^2} = 0$ (3)

In equation (1) $y = 0 \Rightarrow x = -2$

In equation (2) $\dfrac{dy}{dx} = +3$

In equation (3) $\dfrac{d^2y}{dx^2} = +20.25$

11

∫ Integration

AS Level			A Level			Topic	Date attempted	Date completed	Self-assessment
CORE	MODULAR	TRADITIONAL	CORE	MODULAR	TRADITIONAL				
✓	✓	✓	✓	✓	✓	**Concept; standard results**			
	✓	✓	✓	✓	✓	**By substitution**			
		✓	✓	✓	✓	**By parts; by partial fractions**			
✓	✓	✓	✓	✓	✓	**Definite integrals**			

Fact Sheet

Standard Results (Arbitrary Constants Omitted)

$f(x)$	$\int f(x)\, dx$
$x^n \ (n \neq -1)$	$\dfrac{x^{n+1}}{n+1}$
$\dfrac{1}{x}$	$\ln \lvert x \rvert$
e^x	e^x
$\sin x$	$-\cos x$
$\cos x$	$\sin x$
$\tan x$	$\ln \lvert \sec x \rvert$ or $-\ln \lvert \cos x \rvert$
$\cot x$	$-\ln \lvert \operatorname{cosec} x \rvert$ or $\ln \lvert \sin x \rvert$
$\sec^2 x$	$\tan x$
$\operatorname{cosec}^2 x$	$-\cot x$
$\sec x$	$\ln \lvert \sec x + \tan x \rvert$ or $\ln \left\lvert \tan\left(\dfrac{x}{2} + \dfrac{\pi}{4}\right) \right\rvert$
$\operatorname{cosec} x$	$-\ln \lvert \operatorname{cosec} x + \cot x \rvert$ or $\ln \left\lvert \tan \dfrac{x}{2} \right\rvert$
$\sec x \tan x$	$\sec x$
$\operatorname{cosec} x \cot x$	$-\operatorname{cosec} x$
$\dfrac{1}{a^2 + x^2}$	$\dfrac{1}{a} \tan^{-1} \dfrac{x}{a}$
$\dfrac{1}{\sqrt{(a^2 - x^2)}}$	$\sin^{-1} \dfrac{x}{a}$

Integration as the Inverse of Differentiation

$$\int f'(x)\, dx = f(x)$$

$$\int \{f(x)\}^n f'(x)\, dx = \frac{\{f(x)\}^{n+1}}{n+1}$$

$$\int \frac{f'(x)}{f(x)}\, dx = \ln\lvert f(x) \rvert$$

$$\int \{f'(x)\}\{e^{f(x)}\}\, dx = e^{f(x)}$$

Integration by Parts

$$\int u\, \frac{dv}{dx}\, dx = uv - \int v\, \frac{du}{dx}\, dx$$

(Or, more briefly, $\int u\ dv = uv - \int v\ du$, where u and v are functions of x.)

Choose the u and $\dfrac{dv}{dx}$ so that $v\,\dfrac{du}{dx}$ is easier to integrate than $u\,\dfrac{dv}{dx}$.

General Techniques for Integrating Trigonometric Functions

(i) $\sin^n x$ (and $\cos^n x$) where n is an even integer: use the $\cos 2x$ formulae.

(ii) $\sin^n x$ (and $\cos^n x$) where n is an odd integer: write $\sin^n x = \sin^{n-1} x \sin x$ and then change the $\sin^{n-1} x$ into terms in $\cos x$ using $\sin^2 x \equiv 1 - \cos^2 x$.

(iii) $\tan^n x$ where n is an integer and $n > 1$: use $\tan^2 x \equiv \sec^2 x - 1$.

(iv) $\sin ax \cos bx$ and similar terms: use the product formulae.

(v) $\int \sin^n \theta \cos \theta \, d\theta$ let $s = \sin \theta$; then $ds = \cos \theta \, d\theta$.

$\int \cos^n \theta \sin \theta \, d\theta$ let $c = \cos \theta$; then $dc = -\sin \theta \, d\theta$.

$\int \tan^n \theta \sec^2 \theta \, d\theta$ let $t = \tan \theta$; then $dt = \sec^2 \theta \, d\theta$.

$\int \sec^n \theta \tan \theta \, d\theta$ let $s = \sec \theta$; then $ds = \sec \theta \tan \theta \, d\theta$.

Applications

- If $y = f(x)$ cuts the x-axis at $x = c$ $(a < c < b)$,

  shaded area $= A_1 + A_2 = \int_a^c f(x) \, dx - \int_c^b f(x) \, dx$

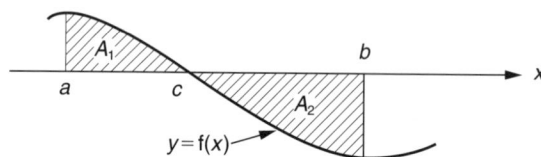

- *Area between a curve and the y-axis*
  If $x = g(y) \geqslant 0$ for $c \leqslant y \leqslant d$, area of elemental strip $\approx x \, \delta y = g(y) \, \delta y$.

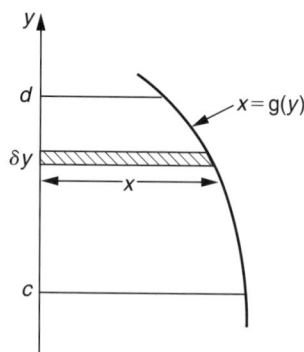

  Area $= \int_c^d x \, dy = \int_c^d g(y) \, dy$

Calculation of Volumes of Revolution

(i) *About the x-axis.* If $y = f(x)$, for $a \leqslant x \leqslant b$, volume of element disc $\approx \pi y^2 \, \delta x$.

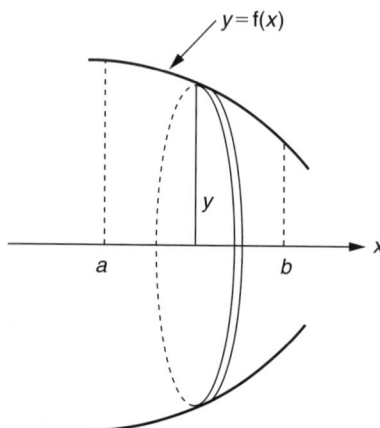

$$\text{Volume} = \int_a^b \pi\, y^2 \, dx$$

$$= \int_a^b \pi\, \{f(x)\}^2 \, dx$$

(ii) *About the y-axis.* If $x = g(y)$ for $c \leqslant y \leqslant d$, volume of elemental disc $\approx \pi x^2 \, \delta y$.

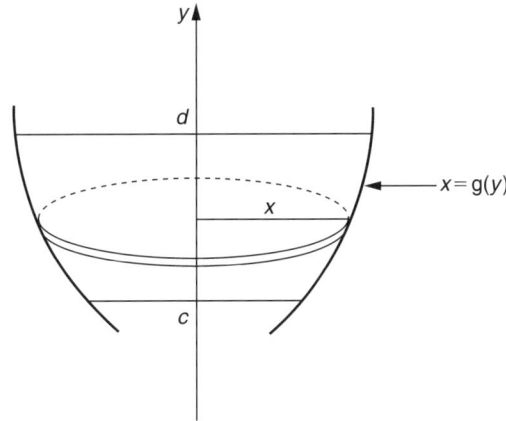

$$\text{Volume} = \int_c^d \pi\, x^2 \, dy$$

$$= \int_c^d \pi\, \{g(y)\}^2 \, dy$$

Worked Examples

1 Find $\int \sqrt{1 + 3x} \, dx$ and hence find the exact value of

$$\int_0^1 \sqrt{1 + 3x} \, dx. \qquad \text{(SEB, 1993)}$$

• $\sqrt{1 + 3x} = (1 + 3x)^{1/2}$

$\int (1 + 3x)^{1/2} \, dx = \frac{1}{3} \cdot \frac{2}{3}(1 + 3x)^{3/2} + C$
$= \frac{2}{9}(1 + 3x)^{3/2} + C$

or, use a substitution $u^2 = 1 + 3x$
Let $u^2 = 1 + 3x$. Then $2u\,du = 3\,dx$.

$$I = \int \sqrt{1 + 3x} \, dx \quad \Rightarrow \quad I = \int u \cdot \frac{2u}{3} \, du$$

$$= \frac{2}{3} \cdot \frac{u^3}{3} + C \qquad \qquad *$$

$$= \frac{2}{9}(1 + 3x)^{3/2} + C$$

6.1

$$\int_0^1 \sqrt{1 + 3x} \, dx = [\tfrac{2}{9}(1 + 3x)^{3/2}]_0^1$$

$$= \tfrac{16}{9} - \tfrac{2}{9}$$

$$= \tfrac{14}{9}$$

If the substitution method has been used then change the limits of the integral to values of *u* and apply at stage *

$$\int_0^1 \sqrt{1 + 3x} \, dx = \int_1^2 \tfrac{2}{3}u^2 \, du = [\tfrac{2}{9}u^3]_1^2 \quad \boxed{6.2}$$

$$= \tfrac{16}{9} - \tfrac{2}{9} = \tfrac{14}{9}$$

2 By using the substitution $x = 1 + u$, or otherwise, show that

$$\int_2^5 \frac{4x^2}{(x - 1)^2} \, dx = 15 + 16 \ln 2 \qquad \text{(NICCEA, 1993)}$$

• If $x = 1 + u$ then $dx = du$; $x = 2, u = 1; x = 5, u = 4$.

$$I = \int_1^4 \frac{4(1 + u)^2}{u^2} \, du$$

$$= 4 \int_1^4 \frac{1 + 2u + u^2}{u^2} \, du$$

$$= 4 \int_1^4 \frac{1}{u^2} + \frac{2}{u} + 1 \, du$$

$$= 4\left[-\frac{1}{u} + 2\ln u + u\right]_1^4 \qquad \boxed{6.1}$$

$$= 4[-\tfrac{1}{4} + 2\ln 4 + 4 + 1 - 2\ln 1 - 1]$$
$$= 15 + 8\ln 4 \qquad\qquad 4 = 2^2$$
$$= 15 + 16\ln 2$$

3 By means of the substitution $u = 1 + \sqrt{x}$, or otherwise, find

$$\int \frac{1}{1 + \sqrt{x}}\,dx$$

giving your answer in terms of x. (UCLES, 1991)

• Remember to substitute for dx

If $u = 1 + \sqrt{x}$ then

$$\sqrt{x} = u - 1 \;\Rightarrow\; x = (u - 1)^2$$
$$dx = 2(u - 1)\,du$$

$$\int \frac{1}{1 + \sqrt{x}}\,dx = \int \frac{1}{u}\cdot 2(u - 1)\,du$$

$$= \int 2 - \frac{2}{u}\,du$$

$$= 2u - 2\ln u + C$$

$$= 2(1 + \sqrt{x}) - 2\ln(1 + \sqrt{x}) + C$$
$$\text{put } 2 + C = C_1$$

$$I = 2\sqrt{x} - 2\ln(1 + \sqrt{x}) + C_1 \qquad \boxed{6.1}$$

An alternative substitution would be $u = \sqrt{x}$.
If $u^2 = x$ then $2u\,du = dx$.

$$\int \frac{1}{1 + \sqrt{x}}\,dx = \int \frac{2u}{1 + u}\,du$$

$$= 2\int 1 - \frac{1}{1 + u}\,du \qquad \text{by division}$$

$$= 2(u - \ln(1 + u)) + C$$
$$= 2(\sqrt{x} - \ln((1 + \sqrt{x})) + C$$

4 Express $\dfrac{5x - 4}{x^2 - 4x}$ in partial fractions.

Hence find the indefinite integral $\displaystyle\int \frac{5x - 4}{x^2 - 4x}\,dx$.

(SEB, 1993)

•

$$\frac{5x - 4}{x(x - 4)} = \frac{A}{x} + \frac{B}{x - 4}$$

$$= \frac{A(x - 4) + Bx}{x(x - 4)}$$

Compare numerators: $x = 0 \;\Rightarrow\; -4 = A(-4)$
$$\Rightarrow\; A = 1$$
Coefficient of x: $A + B = 5 \;\Rightarrow\; B = 4$

Hence $\dfrac{5x - 4}{x(x - 4)} = \dfrac{1}{x} + \dfrac{4}{x - 4}$ $\qquad \boxed{4.3}$

$$\int \frac{5x - 4}{x^2 - 4x}\,dx = \int \left(\frac{1}{x} + \frac{4}{x - 4}\right)dx$$

$$= \ln|x| + 4\ln|x - 4| + C$$

$$= \ln|x(x - 4)^4| + C \qquad \boxed{6.1}$$

Cover over method for partial fractions – use for linear denominators only.

$$\frac{5x - 4}{x(x - 4)} = \frac{A}{x} + \frac{B}{x - 4}$$

Find A by covering over the factor x in the LHS and substitute x = 0 into the residue $\dfrac{5x - 4}{x - 4} = 1$.

Find B by covering over the factor (x − 4) in the LHS and substituting x = 4 in the residue $\dfrac{5x - 4}{x} = 4$.

This is very quick to use.

5 The figure shows part of the curve with equation

$$y = 6\sqrt{x} + \frac{9}{x}.$$

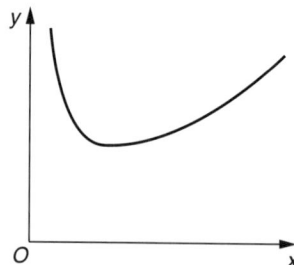

The points A and B lie on the curve and have x-coordinates 1 and 9 respectively.

(a) Determine the equation of the straight line which passes through A and B.

(b) Show that the curve has a point of inflexion when $x = 12^{2/3}$.

(c) Prove that the finite area enclosed by the curve and the chord AB is $32 - 18\ln 3$. (AEB, 1994)

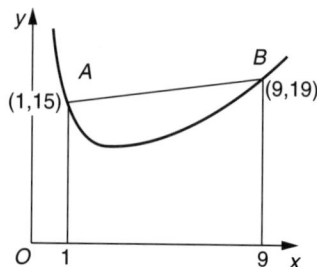

(a) When $x = 1, y = 6 + 9 = 15$
$\quad\quad\quad x = 9, y = 6 \times 3 + \frac{9}{9} = 19$

Gradient of line $= \dfrac{19 - 15}{9 - 1} = \frac{1}{2}$

Equation of AB is $y = \frac{1}{2}(x + 29)$

(b) $y = 6\sqrt{x} + \dfrac{9}{x} \Rightarrow \dfrac{dy}{dx} = \dfrac{3}{\sqrt{x}} - \dfrac{9}{x^2}$

$\Rightarrow \dfrac{d^2y}{dx^2} = -\dfrac{3}{2x^{3/2}} + \dfrac{18}{x^3}$

$\dfrac{d^2y}{dx^2} = 0$ when $x^3 = 12x^{3/2}$

$\Rightarrow x^{3/2} = 12$
$\Rightarrow x = 12^{2/3}$

If $\dfrac{d^2y}{dx^2} = 0$ but $\dfrac{dy}{dx} \neq 0$ there is a point of inflexion

which is *not a stationary point* P ⌐

If in doubt, test the sign of $\dfrac{d^2y}{dx^2}$ on each side of P

When $x = 12^{2/3}, \dfrac{dy}{dx} = \dfrac{3}{12^{1/3}} - \dfrac{9}{12^{4/3}} \neq 0.$

Hence when $x = 12^{2/3}$ the curve has a point of inflexion.

(c) Area enclosed is area of trapezium – area under the curve.

$A = \frac{8}{2}(15 + 19) - \int_1^9 \left(6\sqrt{x} + \dfrac{9}{x}\right) dx$

$= 136 - [\frac{12}{3}x^{3/2} + 9 \ln x]_1^9$ 6.1
$= 136 - 108 - 9 \ln 9 + 4 + 9 \ln 1$
$\quad\quad\quad\quad\quad$ In 1 = 0, In 9 = 2 In 3
$= 32 - 18 \ln 3$

6 The curve with equation $y = \sqrt{x} + \dfrac{3}{\sqrt{x}}$ is sketched for $x > 0$.

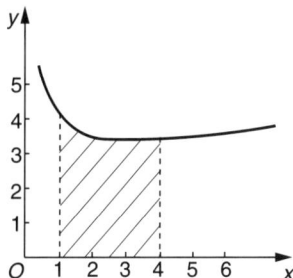

The region R, shaded in the diagram, is bounded by the curve, the x-axis and the lines $x = 1$ and $x = 4$. Determine:

(a) the area of R

(b) the volume of the solid formed when R is rotated through 360° about the x-axis.   (OLE, 1992)

• (a) Area $= \int_1^4 (x^{1/2} + 3x^{-1/2}) dx = \left[\dfrac{2x^{3/2}}{3} + 6x^{1/2}\right]_1^4$ 6.1

$= \frac{16}{3} + 12 - \frac{2}{3} - 6$
$= 10\frac{2}{3}$ 6.2

(b) Volume of revolution is $\int \pi y^2 dx$

Volume of revolution about the x-axis

$= \pi\int_1^4 \left(\sqrt{x} + \dfrac{3}{\sqrt{x}}\right)^2 dx$

$= \pi\int_1^4 \left(x + 6 + \dfrac{9}{x}\right) dx$

$= \pi\left[\dfrac{x^2}{2} + 6x + 9 \ln x\right]_1^4$ 6.1

$= \pi[8 + 24 + 9 \ln 4 - \frac{1}{2} - 6 - 9 \ln 1]$
$\quad\quad\quad\quad\quad$ In 1 = 0 6.2
$= 119.3$

7 For each non-negative integer n let

$I_n = \int_1^e x(\log_e x)^n dx$

(i) Use integration by parts to evaluate I_1.
(ii) Prove that, for each $n \geqslant 1, I_n = \frac{1}{2}e^2 - \frac{1}{2}nI_{n-1}$.
(iii) Hence, or otherwise, deduce that $I_4 + 1 = I_1$.
$\quad\quad\quad\quad\quad$ (NICCEA, 1993)

• (i) $I_1 = \int_1^e x \log_e x\, dx$

By parts: $u = \ln x, \dfrac{dv}{dx} = x, \dfrac{du}{dx} = \dfrac{1}{x}, v = \dfrac{x^2}{2}$

$I_1 = \left[\dfrac{x^2}{2} \ln x\right]_1^e - \int_1^e \dfrac{x^2}{2} \cdot \dfrac{1}{x} dx$

$= \left[\dfrac{x^2}{2} \ln x - \dfrac{x^2}{4}\right]_1^e$ In e = 1 6.1

$= \dfrac{e^2}{2} - \dfrac{e^2}{4} + \dfrac{1}{4}$

$= \frac{1}{4}(e^2 + 1)$

(ii) $I_n = \int_1^e x(\ln x)^n dx$

$u = (\ln x)^n, \dfrac{dv}{dx} = x \Rightarrow v = \dfrac{x^2}{2}$

$$\frac{du}{dx} = n(\ln x)^{n-1} \cdot \frac{1}{x}$$

$$I_n = \left[\frac{x^2}{2}(\ln x)^n\right]_1^e - \int_1^e n\frac{x}{2}(\ln x)^{n-1}\,dx$$

$$= \frac{e^2}{2} - \frac{n}{2}\int_1^e x(\ln x)^{n-1}\,dx$$

$$= \frac{e^2}{2} - \frac{n}{2}I_{n-1}$$

(iii) $I_4 = \dfrac{e^2}{2} - \dfrac{4}{2}I_3$ putting $n = 4$ in I_n

$$= \frac{e^2}{2} - 2\left(\frac{e^2}{2} - \frac{3}{2}I_2\right) \qquad \text{putting } n = 3 \text{ in } I_n$$

$$= -\frac{e^2}{2} + 3I_2$$

$$= -\frac{e^2}{2} + 3\left(\frac{e^2}{2} - I_1\right)$$

$$I_4 = e^2 - 3I_1$$

But $e^2 = 4I_1 - 1$ from (i). Hence

$$I_4 = I_1 - 1 \text{ or } I_4 + 1 = I_1$$

8 By expanding both sides of the identity

$$5\cos\left(\theta - \frac{\pi}{3}\right) - 3\cos\left(\theta + \frac{\pi}{3}\right) \equiv R\cos(\theta - \alpha)$$

show that $R = 7$ and find the acute angle α in radians correct to 3 d.p.

(i) Find all the values of θ in the range $[0, 2\pi]$ satisfying the equation

$$5\cos\left(\theta - \frac{\pi}{3}\right) - 3\cos\left(\theta + \frac{\pi}{3}\right) = 4.$$

(ii) Show that

$$\int_0^{\pi/2} \frac{d\theta}{5\cos\left(\theta - \frac{\pi}{3}\right) - 3\cos\left(\theta + \frac{\pi}{3}\right)}$$

$$= \frac{1}{7}\ln\left(\frac{\operatorname{cosec}\alpha + \cot\alpha}{\sec\alpha - \tan\alpha}\right) \qquad \text{(WJEC, 1992)}$$

• $\text{LHS} = 5\left(\cos\theta\cos\dfrac{\pi}{3} + \sin\theta\sin\dfrac{\pi}{3}\right)$

$$- 3\left(\cos\theta\cos\frac{\pi}{3} - \sin\theta\sin\frac{\pi}{3}\right)$$

$$= 2\cos\theta\cos\frac{\pi}{3} + 8\sin\theta\sin\frac{\pi}{3}$$

$$= \cos\theta + 4\sqrt{3}\sin\theta$$

$$\text{RHS} = R(\cos\theta\cos\alpha + \sin\theta\sin\alpha)$$

Comparing terms in $\cos\theta$ and $\sin\theta$:

$$R\cos\alpha = 1 \quad R\sin\alpha = 4\sqrt{3}$$

$$\sin^2\alpha + \cos^2\alpha \equiv 1, \frac{\sin\alpha}{\cos\alpha} = \tan\alpha$$

$$\Rightarrow \quad R^2 = 1 + 48 = 49, \text{ i.e. } R = 7$$

$\tan\alpha = 4\sqrt{3} \quad \Rightarrow \quad \alpha = 1.427$ radians   9.2

(i) $5\cos\left(\theta - \dfrac{\pi}{3}\right) - 3\cos\left(\theta + \dfrac{\pi}{3}\right) = 7\cos(\theta - \alpha)$
$$= 4$$

$$\therefore \cos(\theta - \alpha) = \tfrac{4}{7}$$
$$\theta - \alpha = (\pm 0.9626) + 2\pi n$$
$$\theta = 0.465 \text{ or } 2.390$$

(ii) $I = \displaystyle\int_0^{\pi/2} \frac{d\theta}{5\cos\left(\theta - \dfrac{\pi}{3}\right) - 3\cos\left(\theta + \dfrac{\pi}{3}\right)}$

$$= \int_0^{\pi/2} \frac{d\theta}{7\cos(\theta - \alpha)}$$

$$= \tfrac{1}{7}\int_0^{\pi/2} \sec(\theta - \alpha)\,d\theta$$

$$= \tfrac{1}{7}[\ln(\sec(\theta - \alpha) + \tan(\theta - \alpha))]_0^{\pi/2}$$

$$= \tfrac{1}{7}\ln\left(\sec\left(\frac{\pi}{2} - \alpha\right) + \tan\left(\frac{\pi}{2} - \alpha\right)\right)$$
$$-\tfrac{1}{7}\ln(\sec(-\alpha) + \tan(-\alpha))$$

$$\sec\left(\frac{\pi}{2} - \alpha\right) = \operatorname{cosec}\alpha$$

$$\tan\left(\frac{\pi}{2} - \alpha\right) = \cot\alpha$$

$$\sec(-\alpha) = \sec\alpha$$
$$\tan(-\alpha) = \tan\alpha$$

$$\therefore I = \tfrac{1}{7}\ln(\operatorname{cosec}\alpha + \cot\alpha) - \tfrac{1}{7}\ln(\sec\alpha - \tan\alpha)$$

$$= \tfrac{1}{7}\ln\left(\frac{\operatorname{cosec}\alpha + \cot\alpha}{\sec\alpha - \tan\alpha}\right)$$

9 On the same diagram, sketch the curve C_1 with polar equation

$$r = 2\cos 2\theta, \quad -\frac{\pi}{4} < \theta \leqslant \frac{\pi}{4}$$

and the curve C_2 with polar equation $\theta = \dfrac{\pi}{12}$.

Find the area of the smaller region bounded by C_1 and C_2.   (L, 1992)

•

$$\text{Area} = \tfrac{1}{2}\int_{\pi/12}^{\pi/4} r^2 \, d\theta$$

$$= \int_{\pi/12}^{\pi/4} 2 \cos^2 2\theta \, d\theta$$

$$= \int_{\pi/12}^{\pi/4} \cos 4\theta + 1 \, d\theta$$

$$= [\tfrac{1}{4} \sin 4\theta + \theta]_{\pi/12}^{\pi/4}$$

$$= \frac{\pi}{4} - \frac{\pi}{12} - \frac{1}{4} \cdot \frac{\sqrt{3}}{2}$$

$$= \frac{\pi}{6} - \frac{\sqrt{3}}{8}$$

10 (i) Complete the square in, and also factorize, the quadratic function

$$1 - 2x - 3x^2$$

(ii) By using the substitution $u = x + \tfrac{1}{3}$, or otherwise, show that

$$\int_{-1/3}^{0} \frac{dx}{\sqrt{1 - 2x - 3x^2}} = \frac{\pi}{6\sqrt{3}}$$

(iii) By using partial fractions, or otherwise, show that

$$\int_{-1/3}^{0} \frac{dx}{1 - 2x - 3x^2} = \frac{1}{4} \log_e 3$$
(NICCEA, 1992)

• (i) $1 - 2x - 3x^2 = -3(x^2 + \tfrac{2}{3}x - \tfrac{1}{3})$
$$= -3((x + \tfrac{1}{3})^2 - \tfrac{4}{9})$$
$$= \tfrac{4}{3} - 3(x + \tfrac{1}{3})^2 \qquad \boxed{4.2}$$

$$= (1 - 3x)(1 + x) \qquad \boxed{4.1}$$

(ii)
$$u = x + \tfrac{1}{3} \;\Rightarrow\; 1 - 2x - 3x^2 = \tfrac{4}{3} - 3u^2; \text{ and}$$
$$du = dx$$

$$\int_{-1/3}^{0} \frac{dx}{\sqrt{1 - 2x - 3x^2}} = \int_{0}^{1/3} \frac{du}{\sqrt{\tfrac{4}{3} - 3u^2}}$$

$$= \sqrt{3} \int_{0}^{1/3} \frac{du}{\sqrt{4 - 9u^2}}$$

Let $3u = z$, then $3\,du = dz$; $u = 0$, $z = 0$; $u = \tfrac{1}{3}$, $z = 1$.

$$I = \frac{\sqrt{3}}{3} \int_{0}^{1} \frac{dz}{\sqrt{4 - z^2}} = \frac{1}{\sqrt{3}} \left[\sin^{-1} \frac{z}{2} \right]_{0}^{1} \qquad \boxed{6.1}$$

$$= \frac{\pi}{6\sqrt{3}}$$

(iii) $\displaystyle\int_{-1/3}^{0} \frac{dx}{1 - 2x - 3x^2} = \int_{-1/3}^{0} \frac{1}{(1 + x)(1 - 3x)} \, dx$

Since the factors are linear, use the 'cover over' method for partial fractions

$$= \int_{-1/3}^{0} \frac{1}{4(1 + x)} + \frac{3}{4(1 - 3x)} \, dx \qquad \boxed{4.3}$$

$$= [\tfrac{1}{4} \ln (1 + x) - \tfrac{1}{4} \ln (1 - 3x)]_{-1/3}^{0} \qquad \boxed{6.1}$$
$$= \tfrac{1}{4} \ln (1) - \tfrac{1}{4} \ln (1) - \tfrac{1}{4} \ln (\tfrac{2}{3}) + \tfrac{1}{4} \ln 2$$
$$= \tfrac{1}{4} \log_e 3$$

11 (i) Write in partial fractions $\dfrac{2x^3 + 7x + 3}{x^3 + 3x}$.

(ii) Find $\displaystyle\int \frac{2x^3 + 7x + 3}{x^3 + 3x} \, dx$. (NICCEA, 1994)

• Since the fraction is an improper fraction, carry out long division first

$$\begin{array}{r} 2 \\ x^3 + 3x \overline{\smash{\big)}\ 2x^3 + 0x^2 + 7x + 3} \\ \underline{2x^3 \qquad\quad + 6x} \\ x + 3 \end{array}$$

$$\frac{2x^2 + 7x + 3}{x^3 + 3x} = 2 + \frac{x + 3}{x(x^2 + 3)}$$

$$\frac{x + 3}{x(x^2 + 3)} \equiv \frac{A}{x} + \frac{Bx + C}{x^2 + 3}$$

$$\equiv \frac{A(x^2 + 3) + (Bx + C)x}{x(x^2 + 3)}$$

$$\therefore x + 3 \equiv A(x^2 + 3) + Bx^2 + Cx \qquad (1)$$

Coefficient of x^2: $\quad 0 \equiv A + B$
Coefficient of x: $\quad 1 \equiv C$
Constant term: $\quad 3 \equiv 3A \;\Rightarrow\; A = 1 \therefore B = -1$

Alternatively substitute suitable values of x such as $x = 0$ into equation 1

$$\frac{2x^3 + 7x + 3}{x^3 + 3x} \equiv 2 + \frac{1}{x} - \frac{x}{x^2 + 3} + \frac{1}{x^2 + 3}$$

4.3

(ii) $\displaystyle\int \frac{2x^3 + 7x + 3}{x^3 + 3x}\, dx = I$

$$I = \int\left(2 + \frac{1}{x} - \frac{x}{x^2 + 3} + \frac{1}{x^2 + 3}\right) dx$$

$$= 2x + \ln|x| - \tfrac{1}{2}\ln(x^2 + 3)$$

$$+ \frac{1}{\sqrt{3}}\tan^{-1}\frac{x}{\sqrt{3}} + C$$

$$= 2x + \tfrac{1}{2}\ln\frac{x^2}{x^2 + 3} + \frac{1}{\sqrt{3}}\tan^{-1}\frac{x}{\sqrt{3}} + C$$

If the result of an integration is ln f(x) the modulus sign must be used unless the value of f(x) cannot be negative

12 Given that $y = \ln(1 + x^2)$, find the values of

$$\frac{dy}{dx}, \quad \frac{d^2y}{dx^2} \quad\text{and}\quad \frac{d^3y}{dx^3} \quad\text{when}\quad x = 1$$

Hence, show that the Taylor's series expansion of $\ln(1 + x^2)$ about $x = 1$, in ascending powers of $(x - 1)$, is

$$\ln 2 + (x - 1) - \tfrac{1}{6}(x - 1)^3 + \ldots$$

By integrating the given terms of the Taylor's series, obtain an approximate value of

$$I = \int_1^{1.2} \ln(1 + x^2)\, dx$$

giving your answer to **five** significant figures.

Use Simpson's rule with two equal intervals to obtain a second approximate value for I, giving your answer to **five** significant figures. (OLE, 1993)

● $y = \ln(1 + x^2)$ *Chain rule: $u = 1 + x^2$, $y = \ln u$*

$$\frac{dy}{dx} = \frac{2x}{1 + x^2} \qquad\text{\textit{Quotient rule}}$$

$$\frac{d^2y}{dx^2} = \frac{(1 + x^2)(2) - 2x(2x)}{(1 + x^2)^2} = \frac{2 - 2x^2}{(1 + x^2)^2}$$

5.1

$$\frac{d^3y}{dx^3} = \frac{(1 + x^2)^2(-4x) - 2(1 - x^2)4x(1 + x^2)}{(1 + x^2)^4}$$

$$= \frac{-4x - 4x^3 - 8x + 8x^3}{(1 + x^2)^3}$$

$$= \frac{4x(x^2 - 3)}{(1 + x^2)^3}$$

5.1

When $x = 1$: $y = \ln 2$, $\dfrac{dy}{dx} = 1$, $\dfrac{d^2y}{dx^2} = 0$, $\dfrac{d^3y}{dx^3} = -1$.

Taylor's series expansion is

$$y = y_{(1)} + (x - 1)\left(\frac{dy}{dx}\right)_{x=1} + \frac{(x - 1)^2}{2!}\left(\frac{d^2y}{dx^2}\right)_{x=1}$$

$$+ \frac{(x - 1)^3}{3!}\left(\frac{d^3y}{dx^3}\right)_{x=1} + \ldots$$

$$= \ln 2 + (x - 1) - \frac{(x - 1)^3}{6} + \ldots$$

$$I = \int_1^{1.2} \ln(1 + x^2)\, dx$$

$$\approx \int_1^{1.2}\left(\ln 2 + (x - 1) - \frac{(x - 1)^3}{6}\right) dx$$

$$\approx \left[x\ln 2 + \frac{(x - 1)^2}{2} - \frac{(x - 1)^4}{24}\right]_1^{1.2}$$

6.1

$$\approx 1.2\ln 2 + \frac{(0.2)^2}{2} - \frac{(0.2)^4}{24} - \ln 2$$

6.2

$$\approx 0.158\,56$$

By Simpson's rule with 2 equal intervals, width 0.1:

$$I \approx \frac{0.1}{3}\left(\ln 2 + 4\ln(2.21) + \ln(2.44)\right)$$

$$\approx 0.158\,57$$

6.3

13 Use the Simpson's composite rule with four strips to obtain an estimate of

$$\int_1^2 x^2 \ln x\, dx$$

(Use 5 d.p. arithmetic in your calculation.)

Given that $\dfrac{d^4}{dx^4}(x^2 \ln x) = \dfrac{-2}{x^2}$ for $x > 0$, estimate

the magnitude of the truncation error.

Hence state your estimate of the integral to an appropriate degree of accuracy.

Simpson's rule:

$$\int_a^b f(x)\, dx = \frac{h}{3}\{f_0 + f_{2n} + 4(f_1 + f_3 + \ldots + f_{2n-1})$$

$$+ 2(f_2 + f_4 + \ldots + f_{2n-2})\} + E$$

$$|E| = \frac{b - a}{180} h^4 M \text{ with } |f^{(iv)}(x)| \leqslant M \text{ for } a \leqslant x \leqslant b$$

(SEB, 1994)

• $\int_1^2 x^2 \ln x \, dx$ with 4 strips

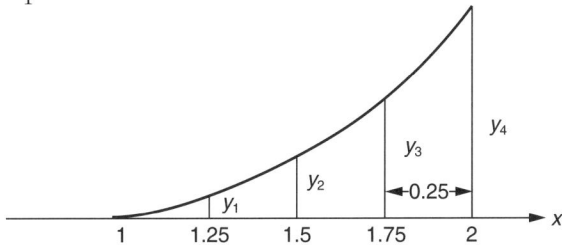

Simpson's rule:

$$\int_1^2 x^2 \ln x \, dx = \tfrac{1}{12}\{1 \ln 1 + 4 \ln 2 + 4[(\tfrac{5}{4})^2 \ln \tfrac{5}{4}$$
$$+ (\tfrac{7}{4})^2 \ln \tfrac{7}{4}] + 2(\tfrac{3}{2})^2 \ln \tfrac{3}{2}\}$$

$$= \tfrac{1}{12}\{12.847\ 12\}$$
$$= 1.070\ 59$$

$$f^{iv}(x) = \frac{-2}{x^2}$$

In the interval $1 \le x \le 2$, $|\, f^{iv}(x)\,|_{max} = 2$.

$$\text{Error} = \frac{2-1}{180}(\tfrac{1}{4})^4 \cdot 2 = 4.34 \times 10^{-5}$$
$$= 0.000\ 04$$

Hence value is in the interval $1.070\ 55 \rightarrow 1.070\ 63$. Thus

$$\int_1^2 x^2 \ln x \, dx = 1.0706$$

Exercises

1 Find (i) $\int(3x^3 + 4x)\,dx$ (ii) $\int \sin^2 x \cos x \, dx$
(SEB, 1994)

2 Use the identity $\sin 2x = 2 \sin x \cos x$, or otherwise, to evaluate

$$\int_0^{\pi/2} \sin x \cos x \, dx \qquad \text{(SMP, 1992)}$$

3 Given that $\dfrac{x^4}{1 + x^2} = ax^2 + b + \dfrac{c}{1 + x^2}$ find the numerical values of the constants a, b and c. Hence show that

$$\int_0^1 \frac{x^4}{1 + x^2}\,dx = \frac{\pi}{4} - \frac{2}{3}$$

Using this result in conjunction with integration by parts, evaluate the integral $I = \displaystyle\int_0^1 x^3 \tan^{-1} x \, dx$
(WJEC, 1993)

4 A solid wooden plinth 5 cm high is in the shape of the solid formed by rotating part of the curve $y = e^x$ about the x-axis through $360°$.

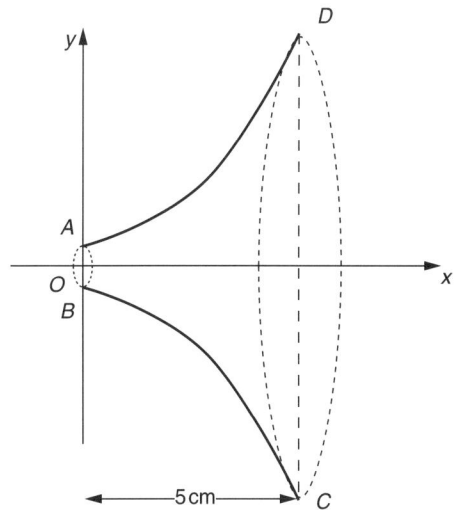

(a) Find the area of cross-section through $ABCD$.
(b) Find its volume.   (JMB, 1996)

5 Find (a) $\int x \cos x \, dx$ (b) $\int \cos^2 y \, dy$.
(L Specimen, 1996)

6 Evaluate $\displaystyle\int_0^1 \frac{x^3}{\sqrt{1 + x^4}}\,dx$.

7 Evaluate (a) $\displaystyle\int_0^3 \frac{dx}{(9 + x^2)^{3/2}}$ (b) $\displaystyle\int_1^2 x^3 \ln 2x \, dx$.
(OLE, 1993)

8 (a) Express $\dfrac{1}{x^2(2x - 1)}$ in the form

$$\frac{A}{x} + \frac{B}{x^2} + \frac{C}{2x - 1}.$$

Hence, or otherwise, evaluate $\displaystyle\int_1^2 \frac{1}{x^2(2x - 1)}\,dx$

(b) Find $\int x^3 \ln(4x)\,dx$.

(c) Using the substitution $x = 3 \tan \theta$, evaluate

$$\int_0^3 \frac{1}{(9 + x^2)^2}\,dx. \qquad \text{(AEB, 1992)}$$

9 Using the substitution $t = \tan x$, evaluate

$$\int_0^{\pi/4} \frac{dx}{3 \cos^2 x + \sin^2 x}$$

giving your answer in terms of π.   (L, 1993)

10 The cargo space of a small bulk carrier is 60 m long. The shaded part of the diagram represents the uniform cross-section of this space. It is shaped like the parabola with equation $y = \tfrac{1}{4}x^2$, $-6 \le x \le 6$ between the lines $y = 1$ and $y = 9$.

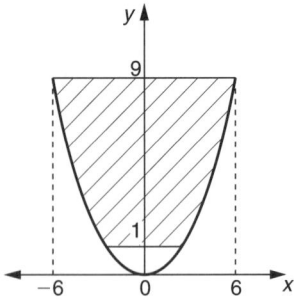

Find the area of this cross-section and hence find the volume of cargo that this ship can carry.
(SEB, 1994)

11 Use the trapezium rule with six ordinates, working to 5 d.p., to estimate the value of the definite integral $\int_0^{1/2} \dfrac{dx}{\sqrt{1 - x^2}}$.

Show that your answer is correct to 3 d.p. by obtaining the exact value of the integral.
(NICCEA, 1993)

12 (a) Show that $(x + 1)$, $(x - 2)$ and $(x - 3)$ are factors of

$$f(x) \equiv x^3 - 4x^2 + x + 6$$

(b) Sketch the curve C with equation $y = f(x)$ and write on your sketch the coordinates of the points where the curve crosses the coordinate axes.

(c) Determine the area of the finite region bounded by the curve C and the x-axis for $-1 \leqslant x \leqslant 2$.

The tangent to the curve C at the point $(1, 4)$ meets the curve again at the point A.

(d) Determine the coordinates of A. (L, 1993)

13 The concrete on the 20 feet by 28 feet rectangular facing of the entrance to an underground cavern is to be repainted.

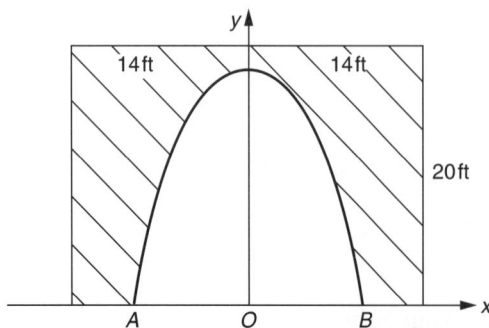

Coordinate axes are chosen as shown in the diagram with a scale of 1 unit equal to 1 foot. The roof is in the form of a parabola with equation $y = 18 - \frac{1}{8}x^2$.

(a) Find the coordinates of the points A and B.

(b) Calculate the total cost of repainting the facing at £3 per square foot. (SEB, 1993)

14 (i) Show that $(x + 1)$ is a factor of $x^3 + 4x^2 + 5x + 2$ and find the other factors.

(ii) Use calculus to find the turning points on the curve $y = x^3 + 4x^2 + 5x + 2$ and determine the nature of each.

(iii) Sketch the graph of this curve, showing the turning points and the points where the curve crosses the axes.

(iv) Calculate the area of the region bounded by the curve, the x-axis and the lines $x = -3$ and $x = -1$. (NICCEA, 1994)

Brief Solutions

1 (i) $\dfrac{3x^4}{4} + 2x^2 + C$ 6.1

(ii) *See Fact Sheet 'General Techniques' or substitute* $s = \sin x$

$$I = \frac{\sin^3 x}{3} + C \qquad \boxed{6.1}$$

2 As in Q1(ii): $I = \left[\dfrac{\sin^2 x}{2} \right]_0^{\pi/2} = \frac{1}{2}$

or $I = \frac{1}{2}\int \sin 2x \, dx = [-\frac{1}{4} \cos 2x]_0^{\pi/2} = \frac{1}{4} + \frac{1}{4} = \frac{1}{2}$ 6.2

3 By long division $\dfrac{x^4}{1 + x^2} = x^2 - 1 + \dfrac{1}{1 + x^2}$ 4.3

$$\int_0^1 \left(x^2 - 1 + \frac{1}{1 + x^2} \right) dx = \left[\frac{x^3}{3} - x + \tan^{-1} x \right]_0^1$$

$$= \frac{\pi}{4} - \frac{2}{3} \qquad \boxed{6.1}$$

$$I = \int_0^1 x^3 \tan^{-1} x \, dx = \left[\frac{x^4}{4} \tan^{-1} x \right]_0^1 - \int_0^1 \frac{x^4}{4(x^2 + 1)} \, dx$$

$$= \frac{\pi}{16} - \frac{1}{4}\left(\frac{\pi}{4} - \frac{2}{3} \right) = \frac{1}{6}$$

4 (a) Area of cross-section $= 2\int_0^5 e^x \, dx = 295 \text{ cm}^2$.   6.2

(b) Volume $= \pi\int_0^5 (e^x)^2 \, dx = \pi\int_0^5 e^{2x} \, dx = \dfrac{\pi}{2} [e^{2x}]_0^5$

$$= 34\,600 \text{ cm}^3 \text{ to 3 s.f.} \qquad \boxed{6.2}$$

5 (a) $\int x \cos x \, dx$ by parts $\Rightarrow x \sin x - \int \sin x \, dx$

$\Rightarrow x \sin x + \cos x + C$ $\boxed{6.1}$

(b) For $\int \cos^2 y \, dy$

Use the identity $\cos 2y = 2 \cos^2 y - 1$

$\boxed{6.1}$

$\frac{1}{2} \int (\cos 2y + 1) \, dy = \frac{1}{2} \left(\frac{\sin 2y}{2} + y \right) + C.$

6 $\int_0^1 \frac{x^3}{\sqrt{1 + x^4}} \, dx$

Substitute $u^2 = 1 + x^4 \quad 2u \, du = 4x^3 \, dx$

$x = 0, u = 1; x = 1, u = \sqrt{2}$

$\frac{1}{2} \int_1^{\sqrt{2}} \frac{u \, du}{u} = \frac{\sqrt{2} - 1}{2} = 0.207$ $\boxed{6.2}$

7 (a) $\int_0^3 \frac{dx}{(9 + x^2)^{3/2}}$ Let $x = 3 \tan \theta$:

$dx = 3 \sec^2 \theta \cdot d\theta; x = 0, \theta = 0; x = 3, \theta = \frac{\pi}{4}$

$I = \int_0^{\pi/4} \frac{3 \sec^2 \theta}{27 \sec^3 \theta} \, d\theta = \frac{1}{9} \int_0^{\pi/4} \cos \theta \, d\theta = \frac{1}{9\sqrt{2}}$

$\boxed{6.1 \text{ and } 6.2}$

(b) $\int_1^2 x^3 \ln 2x \, dx$ By parts:

$I = \frac{x^4}{4} \ln 2x - \int \frac{x^3}{4} \, dx = \left[\frac{x^4}{4} \ln 2x - \frac{x^4}{16} \right]_1^2$

$= \frac{31}{4} \ln 2 - \frac{15}{16} \, (= 4.43)$

8 (a) $\frac{1}{x^2(2x - 1)} = \frac{-2}{x} - \frac{1}{x^2} + \frac{4}{2x - 1}$ $\boxed{4.3}$

$\int_1^2 \frac{1}{x^2(2x - 1)} \, dx = \left[-2 \ln x + \frac{1}{x} + 2 \ln (2x - 1) \right]_1^2$

$\boxed{6.1 \text{ and } 6.2}$

$= 2 \ln \frac{3}{2} - \frac{1}{2} \quad (= 0.311)$

(b) $\int x^3 \ln (4x) \, dx$ By parts

$= \frac{x^4}{4} \ln (4x) - \int \frac{x^3}{4} \, dx = \frac{x^4}{4} \ln (4x) - \frac{x^4}{16} + C$

$\boxed{6.1}$

(c) $\int_0^3 \frac{1}{(9 + x^2)^2} \, dx$ By substitution

$x = 3 \tan \theta, dx = 3 \sec^2 \theta \, d\theta;$

$x = 0, \theta = 0; x = 3, \theta = \frac{\pi}{4}$

$\int_0^{\pi/4} \frac{3 \sec^2 \theta \, d\theta}{81 \sec^4 \theta} = \frac{1}{27} \int_0^{\pi/4} \cos^2 \theta \, d\theta$

$= \frac{1}{27} \left[\frac{\theta}{2} + \frac{\sin 2\theta}{4} \right]_0^{\pi/4} = \frac{1}{27} \left[\frac{\pi}{8} + \frac{1}{4} \right]$ or 0.0238

$\boxed{6.1 \text{ and } 6.2}$

9 $\int_0^{\pi/4} \frac{dx}{3 \cos^2 x + \sin^2 x}$

$t = \tan x, dt = \sec^2 x \, dx; x = 0, t = 0; x = \frac{\pi}{4}, t = 1$

Multiply numerator and denominator by $\sec^2 x$.

$I = \int_0^1 \frac{dt}{3 + t^2} = \left[\frac{1}{\sqrt{3}} \tan^{-1} \frac{t}{\sqrt{3}} \right]_0^1 = \frac{\pi}{6\sqrt{3}}$

$\boxed{6.1}$

10 Area of element $= 2x \delta y$ where $x = 2\sqrt{y}$

Area $= \int_1^9 4\sqrt{y} \, dy = \frac{208}{3}$ $\boxed{6.2}$

Volume $= 60 \times \frac{208}{3} = 4160 \text{ m}^3$

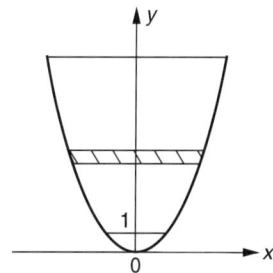

11 $\int_0^{1/2} \frac{dx}{\sqrt{1 - x^2}}$ gives the area under the curve.

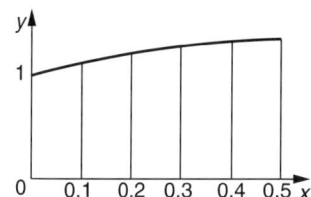

By the trapezium rule, area is:

$\frac{0.1}{2} \left[\frac{1}{\sqrt{1}} + \frac{1}{\sqrt{0.75}} + 2 \left(\frac{1}{\sqrt{0.99}} + \frac{1}{\sqrt{0.96}} \right. \right.$

$\left. \left. + \frac{1}{\sqrt{0.91}} + \frac{1}{\sqrt{0.84}} \right) \right] = 0.524 \text{ to 3 d.p.}$

$$\int_0^{1/2} \frac{1}{\sqrt{1-x^2}}\, dx = [\sin^{-1} x]_0^{1/2} = \frac{\pi}{6} = 0.524 \text{ to 3 d.p.}$$

6.1 and 6.2

12 (a) f(−1) = 0, f(2) = 0, f(3) = 0 ∴ (x + 1), (x − 2) and (x − 3) are factors

(b)

7.1

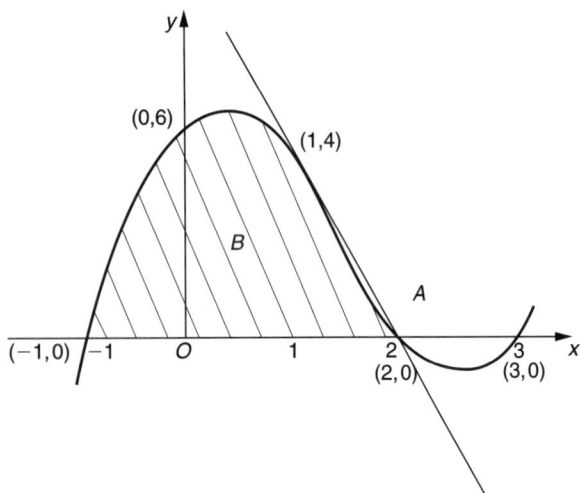

(c) Area $B = \int_{-1}^{2} (x^3 - 4x^2 + x + 6)\, dx$

$$= 11\tfrac{1}{4}$$

(d) $\dfrac{dy}{dx} = 3x^2 - 8x + 1$ At (1, 4), $\dfrac{dy}{dx} = -4$

Tangent at (1, 4) is y − 4 = −4(x − 1)
⇒ y = 8 − 4x. It meets the curve when
8 − 4x = x³ − 4x² + x + 6, that is

$$x^3 - 4x^2 + 5x - 2 = 0$$

A repeated factor of (x − 1) comes from the *tangent* at x = 1:

(x − 1)(x − 1)(x − 2) = 0   A(2, 0)

13 (a) Equation of the parabola is $y = 18 - \tfrac{1}{8}x^2$.
At A and B: y = 0, x = ± 12.

Area under the parabola = $2\displaystyle\int_0^{12} (18 - \tfrac{1}{8}x^2)\, dx$

$$= 288 \text{ sq. ft.}$$ 6.2

(b) Area of rectangle = 560 sq. ft   ⇒   area to be painted is 272 sq. ft.
Cost £816

14 f(x) = x³ + 4x² + 5x + 2
f(−1) = 0 ⇒ (x + 1) is a factor
f(−2) = 0 ⇒ (x + 2) is a factor   4.1
(i) f(x) = (x + 1)(x + 2)(x + 1)

(ii) $\dfrac{dy}{dx} = 3x^2 + 8x + 5 = 0$

when $x = -\tfrac{5}{3}$ or −1, $y = \tfrac{4}{27}$ or 0

Stationary points $(-\tfrac{5}{3}, \tfrac{4}{27})$ or (−1, 0)

$\dfrac{d^2y}{dx^2} = 6x + 8.$

When $x = -\tfrac{5}{3}, \dfrac{d^2y}{dx^2} = -2$ ⇒ maximum point.

When $x = -1, \dfrac{d^2y}{dx^2} = 2$ ⇒ minimum point.

(iii)

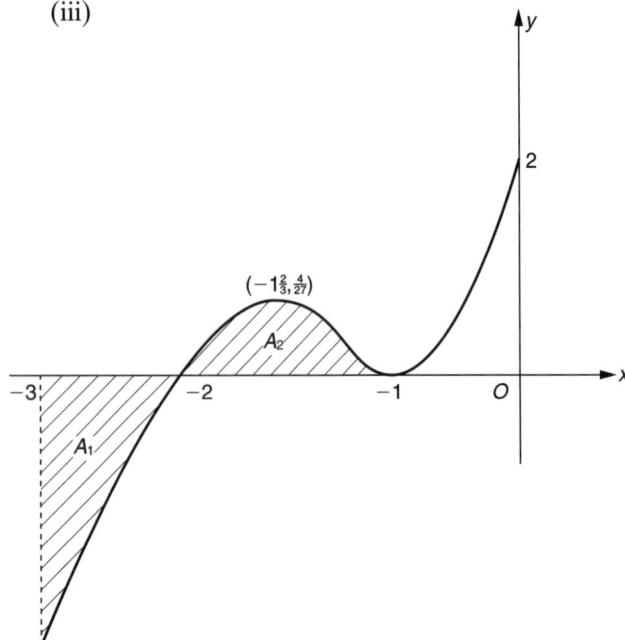

(iv) Area $A_1 = \left| \displaystyle\int_{-3}^{-2} (x^3 + 4x^2 + 5x + 2)\, dx \right| = \tfrac{17}{12}$

6.2

Area $A_2 = \displaystyle\int_{-2}^{-1} f(x)\, dx = \tfrac{1}{12}$

$A_1 + A_2 = 1\tfrac{1}{2}$

12

Differential equations

AS Level			A Level			Topic	Date attempted	Date completed	Self-assessment
CORE	MODULAR	TRADITIONAL	CORE	MODULAR	TRADITIONAL				
	✓	✓	✓	✓	✓	**First order**			
				✓	✓	**Second order**			
				✓	✓	**Particular solutions**			
			✓	✓	✓	**Numerical solutions**			

Fact Sheet

First Order Differential Equations

First order differential equations contain first derivatives only. For example:

$$x^2 \frac{dy}{dx} = \sin x \qquad y^3 \frac{dy}{dx} + x = y, \qquad \text{etc.}$$

Methods of Solution

Direct Integration

$$\frac{dy}{dx} = f(x) \quad \Rightarrow \quad y = \int f(x) \, dx + \text{constant}$$

$$\frac{dy}{dx} = g(y) \quad \Rightarrow \quad \int \frac{1}{g(y)} \, dy = \int x \, dx + \text{constant}$$

Variables Separable

$$\frac{dy}{dx} = f(x) \cdot g(y) \quad \Rightarrow \quad \int \frac{dy}{g(y)} = \int f(x) \, dx + C$$

Integrating Factor

$$\frac{dy}{dx} + f(x)y = g(x)$$

Multiply throughout by $e^{\int f(x) \, dx}$.

Example

$$\frac{dy}{dx} + 2xy = x$$

Multiply by e^{x^2}

$$e^{x^2} \frac{dy}{dx} + 2xe^{x^2}y = xe^{x^2}$$

The LHS is the result of differentiating $e^{x^2}y$. Integrating gives

$$e^{x^2}y = \tfrac{1}{2}e^{x^2} + C$$

Substitutions – Homogeneous Equations

$$\frac{dy}{dx} = \frac{2x + y}{x}$$

Since all terms on the RHS are of the same order, substitute $y = vx$.

If $y = vx$ then $\dfrac{dy}{dx} = x\dfrac{dv}{dx} + v$.

$$\therefore x\frac{dv}{dx} + v = \frac{2x + vx}{x} = 2 + v$$

$$\therefore x\frac{dv}{dx} = 2$$

Separate the variables: $\displaystyle\int dv = \int \frac{2}{x}\,dx$

$$v = 2\ln|x| + C$$

Numerical Methods

$$\frac{dy}{dx} \approx \frac{\delta y}{\delta x} \quad\Rightarrow\quad \delta y \approx \frac{dy}{dx}\,\delta x$$

Hence choosing small repeated increments of δx will give approximate values for δy and hence of y.

Second Order Linear Differential Equations

$$a\frac{d^2y}{dx^2} + b\frac{dy}{dx} + cy = f(x)$$

Type 1

$$b = c = 0, \text{ i.e. } a\frac{d^2y}{dx^2} = f(x)$$

Integrate directly

Type 2

$$f(x) = 0, \text{ i.e. } a\frac{d^2y}{dx} + b\frac{dy}{dx} + cy = 0$$

There are three subgroups depending on the roots of the equation $am^2 + bm + c = 0$ (known as the auxiliary equation):

(i) roots real and distinct m_1 and m_2

$$y = Ae^{m_1x} + Be^{m_2x}$$

(ii) roots real and equal

$$y = e^{mx}(Ax + B)$$

(iii) roots complex $m = a \pm ib$

$$y = e^{ax}(A\sin bx + B\cos bx)$$

Type 3

$$f(x) \neq 0$$

The solutions are:

$$y = \text{complementary function (CF)} + \text{particular integral (PI)}$$

The CF is obtained as in type 2, assuming $f(x) = 0$.
The PI is a function of the same general form as $f(x)$.

Example

$$\frac{d^2y}{dx^2} + 6\frac{dy}{dx} + 5y = 32e^{3x}$$

Auxiliary equation is $m^2 + 6m + 5 = 0 \quad\Rightarrow\quad m = -1 \text{ or } -5$.

$$\text{CF}\quad y_1 = Ae^{-x} + Be^{-5x}$$

$$\text{PI}\quad y_2 = Ce^{3x} \quad \frac{dy_2}{dx} = 3Ce^{3x} \quad \frac{d^2y_2}{dx^2} = 9Ce^{3x}$$

Substitute into the differential equation:

$$Ce^{3x}(9 + 18 + 5) \equiv 32e^{3x}$$
$$\therefore C = 1$$

The general solution is $y = Ae^{-x} + Be^{-5x} + e^{3x}$.

Note: Although the PI satisfies the d.e. it is *not* a general solution. The CF does *not* satisfy the d.e. unless $f(x) = 0$ but it is always part of the general solution.

Worked Examples

1 (a) Solve the differential equation $\dfrac{dy}{dx} = xy$ when $y = 1$ when $x = 0$.

(b) Find a solution of the differential equation

$\dfrac{d^2y}{dx^2} = y$ in the form $y = f(x)$ where $f(0) = 1$

and, for all x, $f(x) = f(-x)$.   (O&C, 1991)

• (a) *Variables are separable*

$$\frac{dy}{dx} = xy \quad \Rightarrow \quad \frac{1}{y}\frac{dy}{dx} = x$$

$$\int_1^y \frac{1}{y}\,dy = \int_0^x x\,dx \qquad \boxed{6.1}$$

$$[\ln y]_1^y = \left[\frac{x^2}{2}\right]_0^x$$

$$\ln y = \frac{x^2}{2} \quad \Rightarrow \quad y = e^{x^2/2}$$

(b) *With second order d.e.'s, bring all terms containing y to the LHS of the equation*

$$\frac{d^2y}{dx^2} - y = 0$$

Auxiliary equation $m^2 - 1 = 0 \quad \Rightarrow \quad m = \pm 1$
General solution $y = Ae^x + Be^{-x}$
Conditions:
$f(0) = 1$ $\therefore A + B = 1$
$f(x) = f(-x)$, i.e. $Ae^x + Be^{-x} = Ae^{-x} + Be^x$
                    $\Rightarrow \quad A = B$

$$\therefore A = B = \tfrac{1}{2}$$

Hence $y = \tfrac{1}{2}(e^x + e^{-x}) = \cosh x$

2 Find the general solution of the differential equation

$$\frac{d^2y}{dx^2} - 4\frac{dy}{dx} + 8y = 4\sin 2x \qquad \text{(NICCEA, 1993)}$$

• The auxiliary equation is $m^2 - 4m + 8 = 0$:

$$m = \frac{4 \pm \sqrt{16 - 32}}{2} = 2 \pm 2i$$

The CF is $y_1 = e^{2x}(A\sin 2x + B\cos 2x)$
The PI is y_2
The general form will be $C\sin 2x + D\cos 2x$

$$y_2 = C\sin 2x + D\cos 2x$$

$$\frac{dy_2}{dx} = 2C\cos 2x - 2D\sin 2x$$

$$\frac{d^2y_2}{dx^2} = -4C\sin 2x - 4D\cos 2x$$

Substitute into the d.e. and collect like terms together:

$$-4C\sin 2x - 4D\cos 2x - 8C\cos 2x + 8D\sin 2x$$
$$+ 8C\sin 2x + 8D\cos 2x \equiv 4\sin 2x$$

$$\sin 2x(-4C + 8D + 8C) \equiv 4\sin 2x$$
$$\Rightarrow \quad C + 2D = 1 \qquad\qquad (1)$$
$$\cos 2x(-4D - 8C + 8D) \equiv 0$$
$$\Rightarrow \quad 4D - 8C = 0 \quad \Rightarrow \quad D = 2C \qquad (2)$$

From equation 1: $C + 4C = 1$
$\Rightarrow \quad C = \tfrac{1}{5}, \quad D = \tfrac{2}{5}$ $\boxed{\text{2.1b and 2.2b}}$
The PI is
$y_2 = \tfrac{1}{5}(\sin 2x + 2\cos 2x)$
The general solution is:

$$y = \tfrac{1}{5}(\sin 2x + 2\cos 2x) + e^{2x}(A\sin 2x + B\cos 2x)$$

3 Obtain the solution of

$$\frac{dy}{dx} + y\tan x = e^{2x}\cos x \quad \left(0 \leqslant x < \frac{\pi}{2}\right)$$

for which $y = 2$ at $x = 0$, giving your answer in the form $y = f(x)$ (L, 1993)

• *This first order d.e. requires an integrating factor of* $e^{\int \tan x \, dx}$

$$e^{\int \tan x \, dx} = e^{\ln \sec x} = \sec x$$

Since $0 \leqslant x < \dfrac{\pi}{2}$, *modulus signs are not needed*

The d.e. becomes:

$$\sec x \frac{dy}{dx} + y\sec x \tan x = e^{2x}$$

Integrating:　　　$y \sec x = \frac{1}{2}e^{2x} + C$

$y = 2, x = 0$:　　　$2 = \frac{1}{2} + C \implies C = \frac{3}{2}$

$$\therefore y = \frac{\cos x}{2}(e^{2x} + 3)$$

4 (a) Solve the differential equation

$$\frac{d^2y}{dx^2} - 6\frac{dy}{dx} + 9y = e^{5x}$$

given that $y = 0 = \dfrac{dy}{dx}$ when $x = 0$.

(b) (i) Given that x and v are positive, obtain the general solution of the differential equation

$$x\frac{dv}{dx} + 2v = x$$

(ii) If $v = \dfrac{dx}{dt}$, express x as a function of t,

given that $x = 1$ and $v = \frac{1}{3}$ when $t = 1$.

(NICCEA, 1994)

• (a) $\dfrac{d^2y}{dx^2} - 6\dfrac{dy}{dx} + 9y = e^{5x}$

The auxiliary equation is $m^2 - 6m + 9 = 0$
　　$\implies m = 3$ (repeated).
Hence the CF is

$y_1 = e^{3x}(Ax + B)$

The PI is

$y_2 = Ce^{5x}$

$$\frac{dy}{dx} = 5Ce^{5x} \qquad \frac{d^2y}{dx^2} = 25Ce^{5x}$$

Substituting into the d.e.:
$Ce^{5x}(25 - 30 + 9) \equiv e^{5x} \implies C = \frac{1}{4}$

Hence the general solution is

$y = e^{3x}(Ax + B) + \frac{1}{4}e^{5x}$

$x = 0, y = 0$:　　$0 = B + \frac{1}{4} \implies B = -\frac{1}{4}$

$x = 0, \dfrac{dy}{dx} = 0$:　$\dfrac{dy}{dx} = e^{3x}(A + 3Ax + 3B) + \frac{5}{4}e^{5x}$

　　　　$0 = A + 3B + \frac{5}{4} \implies A = -\frac{1}{2}$

The required solution is $y = \frac{1}{4}(e^{5x} - e^{3x}(2x + 1))$

(b) (i) $x\dfrac{dv}{dx} + 2v = x$

Divide by x:　$\dfrac{dv}{dx} + \dfrac{2}{x}v = 1$　　　(1)

Integrating factor is $e^{\int (2/x)\,dx} = e^{2\ln x} = x^2$

Multiplying by x^2:　$x^2\dfrac{dv}{dx} + 2xv = x^2$

Integrating:　　$x^2v = \dfrac{x^3}{3} + C$　　　(2)

This is the general solution
When $x = 1$, $v = \frac{1}{3}$　$\therefore C = 0$

$$\therefore x^2v = \frac{x^3}{3} \quad \text{or} \quad v = \frac{x}{3}$$

(ii) If $v = \dfrac{dx}{dt}$ then $\dfrac{dx}{dt} = \dfrac{x}{3}$

$$\int \frac{3}{x}\,dx = \int dt, \quad x = 1 \text{ when } t = 1$$

Use these given conditions to express the integrals as definite integrals – it usually helps to simplify the constants of integration

$$\int_1^x \frac{3}{x}\,dx = \int_1^t 1\,dt$$

$[3\ln x]_1^x = [t]_1^t$ 　　　| 6.1 |
$\therefore \ln x - \ln 1 = \frac{1}{3}(t - 1)$
　　　$x = e^{\frac{1}{3}(t-1)}$

5 A curve has the equation $y = (2x + 1)e^{-2x}$.

(a) Find $\dfrac{dy}{dx}$ and show that $\dfrac{d^2y}{dx^2} + 4\dfrac{dy}{dx} + 4y = 0$.

(b) Calculate the coordinates of the turning point of the curve and determine its nature.

(OLE, 1992)

• (a) Differentiate $y = (2x + 1)e^{-2x}$:

$$\frac{dy}{dx} = 2e^{-2x} - 2(2x + 1)e^{-2x}$$

$$= 2e^{-2x}(1 - 2x - 1) \qquad | 5.1 |$$
$$= -4xe^{-2x} \qquad (1)$$

For repeated differentiation it is often easier with terms such as e^{-2x} to multiply throughout by e^{2x} before differentiating

$e^{2x}y = 2x + 1$

$e^{2x}\dfrac{dy}{dx} + 2e^{2x}y = 2$

$e^{2x}\dfrac{d^2y}{dx^2} + 2e^{2x}\dfrac{dy}{dx} + 2e^{2x}\dfrac{dy}{dx} + 4e^{2x}y = 0$

Divide by e^{2x}:

$$\frac{d^2y}{dx^2} + 4\frac{dy}{dx} + 4y = 0 \qquad (2)$$

(b) $\dfrac{dy}{dx} = 0$ when $x = 0 \Rightarrow y = 1$.

In equation 2, $\dfrac{d^2y}{dx^2} = -4$.

Therefore the turning point $(0, 1)$ is a maximum point.

6 The motion of a piece of apparatus in a laboratory experiment can be modelled by the differential equation

$$\frac{d^2x}{dt^2} + k\frac{dx}{dt} + 4x = 0$$

where k is a parameter whose value can be adjusted, x represents a displacement and t represents time. The initial conditions for the experiment are

$x = 0$ and $\dfrac{dx}{dt} = k$ when $t = 0$.

(i) Given that $k = 2$, find x in terms of t, and state the time interval between successive occasions when $x = 0$.
(ii) Given that $k = 5$, show that x is never zero for $t > 0$.   (UCLES Specimen Paper, 1994)

• (i) $\dfrac{d^2x}{dt^2} + 2\dfrac{dx}{dt} + 4x = 0$

Auxiliary equation is
$m^2 + 2m + 4 = 0 \Rightarrow m = -1 \pm \sqrt{3}\text{i}.$

2.2a

General solution is

$x = \text{e}^{-t}(A \sin \sqrt{3}t + B \cos \sqrt{3}t)$ (1)

Initial condition: $x = 0$ when $t = 0 \Rightarrow B = 0$

$\therefore x = A\text{e}^{-t} \sin \sqrt{3}t$

$\dfrac{dx}{dt} = A\text{e}^{-t}(-\sin \sqrt{3}t + \sqrt{3} \cos \sqrt{3}t)$

Initial condition $\dfrac{dx}{dt} = 2$ when $t = 0$

$\Rightarrow A\sqrt{3} = 2 \Rightarrow A = \dfrac{2}{\sqrt{3}}$

$\therefore x = \dfrac{2}{\sqrt{3}} \text{e}^{-t} \sin \sqrt{3}t$

For $x = 0$, $\sqrt{3}t = n\pi \Rightarrow t = \dfrac{n\pi}{\sqrt{3}}$

Thus the time interval between successive occasions when $x = 0$ is $\dfrac{\pi}{\sqrt{3}}$.

(ii) When $k = 5$ the auxiliary equation is
$m^2 + 5m + 4 = 0 \Rightarrow m = -1$ or -4.

4

The general solution is

$x = A\text{e}^{-t} + B\text{e}^{-4t}$

$t = 0, x = 0:$ $0 = A + B$ (2)

$t = 0, \dfrac{dx}{dt} = 5:$ $5 = -A - 4B$ (3)

Adding equations 2 and 3:

$5 = -3B \Rightarrow B = -\tfrac{5}{3}, A = \tfrac{5}{3}$

$\therefore x = \tfrac{5}{3}(\text{e}^{-t} - \text{e}^{-4t}) = \tfrac{5}{3}\text{e}^{-t}(1 - \text{e}^{-3t})$

For $t > 0$, $\text{e}^{-t} > 0$ and $0 < \text{e}^{-3t} < 1$

$\therefore x > 0$ for all $t > 0$

7 The rate at which a rumour spreads in a crowd of people can be modelled with the differential equation

$$\frac{dn}{dt} = 0.0002n(N - n)$$

where N is the number in the crowd and n the number who have heard the rumour at time t minutes. There are 5000 people in the crowd and initially 10 of them have heard the rumour.

(a) Copy and complete the table below for a step-by-step solution of the differential equation and estimate the number of people who have heard the rumour after 10 minutes.

n	t	$\dfrac{dn}{dt}$	dt	dn
10	0		5	
	5		5	
	10			

(b) State a means by which your estimate could be improved.
(c) Find the rate of spread of the rumour when (i) $n = 2000$, (ii) $n = 4000$.
(d) State what happens to $\dfrac{dn}{dt}$ as n approaches 5000.
(e) Using your answers to (c) and (d), sketch a graph to show how the number of people who have heard the rumour changes with time.
(NEAB SMP, 1993)

(a)

		(i)	(ii)	
n	t	$\dfrac{dn}{dt}$	dt	dn
10	0	9.98	5	50
60	5	59.3	5	296
356	10			

(iii) [row label at 60]

(i) use $\dfrac{dn}{dt} = 0.0002n(N - n)$

(ii) multiply $\dfrac{dn}{dt}$ by dt

(iii) add dn to previous value of n

(b) Reduce dt to 1 (say).
The increments in t are too large to give a reasonable answer. Try it again with increments of 1

(c) (i) $n = 2000, \dfrac{dn}{dt} = 1200$

(ii) $n = 4000, \dfrac{dn}{dt} = 800$

(d) $\dfrac{dn}{dt} \to 0$ as $n \to 5000$

(e)

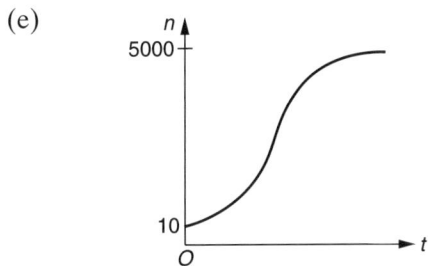

8 (a) Find the general solution of the differential equation

$$\frac{dy}{dx} = (1 - y)^2$$

expressing y in terms of x.

Sketch the solution curve for which $y = 0$ when $x = 0$.

(b) Find the solution of the differential equation

$$\frac{d^2y}{dx^2} - 2\frac{dy}{dx} = e^{-3x}$$

given that $y = 0$ and $\dfrac{dy}{dx} = 0$ when $x = 0$.

(UCLES, 1993)

• (a) $\dfrac{dy}{dx} = (1 - y)^2 \Rightarrow \displaystyle\int \frac{1}{(1-y)^2}\, dy = \int dx$

$\Rightarrow \dfrac{1}{1 - y} = x + C$ $\boxed{6.1}$

$\Rightarrow 1 - y = \dfrac{1}{x + C}$

$\Rightarrow y = 1 - \dfrac{1}{x + C} = \dfrac{x + C - 1}{x + C}$

The general solution is $y = \dfrac{x + C - 1}{x + C}$.

When $x = 0, y = 0 \Rightarrow C = 1$.

The particular solution is $y = \dfrac{x}{x + 1}$.

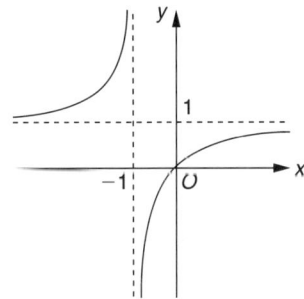

(b) $\dfrac{d^2y}{dx^2} - 2\dfrac{dy}{dx} = e^{-3x}$

Auxiliary equation is $m^2 - 2m = 0$
$\Rightarrow m = 0$ or 2.
∴ the complementary function is $y_1 = Ae^{2x} + B$.
The particular integral is

$$y_2 = Ce^{-3x} \quad \frac{dy_2}{dx} = -3Ce^{-3x} \quad \frac{d^2y_2}{dx^2} = 9Ce^{-3x}$$

Substituting into the d.e.:
$(9C + 6C)e^{-3x} \equiv e^{-3x}$
∴ $15C = 1 \Rightarrow C = \frac{1}{15}$

The general solution of the d.e. is

$y = Ae^{2x} + B + \frac{1}{15}e^{-3x}$

$y = 0, x = 0 \qquad\qquad \Rightarrow\quad 0 = A + B + \frac{1}{15}$

$\dfrac{dy}{dx} = 0, x = 0 \qquad \Rightarrow\quad 0 = 2A - \frac{1}{5}$

∴ $A = \frac{1}{10}, B = -\frac{1}{6}$

∴ $y = \frac{1}{10}e^{2x} + \frac{1}{15}e^{-3x} - \frac{1}{6}$

9 The spread of a disease in a large population can be modelled by means of a differential equation. The proportion x of the population infected with the disease after t days satisfies

$$\frac{dx}{dt} = \tfrac{1}{2}x - \tfrac{1}{2}x^2 \quad \text{for } t \geqslant 0$$

(i) Given that $x = \frac{1}{500}$ when $t = 0$, find x in terms of t.

(ii) Verify that about 6% of the population was infected after seven days.

(iii) How long will it take for 25% of the population to become infected?

(SEB, 1994)

• (i) $\dfrac{\mathrm{d}x}{\mathrm{d}t} = \frac{1}{2}x - \frac{1}{2}x^2 = \dfrac{x - x^2}{2}$

$\therefore \dfrac{2}{x - x^2}\dfrac{\mathrm{d}x}{\mathrm{d}t} = 1$

$\dfrac{2}{x(1 - x)} = \dfrac{2}{x} + \dfrac{2}{1 - x}$ *Cover over method*

$\boxed{4.3}$

$\therefore \displaystyle\int_{1/500}^{x}\left(\dfrac{2}{x} + \dfrac{2}{1 - x}\right)\mathrm{d}x = \int_{0}^{t} t\,\mathrm{d}t$

$\therefore \left[2\ln x - 2\ln(1 - x)\right]_{1/500}^{x} = t$ $\boxed{6.1}$

$2\ln\left(\dfrac{x}{1 - x}\right) - 2\ln\left(\dfrac{1}{499}\right) = t$

$\ln\left(\dfrac{499x}{1 - x}\right) = \dfrac{t}{2}$

$\dfrac{499x}{1 - x} = \mathrm{e}^{t/2}$

$499x = \mathrm{e}^{t/2} - x\mathrm{e}^{t/2}$

$\therefore x = \dfrac{\mathrm{e}^{t/2}}{499 + \mathrm{e}^{t/2}}$

(ii) When $t = 7$, $x = 0.0622$.
That is, 6% of the population was infected after seven days.

(iii) When $x = 0.25$

$\mathrm{e}^{t/2} = \dfrac{499(0.25)}{0.75}$

$\qquad = 166.3$
$\quad t = 10.2$ days

About 25% of the population was infected after 10 days.

Exercises

1 (a) Solve the differential equation

$\dfrac{\mathrm{d}^2x}{\mathrm{d}t^2} + 5\dfrac{\mathrm{d}x}{\mathrm{d}t} + 6x = \mathrm{e}^{2t}$

(b) The variables u and x are related by

$x\dfrac{\mathrm{d}u}{\mathrm{d}x} = 3(4 + u^2x^6 - u)$

By making the substitution $y = ux^3$, show that

$\dfrac{\mathrm{d}y}{\mathrm{d}x} = 3x^2(4 + y^2)$

Solve this equation and hence find u in terms of x given that $u = 0$ when $x = 1$. (UCLES, 1992)

2 Solve the differential equation

$x\dfrac{\mathrm{d}y}{\mathrm{d}x} - y = x^3 + 1$

given that $y = 1$ when $x = 1$.   (NICCEA, 1993)

3 A bottle is shaped so that when the depth of water is x cm, the volume of water in the bottle is $(x^2 + 4x)$ cm^3, $x \geqslant 0$. Water is poured into the bottle so that at time t s after pouring commences, the depth of water is x cm and the rate of increase of the volume of the water is $(x^2 + 25)$ cm^3 s^{-1}.

(a) Show that $\dfrac{\mathrm{d}x}{\mathrm{d}t} = \dfrac{x^2 + 25}{2x + 4}$.

(b) Given that the bottle was empty at $t = 0$, solve this differential equation to obtain t in terms of x.   (L, 1994)

4 If $x < 1$ solve the differential equation

$(1 - x)\dfrac{\mathrm{d}y}{\mathrm{d}x} = (1 + x)y$

given that $y = 2$ when $x = 0$.   (NICCEA, 1994)

5 Determine the solution of the differential equation

$\dfrac{\mathrm{d}^2y}{\mathrm{d}x^2} - 4y = 16$

satisfying $y = 1$ and $\dfrac{\mathrm{d}y}{\mathrm{d}x} = 2$ when $x = 0$.

Show that the series expansion of the solution is

$y = 1 + 6x + 10x^2 + cx^3 + \ldots$

and state the value of the constant c.

6 (a) Solve the differential equation $\dfrac{\mathrm{d}y}{\mathrm{d}x} = 2xy$ and sketch one of the solution curves, other than $y = 0$.

(b) The differential equation

$\dfrac{\mathrm{d}^2x}{\mathrm{d}t^2} + 5\dfrac{\mathrm{d}x}{\mathrm{d}t} + 6x = C\mathrm{e}^{-t}$

where C is a constant, has a particular integral

$x = 4\mathrm{e}^{-t}.$

(i) Find the value of C.
(ii) Find the solution of the differential

equation for which $x = 1$ and $\dfrac{dx}{dt} = 2$ when $t = 0$.

(UCLES, 1991)

7 Given that $x = e^t$, show that

$$\frac{d^2y}{dx^2} = e^{-2t}\left(\frac{d^2y}{dt^2} - \frac{dy}{dt}\right)$$

Hence, show that the substitution $x = e^t$ transforms the d.e.

$$x^2\frac{d^2y}{dx^2} - 4x\frac{dy}{dx} + 6y = 3 \qquad \text{(i)}$$

into

$$\frac{d^2y}{dt^2} - 5\frac{dy}{dt} + 6y = 3 \qquad \text{(ii)}$$

Hence find the general solution of the differential equation (i). (L Specimen Paper, 1996)

8 Determine the values of the constants a, b and c for which $a + b \sin 3\theta + \cos 3\theta$ is a particular integral of the differential equation

$$\frac{d^2r}{d\theta^2} + 10\frac{dr}{d\theta} + 9r = 9 - 30 \sin 3\theta$$

Hence find the general solution of the differential equation. (AEB, 1993)

9 During a period of acceleration, the motion of a canal barge may be modelled by the differential equation

$$10\,000\,\frac{dv}{dt} = \frac{1}{v}(6250 - kv^2)$$

where v is the speed in m s^{-1}, t is the time in seconds and k is a constant.

The terminal speed of the barge is predicted to be 2.5 m s^{-1} in these circumstances.
(i) Find the value of k.
(ii) Solve the differential equation for t in terms of v given that $v = 1$ when $t = 0$.
(iii) How long does it take for the barge to accelerate from 1 m s^{-1} to 2 m s^{-2}?
(iv) By rearranging your expression in (ii) show that the distance travelled by the barge between $t = 0$ and $t = T$ is given by

$$x = \int_0^T \tfrac{1}{2}\sqrt{25 - 21e^{-t/5}}\,dt \qquad \text{(MEI, 1993)}$$

10 Find the solution of $\dfrac{d^2p}{dt^2} + 2\dfrac{dp}{dt} + 17p = 34$ such that $p = 2.5$ and $\dfrac{dp}{dt} = 0$ when $t = 0$.

In an economic model the demand D (the amount required) for a particular item is given by

$$D = 6\frac{d^2p}{dt^2} + 7\frac{dp}{dt} + 6p + 16$$

where p is the price in pounds of the item at time t months.

Similarly the supply S (the amount available) is given by

$$S = 7\frac{d^2p}{dt^2} + 9\frac{dp}{dt} + 23p - 18$$

Given that $S = D$ and that, at $t = 0$, $p = 2.5$ and

$$\frac{dp}{dt} = 0 \text{ find}$$

(i) the price after a long time
(ii) the lowest price at which the item can be bought and the time when this would be possible. (WJEC, 1992)

11 Find the general solution of the differential equation

$$\frac{dy}{dx} + 2y = e^{-2x}$$

giving y explicitly in terms of x in your answer.

Find also the particular solution for which $y = 1$ when $x = 0$. (UCLES Specimen Paper)

12 Obtain the general solution of the differential equation

$$\frac{d^2y}{dx^2} + 4\frac{dy}{dx} + 13y = e^{-3x} \qquad \text{(L, 1994)}$$

13 The current I amperes in an electronic circuit (consisting of inductance and resistance in series with a sinusoidal power source) is given by the differential equation

$$\frac{dI}{dt} + 2I = \sin 3t$$

where t is elapsed time in seconds.
Find
(i) the complementary function
(ii) the particular integral
(iii) the current I as a function of time, given that initially $I = 0$
(iv) the amplitude of the oscillations after a long time has elapsed. (MEI, 1993)

14 Taking steps $\delta x = 0.2$ over the interval $1 < x < 2$, estimate the solution of the equation $\dfrac{dy}{dx} = \dfrac{1}{x}$ which passes through the point $(1, 1)$, when $x = 2$.

By integrating the differential equation find the percentage error in the value of y at $x = 2$.

Brief Solutions

1 (a) Auxiliary equation
$$m^2 + 5m + 6 = 0 \quad \Rightarrow \quad m = -2 \text{ or } -3$$

$\boxed{2.2a}$

CF: $x_1 = Ae^{-2t} + Be^{-3t}$
PI: $x_2 = \frac{1}{20}e^{2t}$
General solution: $x = \frac{1}{20}e^{2t} + Ae^{-2t} + Be^{-3t}$

(b) $x\dfrac{\mathrm{d}u}{\mathrm{d}x} = 3(4 + u^2x^6 - u)$

$y = ux^3 \quad \Rightarrow \quad \dfrac{\mathrm{d}y}{\mathrm{d}x} = x^3\dfrac{\mathrm{d}u}{\mathrm{d}x} + 3x^2u$

$x^3\dfrac{\mathrm{d}u}{\mathrm{d}x} = 3x^2(4 + u^2x^6 - u)$

$\dfrac{\mathrm{d}y}{\mathrm{d}x} - 3x^2u = 3x^2(4 + y^2 - u)$

$\Rightarrow \quad \dfrac{\mathrm{d}y}{\mathrm{d}x} = 3x^2(4 + y^2)$

Separate the variables and integrate $\boxed{6.1}$

$\frac{1}{2}\tan^{-1}\dfrac{y}{2} = x^3 - 1 \quad \Rightarrow \quad \dfrac{y}{2} = \tan 2(x^3 - 1)$

$\Rightarrow \quad u = \dfrac{2}{x^3}\tan(2x^3 - 2)$

2 Use the integrating factor $e^{\int -\frac{1}{x}\,\mathrm{d}x} = e^{-\log_e x} = \dfrac{1}{x}$

$\dfrac{\mathrm{d}y}{\mathrm{d}x} - y\left(\dfrac{1}{x}\right) = x^2 + \dfrac{1}{x} \quad \Rightarrow \quad \dfrac{1}{x}\dfrac{\mathrm{d}y}{\mathrm{d}x} - y\dfrac{1}{x^2} = x + \dfrac{1}{x^2}$

Integrating

$\dfrac{y}{x} = \dfrac{x^2}{2} - \dfrac{1}{x} + \dfrac{3}{2} \text{ or } y = \dfrac{x^3}{2} + \dfrac{3}{2}x - 1$

3 $V = x^2 + 4x \quad \dfrac{\mathrm{d}V}{\mathrm{d}x} = 2x + 4 \quad \dfrac{\mathrm{d}V}{\mathrm{d}t} = x^2 + 25$

(a) $\dfrac{\mathrm{d}x}{\mathrm{d}t} = \dfrac{\mathrm{d}x}{\mathrm{d}V}\dfrac{\mathrm{d}V}{\mathrm{d}t} = \dfrac{x^2 + 25}{2x + 4}$

(b) $\Rightarrow \quad \displaystyle\int_0^x \dfrac{2x + 4}{x^2 + 25}\,\mathrm{d}x = \int_0^t 1\,\mathrm{d}t$

Split $\dfrac{2x + 4}{x^2 + 25}$ into $\dfrac{2x}{x^2 + 25} + \dfrac{4}{x^2 + 25}$ (which are

standard integrals)

$\left[\ln(x^2 + 25) + \frac{4}{5}\tan^{-1}\dfrac{x}{5}\right]_0^x = t$

$= \ln\left(\dfrac{x^2 + 25}{25}\right) + \frac{4}{5}\tan^{-1}\dfrac{x}{5}$

4 $\displaystyle\int_2^y \dfrac{1}{y}\,\mathrm{d}y = \int_0^x \dfrac{1 + x}{1 - x}\,\mathrm{d}x = \int_0^x \dfrac{2}{1 - x} - 1\,\mathrm{d}x$

$\boxed{6.1}$

$\ln\dfrac{y}{2} + 2\ln(1 - x) = -x \quad \Rightarrow \quad \dfrac{y(1 - x)^2}{2} = e^{-x}$

$y = \dfrac{2e^{-x}}{(1 - x)^2}$

5 CF: $\quad y_1 = Ae^{2x} + Be^{-2x}$
PI: $\quad y_2 = -4$
General solution: $\quad y = Ae^{2x} + Be^{-2x} - 4$
Substituting given conditions gives
$y = 4e^{2x} + e^{-2x} - 4$
Expanding:

$y = 4\left(1 + 2x + \dfrac{4x^2}{2!} + \dfrac{8x^3}{3!}\cdots\right)$

$+ \left(1 - 2x + \dfrac{4x^2}{2!} - \dfrac{8x^3}{3!}\cdots\right) - 4$

$= 1 + 6x + 10x^2 + 4x^3 + \ldots \quad c = 4$

6 (a) $\displaystyle\int \dfrac{1}{y}\,\mathrm{d}y = \int 2x\,\mathrm{d}x \quad \Rightarrow \quad \ln|y| = x^2 + C$

$\Rightarrow \quad |y| = e^{x^2 + C} = Ae^{x^2}$
$\Rightarrow \quad y = Ae^{x^2}$

(b) Auxiliary equation is $m^2 + 5m + 6 = 0$

$\boxed{2.2a}$

(i) CF: $x_1 = Ae^{-2t} + Be^{-3t}$
PI: $x_2 = 4e^{-t} \quad \Rightarrow \quad C = 8$

(ii) $x = Ae^{-2t} + Be^{-3t} + 4e^{-t}$

Substituting the given conditions gives
$A = -3, B = 0 \quad \Rightarrow \quad x = 4e^{-t} - 3e^{-2t}$

7 $x = e^t \quad \Rightarrow \quad \dfrac{\mathrm{d}x}{\mathrm{d}t} = e^t, \dfrac{\mathrm{d}y}{\mathrm{d}x} = \dfrac{\mathrm{d}y}{\mathrm{d}t}\dfrac{\mathrm{d}t}{\mathrm{d}x} = e^{-t}\dfrac{\mathrm{d}y}{\mathrm{d}t}$

$\dfrac{\mathrm{d}}{\mathrm{d}x}\left(\dfrac{\mathrm{d}y}{\mathrm{d}x}\right) = \dfrac{\mathrm{d}}{\mathrm{d}t}\left(e^{-t}\dfrac{\mathrm{d}y}{\mathrm{d}t}\right)\dfrac{\mathrm{d}t}{\mathrm{d}x}$

$= \left(-e^{-t}\dfrac{\mathrm{d}y}{\mathrm{d}t} + e^{-t}\dfrac{\mathrm{d}^2y}{\mathrm{d}t^2}\right)e^{-t} = e^{-2t}\left(\dfrac{\mathrm{d}^2y}{\mathrm{d}t^2} - \dfrac{\mathrm{d}y}{\mathrm{d}t}\right)$

Hence

$x^2\dfrac{\mathrm{d}^2y}{\mathrm{d}x^2} - 4x\dfrac{\mathrm{d}y}{\mathrm{d}x} + 6y = 3$

$$\Rightarrow \quad \frac{d^2y}{dt^2} - \frac{dy}{dt} - 4\frac{dy}{dt} + 6y = 3 \qquad \boxed{2.2a}$$

Auxiliary equation is $m^2 - 5m + 6 = 0$
$\Rightarrow \quad m = 2$ or 3.
CF: $\quad y_1 = Ae^{2t} + Be^{3t}$
PI: $\quad y_2 = \frac{1}{2}$
General solution:
$y = Ae^{2t} + Be^{3t} + \frac{1}{2} = Ax^2 + Bx^3 + \frac{1}{2}$

8 $\quad r = a + b\sin 3\theta + c\cos 3\theta$

$$\frac{dr}{d\theta} = 3b\cos 3\theta - 3c\sin 3\theta$$

$$\frac{d^2r}{d\theta^2} = -9b\sin 3\theta - 9c\cos 3\theta$$

Substituting:

$$\frac{d^2r}{d\theta^2} + 10\frac{dr}{d\theta} + 9r = 9a - 30c\sin 3\theta + 30b\cos 3\theta$$

$$= 9 - 30\sin 3\theta$$
$$\Rightarrow \quad a = 1, c = 1, b = 0$$

PI is $r_1 = 1 + \cos 3\theta$ $\qquad \boxed{2.2a}$
Auxiliary equation is
$m^2 + 10m + 9 = 0 \quad \Rightarrow \quad m = -9$ or -1.
General solution is $r = 1 + \cos 3\theta + Ae^{-9\theta} + Be^{-\theta}$

9 (i) $\quad 10\,000\dfrac{dv}{dt} = \dfrac{1}{v}(6250 - kv^2)$

$$\frac{dv}{dt} = 0 \text{ when } v = 2.5 \quad \Rightarrow \quad k = 1000$$

$\boxed{6.1}$

(ii) $\displaystyle \int_1^v \frac{10\,000v}{6250 - 1000v^2}\,dv = \int_0^t dt$

$$\Rightarrow \quad \int_1^v \frac{40v}{25 - 4v^2}\,dv = \left[-5\ln\left(\frac{25 - 4v^2}{21}\right)\right] = t$$

$$\Rightarrow \quad t = 5\ln\left(\frac{21}{25 - 4v^2}\right)$$

(iii) When $v = 2$, $t = 5\ln\frac{21}{9} = 4.24$ s

(iv) From (ii), $e^{t/5} = \dfrac{21}{25 - 4v^2}$

$$\Rightarrow \quad 4v^2 = 25 - 21e^{-t/5}$$
$$v = \tfrac{1}{2}\sqrt{25 - 21e^{-t/5}}$$

But $v = \dfrac{dx}{dt} \quad \Rightarrow \quad x = \displaystyle\int_0^T \tfrac{1}{2}\sqrt{25 - 21e^{-t/5}}\,dt$

10 Auxiliary equation is
$m^2 + 2m + 17 = 0 \quad \Rightarrow \quad m = -1 \pm 4i$
CF: $\quad p_1 = e^{-t}(A\sin 4t + B\cos 4t)$
PI: $\quad p_2 = 2$

General solution: $\quad p = 2 + e^{-t}(A\sin 4t + B\cos 4t)$
Substituting in the d.e. gives $A = \frac{1}{8}$, $B = \frac{1}{2}$

$$p = 2 + \frac{e^{-t}}{8}(\sin 4t + 4\cos 4t)$$

When $S = D$, $S - D = 0$ which gives the d.e. above.
(i) The price after a long time is £2.

(ii) $\dfrac{dp}{dt} = \dfrac{e^{-t}}{8}(-\sin 4t - 4\cos 4t + 4\cos 4t$

$$- 16\sin 4t)$$

$$= \frac{e^{-t}}{8}(-17\sin 4t)$$

$$\frac{dp}{dt} = 0 \text{ when } \sin 4t = 0 \quad \Rightarrow \quad t = n\pi/4$$

$$\Rightarrow \quad p = 2 + \frac{e^{-n\pi/4}}{2}\cos n\pi$$

When n is odd, $p = 2 - \dfrac{e^{-n\pi/4}}{2}$ and $e^{-n\pi/4}$ is

positive and decreasing.

$\therefore$ The minimum price paid is when $t = \dfrac{\pi}{4}$

$$\Rightarrow \quad p = £1.77$$

11 Integrating factor is $e^{\int 2\,dx} = e^{2x}$

$$e^{2x}\frac{dy}{dx} + 2e^{2x}y = 1 \quad \Rightarrow \quad e^{2x}y = x + C$$

$$\therefore y = (x + C)e^{-2x}$$
When $x = 0$, $y = 1 \quad \therefore C = 1 \quad \Rightarrow \quad y = (x + 1)e^{-2x}$

12 Auxiliary equation is
$m^2 + 4m + 13 = 0 \quad \Rightarrow \quad m = -2 \pm 3i$

$\boxed{2.2a}$

CF: $\quad y_1 = e^{-2x}(A\sin 3x + B\cos 3x)$
PI: $\quad y_2 = Ce^{-3x} \quad \Rightarrow \quad 10C = 1 \quad \Rightarrow \quad C = \frac{1}{10}$
General solution:
$y = \frac{1}{10}e^{-3x} + e^{-2x}(A\sin 3x + B\cos 3x)$

13 Auxiliary equation is $m + 2 = 0 \quad \Rightarrow \quad m = -2$.
CF: $\quad Ae^{-2t}$
PI: $\quad B\sin 3t + C\cos 3t \quad \Rightarrow \quad B = \frac{2}{13}, C = -\frac{3}{13}$
General solution:
$I = \frac{1}{13}(2\sin 3t - 3\cos 3t) + Ae^{-2t}$
$t = 0, I = 0 \quad \Rightarrow \quad A = \frac{3}{13}$
$\Rightarrow \quad I = \frac{1}{13}(2\sin 3t - 3\cos 3t + 3e^{-2t})$

After a long time
$e^{-2t} \to 0 \quad \Rightarrow \quad I \approx \frac{1}{13}(\sqrt{13}\sin(3t - \alpha))$ where
$\tan \alpha = \frac{3}{2}$.

Amplitude of I is $\dfrac{\sqrt{13}}{13}$.

14

x	y	δx	$\dfrac{\delta y}{\delta x}$	δy
1	1	0.2	1	0.2
1.2	1.2	0.2	0.833	0.167
1.4	1.367	0.2	0.714	0.143
1.6	1.510	0.2	0.625	0.125
1.8	1.635	0.2	0.556	0.111
2.0	1.746			

$$\frac{dy}{dx} = \frac{1}{x} \quad \Rightarrow \quad y = \ln x + 1$$

When $x = 2$, $y = 1.693$

Error is $1.746 - 1.693 = 0.053$

% error = 3.1%

13

Miscellaneous pure topics

AS Level			A Level			Topic	Date attempted	Date completed	Self-assessment
CORE	MODULAR	TRADITIONAL	CORE	MODULAR	TRADITIONAL				
				✓		**Polar coordinates**			
		✓		✓	✓	**Complex numbers**			
			✓	✓	✓	**Iterative methods**			
			✓	✓	✓	**Numerical integration**			
				✓	✓	**Volumes of revolution**			
					✓	**Matrices**			
				✓	✓	**Vector planes**			

Fact Sheet

Polar Coordinates (r, θ)

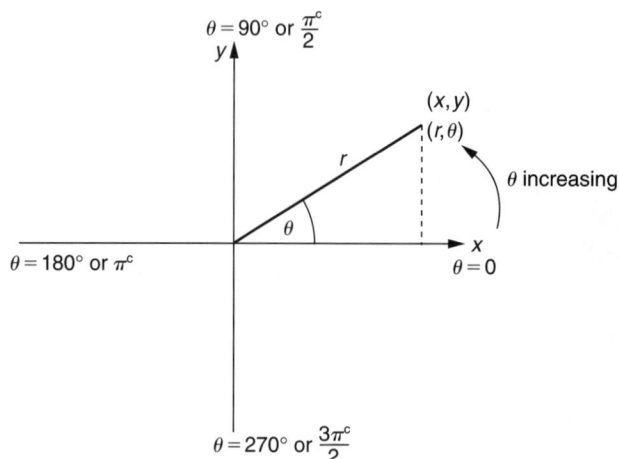

$x = r \cos \theta$

$y = r \sin \theta$

$x^2 + y^2 = r^2$

$\tan \theta = \dfrac{y}{x}$

Polar Equations of Curves $r = \mathrm{f}(\theta)$

(i) $\mathrm{f}(\theta)$ is a function of $\sin \theta$ – the curve is symmetrical about the $\theta = 90°$ and 270° lines.

(ii) $\mathrm{f}(\theta)$ is a function of $\cos \theta$ – the curve is symmetrical about the $\theta = 0°$, 180° lines.

(iii) If $\mathrm{f}(\theta)$ is negative the point is plotted as $(\theta + 180°, -\mathrm{f}(\theta))$.

(iv) Angles should be measured in *radians* but think as degrees for plotting.

(v) $r = k\theta$ or $r = \mathrm{e}^{k\theta}$ are both spirals; θ *must* be in radians.

Polar → Cartesian Equations

Use the conditions above.

Example

$r = \sin 2\theta$ ($= 2 \sin \theta \cos \theta$) becomes

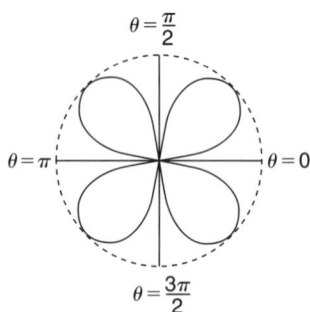

$$\sqrt{x^2 + y^2} = 2 \frac{y}{r} \frac{x}{r} \quad \Rightarrow \quad (x^2 + y^2)^{3/2} = 2xy$$

$$\text{Area} = \tfrac{1}{2}\int r^2 \, \mathrm{d}\theta$$

Complex Numbers

$$\mathrm{i} = \sqrt{(-1)} \quad \text{or} \quad j = \sqrt{(-1)}$$

- $\mathrm{i}^2 = -1 \quad \mathrm{i}^3 = -\mathrm{i} \quad \mathrm{i}^4 = 1 \quad \dfrac{1}{\mathrm{i}} = -\mathrm{i}$

- $\mathrm{i}^{4n} = 1 \quad \mathrm{i}^{4n+1} = \mathrm{i} \quad \mathrm{i}^{4n+2} = -1 \quad \mathrm{i}^{4n+3} = -\mathrm{i}$ for all integer values of n with similar results for j

Real and Imaginary Numbers

- If $z = x + iy$ where x and y are real numbers:
  $x = R(z)$ or $Re(z)$: the real part of z
  $y = I(z)$ or $Im(z)$: the imaginary part of z

Complex Conjugate

- If $z = x + iy$, $\bar{z}$ (or $z*$) $= x - iy$ is the complex conjugate of z.

Algebra

- $(a + ib) + (c + id) = (a + c) + i(b + d)$
  $z + z* = 2x = 2Re(z)$
- $(a + ib) - (c + id) = (a - c) + i(b - d)$
  $z - z* = 2iy = 2Im(z)$
- $(a + ib)(c + id) = (ac - bd) + i(bc + ad)$
  $zz* = x^2 + y^2$ (real)
- $\dfrac{a + ib}{c + id} = \dfrac{(a + ib)(c - id)}{(c + id)(c - id)} = \dfrac{(ac + bd)}{(c^2 + d^2)} + \dfrac{i(bc - ad)}{(c^2 + d^2)}$

Argand Diagram

- $z = x + iy$ can be regarded as an ordered pair (x, y) and can be plotted on cartesian axes (called an Argand diagram).
- From the sketch $x = r\cos\theta$, $y = r\sin\theta$,

  $z = x + iy = r(\cos\theta + i\sin\theta)$ or r cis θ.

This is the trigonometric form of a complex number.

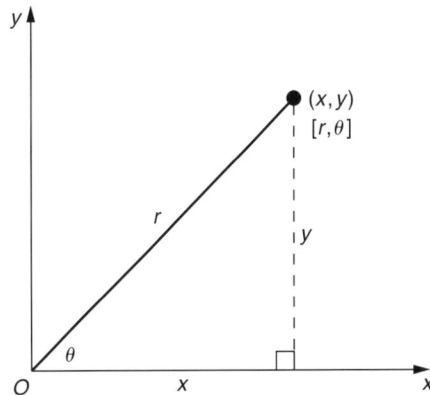

- $r = \sqrt{(x^2 + y^2)}$ is called the *modulus* of z, written $|z| = r = \sqrt{(x^2 + y^2)}$.
- From the previous section it can be seen that $|z|^2 = zz*$.
- $\theta = \arctan\left(\dfrac{y}{x}\right)$ is called the *argument* of z, written arg $z = \theta = \arctan\left(\dfrac{y}{x}\right)$.
- Since arctan is multivalued, the value of θ in the interval $-\pi < \theta \leq \pi$ is called the *principal value*.
- z can also be regarded as the ordered pair $[r, \theta]$ (on an Argand diagram).

Exponential Form

It can be shown that $\cos\theta + i\sin\theta = e^{i\theta}$
so $z = x + iy = r(\cos\theta + i\sin\theta) = re^{i\theta}$.
If $z_1 = r_1e^{i\theta_1}$ and $z_2 = r_2e^{i\theta_2}$ then

(i) $z_1z_2 = r_1r_2e^{i(\theta_1 + \theta_2)}$

Thus, in a product, moduli are multiplied and arguments are added, i.e. multiplication causes an enlargement factor r followed by a rotation θ on an Argand diagram.

(ii) $\dfrac{z_1}{z_2} = \dfrac{r_1}{r_2}\, e^{i(\theta_1 - \theta_2)}$

Thus, in a quotient, moduli are divided and arguments are subtracted.

De Moivre's Theorem

If $z = (\cos\theta + i\sin\theta)$ then $z^n = (\cos\theta + i\sin\theta)^n = \cos n\theta + i\sin n\theta$.
From this $\cos n\theta = \mathrm{Re}[(\cos\theta + i\sin\theta)^n]$

$$= \cos^n\theta - \binom{n}{2}\cos^{n-2}\theta\sin^2\theta + \binom{n}{4}\cos^{n-4}\theta\sin^4\theta\ldots$$

$$\sin n\theta = \mathrm{Im}[(\cos\theta + i\sin\theta)^n]$$

$$= \binom{n}{1}\cos^{n-1}\theta\sin\theta - \binom{n}{3}\cos^{n-3}\theta\sin^3\theta + \ldots$$

Numerical Methods of Integration

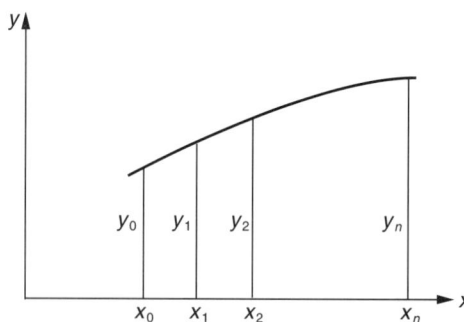

$y_0, y_1, y_2, \ldots, y_n$ are the *ordinates* at $x_0, x_1, \ldots, x_n$ which are equally spaced:

$$x_1 - x_0 = x_2 - x_1 = \ldots = h$$

Trapezium Rule

$$\text{Area} \approx \frac{h}{2}\left[(y_0 + y_n) + 2(y_1 + y_2 + \ldots + y_{n-1})\right]$$

Simpson's Rule

$$\text{Area} \approx \frac{h}{3}\left[(y_0 + y_n) + 4(y_1 + y_3 + y_5 \ldots) + 2(y_2 + y_4 + \ldots)\right]$$

Simpson's rule requires an *odd* number of ordinates, an *even* number of intervals.

Mid Ordinate Rule

$$\text{Area} \approx \sum\left(h \times f\left(\frac{x_r + x_{r+1}}{2}\right)\right) \quad \text{or} \quad h\sum f\left(\frac{x_r + x_{r+1}}{2}\right)$$

i.e. the sum of the areas of rectangles (as shown in the following diagram).

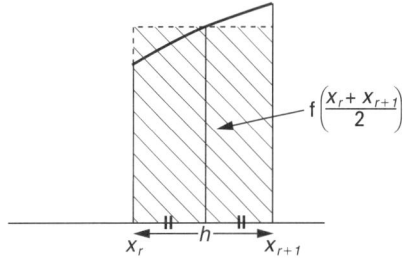

$$f\left(\frac{x_r + x_{r+1}}{2}\right)$$

Numerical Methods of Solving Equations

Newton–Raphson

If x_0 is an approximate root of an equation $y = f(x)$, then $x_1 = x_0 - \dfrac{f(x_0)}{f'(x_0)}$

is a better approximation providing $f(x)$ is a continuous function in the neighbourhood of x_0.

Iterative Methods

(i) $x^3 - 3x^2 + 7 = 0 \qquad x^2(x - 3) = -7$

$$x_i = \sqrt{\frac{-7}{x_{i-1} - 3}}$$

$x_0 = -1, x_1 = -1.323, x_2 = -1.272, x_3 = -1.280, x_4 = -1.279, x_5 = -1.279$

(ii) $x^3 - 3x^2 + 7 = 0 \quad x = \sqrt[3]{3x^2 - 7}$

$x_0 = -1, x_1 = -1.587, x_2 = 0.824, x_3 = -1.706$

These are not converging, therefore this method is unsuitable.

(iii) $x^5 = 10 \quad \Rightarrow \quad x^6 = 10x \quad x = \sqrt[3]{\sqrt{10x}}$

$x_0 = 2, x_1 = 1.648, x_2 = 1.595, x_3 = 1.587, x_4 = 1.585, x_5 = 1.585$

Volumes of Revolution

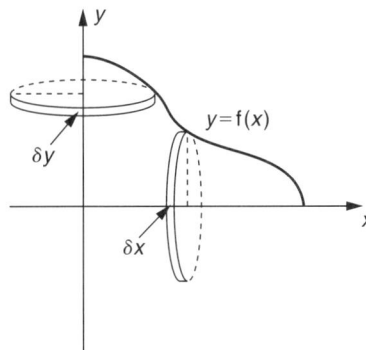

(i) About the x-axis: $V_x = \displaystyle\int_a^b \pi y^2 \, dx$

(ii) About the y-axis: $V_y = \displaystyle\int_a^b \pi x^2 \, dy$

Square Matrices

2×2 matrices are used for transformations of the (x, y) plane.

Standard Transformation Matrices

- $\begin{pmatrix} a & 0 \\ 0 & a \end{pmatrix}$ is an enlargement, factor a.

- $\begin{pmatrix} -1 & 0 \\ 0 & 1 \end{pmatrix}$ is a reflection in the y-axis.

- $\begin{pmatrix} 1 & 0 \\ 0 & -1 \end{pmatrix}$ is a reflection in the x-axis.

- $\begin{pmatrix} \cos\theta & -\sin\theta \\ \sin\theta & \cos\theta \end{pmatrix}$ is an anticlockwise rotation of θ about the origin.

Identity Matrix

The identity matrix is a square matrix with elements on the leading diagonal $= 1$ and all other elements $= 0$.

Inverse Matrix

If $AA^{-1} = I$, then A^{-1} is the inverse matrix of A and A is the inverse matrix of A^{-1}.

Multiplication of Matrices

$$\begin{pmatrix} a & b \\ c & d \end{pmatrix}\begin{pmatrix} x \\ y \end{pmatrix} = \begin{pmatrix} e \\ f \end{pmatrix} \Rightarrow ax + by = e \quad cx + dy = f$$

Multiply the elements in the first row of the left-hand matrix by the corresponding elements in the first column of the right-hand matrix to get the element in the first row first column of the resulting matrix

row i $\begin{pmatrix} x_1 & x_2 & x_3 & \ldots & x_n \end{pmatrix} \times \begin{pmatrix} y_1 \\ y_2 \\ \vdots \\ y_n \end{pmatrix} = \begin{pmatrix} z_{ij} \end{pmatrix}$

column j z_{ij} is the element in *row i, column j*

$$z_{ij} = x_1 y_1 + x_2 y_2 + x_3 y_3 + \ldots + x_n y_n$$

A matrix with n columns can only pre-multiply a matrix with n rows.

If A has p rows and q columns, and B has q rows and r columns, then AB has p rows and r columns.

Eigenvalues of A are the values of λ which make $\text{Det}(A - \lambda I) = 0$. The eigenvector $\begin{bmatrix} X_1 \\ X_2 \end{bmatrix}$ is the non-trivial solution of $[A - \lambda I]\begin{bmatrix} X_1 \\ X_2 \end{bmatrix} = \begin{bmatrix} 0 \\ 0 \end{bmatrix}$

Vector Planes

The Equation of the Plane

(i) A plane is defined by 3 non-collinear points.

If the points A, B and C have position vectors $\mathbf{a}$, $\mathbf{b}$ and $\mathbf{c}$ then the equation of the plane is $\mathbf{r} = \mathbf{a} + \lambda\mathbf{b} + \mu\mathbf{c}$.

(ii) A plane is defined by one point and a direction vector perpendicular to the plane.

If **a** is the position vector of A and **n** is a vector perpendicular to the plane then the equation is $\mathbf{r} \cdot \mathbf{n} = \mathbf{a} \cdot \mathbf{n}$.

If the unit vector $\hat{\mathbf{n}}$ is used instead of **n**, $\mathbf{a} \cdot \hat{\mathbf{n}}$ is the perpendicular distance of the plane from the origin.

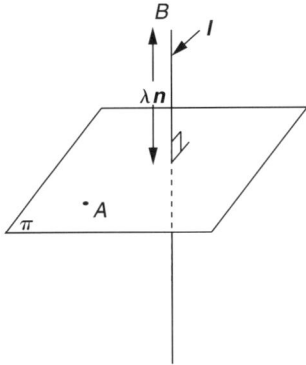

Distance of a Point from a Plane π

The equation of the line l perpendicular to π passing through B (with position vector **b**) is $\mathbf{r} = \mathbf{b} + \lambda\mathbf{n}$.

Substitute the coordinates of a general point on the line l into $\mathbf{r} \cdot \mathbf{n} = \mathbf{a} \cdot \mathbf{n}$ to find the value of λ.

The distance is $|\lambda\mathbf{n}|$.

The Angle Between a Line l_1 and a Plane π

Equation of the line l is $\mathbf{r} = \mathbf{a} + \lambda\mathbf{d}$ where **d** is the direction vector. Equation of the plane π is $\mathbf{r} \cdot \mathbf{n} = $ constant.

$$\sin \theta = \frac{\mathbf{n} \cdot \mathbf{d}}{|\mathbf{n}||\mathbf{d}|}$$

Line of Intersection of Two Planes

Put $x = 0$ into the equations of both planes and solve giving $(0, y_1, z_1)$.
Put $y = 0$ into the equations of both planes and solve giving $(x_2, 0, z_2)$.

The line of intersection passes through these points.

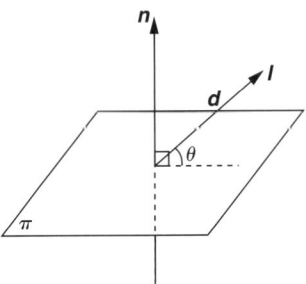

The Angle Between Two Planes

The angle ϕ between two planes is given by

$$\cos \phi = |\hat{\mathbf{n}}_1| \cdot |\hat{\mathbf{n}}_2|$$

Worked Examples

1 (a) (i) Express $12 \cos \theta + 5 \sin \theta$ in the form $A \cos (\theta + \alpha)$, stating clearly the values of A and α.

Solve the equation

$12 \cos \theta + 5 \sin \theta = 7$ for $0° \leqslant \theta \leqslant 360°$

(ii) Sketch the curve with polar equation

$r = 13(1 + \cos \theta)$ for $0° \leqslant \theta \leqslant 360°$

Hence, or otherwise, sketch the curve with polar equation

$r = 13 + 12 \cos \theta + 5 \sin \theta$
for $0° \leqslant \theta \leqslant 360°$

(O&C, 1992)

• (i) $a \cos \theta + b \sin \theta \equiv \sqrt{a^2 + b^2} \cos (\theta - \alpha)$

where $\tan \alpha = \dfrac{b}{a}$

$\boxed{9.2}$

Hence $12 \cos \theta + 5 \sin \theta = 13 \cos (\theta - \alpha)$
where $\tan \alpha = \frac{5}{12} \Rightarrow \alpha = 22.6°, A = 13$

$12 \cos \theta + 5 \sin \theta = 7$
$\Rightarrow 13 \cos (\theta - 22.6°) = 7$
$\cos (\theta - 22.6°) = \frac{7}{13}$
$\therefore \theta - 22.6° = [57.4, -57.4] + 360n$
$\theta = (80°, -34.8°) + 360n$
For $0° \leqslant \theta \leqslant 360°$ $\theta = 80°, 325°$

(ii) $r = 13(1 + \cos \theta)$

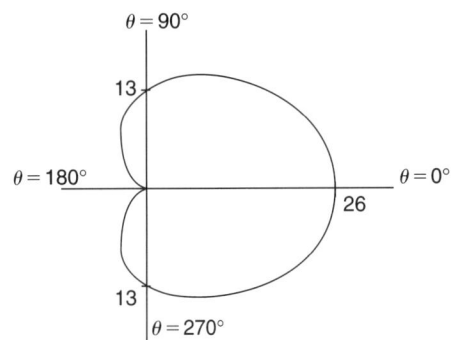

$r = 13(1 + \cos(\theta - 22.6))$
Rotation anticlockwise of 22.6°

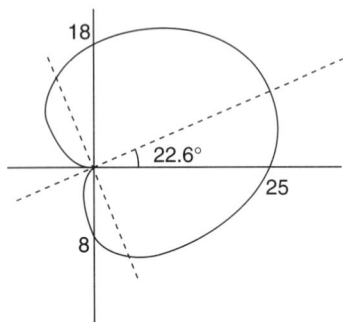

7.1

2 A curve C has polar equation

$$r = a(3 + 2\cos 2\theta) \quad 0 \le \theta < 2\pi$$

where a is a positive constant.

(i) State all the symmetries of C and find the maximum and minimum values of r, stating the values of θ where they are attained. Sketch C.

(ii) Show that the cartesian equation of C is
$(x^2 + y^2)^{3/2} = a(5x^2 + y^2)$.

(iii) Find the area of the sector enclosed by C and

the lines $\theta = 0$ and $\theta = \dfrac{\pi}{4}$.   (UCLES, 1992)

• $r = a(3 + 2\cos 2\theta)$ 7.1

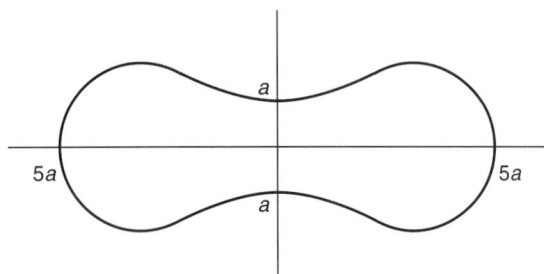

Minimum value of r is a at $\theta = \dfrac{\pi}{2}$ and $\dfrac{3\pi}{2}$.

Maximum value of r is $5a$ at $\theta = 0, \pi$.

(i) C is symmetrical about $\theta = 0$ and $\theta = \dfrac{\pi}{2}$

(reflective) and has rotational symmetry of π radians about the pole.

(ii) Using $x = r\cos\theta$, $y = r\sin\theta$ and $\cos 2\theta = \cos^2\theta - \sin^2\theta$, the cartesian equation is given by:

$$r = a(3 + 2(\cos^2\theta - \sin^2\theta))$$
Multiply by r^2:
$$r^3 = a(3r^2 + 2(r^2\cos^2\theta - r^2\sin^2\theta))$$
$$(x^2 + y^2)^{3/2} = a(3(x^2 + y^2) + 2x^2 - 2y^2))$$
$$(x^2 + y^2)^{3/2} = a(5x^2 + y^2)$$

(iii) Area of the sector

$$= \int_0^{\pi/4} \tfrac{1}{2}r^2 \, d\theta$$

$$= \tfrac{1}{2}a^2 \int_0^{\pi/4} (3 + 2\cos 2\theta)^2 \, d\theta$$

$$= \tfrac{1}{2}a^2 \int_0^{\pi/4} 9 + 12\cos 2\theta + 2(\cos 4\theta + 1) \, d\theta$$

$$= \tfrac{1}{2}a^2 \left[11\theta + 6\sin 2\theta + \tfrac{1}{2}\sin 4\theta \right]_0^{\pi/4}$$

6.1

$$= \tfrac{1}{2}a^2 \left[\frac{11\pi}{4} + 6 \right]$$

3 (a) By comparing real and imaginary parts, find the real numbers x and y which satisfy the equation

$$(1 + 4i)x + (1 + 2i)^2 y + 12 = 0$$

Determine the modulus and argument of the complex number $x + iy$.

(b) Express $\sqrt{3} + i$ in the form $r(\cos\theta + i\sin\theta)$, where $r > 0$ and θ is in radians.

(i) Using de Moivre's theorem, simplify

$$\frac{(\sqrt{3} + i)^{11}}{512} - \frac{128}{(\sqrt{3} + i)^5}$$

(ii) Illustrate, by shading on the Argand diagram, the region R where the complex number z satisfies both

$$5 \le |z| \le 20$$

and $\tfrac{1}{4}\pi \le \arg z \le \tfrac{3}{4}\pi$

Determine for which integer values of n the complex number $(\sqrt{3} + i)^n$ lies in the region R.   (O&C, 1993)

•   (a) $(1 + 4i)x + (1 + 2i)^2 y + 12 = 0$
$\Rightarrow (1 + 4i)x + (1 - 4 + 4i)y + 12 = 0$
Re: $x - 3y + 12 = 0$ (1)
Im: $4x + 4y = 0$ (2)
Hence $x = -y \Rightarrow x = -3, y = 3$

$z = -3 + 3i \Rightarrow |z| = 3\sqrt{2}, \arg z = \dfrac{3\pi}{4}$

(b) $|\sqrt{3} + i| = 2 \Rightarrow \arg(\sqrt{3} + i) = \dfrac{\pi}{6}$

$$\sqrt{3} + i = 2\left(\cos\frac{\pi}{6} + i\sin\frac{\pi}{6}\right)$$

by De Moivre's theorem

(i) $(\sqrt{3} + i)^{11} = 2^{11}\left(\cos\dfrac{11\pi}{6} + i\sin\dfrac{11\pi}{6}\right)$

$$(\sqrt{3} + i)^5 = 2^5\left(\cos\frac{5\pi}{6} + i\sin\frac{5\pi}{6}\right)$$

$$\frac{1}{(\sqrt{3} + i)^5} = \frac{1}{2^5}\left(\cos\frac{-5\pi}{6} + i\sin\frac{-5\pi}{6}\right)$$

$$\frac{(\sqrt{3} + i)^{11}}{512} - \frac{128}{(\sqrt{3} + i)^5}$$

$$= 4\left(\cos\frac{11\pi}{6} + i\sin\frac{11\pi}{6}\right)$$

$$- 4\left(\cos\frac{-5\pi}{6} + i\sin\frac{-5\pi}{6}\right)$$

$$= 4\left(\frac{\sqrt{3}}{2} - \tfrac{1}{2}i\right) - 4\left(-\frac{\sqrt{3}}{2} - \tfrac{1}{2}i\right)$$

$$= 4\sqrt{3}$$

(ii)

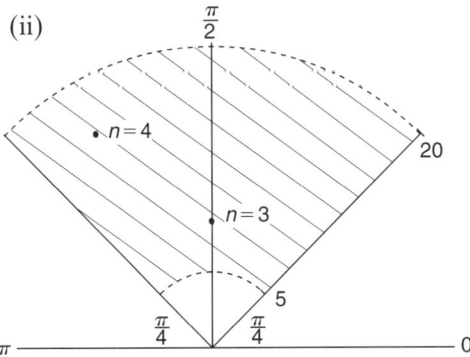

$$(\sqrt{3} + i)^n = 2^n e^{\pi n/6}$$

$5 > n > 2$ for $|z^n|$ $5 > n > 1$ for $\arg(z^n)$
$\therefore n = 3$ or 4.

$n = 3$: $(\sqrt{3} + i)^3 = 8e^{i\pi/2}$

$n = 4$: $(\sqrt{3} + i)^4 = 16e^{i2\pi/3}$

4 Let z be the complex number $1 + i$.

(a) Find the real numbers a and b such that

$$\frac{a}{b + z} = -3 + i$$

(b) (i)   Find the modulus and argument of z^3.
    (ii) In an Argand diagram O is the origin, A represents the complex number z, B represents the complex number z^3 and the point C is chosen so that $OABC$ is a parallelogram.
Determine, in the form $p + qi$, the complex number represented by C.

(AEB, 1994)

• (a) $\dfrac{a}{b + z} = -3 + i$

$$\Rightarrow a = (-3 + i)(b + 1 + i)$$
$$a = -3b + i^2 - 3 + i(-3 + b + 1)$$

$$\left.\begin{array}{l}\text{Re: } a = -3b - 4\\ \text{Im: } 0 = b - 2\end{array}\right\} \Rightarrow b = 2, a = -10$$

(b) (i) $|z| = \sqrt{2}$ $\arg z = \dfrac{\pi}{4}$

$|z^3| = 2\sqrt{2}$ $\arg z^3 = \dfrac{3\pi}{4}$

(ii) B represents $-2 + 2i$
CB is parallel to OA
$\overline{CB} = 1 + i$
$\Rightarrow C$ is $-3 + i$

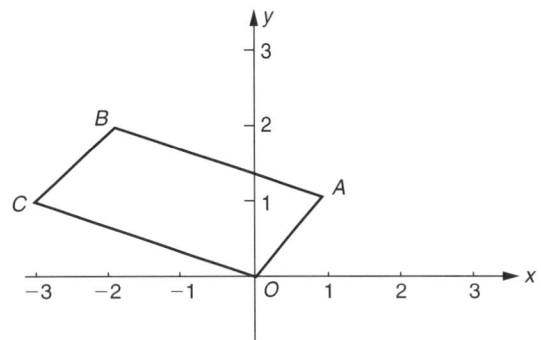

5 (a) Writing

$$\tan x = \frac{\sin x}{\cos x}$$

and using appropriate rules of differentiation and integration, show that

(i) $\dfrac{d}{dx}(\tan x) = \sec^2 x$

and

(ii) $\int\tan x\, dx = -\ln|\cos x| + C$

(b) For values of x in the interval $\left[0, \dfrac{\pi}{4}\right]$, sketch the curve

$$y = (x + 1)\sec^2 x$$

The region enclosed by the curve, the x-axis and the ordinates $x = 0$ and $x = \dfrac{\pi}{4}$ is denoted by R.

(i)   Express the area of R as a definite integral. Use integration by parts to find its value.

(ii) R is rotated through four right-angles about the x-axis. Express the volume generated as a definite integral. Use the trapezium rule with 5 ordinates and an interval of $\dfrac{\pi}{16}$ to find an approximate value of this integral, giving your answer correct to 3 significant figures.

(WJEC, 1994)

• (a) $\tan x = \dfrac{\sin x}{\cos x}$

(i) Using the quotient rule

$$\frac{d}{dx}(\tan x) = \frac{\cos x(\cos x) - \sin x(-\sin x)}{\cos^2 x}$$

$$= \frac{\cos^2 x + \sin^2 x}{\cos^2 x}$$

$$= \sec^2 x$$

(ii) Since $-\sin x$ is the result of differentiating $\cos x$, in $\int \tan x \, dx$ let $u = \cos x$, $du = -\sin x \, dx$.

$$\int \tan x \, dx = \int \frac{-du}{u}$$

$$= -\ln |u| + C$$

$$= -\ln |\cos x| + C$$

(b) $y = (x + 1)\sec^2 x$

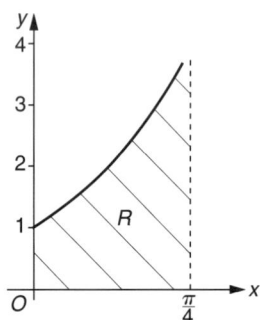

7.1

(i) Area of $R = \int_0^{\pi/4} (x + 1)\sec^2 x \, dx$

Using integration by parts:

$$u = (x + 1), \frac{dv}{dx} = \sec^2 x$$

$$\Rightarrow \quad \frac{du}{dx} = 1, v = \tan x$$

$$R = \left[(x + 1)\tan x\right]_0^{\pi/4} - \int_0^{\pi/4} \tan x \, dx$$

$$= [(x + 1)\tan x + \ln |\cos x|]_0^{\pi/4}$$

6.1 and 6.2

$$= \left(1 + \frac{\pi}{4}\right) + \ln \frac{1}{\sqrt{2}}$$

$$= 1.44 \text{ (to 3 s.f.)}$$

(ii) Volume generated $= \pi \int_0^{\pi/4} y^2 \, dx$

$$= \pi \int_0^{\pi/4} (x + 1)^2 \sec^4 x \, dx$$

x	0	$\dfrac{\pi}{16}$	$\dfrac{\pi}{8}$	$\dfrac{3\pi}{16}$	$\dfrac{\pi}{4}$
$(x + 1)^2 \sec^4 x$	1	1.546 75	2.662 28	5.283 10	12.750 59

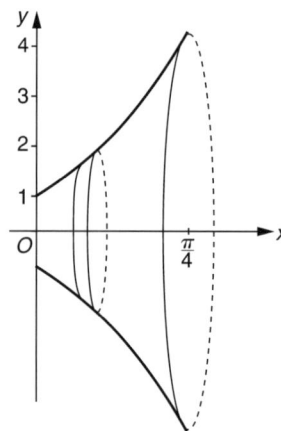

6.3

By the trapezium rule

$$V = \pi \cdot \frac{\pi}{32}[1 + 12.750\ 59$$

$$+ 2(1.546\ 75 + 2.662\ 28 + 5.283\ 10)]$$

$$= 10.096$$

$$= 10.1 \text{ to 3 s.f.}$$

6 The region R is bounded by the x-axis and the part of the curve $y = \sin 2x$ between $x = 0$ and $x = \frac{1}{2}\pi$. Use integration to find the exact values of

(i) the area of R

(ii) the volume of the solid formed when R is rotated completely about the x-axis.

(UCLES, 1993)

• $y = \sin 2x$

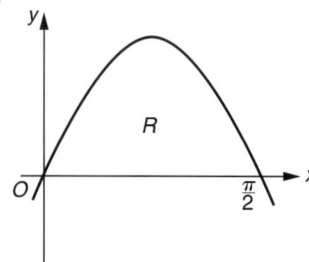

(i) Area of $A = \int_0^{\pi/2} \sin 2x \, dx = \left[\frac{-\cos 2x}{2}\right]_0^{\pi/2}$

$$= \frac{1}{2} + \frac{1}{2}$$

$$= 1$$

6.1

(ii) Volume $= \pi \int_0^{\pi/2} \sin^2 2x \, dx$

$$= \frac{\pi}{2} \int_0^{\pi/2} (1 - \cos 4x) \, dx$$

$$= \frac{\pi}{2}\left[x - \frac{\sin 4x}{4}\right]_0^{\pi/2}$$

6.1

$$= \frac{\pi}{2}\left[\frac{\pi}{2}\right]$$

$$= \frac{\pi^2}{4}$$

7 Show that the equation $x^3 + 5x^2 - 9 = 0$ can be expressed as $x = \sqrt{\dfrac{a}{x + b}}$

Hence, using the iteration formula

$x_{n+1} = \sqrt{\dfrac{a}{x_n + b}}$ and $x_0 = 1$, find the approximate

solution to the equation correct to 4 d.p.

• $x^3 + 5x^2 - 9 = 0 \;\Rightarrow\; x^3 + 5x^2 = 9$

$\qquad x^2(x + 5) = 9$

$\qquad x^2 = \dfrac{9}{x + 5}$

$\qquad x = \sqrt{\dfrac{9}{x + 5}}$

If $x_0 = 1$, $x_1 = \sqrt{\dfrac{9}{6}} = 1.224\,74$

$x_2 = \sqrt{\dfrac{9}{6.224\,74}} = 1.202\,43$ 2.2a

$x_3 = \sqrt{\dfrac{9}{6.202\,43}} = 1.204\,59$

$x_4 = \sqrt{\dfrac{9}{6.204\,59}} = 1.204\,38$

$x_5 = \sqrt{\dfrac{9}{6.204\,38}} = 1.204\,40$

Root is 1.2044 to 4 d.p.

8 Use a numerical method to obtain an approximate value for the gradient of $\dfrac{2\sqrt{x}}{1 + x}$ at the point on the curve where $x = 2$. (Oxford)

• $f(2.1) = 0.934\,93$
$f(1.9) = 0.950\,62$

Gradient $\approx \dfrac{0.934\,93 - 0.950\,62}{0.2} = -0.0785$

Taking a smaller interval would give a more accurate answer
$f(2.05) = 0.938\,87$
$f(1.95) = 0.946\,73$

Gradient $\approx \dfrac{0.938\,87 - 0.946\,73}{0.1} = -0.0786$

(O&C, 1993)

9 A golden rectangle has one side of length 1 unit and a shorter side of length ψ units, where ψ is called the golden section.

ψ can be found by using the iterative formula $x_{n+1} = \sqrt[3]{x_n(1 - x_n)}$. Choosing a suitable value for x_1 and showing intermediate values, use this iterative formula to find the value of ψ to 2 decimal places.
(NEAB SMP, 1993)

x_n	$1 - x_n$	x_{n+1}
0.5	0.5	0.629 96
0.629 96	0.370 04	0.615 44
0.615 44	0.384 56	0.618 56
0.618 56	0.381 44	0.617 92

$\psi = 0.62$ to 2 d.p.   2.2a

10 Shade on a sketch the finite region R in the first quadrant bounded by the x-axis, the curve $y = \ln x$ and the line $x = 5$.

By means of integration, calculate the area of R.

The region R is rotated completely about the x-axis to form a solid of revolution S.

x	1	2	3	4	5
$(\ln x)^2$	0	0.480	1.207	1.922	2.590

Use the given table of values and apply the trapezium rule to find an estimate of the volume of S, giving your answer to one decimal place.
(AEB, 1987)

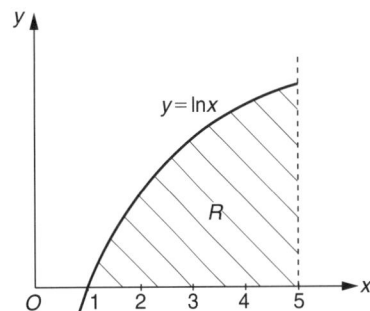

Area of $R = \displaystyle\int_1^5 \ln x \, dx$

Let $u = \ln x, \dfrac{dv}{dx} = 1 \;\Rightarrow\; \dfrac{du}{dx} = \dfrac{1}{x}, v = x$

$\displaystyle\int_1^5 \ln x \, dx = [x \ln x]_1^5 - \int_1^5 1 \, dx$

$= [x \ln x - x]_1^5$
$= 3.047 + 1$
$= 4.047$

Volume of $S = \pi \displaystyle\int_1^5 (\ln x)^2 \, dx$

$$\int_1^5 (\ln x)^2 \, dx \approx \tfrac{1}{2}[(0 + 2.590) + 2(0.480 + 1.207 + 1.922)]$$

$$= 4.904$$

Volume of $S = \pi(4.904) = 15.4$

11 A transformation of the plane is defined by

$$\begin{pmatrix} x \\ y \end{pmatrix} \to M\begin{pmatrix} x \\ y \end{pmatrix} \text{ where } M = \begin{pmatrix} -1 & 4 \\ -1 & 3 \end{pmatrix}$$

(i) Prove that every point of the line l with equation $x - 2y = 0$ is left unchanged by the transformation.

(ii) By induction, or otherwise, prove that, for any positive integer n,

$$M^n = \begin{pmatrix} 1 - 2n & 4n \\ -n & 1 + 2n \end{pmatrix} \qquad \text{(O\&C, 1992)}$$

• Any point on the line $x - 2y = 0$ has coordinates $(2t, t)$

$$M\begin{pmatrix} 2t \\ t \end{pmatrix} = \begin{pmatrix} -1 & 4 \\ -1 & 3 \end{pmatrix}\begin{pmatrix} 2t \\ t \end{pmatrix}$$

$$= \begin{pmatrix} -2t + 4t \\ -2t + 3t \end{pmatrix}$$

$$= \begin{pmatrix} 2t \\ t \end{pmatrix}$$

i.e. all points on the line $x - 2y = 0$ are unchanged by the transformation.

Assume that $M^k = \begin{pmatrix} 1 - 2k & 4k \\ -k & 1 + 2k \end{pmatrix}$

Pre-multiply by $M \Rightarrow$

$$M^{k+1} = \begin{pmatrix} -1 & 4 \\ -1 & 3 \end{pmatrix}\begin{pmatrix} 1 - 2k & 4k \\ -k & 1 + 2k \end{pmatrix}$$

$$= \begin{pmatrix} -1 + 2k - 4k & -4k + 4 + 8k \\ -1 + 2k - 3k & -4k + 3 + 6k \end{pmatrix}$$

$$= \begin{pmatrix} 1 - 2(k + 1) & 4(k + 1) \\ -(k + 1) & 1 + 2(k + 1) \end{pmatrix}$$

Hence if valid for $n = k$ it is valid for $n = k + 1$. But when $k = 1$,

$$\begin{pmatrix} -1 & 4 \\ -1 & 3 \end{pmatrix} = \begin{pmatrix} 1 - 2(1) & 4(1) \\ -(1) & 1 + 2(1) \end{pmatrix}$$

Hence valid for all integer values of k.

12 (a) The transformation T_1 is represented by the matrix

$$A = \begin{bmatrix} -\tfrac{3}{5} & \tfrac{4}{5} \\ \tfrac{4}{5} & \tfrac{3}{5} \end{bmatrix}$$

(i) Find the eigenvalues and eigenvectors of A.

(ii) State the equation of the line of invariant points and describe the transformation T_1 geometrically.

(b) Find the 2×2 matrix B which represents the transformation T_2, a rotation about the origin through $\tan^{-1}(\tfrac{3}{4})$ anticlockwise.

(c) The transformation $T_3 = T_1 T_2 T_1$. Find the 2×2 matrix C which represents T_3 and hence describe T_3 geometrically. (OLE, 1992)

• (a) (i) $A - \lambda I = \begin{bmatrix} -\tfrac{3}{5} - \lambda & \tfrac{4}{5} \\ \tfrac{4}{5} & \tfrac{3}{5} - \lambda \end{bmatrix}$

$\text{Det} | A - \lambda I | = 0$ for the eigenvalues

$\text{Det}(A - \lambda I) = (-\tfrac{3}{5} - \lambda)(\tfrac{3}{5} - \lambda) - (\tfrac{4}{5})^2$
$= -\tfrac{9}{25} + \lambda^2 - \tfrac{16}{25} = \lambda^2 - 1$

$\text{Det}(A - \lambda I) = 0$ when $\lambda = \pm 1$ (the eigenvalues)

When $\lambda = 1$

$$A - I\lambda = \begin{bmatrix} -\tfrac{8}{5} & \tfrac{4}{5} \\ \tfrac{4}{5} & -\tfrac{2}{5} \end{bmatrix}$$

For the eigenvector $\begin{vmatrix} X_1 \\ X_2 \end{vmatrix}$, $[A - \lambda I]\begin{bmatrix} X_1 \\ X_2 \end{bmatrix} = 0$

$\Rightarrow -\tfrac{8}{5}X_1 + \tfrac{4}{5}X_2 = 0 \Rightarrow X_2 = 2X_1$

When $\lambda = -1$

$$A - I\lambda = \begin{bmatrix} \tfrac{2}{5} & \tfrac{4}{5} \\ \tfrac{4}{5} & \tfrac{8}{5} \end{bmatrix}$$

$[A - I\lambda]\begin{bmatrix} X_1 \\ X_2 \end{bmatrix} = 0 \Rightarrow \tfrac{2}{5}X_1 + \tfrac{4}{5}X_2 = 0$
$\Rightarrow 2X_2 = -X_1$

Hence the corresponding eigenvectors are

$$\begin{pmatrix} 1 \\ 2 \end{pmatrix} \text{ and } \begin{pmatrix} 2 \\ -1 \end{pmatrix}.$$

(ii) Testing $[A]\begin{bmatrix} 1a \\ 2a \end{bmatrix}$ gives $\begin{bmatrix} 1a \\ 2a \end{bmatrix}$

$\Rightarrow$ invariant points.

Testing $[A]\begin{bmatrix} -2 \\ 1 \end{bmatrix}$ gives $\begin{bmatrix} 2 \\ -1 \end{bmatrix}$

$\Rightarrow$ invariant line.

Hence the line of invariant points is
$y = 2x$.
T_1 is a reflection in the line $y = 2x$.

(b) *The coordinates of A' and C' give the transformation matrix for a rotation of θ of the unit square OABC*

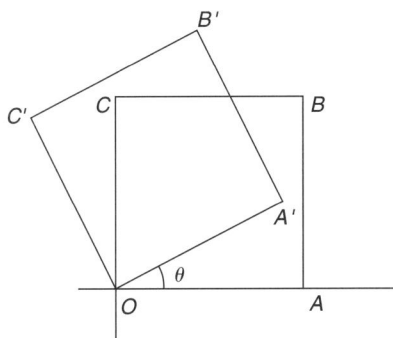

T_2: $\begin{bmatrix} \frac{4}{5} & -\frac{3}{5} \\ \frac{3}{5} & \frac{4}{5} \end{bmatrix}$

(c) $T_1 T_2 T_1 - \begin{bmatrix} -\frac{3}{5} & \frac{4}{5} \\ \frac{4}{5} & \frac{3}{5} \end{bmatrix}\begin{bmatrix} \frac{4}{5} & -\frac{3}{5} \\ \frac{3}{5} & \frac{4}{5} \end{bmatrix}\begin{bmatrix} -\frac{3}{5} & \frac{4}{5} \\ \frac{4}{5} & \frac{3}{5} \end{bmatrix}$

$= \begin{bmatrix} 0 & 1 \\ 1 & 0 \end{bmatrix}\begin{bmatrix} -\frac{3}{5} & \frac{4}{5} \\ \frac{4}{5} & \frac{3}{5} \end{bmatrix}$

$= \begin{bmatrix} \frac{4}{5} & \frac{3}{5} \\ -\frac{3}{5} & \frac{4}{5} \end{bmatrix}$

$\Rightarrow$ a rotation of $\tan^{-1}\left(\frac{3}{4}\right)$ clockwise

13 Planes Π_1, Π_2, Π_3 have equations
$\Pi_1: x + y + 2z = 1$
$\Pi_2: 2x - y + z = 5$
$\Pi_3: x - 2y - z = 8$

(i) Line L is the intersection of Π_1 and Π_2. Find the coordinates of the point P on L with x-coordinate equal to zero and also the coordinates of the point Q on L with z-coordinate equal to zero. Hence, or otherwise, find the equation of L in the form $\mathbf{r} = \mathbf{p} + \lambda(\mathbf{q} - \mathbf{p})$, where $\mathbf{p}, \mathbf{q}$ are the position vectors of P and Q respectively with respect to the origin.

(ii) Using your equation for L, show that the perpendicular distance of every point of L from π_3 is $\dfrac{4}{\sqrt{6}}$. Deduce that the three planes form a prism.

Find the angle between each pair of planes and deduce the shape of cross-section of the prism. Find the area of this cross-section.

(iii) The plane π_0 is perpendicular to π_1, π_2, π_3 and passes through the origin. Find the equation of π_0 in the form $\mathbf{r} \cdot \mathbf{n} = k$, and also find the

perpendicular distance of the line L from the origin.

(O&C, 1993)

$\bullet$ $\pi_1: x + y + 2z = 1$
    $\pi_2: 2x - y + z = 5$

(i) At $P, x = 0 \Rightarrow y + 2z = 1$
                        $-y + z = 5$
    Adding, $3z = 6 \Rightarrow z = 2, y = -3$
                            $\Rightarrow P(0, -3, 2)$
    At $Q, z = 0 \Rightarrow x + y = 1$
                        $2x - y = 5$
                $\Rightarrow Q(2, -1, 0)$

Equation of line PQ:

$$\mathbf{r} = \begin{pmatrix} 0 \\ -3 \\ 2 \end{pmatrix} + \lambda\begin{pmatrix} 2 \\ 2 \\ -2 \end{pmatrix} = L$$

(ii) $\pi_3: \mathbf{r} \cdot \begin{pmatrix} 1 \\ -2 \\ 1 \end{pmatrix} = 8$

Direction vector of $PQ = \begin{pmatrix} 1 \\ 1 \\ -1 \end{pmatrix}$

Normal vector of $\pi_3 = \begin{pmatrix} 1 \\ -2 \\ -1 \end{pmatrix}$

Scalar product of $\begin{pmatrix} 1 \\ 1 \\ -1 \end{pmatrix}$ and $\begin{pmatrix} 1 \\ -2 \\ -1 \end{pmatrix} = 0$

$\Rightarrow$ line PQ is parallel to the plane π_3, i.e. every point of PQ is the same distance from π_3.

Line through $P(0, -3, 2)$ which is perpendicular to π_3 is

$$\mathbf{r} = \begin{pmatrix} 0 \\ -3 \\ 2 \end{pmatrix} + \mu\begin{pmatrix} 1 \\ -2 \\ -1 \end{pmatrix}.$$

This meets the plane π_3 when

$$\begin{pmatrix} \mu \\ -3 - 2\mu \\ 2 - \mu \end{pmatrix} \cdot \begin{pmatrix} 1 \\ -2 \\ -1 \end{pmatrix} = 8$$

$\Rightarrow \mu + 6 + 4\mu - 2 + \mu = 8 \qquad \mu = \dfrac{4}{6}$

Hence the distance of point $(0, -3, 2)$ from π_3

is $\frac{2}{3}\sqrt{1^2 + 2^2 + 1^2} = \dfrac{4}{\sqrt{6}}$.

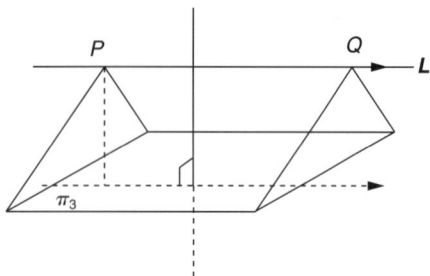

Hence the planes form a prism.

The angle between π_1 and π_2 is α where

$$\cos \alpha = \frac{\begin{pmatrix} 1 \\ 1 \\ 2 \end{pmatrix} \cdot \begin{pmatrix} 2 \\ -1 \\ 1 \end{pmatrix}}{\sqrt{6} \ \sqrt{6}} = \frac{1}{2}$$

$\Rightarrow \quad \alpha = 60°$.

The angle between π_1 and π_3 is β where

$$\cos \beta = \frac{\left| \begin{pmatrix} 1 \\ 1 \\ 2 \end{pmatrix} \cdot \begin{pmatrix} 1 \\ -2 \\ -1 \end{pmatrix} \right|}{\sqrt{6} \ \sqrt{6}} = \frac{1}{2}$$

$\Rightarrow \quad \beta = 60°$.

The shape of the cross-section is an equilateral triangle of height $\dfrac{4}{\sqrt{6}}$.

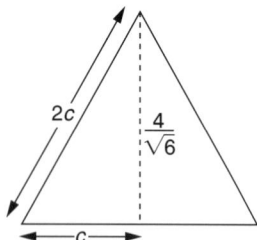

Each side of the triangle is $\dfrac{2}{\sqrt{3}} \cdot \dfrac{4}{\sqrt{6}} = \dfrac{8}{3\sqrt{2}}$

Area $= \dfrac{4}{\sqrt{6}} \cdot \dfrac{4}{3\sqrt{2}} = \dfrac{8}{3\sqrt{3}}$

(iii) If π_0 is perpendicular to π_1, π_2 and π_3 it has L as a normal to the plane $\Rightarrow$ equation of π_0 is

$$\mathbf{r} \cdot \begin{pmatrix} 1 \\ 1 \\ -1 \end{pmatrix} = 0.$$

Line L has general point $(2\lambda, -3 + 2\lambda, 2 - 2\lambda)$.
This meets π_0: $x + y - z = 0$ when
$2\lambda - 3 + 2\lambda - 2 + 2\lambda = 0$
i.e. $\lambda = \frac{5}{6}$

Hence L meets π_0 at the point $(\frac{10}{6}, -\frac{8}{6}, \frac{2}{6})$, which is $\frac{1}{3}\sqrt{25 + 16 + 1}$ from the origin

i.e. $\sqrt{\frac{14}{3}}$ from the origin

Exercises

1 (a) The equation of a curve C_1 is given in polar coordinates by

$$r = a(1 + \sin 2\theta)$$

where a is a positive constant and $0 \leqslant \theta \leqslant 2\pi$.
(i) Sketch C_1 making clear the form near the pole and the symmetries of the curve.
(ii) Find the area enclosed by one loop of C_1.
(b) The equation of a curve C_2 is given in polar coordinates by

$$r^2 = a^2(1 + \sin 2\theta)$$

where a is a positive constant, $r \geqslant 0$ and $0 \leqslant \theta < 2\pi$. Show that the equation of C_2 in cartesian coordinates may be expressed as

$$(x^2 + y^2)^2 = a^2(x + y)^2 \qquad \text{(UCLES, 1991)}$$

2 On the same diagram, sketch the curve C_1, with polar equation

$$r = 2 \cos 2\theta \qquad -\frac{\pi}{4} < \theta \leqslant \frac{\pi}{4}$$

and the curve C_2 with polar equation $\theta = \dfrac{\pi}{12}$.

Find the area of the smaller region bounded by C_1 and C_2. (L, 1992)

3 Find the modulus and argument of the complex number $\dfrac{5 + i}{3 - 2i}$

Hence solve the equation $z^4 = \dfrac{5 + i}{3 - 2i}$ giving your answers in the form $re^{i\theta}$, where $r > 0$ and $-\pi < \theta \leqslant \pi$. (L, 1992)

4 (a) One root of the equation $z^2 - 4z + 8 = 0$ is $z_1 = a + bj$ where a and b are positive integers.
(i) Find z_1 and write down the other root, z_2, of this equation. Illustrate z_1 and z_2 on an Argand diagram.
(ii) Calculate $|z_2|$ and arg z_2. Show clearly the meaning of $|z_2|$ and arg z_2 on your diagram.
(b) Find the complex number w in the form $X + Yj$, given that

$$\frac{1}{w} = \frac{1}{1 + 3j} + \frac{1}{1 - 2j} \qquad \text{(O\&C MEI, 1993)}$$

5 Given that $(8 - 6i)^{1/2} = x + iy$, where x, y are real, square both sides of this equation and hence obtain a pair of simultaneous equations for x and y.

Solve these equations and deduce two square roots of $8 - 6i$. (WJEC, 1994)

6

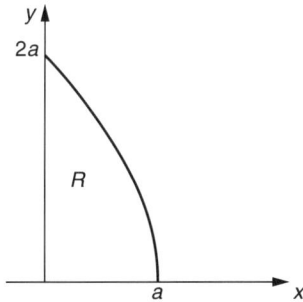

The diagram shows the region R, which is bounded by the axes and the part of the curve $y^2 = 4a(a - x)$ lying in the first quadrant. Find, in terms of a,

(i) the area of R

(ii) the volume, V_x, of the solid formed when R is rotated completely about the x-axis.

The volume of the solid formed when R is rotated completely about the y-axis is V_y. Show that $V_y = \frac{8}{15} V_x$.

The region S, lying in the first quadrant, is bounded by the curve $y^2 = 4a(a - x)$ and the lines $x = a$ and $y = 2a$. Find, in terms of a, the volume of the solid formed when S is rotated completely about the y-axis. (UCLES, 1992)

7 Sketch the curve $y = 9 - x^2$, stating the coordinates of the turning point and of the intersections with the axes.

The finite region bounded by the curve and the x-axis is denoted by R.

(i) Find the area of R.

(ii) Hence, or otherwise, find $\int_0^9 \sqrt{9 - y} \, dy$.

(iii) Find the volume of the solid of revolution obtained when R is rotated through 2π radians about the x-axis.

(iv) Find the volume of the solid of revolution obtained when R is rotated through π radians about the y-axis. (UCLES, 1993)

8 A sketch graph of $y = x^2 + px + q$ is shown. The vertex of the graph is $(2, -3)$.

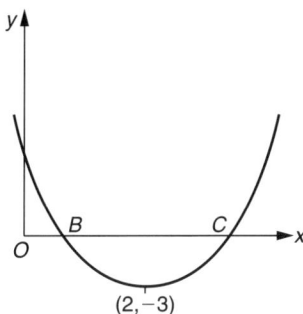

(a) Find p and q. Hence calculate the x-coordinates of the points B and C.

(b) Draw a sketch graph to show that the equation $x^2 + px + q = \sqrt{x}$ has two solutions.

(c) Use the iterative formula $x_{n+1} = \frac{1}{4}(x_n^2 - \sqrt{x_n} + 1)$ and a start value of 0.2 to find the smaller of the two roots correct to 2 decimal places showing all intermediate values. (NEAB SMP, 1994)

9 Show that the equation $x^3 = 5$ can be arranged to the form

$$x = \sqrt{\sqrt{\sqrt{5x}}}$$

Use an iterative process based on this form, starting with $x_1 = 1$ and showing intermediate values, to obtain the positive solution for $x^3 = 5$ correct to 2 decimal places. (NEAB SMP Specimen Paper)

10 (a) Part of the graphs for $y = x$ and $y = \cos x$ are shown. Taking $x = 0.8$ as the x-coordinate of the point A, calculate the area OAB shaded.

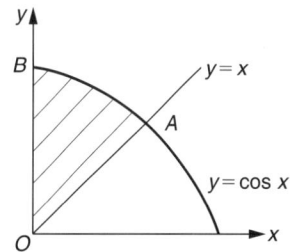

(b) Given that 0.8 is an approximate solution to the equation $x = \cos x$, use a single application of the Newton–Raphson method to obtain a better solution.

(c) Draw a sketch which illustrates that the equation $x = \cos x$ has only one solution. (SMP, 1993)

11 Use the trapezium rule with 5 ordinates and interval width 0.25 to find an approximate value of the integral

$$\int_1^2 \frac{e^x}{x} \, dx$$

Give your answer correct to 2 decimal places. (WJEC, 1992)

12

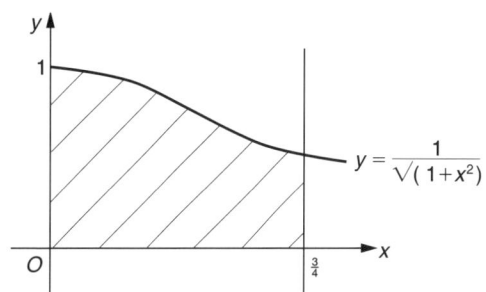

The area A of the shaded region in the figure is

$$\int_0^{3/4} \frac{1}{\sqrt{(1+x^2)}} \, dx$$

(a) By evaluating the integral, show that $A = \ln 2$.
(b) By applying Simpson's rule, with 5 equally spaced ordinates, to the integral, obtain an approximate value of $\ln 2$, giving your answer to 4 decimal places. (L, 1994)

13 A 2×2 matrix A is such that

$$A^2 - 4A + 3I = 0$$

(i) Show that $(A + 3I)(A - 7I) = kI$ where k is a constant, stating the value of k. Hence describe the transformation given by

$$\begin{pmatrix} x \\ y \end{pmatrix} \rightarrow (A + 3I)(A - 7I) \begin{pmatrix} x \\ y \end{pmatrix}$$

(ii) Prove that $A - 2I$ is its own inverse.

(iii) Show that the matrix $\begin{pmatrix} -3 & -4 \\ 6 & 7 \end{pmatrix}$ satisfies the equation in (i). (O&C, 1993)

14 (a) Two square matrices $\mathbf{M}$ and $\mathbf{N}$ (of the same size) have inverses $\mathbf{M}^{-1}$ and $\mathbf{N}^{-1}$ respectively. Show that the inverse of $\mathbf{MN}$ is $\mathbf{N}^{-1}\mathbf{M}^{-1}$.

(b) $\mathbf{A} = \begin{pmatrix} 1 & 7 & 4 \\ 0 & 1 & 2 \\ 0 & 0 & 1 \end{pmatrix}$, $\mathbf{B} = \begin{pmatrix} 1 & 0 & 0 \\ 3 & 1 & 0 \\ -1 & -4 & 1 \end{pmatrix}$

and $\mathbf{C} = \mathbf{AB}$.

(i) Evaluate the matrix $\mathbf{C}$.
(ii) Work out the matrix product

$$\mathbf{A} \begin{pmatrix} 1 & a & b \\ 0 & 1 & c \\ 0 & 0 & 1 \end{pmatrix}.$$

(iii) By equating the product in (ii) to

$$\begin{pmatrix} 1 & 0 & 0 \\ 0 & 1 & 0 \\ 0 & 0 & 1 \end{pmatrix}, \text{find } \mathbf{A}^{-1}.$$

(iv) Using a similar method, or otherwise, find $\mathbf{B}^{-1}$.
(v) Using your results from (iii) and (iv), find $\mathbf{C}^{-1}$.
(O&C MEI, 1993)

15 Planes π_1 and π_2 have equations

$$\mathbf{r} \cdot (\mathbf{i} - \mathbf{j} + 2\mathbf{k}) = 5 \text{ and } 3x - z = 2$$

respectively.

Find the size of the acute angle between π_1 and π_2 giving your answer to the nearest 0.1 degree.
(L, 1992)

16 $OABCDEFG$ is a unit cube with M the mid-point of EF. O is taken as the origin of a coordinate system with x, y, z axes as shown.

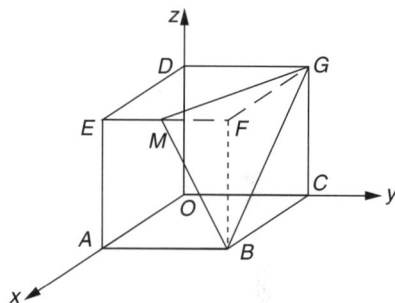

(a) Find the equation of plane BGM.
(b) Find the coordinates of the point where the line OF intersects the plane BGM.
(c) Calculate the angle between the line OF and the plane BGM. (JMB)

17 The line l_1 has vector equation
$\mathbf{r} = (-7\mathbf{i} + 5\mathbf{j}) + \lambda(8\mathbf{i} - 3\mathbf{j} + \mathbf{k})$ and the plane π_1 has equation $\mathbf{r} \cdot (\mathbf{i} - 2\mathbf{j} - \mathbf{k}) = 9$.

(a) Calculate, to the nearest degree, the acute angle between the line l_1 and the plane π_1.
(b) Find the coordinates of the point of intersection of the line l_1 and the plane π_1.
(c) Evaluate the vector product $(8\mathbf{i} - 3\mathbf{j} + \mathbf{k}) \times (\mathbf{i} - 2\mathbf{j} - \mathbf{k})$ and hence, or otherwise, determine an equation of the plane π_2 which contains l_1 and which is perpendicular to π_1.
(d) Determine a vector equation of the line of intersection of the planes π_1 and π_2.
(AEB, 1993)

Brief Solutions

1 (a) $C_1: r = a(1 + \sin 2\theta)$

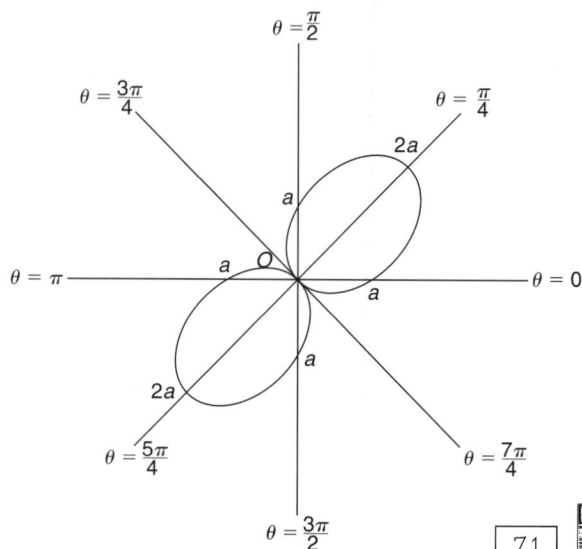

7.1

The curve is symmetrical about the lines

$$\theta = \frac{\pi}{4} + n\left(\frac{\pi}{2}\right), n = 0, 1, 2, 3$$

and has rotational symmetry of π about the pole.

$$\text{Area of one loop} = 2\int_{-\pi/4}^{\pi/4} \tfrac{1}{2}r^2 \, \mathrm{d}\theta$$
$$= a^2\int_{-\pi/4}^{\pi/4} (1 + 2\sin 2\theta + \sin^2 2\theta)\mathrm{d}\theta = \frac{3\pi a^2}{4}$$

6.1 and 6.2

(b) $C_2: r^2 = a^2(1 + \sin 2\theta) \;\Rightarrow\; r^4 = a^2r^2 + a^2 2xy$
$\;\Rightarrow\; (x^2 + y^2)^2 = a^2(x^2 + y^2) + 2a^2xy$
$\;\Rightarrow\; (x^2 + y^2)^2 = a^2(x + y)^2$

2

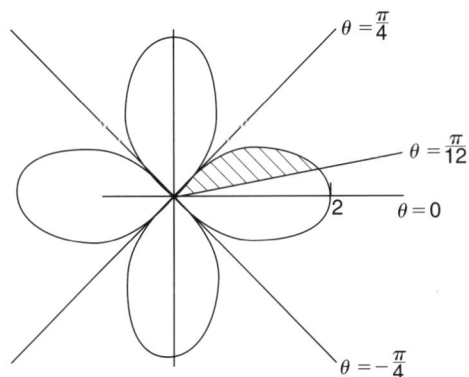

$\theta = \frac{\pi}{4}$

$\theta = \frac{\pi}{12}$

2 $\theta = 0$

$\theta = -\frac{\pi}{4}$

7.1

Area of the shaded region is

$$A = \tfrac{1}{2}\int_{\pi/12}^{\pi/4} r^2 \, \mathrm{d}\theta$$

$$= 2\int_{\pi/12}^{\pi/4} \cos^2 2\theta \, \mathrm{d}\theta$$

6.1 and 6.2

$$= \frac{\pi}{6} - \frac{\sqrt{3}}{8} \; (\approx 0.3071)$$

3 $\dfrac{5 + i}{3 - 2i} = \dfrac{(5 + i)(3 + 2i)}{(3 - 2i)(3 + 2i)} = 1 + i$

Modulus $\sqrt{2}$ Arg $\dfrac{\pi}{4}$

$z^4 = 2^{1/2} \, \mathrm{e}^{\mathrm{i}(\pi/4 + 2n\pi)}$
$z_1 = 2^{1/8} \, \mathrm{e}^{\mathrm{i}\pi/16}$ $z_2 = 2^{1/8} \, \mathrm{e}^{\mathrm{i}9\pi/16}$ $z_3 = 2^{1/8} \, \mathrm{e}^{\mathrm{i}(-7\pi/16)}$
$z_4 = 2^{1/8} \, \mathrm{e}^{\mathrm{i}(-15\pi/16)}$

4 (a) $z^4 - 4z + 8 = 0 \;\Rightarrow\; z = 2 \pm 2j$
$z_1 = 2 + 2j$ $z_2 = 2 - 2j$

Arg $z_2 = \dfrac{-\pi}{4}$

$|z_2| = \sqrt{2^2 + 2^2} = 2\sqrt{2}$

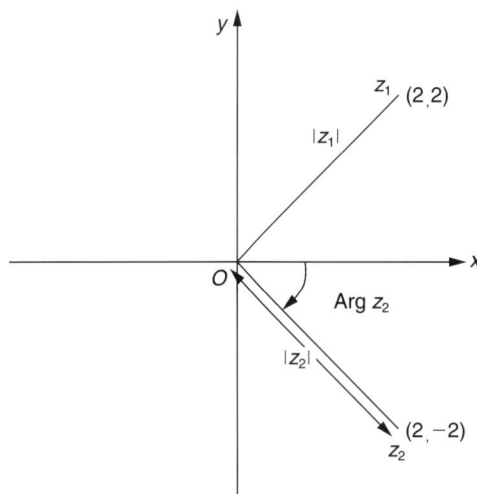

(b) $\dfrac{1}{w} = \dfrac{1}{1 + 3j} + \dfrac{1}{1 - 2j} = \dfrac{1 - 3j}{10} + \dfrac{1 + 2j}{5}$

$$= \frac{3 + j}{10}$$

$$\therefore w = \frac{10}{3 + j} = 3 - j$$

5 $((8 - 6i)^{1/2})^2 = (x + iy)^2$
$\Rightarrow\; 8 = x^2 - y^2 \qquad -6 = 2xy$
Try the factors of 6 before going into algebraic substitution
$x = \pm 3, y = \mp 1 \;\Rightarrow\; (8 - 6i)^{1/2} = 3 - i$ or $-3 + i$

6 (i) Area of $R = \displaystyle\int_0^{2a} x \, \mathrm{d}y = \int_0^{2a} \frac{4a^2 - y^2}{4a} \, \mathrm{d}y = \frac{4a^2}{3}$

(ii) Volume $V_x = \displaystyle\int_0^a \pi y^2 \, \mathrm{d}x = 2\pi a^3$

Volume $V_y = \displaystyle\int_0^{2a} \pi x^2 \, \mathrm{d}y$
$$= \frac{\pi}{16a^2}\int_0^{2a} (4a^2 - y^2)^2 \, \mathrm{d}y = \frac{16a^3\pi}{15}$$

$V_y = \tfrac{8}{15} V_x$

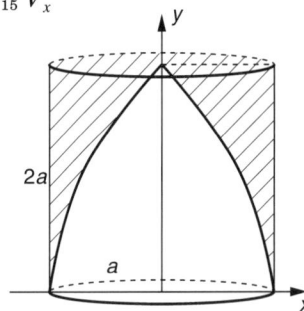

Volume when rectangle is rotated is $\pi(a)^2 2a$
$$= 2\pi a^3$$
Volume when S is rotated is $2\pi a^3 - V_y$
$$= \frac{14\pi a^3}{15}$$

7

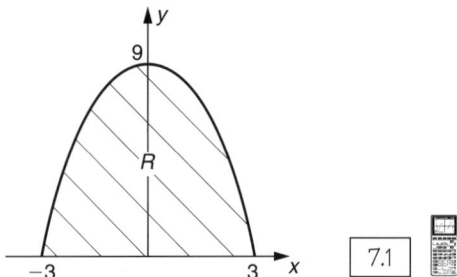

7.1

Points of intersection with the axes are
$(-3, 0), (3, 0), (0, 9)$ (turning point)

(i)   Area $R = 2\int_0^3 (9 - x^2)\,dx = 36$

6.1 and 6.2

(ii) $\int_0^9 \sqrt{9 - y}\,dy = \frac{1}{2}R = 18$

(iii)   Volume $V_x = 2\pi\int_0^3 y^2\,dx = 814.3$

(iv)   Volume $V_y = \pi\int_0^9 x^2\,dy = 127.2$

8 (a) $y = (x - 2)^2 - 3$ (from turning point)
           $= x^2 - 4x + 1$. So $p = -4, q = 1$.
        $y = 0$ when $x = 2 \pm \sqrt{3}$
        $\Rightarrow \quad B(2 - \sqrt{3}, 0), C(2 + \sqrt{3}, 0)$

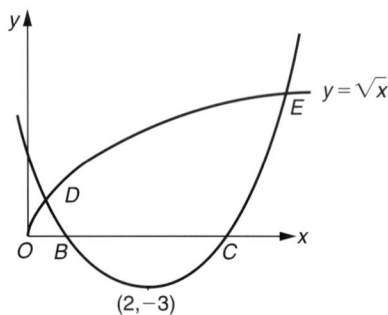

2.4a

(b) At D and E, $\sqrt{x} = x^2 - 4x + 1$
      Rearrange $4x = x^2 - \sqrt{x} + 1$
      $\Rightarrow \quad x = \frac{1}{4}(x^2 - \sqrt{x} + 1)$

2.2a

x_n	x_{n+1}
0.2	0.1482
0.1482	0.1592
0.1592	0.1566
0.1566	0.1572

Hence the smaller root is $x = 0.16$ to 2 d.p.

9 $x^3 = 5 \quad \Rightarrow \quad x^4 = 5x \quad \Rightarrow \quad x^2 = \sqrt{5x}$
   $\Rightarrow \quad x = \sqrt{\sqrt{5x}}$

2.2a

$x_1 = 1 \quad x_2 = 1.4953 \quad x_3 = 1.6536 \quad x_4 = 1.6957$
$x_5 = 1.7064 \quad x_6 = 1.7091$
$x = 1.71$ to 2 d.p.

10 (a) $x = 0.8, y = 0.8$ (from straight line)

6.2

Shaded area $= \int_0^{0.8} \cos x\,dx - \frac{1}{2}(0.8)^2 = 0.397$

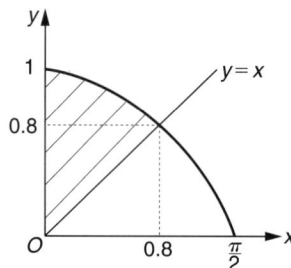

(b) $f(x) = x - \cos x \quad f'(x) = 1 + \sin x$
     $f(0.8) = 0.1033 \quad\;\; f'(0.8) = 1.7174$

By the Newton–Raphson method,

$$x_1 = 0.8 - \frac{0.1033}{1.7174} = 0.74.$$

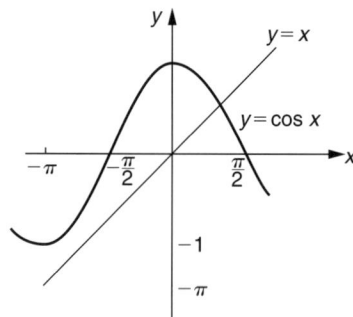

2.4a

(c) From the sketch, $x \approx 0.74$ is the only solution.

11

6.3

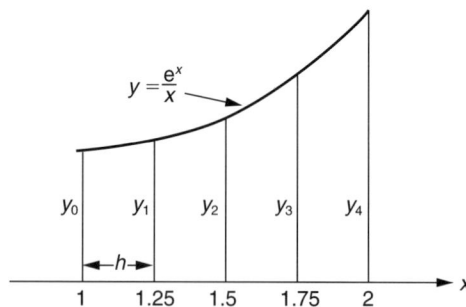

$y_0 = 2.7183$
$y_1 = 2.7923$
$y_2 = 2.9878$
$y_3 = 3.2883$
$y_4 = 3.6945$

Trapezium rule:

$$I = \frac{0.25}{2}(2.7183 + 3.6945$$

$$+ 2(2.7923 + 2.9878 + 3.2883))$$

$$\int_1^2 \frac{e^x}{x}\,dx \approx 3.07$$

12 (a) $A = \int_0^{3/4} \dfrac{1}{\sqrt{1 + x^2}} \, dx$

Use either x = sinh u or x = tan θ as a substitution

$= \ln 2$

(b) Using Simpson's rule with $h = \frac{3}{16}$: $\boxed{6.3}$

$A \approx \frac{1}{16}\left[1 + 4\left(\dfrac{1}{\sqrt{1 + (\frac{3}{16})^2}} + \dfrac{1}{\sqrt{1 + (\frac{9}{16})^2}} \right) \right.$

$\left. + 2\left(\dfrac{1}{\sqrt{1 + (\frac{3}{8})^2}} \right) + \dfrac{1}{\sqrt{1 + (\frac{3}{4})^2}} \right] \approx 0.6931$

Therefore $\ln 2 = 0.6931$ to 4 d.p.

13 (i) $(A + 3I)(A - 7I)$

$= A^2 + 3IA - 7AI - 21I^2 \quad IA = AI = A, \quad I^2 = I$

$= A^2 - 4A - 21I \qquad A^2 - 4A + 3I = O$

$= -24I$

This is an enlargement, factor -24.

(ii) $(A - 2I)(A - 2I) = A^2 - 4A + 4I$

$= (A^2 - 4A + 3I) + I$

$= I$ i.e. self-inverse

(iii) $A = \begin{pmatrix} -3 & -4 \\ 6 & 7 \end{pmatrix} \quad A + 3I = \begin{pmatrix} 0 & -4 \\ 6 & 10 \end{pmatrix}$

$A - 7I = \begin{pmatrix} -10 & -4 \\ 6 & 0 \end{pmatrix}$

$(A + 3I)(A - 7I) = \begin{pmatrix} 0 & -4 \\ 6 & 10 \end{pmatrix}\begin{pmatrix} -10 & -4 \\ 6 & 0 \end{pmatrix}$

$= \begin{pmatrix} -24 & 0 \\ 0 & -24 \end{pmatrix} = -24I$

(i) is satisfied.

14 (a) $\quad \mathbf{MN} \times \mathbf{N^{-1}M^{-1}} = \mathbf{M(NN^{-1})M^{-1}} \quad \mathbf{NN^{-1}} = \mathbf{I}$

$= \mathbf{MM^{-1}}$

$= \mathbf{I}$

(b) (i) $\quad \mathbf{C = AB} = \begin{pmatrix} 18 & -9 & 4 \\ 1 & -7 & 2 \\ -1 & -4 & 1 \end{pmatrix}$

(ii) $\mathbf{A}\begin{pmatrix} 1 & a & b \\ 0 & 1 & c \\ 0 & 0 & 1 \end{pmatrix}$

$= \begin{pmatrix} 1 & a+7 & b+7c+4 \\ 0 & 1 & c+2 \\ 0 & 0 & 1 \end{pmatrix}$

$= I$ if $a = -7, c = -2, b = 10$

Similarly

$\mathbf{B}\begin{pmatrix} 1 & 0 & 0 \\ d & 1 & 0 \\ e & f & 1 \end{pmatrix}$

$= \begin{pmatrix} 1 & 0 & 0 \\ 3+d & 1 & 0 \\ -1-4d+e & -4+f & 1 \end{pmatrix}$

$= I$ if $d = -3, f = 4, e = -11$

(iii) $\mathbf{A^{-1}} = \begin{pmatrix} 1 & -7 & 10 \\ 0 & 1 & -2 \\ 0 & 0 & 1 \end{pmatrix}$

(iv) $\mathbf{B^{-1}} = \begin{pmatrix} 1 & 0 & 0 \\ -3 & 1 & 0 \\ -11 & 4 & 1 \end{pmatrix}$

(v) $\quad \mathbf{C^{-1} = B^{-1}A^{-1}}$

$= \begin{pmatrix} 1 & -7 & 10 \\ -3 & 22 & -32 \\ -11 & 81 & -117 \end{pmatrix}$

15 Normal vectors to π_1 and π_2 are $\begin{pmatrix} 1 \\ -1 \\ 2 \end{pmatrix}$ and $\begin{pmatrix} 3 \\ 0 \\ -1 \end{pmatrix}$.

The acute angle θ is given by

$\cos \theta = \dfrac{\begin{pmatrix} 1 \\ -1 \\ 2 \end{pmatrix} \cdot \begin{pmatrix} 3 \\ 0 \\ -1 \end{pmatrix}}{\sqrt{6}\,\sqrt{10}}$

$\Rightarrow \quad \theta = 82.6°$

16

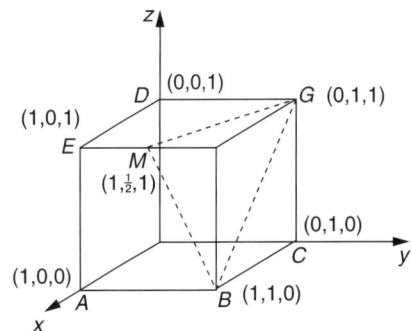

(a) Coordinates of M, B and G are

$M(1, \frac{1}{2}, 1) \quad B(1, 1, 0) \quad G(0, 1, 1)$

Direction vectors:

$\overrightarrow{MG} = \begin{pmatrix} -1 \\ \frac{1}{2} \\ 0 \end{pmatrix} \qquad \overrightarrow{BG} = \begin{pmatrix} -1 \\ 0 \\ 1 \end{pmatrix}$

Normal to plane MBG

$= \begin{pmatrix} -1 \\ \frac{1}{2} \\ 0 \end{pmatrix} \times \begin{pmatrix} -1 \\ 0 \\ 1 \end{pmatrix} = \begin{pmatrix} \frac{1}{2} \\ 1 \\ \frac{1}{2} \end{pmatrix}$

Equation of plane MBG is $\mathbf{r} \cdot \begin{pmatrix} \frac{1}{2} \\ 1 \\ \frac{1}{2} \end{pmatrix} = 1\frac{1}{2}$

or $\quad \frac{1}{2}x + y + \frac{1}{2}z = 1\frac{1}{2}$

(b) Equation of line OF is $\mathbf{r} = \lambda \begin{pmatrix} 1 \\ 1 \\ 1 \end{pmatrix}$

Point of intersection of OF and MBG is given by $\frac{1}{2}\lambda + \lambda + \frac{1}{2}\lambda = 1\frac{1}{2} \implies \lambda = \frac{3}{4}$
Point of intersection is $(\frac{3}{4}, \frac{3}{4}, \frac{3}{4})$

(c) Angle θ between OF and BGM is given by

$$\sin \theta = \frac{\begin{pmatrix} 1 \\ 1 \\ 1 \end{pmatrix} \cdot \begin{pmatrix} 1 \\ 2 \\ 1 \end{pmatrix}}{\sqrt{3}\,\sqrt{6}}$$

$\implies \quad \theta = 70.5°.$

17 (a) The acute angle θ between l_1 and π_1 is given by

$$\sin \theta = \frac{\begin{pmatrix} 8 \\ -3 \\ 1 \end{pmatrix} \cdot \begin{pmatrix} 1 \\ -2 \\ -1 \end{pmatrix}}{\sqrt{74}\,\sqrt{6}}$$

$\implies \quad \theta = 38°$ to the nearest degree

(b) Any point on l_1 is of the form $(-7 + 8\lambda, 5 - 3\lambda, \lambda)$
This lies on π_1 when
$-7 + 8\lambda - 2(5 - 3\lambda) - \lambda = 9 \implies \lambda = 2$
Point of intersection is $(9, -1, 2)$

(c) $\begin{pmatrix} 8 \\ -3 \\ 1 \end{pmatrix} \times \begin{pmatrix} 1 \\ -2 \\ -1 \end{pmatrix} = 5\mathbf{i} + 9\mathbf{j} - 13\mathbf{k}$

This vector is perpendicular to l_1 and parallel to π_1, i.e. is perpendicular to π_2

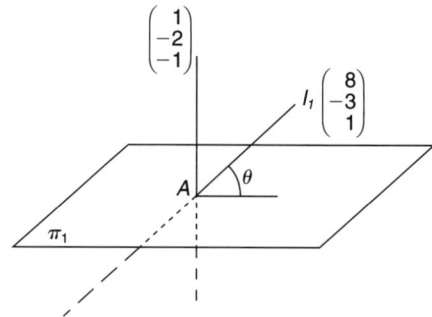

Equation of π_2 is

$$\mathbf{r} \cdot \begin{pmatrix} 5 \\ 9 \\ -13 \end{pmatrix} = \begin{pmatrix} 9 \\ -1 \\ 2 \end{pmatrix} \cdot \begin{pmatrix} 5 \\ 9 \\ -13 \end{pmatrix}$$

$\implies \quad 5x + 9y - 13z = 10$

The common line of π_1 and π_2 is perpendicular to the normal vectors to π_1 and π_2 $\implies$ direction of the common line is given by $\mathbf{n}_1 \times \mathbf{n}_2$

$$\mathbf{n}_1 \times \mathbf{n}_2 = \begin{pmatrix} 35 \\ 8 \\ 19 \end{pmatrix}$$

Hence the common line (which passes through A) has equation

$$\mathbf{r} = \begin{pmatrix} 9 \\ -1 \\ 2 \end{pmatrix} + \mu \begin{pmatrix} 35 \\ 8 \\ 19 \end{pmatrix}$$

14

Descriptive statistics

AS Level			A Level			Topic	Date attempted	Date completed	Self-assessment
CORE	MODULAR	TRADITIONAL	CORE	MODULAR	TRADITIONAL				
	✓	✓	✓	✓	✓	**Presentation of data**			
	✓	✓	✓	✓	✓	**Histograms, frequencies**			
	✓	✓	✓	✓	✓	**Mean, median, quartiles**			
	✓	✓	✓	✓	✓	**Variance, standard deviation**			
	✓	✓	✓	✓	✓	**Permutations and combinations**			

Fact Sheet

Representation of Data

Tabular

Data should be divided into different categories or groups. These can then be illustrated on a **bar chart**. Each **bar** (horizontal) or **column** (vertical) must be separate from the others.

Stem and Leaf

This is a method of analysis of raw data (usually numerical). Data is collected in equal class widths and ordered within each class. The raw data is used in the leaf. This method is useful for comparing two sets of data.

```
        0 | 1  1  5  7  9                                    9 8 7 5 3 1 | 3 | 2  4  7  9
        1 | 0  1  2  4  8  8  9            or     Leaf 1  1  6 9 5 1 1 0 | 4 | 1  2  2  4  5  6   Leaf 2
  Stem  2 | 1  1  1  3  7  2  4  1 Leaf                9 8 5 4 2 | 5 | 0  1  2  4  3
        3 | 0  1  2  3                                       1 | 6 | 0  0  1  1
        4 | 1  2                                                 S
                                                                 T
                                                                 E
                                                                 M
```

Pie Charts

The area of each sector of the circle represents the frequency of that class.

The angle is calculated as

$$\frac{\text{frequency of class}}{\text{total frequency}} \times 360°$$

Since the *area* represents frequencies, when comparing two sets of data with total frequencies f_1 and f_2 the radii of the circles should be proportional to $\sqrt{f_1}$ and $\sqrt{f_2}$.

Histograms

The area of each bar represents the frequency of that class.

There are *no* gaps in histograms unless a class has frequency zero.

Frequency A = Frequency B

Frequency A = $\frac{3}{2} \times$ Frequency C

Frequency Polygons

These join the mid-points of the top of each bar in the histogram. The first and final points are on the horizontal axis at the mid-point of the classes with frequency zero which precede and follow the bars plotted on the diagram (see above).

Arithmetic and Geometric Means

$$\text{Arithmetic mean} = \frac{x_1 + x_2 + \ldots + x_n}{n} \text{ or } \frac{\Sigma x_i}{n}$$

$$\text{Geometric mean} = \sqrt[n]{x_1 x_2 x_3 \ldots x_n}$$

Cumulative Frequency Curve (Ogive)

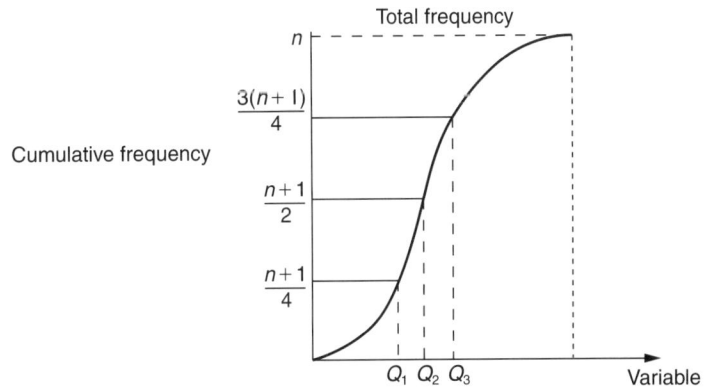

The cumulative frequency is the total frequency up to a particular value of the variable.

The first point is plotted at frequency = 0 and is the lower limit of the first class – remember a 'continuity factor' to avoid gaps between classes.

Q_1 or the **lower quartile** is the value of the variable when the cumulative frequency $= \dfrac{n + 1}{4}$.

Q_2 or the **median** is the value of the variable when the cumulative frequency $= \dfrac{n + 1}{2}$.

Q_3 or the **upper quartile** is the value of the variable when the cumulative frequency $= \dfrac{3(n + 1)}{4}$.

$Q_3 - Q_1$ is the interquartile range.

The cumulative frequency can be cut into deciles ($\frac{1}{10}$) or percentiles ($\frac{1}{100}$) – the median is the 5th decile or 50th percentile.

Expected Values

- Mean $\bar{x}$ or $E(X) = \dfrac{\Sigma x_i}{n}$ where n = number of elements.

Population mean is often denoted μ

If the data is expressed with frequencies f_i then

$$E(X) = \frac{\Sigma f_i x_i}{\Sigma f_i} = \frac{\Sigma f_i x_i}{n}$$

- Variance $s^2 = \frac{1}{n}\Sigma f_i (x_i - \bar{x})^2 = \frac{\Sigma (f_i x_i)^2}{n} - \bar{x}^2 = E(X^2) - (E(X))^2$

- Population variance is often denoted σ^2
- Standard deviation $= \sqrt{\text{variance}}$
- Unbiased estimate of the population mean and variance

$$\text{mean}_{\text{population}} = \text{mean}_{\text{sample}} \quad (\mu \approx \bar{x})$$

$$\text{variance}_{\text{population}} = \hat{s}^2 = \frac{n}{n-1} s^2 \text{ where } s^2 = \text{variance}_{\text{sample size } n}$$

$$(\sigma \approx \hat{s}. \quad \text{If } n \text{ is large } \sigma \approx \hat{s} \approx s)$$

Permutations

Permutations are arrangements where order is important.

n different objects have $n!$ permutations.

The number of permutations of r different objects chosen from n different

objects is written $^nP_r = \dfrac{n!}{(n-r)!}$

The number of permutations of n objects, r of which are identical, is $\dfrac{n!}{r!}$

Combinations

Combinations are groups where order is not important.

The number of groups of size r from n different objects is $^nC_r = \dfrac{n!}{r!(n-r)!}$

Worked Examples

1   The following data relates to the manpower (in thousands) employed in the health and personal social services in 1980.

Type of Employee	Number (thousands)
Medical and dental	46.5
Nursing	466.1
Professional and technical	76.4
Administrative	124.7
Other	274.4
Total	988.1

If these data are to be represented by a pie chart of radius 5 cm, calculate, to the nearest degree, the angles corresponding to each of the five

classifications. (DO NOT DRAW THE PIE CHART.)

If a comparable pie chart were to be drawn for 1981 when there were 1025.6 thousand employees, find the radius of this pie chart. Give your answer to 2 decimal places.   (L Specimen Paper, 1996)

-

	Thousand	Angle
Medical and dental	46.5	17°
Nursing	466.1	170°
Professional and technical	76.4	28°
Administrative	124.7	45°
Other	274.4	100°
Total	988.1	360°

$$\text{Angle} = \frac{\text{manpower}}{\text{total}} \times 360°$$

In 1981 there were 1025.6 thousand employees:

$$\text{Radius} = 5 \times \sqrt{\frac{1025.6}{988.1}} \text{ cm}$$

$$= 5.09 \text{ cm}$$

2 During one week, 159 randomly selected areas of Britain were monitored for the amount of rainfall recorded. The results are given in the table.

Rainfall (mm)	Number of areas
<0.1	6
0.1–0.2	23
0.2–0.3	31
0.3–0.4	50
0.4–0.5	28
0.5–0.6	14
0.6–0.7	5
0.7–0.8	2

(a) Construct a cumulative frequency table and plot its graph.
(b) Using the graph, find:
   (i) the percentage of areas having less than 0.28 mm of rainfall recorded
   (ii) the median rainfall.
(c) Use linear interpolation to find the upper and lower quartiles, and hence write down the interquartile range of the rainfall. (OLE, 1993)

• (a)

Rainfall (mm)	Number of areas	Cumulative frequency
<0.1	6	6
0.1–0.2	23	29
0.2–0.3	31	60
0.3–0.4	50	110
0.4–0.5	28	138
0.5–0.6	14	152
0.6–0.7	5	157
0.7–0.8	2	159

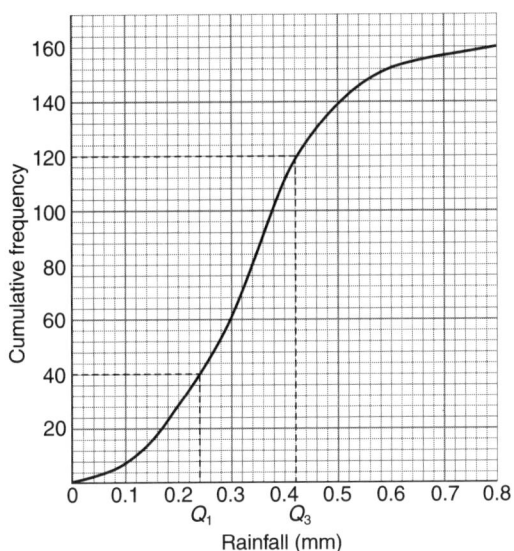

(b) (i) Percentage of areas having less than 0.28 mm rainfall:

$$\frac{53}{159} \times 100\% = 33\%$$

(ii) Median rainfall is 0.351 mm.
(c) Lower quartile = 0.235 mm
Upper quartile = 0.425 mm
IQR = 0.19 mm

3 A railway enthusiast simulates train journeys and records the number of minutes, x, to the nearest minute, trains are late according to the schedule being used. A random sample of 50 journeys gave the following times.

17	5	3	10	4	3	10	5	2	14
3	14	5	5	21	9	22	36	14	34
22	4	23	6	8	15	41	23	13	7
6	13	33	8	5	34	26	17	8	43
24	14	23	4	19	5	23	13	12	10

(a) Construct a stem and leaf diagram to represent these data.
(b) Comment on the shape of the distribution produced by your diagram.
(c) Given that $\Sigma x = 738$ and $\Sigma x^2 = 16\,526$, calculate to 2 decimal places, unbiased estimates of the mean and the variance of the population from which this sample was drawn.
(d) Explain briefly the effect that grouping of these data would have had on your calculations in (c). (L, 1992)

•
```
0 | 5 3 4 3 5 2 3 5 5 9 4 6 8 7 6 8 5 8 4 5
1 | 7 0 0 4 4 4 5 3 3 7 4 9 3 2 0
2 | 1 2 2 3 3 6 4 3 3          Rough sort
3 | 6 4 3 4
4 | 1 3
```

(a) *Stem and leaf*
```
0 | 2 3 3 3 4 4 4 5 5 5 5 5 5 6 6 7 8 8 8 9   20
1 | 0 0 0 2 3 3 3 4 4 4 4 5 7 7 9             15
2 | 1 2 2 3 3 3 3 4 6                          9
3 | 3 4 4 6                                    4
4 | 1 3                                        2
```

(b) Positive skew
(c) Unbiased estimators for mean and variance are

$$\text{mean}_{\text{population}} = \bar{x} = 14.76 \qquad \boxed{10.1}$$

$$s^2 = \frac{\Sigma x^2}{} - \bar{x}^2 = \frac{16\,526}{50} - (14.76)^2 = 112.66$$

$$\text{variance}_{\text{population}} \approx \hat{s}^2 = \frac{ns^2}{n-1}$$

$$= \frac{50}{49} \times 112.66 = 114.96$$

(d) The accuracy would be reduced if the data were grouped.

4 The heights x in centimetres of 42 seedlings were measured and the data summarised in the table below.

Height	Number of seedlings
$0 < x \leqslant 20$	6
$20 < x \leqslant 30$	6
$30 < x \leqslant 35$	6
$35 < x \leqslant 40$	6
$40 < x \leqslant 50$	6
$50 < x \leqslant 70$	6
$70 < x \leqslant 100$	6

(i)   Draw a histogram to represent these data.
(ii) Draw separately a cumulative frequency diagram.
(iii) If one of the seedlings is chosen at random, what is the probability that its height lies between 30 cm and 70 cm?
(iv) Estimate the values of the median and the interquartile range.   (O&C, 1992)

•

Height	Number of seedlings	Frequency /unit width	Cum. freq.
$0 < x \leqslant 20$	6	0.3	6
$20 < x \leqslant 30$	6	0.6	12
$30 < x \leqslant 35$	6	1.2	18
$35 < x \leqslant 40$	6	1.2	24
$40 < x \leqslant 50$	6	0.6	30
$50 < x \leqslant 70$	6	0.3	36
$70 < x \leqslant 100$	6	0.2	42

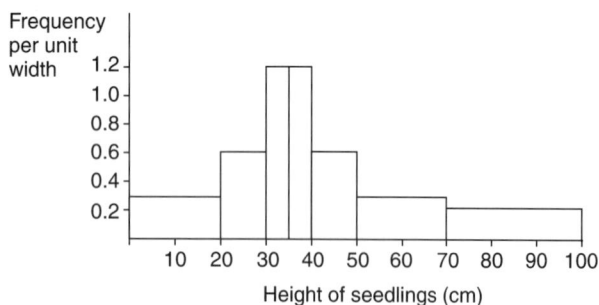

Frequency per unit width — Height of seedlings (cm)

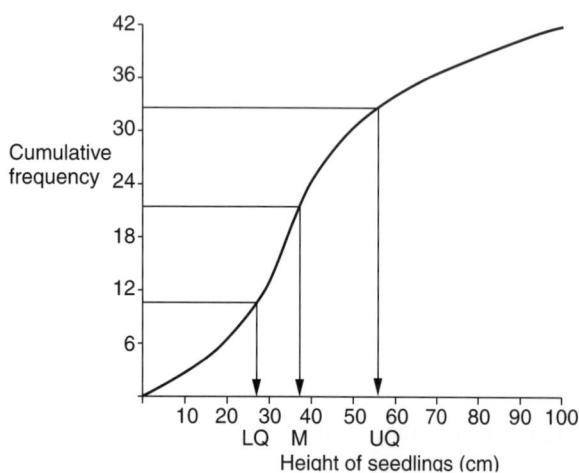

Cumulative frequency — Height of seedlings (cm)

(iii) From the table, 24 seedlings have heights between 30 and 70.

Probability $= \frac{24}{42}$ or $\frac{4}{7}$

(iv) From the cumulative frequency diagram:

Median is 38 cm

measured at cumulative frequency $21\frac{1}{2}$ $\left(\dfrac{n+1}{2}\right)$

Lower Quartile is 28 cm

measured at cumulative frequency $10\frac{3}{4}$ $\left(\dfrac{n+1}{4}\right)$

Upper Quartile is 55 cm

measured at cumulative frequency $32\frac{1}{4}$ $\left(\dfrac{3(n+1)}{4}\right)$

IQR is 27 cm

5 A random sample of 51 people were asked to record the number of miles they travelled by car in a given week. The distances, to the nearest mile, are shown below.

```
67  76  85  42  93  48  93  46
52  72  77  53  41  48  86  78
56  80  70  70  66  62  54  85
60  58  43  58  74  44  52  74
52  82  78  47  66  50  67  87
78  86  94  63  72  63  44  47
57  68  81
```

(a) Construct a stem and leaf diagram to represent these data.
(b) Find the median and quartiles of this distribution.
(c) Draw a vertical box and whisker plot to illustrate these data.   (L Specimen Paper, 1996)

• Do a rough sort of the data before drawing up the stem and leaf diagram

```
40 | 42  48  46  41  48  43  44  47  44  47
50 | 52  53  56  54  58  58  52  52  50  57
60 | 67  66  62  60  66  67  63  63  68
70 | 76  72  77  78  70  70  74  74  78  78  72
80 | 85  86  80  85  82  87  86  81
90 | 93  93  94
```

(a) Stem and leaf diagram:

```
4 | 1  2  3  4  4  6  7  7  8  8
5 | 0  2  2  2  3  4  6  7  8  8
6 | 0  2  3  3  6  6  7  7  8
7 | 0  0  2  2  4  4  6  7  8  8  8
8 | 0  1  2  5  5  6  6  7
9 | 3  3  4
```

(b) Median is the 26th piece of ranked data.
Median = 66
Lower quartile (13th) = 52
Upper quartile = 78

(c) Miles travelled:

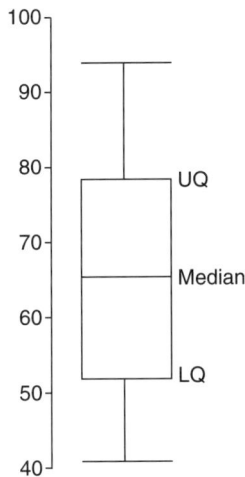

6 Four cards are drawn from a pack of 52 cards. Find the probability that

(i) all are queens
(ii) three are queens and one is an ace
(iii) there is one each of ace, king, queen and jack
(iv) there is ace, king, queen and jack of hearts
(v) there are three clubs and one diamond.

• (i) P(4 queens)

$$= \frac{4}{52} \times \frac{3}{51} \times \frac{2}{50} \times \frac{1}{49}$$

$$= \frac{^4C_4}{^{52}C_4}$$

$$= 0.000\,003\,69 \text{ or } \frac{1}{270\,725} \text{ or } 3.69 \times 10^{-6}$$

(ii) P(3 queens and 1 ace)

$$= \frac{\left(\begin{array}{c}\text{selection of}\\ \text{3 out of 4}\end{array}\right)\left(\begin{array}{c}\text{selection of}\\ \text{1 out of 4}\end{array}\right)}{(\text{selection of 4 out of 52})}$$

$$= \frac{^4C_3 \times {}^4C_1}{^{52}C_4}$$

$$= \frac{4 \times 4}{270\,725} = 0.000\,059\,1$$

(iii) P(1 ace, 1 king, 1 queen, 1 jack)

$$= \frac{(^4C_1)^4}{^{52}C_4} = 0.000\,946$$

(iv) P(ace, king, queen, jack of hearts)

$$= \frac{(^1C_1)^4}{^{52}C_4} = \frac{1}{270\,725} \text{ or } 3.69 \times 10^{-6}$$

(v) P(3 clubs, 1 diamond)

$$= \frac{^{13}C_3 \times {}^{13}C_1}{^{52}C_4} = \frac{3718}{270\,725} = 0.013\,733$$

7 A committee of 5 is to be chosen from 7 men and 5 women. How many different committees can be chosen if

(i) at least one woman must be on the committee
(ii) there must be more men than women?

• (i) Number of possible committees is $^{12}C_5 = 792$.
Number of committees which are all men is $^7C_5 = 21$.
Hence the number of different committees which contain at least one woman $= 792 - 21 = 771$.

(ii) Committees of (5M), (4M, 1W) and (3M, 2W):
Number of committees all male $= 21$
Number of committees (4M, 1W) $= {}^7C_4 \times {}^5C_1 = 175$
Number of committees (3M, 2W) $= {}^7C_3 \times {}^5C_2 = 350$
Hence the number of committees with more men than women is 546.

8 A population consists of the numbers

3, 4, 4, 5, 7, 9, 10, 14

(a) Calculate the mean and variance of this population.
A random sample of size 4 is taken **with** replacement from this population.
(b) Calculate the variance of the sample mean.
A random sample of 4 numbers is taken **without** replacement.
(c) Calculate the variance of the sample mean.

$\left(\begin{array}{l}\text{When sampling without replacement from a}\\ \text{population of size } N, \text{ the variance of the}\end{array}\right.$

$\left.\text{sample mean is } \dfrac{N-n}{N-1} \cdot \dfrac{\sigma^2}{n}.\right)$

The numbers are then split into 2 groups as follows:
    Group 1: 3, 4, 4, 5
    Group 2: 7, 9, 10, 14
(d) Calculate the variance of the numbers in each group.
A sample of size 4 is obtained by independently selecting 2 numbers, at random, **without** replacement, from each of the two groups. The population mean is to be estimated by

$$Y = \tfrac{1}{2}(\bar{X}_1 + \bar{X}_2)$$

where $\bar{X}_1$ and $\bar{X}_2$ are the **sample** means of the two groups.
(e) Find the variance of Y.
(f) State, giving a reason, which of the three samples is likely to give the best estimate of the population mean.   (L, 1993)

- 3, 4, 4, 5, 7, 9, 10, 14: $n = 8$

(a) $\Sigma x = 56 \Rightarrow$ mean value of $x = \frac{56}{8} = 7$

$\boxed{10.1}$

$\Sigma x^2 = 492 \Rightarrow E(X^2) = 61.5$
$\text{Var}(X) = E(X^2) - (E(X))^2 = 61.5 - 49$
$\qquad\qquad\qquad\qquad\qquad = 12.5$

(b) For a sample$_1$, size 4, with replacement:

$\text{Variance} = \dfrac{12.5}{4} = 3.125$

(c) For a sample$_2$, size 4, without replacement:

$\text{Variance} = \dfrac{8-4}{7} \cdot \dfrac{12.5}{4} = 1.79$

(d) Group 1: 3, 4, 4, 5, $\quad \Sigma x_1 = 16, \Sigma x_1^2 = 66$
Group 2: 7, 9, 10, 14, $\Sigma x_2 = 40, \Sigma x_2^2 = 426$

$\text{Variance (Group 1)} = \dfrac{66}{4} - (4)^2 = 0.5$

$\boxed{10.1}$

$\text{Variance (Group 2)} = \dfrac{426}{4} - (10)^2 = 6.5$

(e) Group 1, sample size 2, without replacement:

$\text{Variance} = \dfrac{2}{3} \cdot \dfrac{(0.5)}{2} = \dfrac{0.5}{3}$

Group 2, sample size 2, without replacement:

$\text{Variance} = \dfrac{2}{3} \cdot \dfrac{6.5}{2} = \dfrac{6.5}{3}$

Sample 3: $Y = \frac{1}{2}(X_1 + X_2)$
Variance of $\frac{1}{2}(X_1 + X_2) = \frac{1}{4}(\text{Var } X_1 + \text{Var } X_2)$

$\qquad\qquad\qquad = \dfrac{7}{12} = 0.583$

Sample 3 has the least variance, so giving the best estimate of the population mean.

9 (i) Find the total number of permutations of the word CALCULUS. In how many of these will the 2 C's be together?
(ii) In how many ways can all the vowels come together?
(iii) What is the number of possible arrangements of 3 letters from the word CALCULUS?

- (i) 8 letters: 2 C's, 2 U's, 2 L's, A, S.

$\text{Number of arrangements} = \dfrac{8!}{2!2!2!} = 5040$

If the two C's are together, consider 7 elements:

2 U's, 2 L's, A, S and (CC)

$\text{Number of arrangements} = \dfrac{7!}{2!2!} = 1260$

(ii) If all the vowels are together, then consider 6 elements:

(UUA), 2 L's, 2 C's, S

UUA can be arranged in 3 ways.

$\text{Number of arrangements} = \dfrac{6!}{2!2!} \times 3 = 540$

(iii) Number of different letters = 5.
Choose 3 from 5 and arrange $= 5 \times 4 \times 3$ ways
$\qquad\qquad\qquad\qquad\qquad\qquad = 60$ ways
Choose two the same + one other
$\qquad\qquad\qquad\qquad\qquad\qquad = 3 \times 4 \times 3$ ways
$\qquad\qquad\qquad\qquad\qquad\qquad = 36$
Hence number of arrangements of 3 letters
$\qquad\qquad\qquad\qquad\qquad\qquad = 96.$

Exercises

1 A box contains 4 red balls and 2 blue balls. A random sample of 3 balls is selected from the box. Calculate the probability that 2 red balls and 1 blue ball are selected given that the selection is made

(i) without replacement
(ii) with replacement.

For sampling with replacement, name the distribution of the number of red balls in the sample. Hence, or otherwise, find the mean and variance of the number of red balls in the sample.

(WJEC, 1994)

2 The numbers of half-days the 27 pupils in a certain secondary school class were absent during the last school year are given below. Make a stem-and-leaf plot of these numbers and hence find the median.

24	0	12	8	3	23	44	2	6	32
26	40	0	10	2	62	0	6	10	20
36	2	30	13	20	2	33			

(SEB, 1994)

3 (a) The following are weekly sales, in cases of 24 tins, of Tiddlers Cat Food at 40 retail outlets. Draw a stem and leaf diagram to represent this data.

14	27	51	36	62	60	17	34
53	54	37	40	27	30	26	36
29	41	23	37	51	37	12	36
39	17	40	36	35	47	35	13
21	21	48	54	23	37	43	23

(b) Draw a horizontal box plot to represent the following ordered data.

3	4	5	6	6	7	7	7
8	8	8	9	9	10	11	13
15	16	16	18	19	22	23	23
23	25	27	28	28	29	30	31

(SEB, 1993)

4 In a golf tournament, records are kept of the scores of 128 competitors for the 8th hole on the first day of the tournament. The results are summarized in the following table.

Score	3	4	5	6	7
Frequency	21	83	17	5	2

(i) Find the mode, the mean and the variance of the above distribution.

(ii) Two of these competitors are chosen at random.
  (a) Show that the probability that the sum of their scores for the 8th hole equals 10 is 0.073, correct to three places of decimals.
  (b) Find, correct to three places of decimals, the conditional probability that the difference in their scores for the 8th hole equals 2, given that the sum of their scores equals 10.

(iii) State whether the above data can be used to predict the expected score for the 8th hole by one particular competitor on the last day of the tournament. Give a reason for your answer.
(UCLES, 1992)

5 A group of four boys and six girls are to be seated in a row.

(i) How many possible arrangements are there?
(ii) What is the probability that
  (a) there is a girl at each end
  (b) all the boys are seated together
  (c) all the boys are seated together at one end of the row?

6 Summarized below is the distribution of marks obtained by a group of students in a Geography examination.

Mark range	Frequency
19 or less	2
20–29	14
30–39	21
40–44	34
45–49	39
50–59	42
60–69	13
70–79	9
80–89	4
90 or more	2

Explain how the median and quartiles of a distribution can be used when describing the shape of a distribution.

Use interpolation to estimate the median and quartiles of this distribution. Hence describe its shape. (L Specimen Paper, 1996)

7 A box contains 5 red, 7 blue and 8 white balls. If 3 balls are chosen at random determine the probability that

(i) all are red
(ii) all are the same colour
(iii) there is one of each colour
(iv) the balls are drawn in the order red, blue and white.

8 Twelve people are to travel to the theatre in three cars. Assuming that each car is driven by its owner, in how many ways can the remaining 9 people travel, given that the cars can take 2, 3 and 4 of them respectively?

Brief Solutions

1 4 red balls, 2 blue balls: total of 6 balls

(i) 3 without replacement:
$$P(R, R, B) = \frac{4}{6} \times \frac{3}{5} \times \frac{2}{4} = \frac{1}{5}$$

Number of arrangements of RRB = 3

$$\therefore P(2 \text{ reds and a blue}) = 3 \times \frac{1}{5} = \frac{3}{5}$$

(ii) With replacement:
$$P(R, R, B) = \frac{4}{6} \times \frac{4}{6} \times \frac{2}{6} = \frac{4}{27}$$

$$P(2 \text{ reds, 1 blue}) = \frac{4}{9}$$

With replacement this is a Binomial distribution. $B(n, p) = B(6, \frac{2}{3})$: mean = 4, variance = $\frac{4}{3}$. (See Fact Sheet, Chapter 16, page 198.)

2
```
0 | 0  0  0  2  2  2  2  3  6  6  8
1 | 0  0  2  3
2 | 0  0  3  4  6
3 | 0  2  3  6
4 | 0  4
5 |
6 | 2
```
Median 12

3 (a)
```
1 | 2  3  4  7  7
2 | 1  1  3  3  3  6  7  7  9
3 | 0  4  5  5  6  6  6  6  7  7  7  7  9
4 | 0  0  1  3  7  8
5 | 1  1  3  4  4
6 | 0  2
```

(b) Median = 14, Lower quartile = 7.25, Upper quartile = 23

10.1

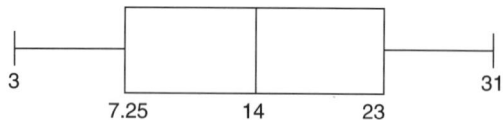

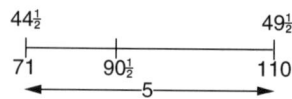

4 (i) $\Sigma f_i = 128$, modal score is 4   most frequent

$$\text{Mean score} = \frac{\Sigma f_i x_i}{128} = 4.094$$ [10.1]

$$\text{Variance} = \frac{\Sigma f_i x_i^2}{n} - (4.094)^2 = 0.585$$ [10.1]

(ii) (a) $P(X_1 + X_2 = 10)$
$2P(3)\,P(7) + 2P(4)P(6) + (P(5))^2 = 0.0734$
or 0.073 to 3 d.p.

(b) P(differ. in scores is 2 | sum of scores is 10)

$$= \frac{2P(4)P(6)}{0.0734} = 0.690$$

(iii) No. A *particular* competitor's score is not influenced by the scores of the other 127.

5 (i) Since people are all different:
$^{10}P_{10} = 10! = 3\,628\,800$ ways

(ii) (a) A girl at each end: $6 \times 5 = 30$ ways.

Remaining 8 are arranged in 8! ways.

$$\text{Probability is } \frac{30 \times 8!}{10!} = \frac{1}{3}.$$

(b) All boys make a group with 4! arrangements. 6 girls + group of boys: 7!4! arrangements.

$$\text{Probability is } \frac{7!\,4!}{10!} = \frac{1}{30}$$

(c) All boys together at one end:
(Group B) 4!   (Group G) 6!
Number of arrangements: $4!6! \times 2$
(BG and GB)

$$\text{Probability} = \frac{4!6! \times 2}{10!} = \frac{1}{105}$$

6 If $Q_2 - Q_1 < Q_3 - Q_2$ positive skew;
if $Q_2 - Q_1 > Q_3 - Q_2$ negative skew
Total number of students = 180.

The median is the score of the $\dfrac{181\text{st}}{2}$ student when ranked.

This comes in the interval $44\frac{1}{2} \to 49\frac{1}{2}$
(must be continuous).

$$\text{Median} = 44\frac{1}{2} + \left(\frac{90\frac{1}{2} - 71}{39}\right) \times 5 = 47$$

Lower quartile is the score of the $\dfrac{181\text{st}}{4}$ student.

$$\text{LQ} = 39\frac{1}{2} + \frac{8\frac{1}{4}}{34} \times 5 = 40.7.$$

Upper quartile is the score of the $\dfrac{3 \times 181\text{st}}{4}$ student ($135\frac{3}{4}$ student).

This comes in the 50–59 range.

$$\text{UQ is } 49\frac{1}{2} + \frac{25\frac{3}{4}}{42} \times 10 = 55.6.$$

$Q_2 - Q_1 = 47 - 40.7 = 6.3$
$Q_3 - Q_2 = 55.6 - 47 = 8.6$

Since these intervals are unequal there is a slight positive skew.
Q_1 = lower quartile; Q_2 = median; Q_3 = upper quartile (see Fact Sheet)

7 (i) $P(R, R, R) = \dfrac{5}{20} \times \dfrac{4}{19} \times \dfrac{3}{18} = \dfrac{1}{114}$

(ii) $P(B, B, B) = \dfrac{7}{20} \times \dfrac{6}{19} \times \dfrac{5}{18} = \dfrac{7}{228}$

$P(W, W, W) = \dfrac{8}{20} \times \dfrac{7}{19} \times \dfrac{6}{18} = \dfrac{14}{285}$

Hence the probability that all are the same colour

is $\dfrac{1}{114} + \dfrac{7}{228} + \dfrac{14}{285} = \dfrac{101}{1140}$ or 0.0886

(iii) P(one of each colour) = $P(R, B, W) \times 3!$

$$= \frac{5}{20} \times \frac{7}{19} \times \frac{8}{18} \times 6$$

$$= \frac{14}{57} = 0.2456$$

(iv) $P(R, B, W) = \dfrac{5}{20} \times \dfrac{7}{19} \times \dfrac{8}{18}$

$$= \frac{7}{171} = 0.0409.$$

8   For the first car choose 2 from 9: $\dfrac{9 \times 8}{2!}$ ways

For the second car choose 3 from 7: $\dfrac{7 \times 6 \times 5}{3!}$ ways

Remaining people go in car 3.

Total number of ways is

$$\dfrac{9 \times 8 \times 7 \times 6 \times 5}{2!3!} = \dfrac{9!}{2!3!4!} = 1260 \text{ ways.}$$

15

Probability

$$\int \frac{1}{1+x^2}\,dx$$

	AS Level			A Level		Topic			
CORE	MODULAR	TRADITIONAL	CORE	MODULAR	TRADITIONAL		Date attempted	Date completed	Self-assessment
	✓	✓	✓	✓	✓	**Probability, Venn diagrams**			
	✓	✓	✓	✓	✓	**Conditional probability**			
	✓	✓	✓	✓	✓	**Independent and/or exclusive events**			
	✓	✓	✓	✓	✓	**Tree diagrams**			

Fact Sheet

For One Event A

(i) $0 \leqslant P(A) \leqslant 1$

(ii) A' or $\overline{A}$ denotes the event 'A does not happen', and $P(A) + P(A') = 1$.

For Two Events, A and B

(i) $P(A$ or $B)$ means A and/or B occurs and is written $P(A \cup B)$.

(ii) $P(A$ and $B)$ means that both A and B occur and is written $P(A \cap B)$.

(iii) $P(A \cup B) = P(A) + P(B) - P(A \cap B)$.

Conditional Probability

The probability of A, given that B has already occurred, is written $P(A \mid B)$, where

$$P(A \mid B) = \frac{P(A \cap B)}{P(B)}, \quad P(B) \neq 0.$$

Independent Events

If $P(A \mid B) = P(A)$ then the occurrence or non-occurrence of event B does not influence the probability of event A. The events are said to be independent and $P(A)P(B) = P(A \cap B)$.

This is the multiplication law for independent events.

Mutually Exclusive Events

If $P(A \cap B) = 0$ then both of A and B cannot occur. The events are said to be mutually exclusive and $P(A \cup B) = P(A) + P(B)$.

This is the addition law for mutually exclusive events.

Exhaustive Events

$P(A \cup B) = 1$

Tree Diagrams

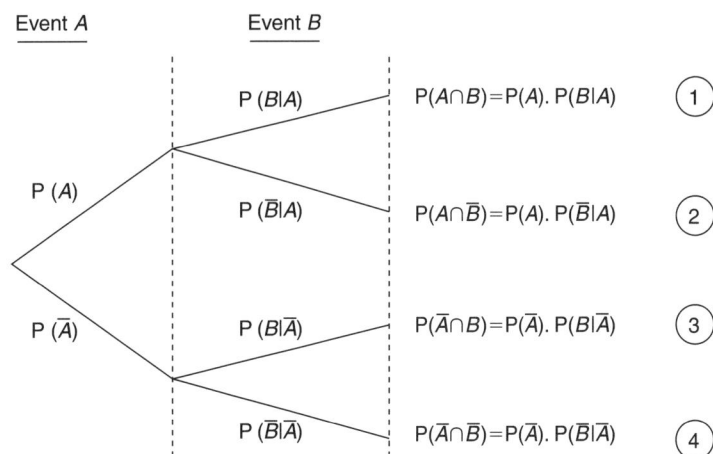

Remember:
(i) The total probability for any one set of 'branches' = 1.
(ii) The sum of final probabilities (intersections) = 1.
(iii) The tree shows conditional probabilities for $(B \mid A)$ etc.
  If $P(A \mid B)$ is required, then:
  (a) Look in the final column for the intersections containing B; in this case
  $P(B) = ① + ③$.
  (b) Find the term giving $A \cap B$, in this case ①.

$$\text{Then } P(A \mid B) = \frac{P(A \cap B)}{P(B)} = \frac{①}{① + ③}.$$

Venn Diagram

A Venn diagram can be used to represent the number of outcomes of an event or the probability of that event.

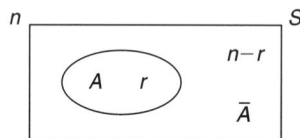

$$P(A) = \frac{r}{n} \qquad P(\overline{A}) = \frac{n - r}{n}$$

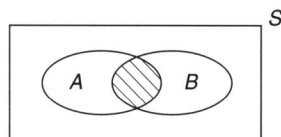

The shaded area represents $A \cap B$.

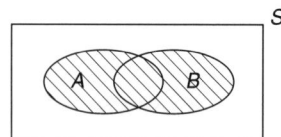

The shaded area represents $A \cup B$.

$$P(A) + P(B) - P(A \cap B) = P(A \cup B)$$

Worked Examples

1 The probability that a fisherman catches a fish is $\frac{7}{10}$ on a cloudy day, and $\frac{1}{5}$ on a clear day. If the probability of a cloudy day is $\frac{3}{5}$, find the probability that the day was cloudy given that he did not catch a fish.

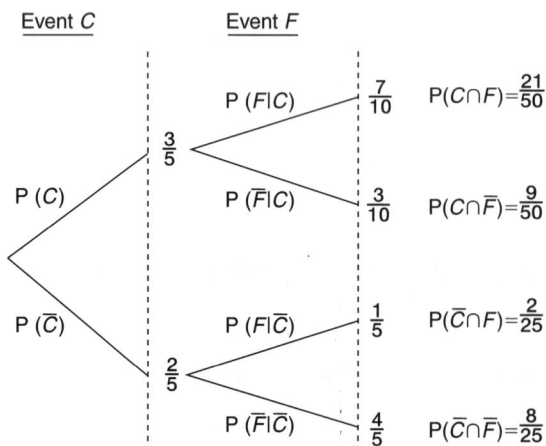

• Let event C be 'the weather was cloudy'.
Let event F be 'he caught a fish'.
To find $P(C \mid \overline{F})$:

$$P(\text{he did not catch a fish}) = P(\overline{F})$$
$$= P(C \cap \overline{F}) + P(\overline{C} \cap \overline{F})$$
$$= \tfrac{9}{50} + \tfrac{8}{25}$$
$$= \tfrac{1}{2}.$$

$$P(C \mid \overline{F}) = \frac{P(C \cap \overline{F})}{P(\overline{F})} = \frac{9/50}{1/2} = \frac{9}{25}.$$

2 Three boxes, X, Y and Z, contain coloured balls. X contains 5 black and 4 white balls, Y contains 7 black and 5 white balls and Z contains 3 black and 5 white balls.
  (a) If balls are withdrawn from box Z, with replacement, find the probability that the third ball drawn is the second white ball.
  (b) One of the boxes is selected at random and a ball is withdrawn from it. Find the probability that
    (i)   box X was chosen and the ball was black
    (ii) a white ball was chosen
    (iii) the ball was selected from box Z, given that it was black.

• (a) Box Z.
  $P(\text{black drawn}) = \frac{3}{8}$, $P(\text{white drawn}) = \frac{5}{8}$.
  For the third ball to be the second white one the possible orders are BWW and WBW.

  The balls are replaced after selection so
  $P(BWW) = (\frac{3}{8})(\frac{5}{8})(\frac{5}{8}) = \frac{75}{512}$
  $P(WBW) = (\frac{5}{8})(\frac{3}{8})(\frac{5}{8}) = \frac{75}{512}$.
  The probability that the third ball is the second white one is $\frac{75}{256}$.

  (b)

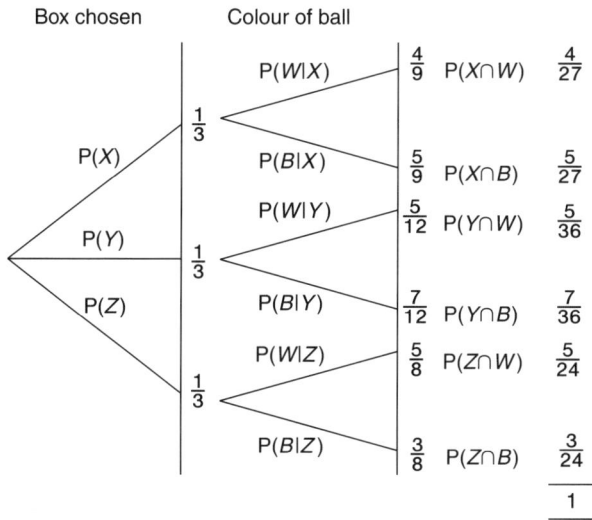

Box chosen / Colour of ball tree diagram

  (i) Prob. of box X and a black ball
    $= P(X \cap B)$
    $= (\frac{1}{3})(\frac{5}{9}) = \frac{5}{27}$.

  (ii) Prob. that a white ball was chosen
    $= P(X \cap W) + P(Y \cap W) + P(Z \cap W)$
    $= (\frac{1}{3})(\frac{4}{9}) + (\frac{1}{3})(\frac{5}{12}) + (\frac{1}{3})(\frac{5}{8}) = \frac{107}{216}$.

  (iii) Prob. that a black ball was chosen
    $= 1 - P(W) = \frac{109}{216}$.
    Prob. of box Z and a black ball $= (\frac{1}{3})(\frac{3}{8}) = \frac{1}{8}$.
    Prob. that the ball was selected from box Z given that it was a black ball
    $$= P(Z \mid B) = \frac{P(Z \cap B)}{P(B)} = \frac{1/8}{109/216} = \frac{27}{109}.$$

3 A card is drawn from a standard pack of 52 playing cards and is not replaced. A second card is then drawn. Find

  (a) the probability that the second card is a club given that the first card is a club
  (b) the probability that the second card is a club
  (c) the expected number of club cards in two draws

  Answer the same questions again if the first card is replaced before the second card is drawn.

• In a pack of 52 cards, 13 are clubs.

  Without replacement:

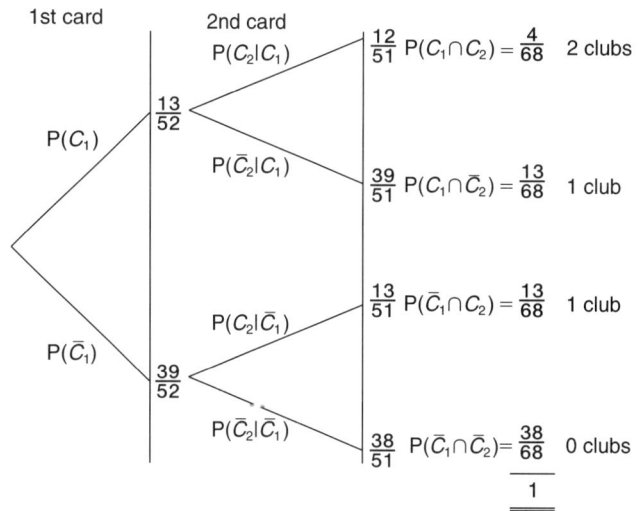

  Let C_1 be the event 'a club is drawn first'.
  Let C_2 be the event 'a club is drawn second'.
  From the tree diagram

  (a) $P(C_2 \mid C_1) = \frac{12}{51} = \frac{4}{17}$.
  (b) $P(C_2) = P(C_1 \cap C_2) + P(\bar{C}_1 \cap C_2)$
    $= (\frac{13}{52})(\frac{12}{51}) + (\frac{39}{52})(\frac{13}{51}) = \frac{17}{68} = \frac{1}{4}$.
    Note: This would give the same answer for C_3, C_4, etc. The probability that any chosen card is a club is $\frac{1}{4}$.
  (c) $P(0 \text{ clubs}) = (\frac{3}{4})(\frac{38}{51}) = \frac{38}{68}$
    $P(1 \text{ club}) = (\frac{1}{4})(\frac{13}{17}) + (\frac{3}{4})(\frac{13}{51}) = \frac{26}{68}$
    $P(2 \text{ clubs}) = (\frac{1}{4})(\frac{4}{17}) = \frac{4}{68}$
    Expected number of clubs
    $= (\frac{38}{68})(0) + (\frac{26}{68})(1) + (\frac{4}{68})(2) = \frac{1}{2}$.

  With replacement:
  (a) $P(C_2 \mid C_1) = \frac{1}{4}$
  (b) $P(C_2) = \frac{1}{4}$
  (c) $P(0 \text{ clubs}) = (\frac{3}{4})(\frac{3}{4}) = \frac{9}{16}$
    $P(1 \text{ club}) = (\frac{13}{52})(\frac{39}{52})(2) = \frac{6}{16}$
    $P(2 \text{ clubs}) = (\frac{1}{4})(\frac{1}{4}) = \frac{1}{16}$
    Expected number of clubs
    $= (\frac{9}{16})(0) + (\frac{6}{16})(1) + (\frac{1}{16})(2) = \frac{1}{2}$.

4 A bag contains 6 white beads, 5 red beads and 4 blue beads. Three are selected at random without replacement. Find the probability that
  (a) the third bead is white
  (b) the third bead is the first white
  (c) the beads were selected white, red and blue in that order

(d) the three beads are different colours
(e) the third bead is red, given that the first was red.

Explain the difference between the expected number of white beads and the most likely number of white beads and find
(f) the expected number of white beads and
(g) the most likely number of white beads.

- Total number of beads = 15.
  (a) The probability that any bead is white = $\frac{6}{15}$.
  (b) The probability that the first white bead is the third bead
  $= $ P(1st not white) $\times$ P(2nd not white)
  $\qquad \times$ P(3rd is white)
  $= (\frac{9}{15})(\frac{8}{14})(\frac{6}{13}) = \frac{72}{455}$.
  (c) P(1st is white) $\times$ P(2nd is red) $\times$ P(3rd is blue)
  $= (\frac{6}{15})(\frac{5}{14})(\frac{4}{13}) = \frac{4}{91}$
  (d) White, red and blue can be arranged in 3! = 6 different ways so

  P(3 different colours) $= (6)(\frac{4}{91}) = \frac{24}{91}$

  (e) If the first bead is red then 6W, 4R and 4B beads are left.
  P(any bead is red) $= \frac{4}{14} = \frac{2}{7}$. Therefore
  $P(R_3 \mid R_1) = \frac{2}{7}$.
  The expected number of white beads is the statistical mean.
  In general $E(X) = \Sigma\{xP(X = x)\}$
  This is not usually an integer, i.e. not a member of the possibility space.
  The most likely number of white beads is that with the highest probability.
  (f) Expected number of white beads $= 3(\frac{6}{15}) = \frac{6}{5}$.
  (g) P(1 white bead)
  $= P(W, \overline{W}, \overline{W}) + P(\overline{W}, W, \overline{W}) + P(\overline{W}, \overline{W}, W)$
  $= (\frac{6}{15})(\frac{9}{14})(\frac{8}{13})(3)$.
  P(2 white beads)
  $= P(W, W, \overline{W}) \times 3$
  $= (\frac{6}{15})(\frac{5}{14})(\frac{9}{13})(3)$.
  Since P(1 W bead) > P(2 W beads), the most likely number of white beads is 1.

5 Two boys play a game taking it in turns to roll a six-sided die. The first boy to obtain a six wins. What is the probability that the boy who rolls first wins on
(i) his first roll
(ii) his second roll
(iii) his third roll?
What is the probability that the boy who rolls first wins?

The rules of the game state that the boy who lost the first game should start the second game. What is the probability that the boy who rolled first in the first game wins both the first and the second game?

- (i) Prob. boy A wins on 1st roll $= \frac{1}{6}$.

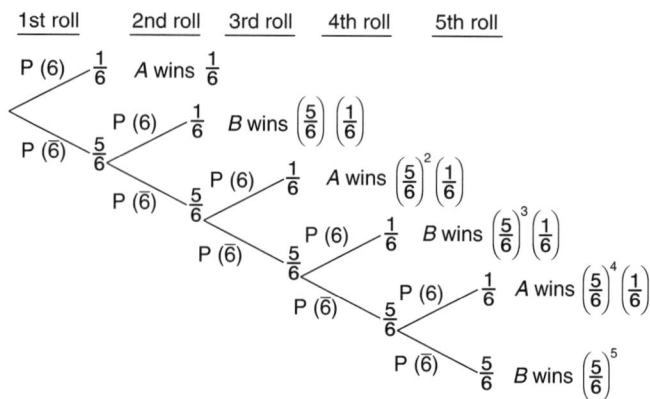

1st roll	2nd roll	3rd roll	4th roll	5th roll

P (6) $\frac{1}{6}$ A wins $\frac{1}{6}$

P (6) $\frac{1}{6}$ B wins $(\frac{5}{6})(\frac{1}{6})$

P ($\overline{6}$) $\frac{5}{6}$

P (6) $\frac{1}{6}$ A wins $(\frac{5}{6})^2(\frac{1}{6})$

P ($\overline{6}$) $\frac{5}{6}$

P (6) $\frac{1}{6}$ B wins $(\frac{5}{6})^3(\frac{1}{6})$

P ($\overline{6}$) $\frac{5}{6}$

P (6) $\frac{1}{6}$ A wins $(\frac{5}{6})^4(\frac{1}{6})$

P ($\overline{6}$) $\frac{5}{6}$

P ($\overline{6}$) $\frac{5}{6}$ B wins $(\frac{5}{6})^5$

(ii) Prob. boy B wins on his 1st roll
$= $ (Prob. A did not win)$(\frac{1}{6}) = (\frac{5}{6})(\frac{1}{6}) = \frac{5}{36}$.
Prob. boy A wins on his 2nd roll
$= (\frac{5}{6})(\frac{5}{6})(\frac{1}{6}) = (\frac{5}{6})^2(\frac{1}{6})$.
(iii) Prob. boy A wins on his 3rd roll $= (\frac{5}{6})^4(\frac{1}{6})$.
Prob. that boy A wins is $\frac{1}{6} + \frac{1}{6}(\frac{5}{6})^2 + (\frac{1}{6})(\frac{5}{6})^4 + \dots$
This is a geometric progression with $a = \frac{1}{6}$, $r = \frac{25}{36}$.

Prob. that A wins $= \dfrac{a}{1-r} = \dfrac{1/6}{11/36} = \dfrac{6}{11}$.

If boy A wins the first game, boy B starts the second game and has a probability of $\frac{6}{11}$ of winning the second game.
Thus boy A has a prob. of $(1 - \frac{6}{11}) = \frac{5}{11}$ of winning the second game.
Probability of boy A winning both games $= (\frac{6}{11})(\frac{5}{11}) = \frac{30}{121}$.

6 (a) Two cards are drawn without replacement from ten cards which are numbered from 1 to 10. Find the probability that
(i) the numbers on both cards are even
(ii) the number on one card is odd and the number on the other card is even
(iii) the sum of the numbers on the two cards exceeds 4.
(b) Events A and C are independent. Probabilities relating to events A, B and C are as follows:

$P(A) = \frac{1}{5}$ $P(B) = \frac{1}{6}$ $P(A \cap C) = \frac{1}{20}$
$P(B \cup C) = \frac{3}{8}$

(a) Evaluate $P(C)$ and show that events B and C are independent.

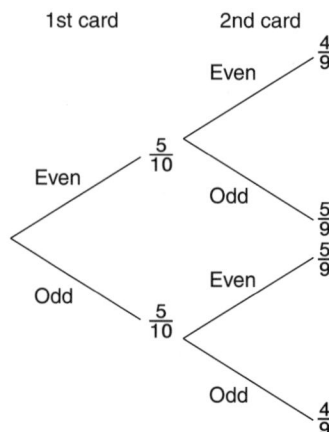

-

1st card	2nd card

Even $\frac{4}{9}$

Even $\frac{5}{10}$

Odd $\frac{5}{9}$

Even $\frac{5}{9}$

Odd $\frac{5}{10}$

Odd $\frac{4}{9}$

(i) Probability of 2 even cards $= (\frac{1}{2})(\frac{4}{9}) = \frac{2}{9}$.

(ii) Probability of 1 odd and 1 even
$= \frac{5}{18} + \frac{5}{18} = \frac{5}{9}$.

(iii) Probability that the sum exceeds 4
$= 1 - \{P(1, 2) + P(2, 1) + P(1, 3) + P(3, 1)\}$
$= 1 - (\frac{1}{10})(\frac{1}{9})(4) = \frac{43}{45}$.

(b) A and C are independent so $P(A \mid C) = P(A)$
or $P(A) \cdot P(C) = P(A \cap C)$.

Venn diagram

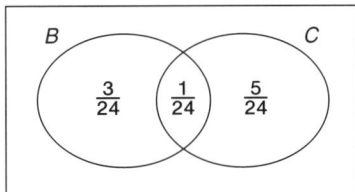

$P(A) = \frac{1}{6}$, $P(A \cap C) = \frac{1}{20}$, thus $P(C) = \frac{1}{4}$.
$P(B) + P(C) = \frac{1}{6} + \frac{1}{4} = \frac{10}{24}$.
$P(B \cup C) = \frac{9}{24}$, therefore $P(B \cap C) = \frac{1}{24}$.

$$P(B \mid C) = \frac{P(B \cap C)}{P(C)} = \frac{1/24}{1/4} = \frac{1}{6}.$$

Therefore $P(B \mid C) = P(B)$. Hence B and C are independent.

7 The events A and B are such that

$$P(A') = \frac{3}{4} \quad P(A \mid B) = \frac{1}{3} \quad P(A \cup B) = \frac{2}{3}$$

where A' denotes the event 'A does not occur'.

Find
(i) $P(A)$
(ii) $P(A \cap B)$
(iii) $P(B)$
(iv) $P(A \mid B')$
where B' denotes the event 'B does not occur'.

Determine whether A and B are independent.
(O&C, 1991)

•

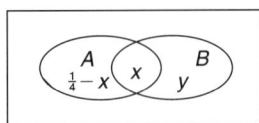

Let $P(A \cap B) = x$ $P(B \cap A') = y$
(i) $P(A') = \frac{3}{4}$ $\Rightarrow$ $P(A) = \frac{1}{4}$
(ii) $P(A \cup B) = \frac{1}{4} - x + x + y = \frac{2}{3}$
$\therefore y = \frac{5}{12}$

$$P(A \mid B) = \frac{P(A \cap B)}{P(B)} = \frac{x}{x + \frac{5}{12}}$$

$$\therefore \frac{x}{x + \frac{5}{12}} = \frac{1}{3} \Rightarrow x = \frac{5}{24}$$

Hence $P(A \cap B) = \frac{5}{24}$

(iii) $P(B) = x + y = \frac{15}{24} = \frac{5}{8}$.

(iv) $P(A \mid B') = \dfrac{P(A \cap B')}{P(B')} = \dfrac{\frac{1}{24}}{\frac{3}{8}} = \dfrac{1}{9}$

If A and B are independent
$P(A \cap B) = P(A) \cdot P(B)$
$P(A \cap B) = \frac{5}{24}$, $P(A) \cdot P(B) = \frac{1}{4} \times \frac{5}{8} = \frac{5}{32}$
Hence A and B are not independent.

8 A lecturer uses a statistical technique to judge the true reception given to her lectures. She asks each student in the lecture-room to toss two coins in the air and note whether or not they land showing two heads. She then tells the students to raise their right arm if one of the following two conditions apply:
I both coins land heads and the student enjoyed the lecture;
II the coins did not land showing two heads and the student did not enjoy the lecture.

Let A denote the event that a randomly selected student obtained two heads, and let B denote the event that a randomly chosen student raises the right arm.

(a) Describe in words the complement, $\bar{A}$, of the event A described above.

(b) If the proportion of the students who actually enjoyed the lecture is p, what are the following probabilities:
(i) $P(A)$
(ii) $P(B$ given $A)$
(iii) $P(A \cap B)$
(iv) $P(\bar{A} \cap B)$?

(c) Hence, or otherwise, demonstrate carefully that

$$P(B) = (3 - 2p)/4$$

(d) What conclusion would you draw if it turned out that 36 out of an audience of 100 raise their right arm? (OLE, 1992)

• (a) $\bar{A}$ is the event that a randomly selected student did not throw 2 heads.

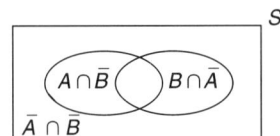

(b) (i) $P(A) = \frac{1}{4}$
(ii) $P(B$ given $A) = p$
(iii) $P(A \cap B) = \frac{1}{4}(p)$
(iv) $P(\bar{A} \cap B) = \frac{3}{4}(1 - p)$

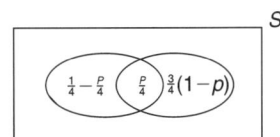

(c) $P(B) = $ (probability of 2H and enjoyed) $+$ (prob. of not 2H and didn't enjoy)
$= \frac{1}{4}p + \frac{3}{4}(1 - p)$
$= \frac{1}{4}(3 - 2p)$

(d) If $P(B) = \frac{36}{100}$ then $\frac{36}{100} = \frac{75}{100} - \frac{1}{2}p \Rightarrow p = \frac{78}{100}$
i.e. 78% enjoyed the lecture

9 The events A and B are such that

$$P(A) = \tfrac{1}{2} \quad P(A' \mid B) = \tfrac{1}{3} \quad P(A \cup B) = \tfrac{3}{5}$$

where A' is the event 'A does not occur'.
Using a Venn diagram, or otherwise, determine
$P(B \mid A')$, $P(B \cap A)$ and $P(A \mid B')$.

The event C is independent of A and
$P(A \cap C) = \tfrac{1}{8}$. Determine $P(C \mid A')$.
State, with a reason in each case, whether
(a) A and B are independent
(b) A and C are mutually exclusive.

• Let $P(A \cap B) = x$ and $P(B \cap A') = y$
$P(A) = \tfrac{1}{2} \qquad P(A') = \tfrac{1}{2}$
$P(B) = x + y \qquad P(B') = 1 - (x + y)$

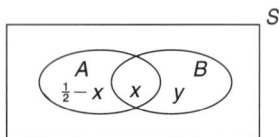

$P(A' \mid B) = \dfrac{y}{x+y} = \tfrac{1}{3} \Rightarrow x = 2y$ (1)

$P(A \cup B) = \tfrac{3}{5} \Rightarrow \tfrac{1}{2} + y = \tfrac{3}{5}$ (2)

From equation 2, $y = \tfrac{1}{10}$
From equation 1, $x = \tfrac{1}{5}$

$P(B \mid A') = P(B \cap A')/P(A') = \dfrac{y}{\tfrac{1}{2}} = \tfrac{1}{5}$

$P(B \cap A) = \tfrac{1}{5}$

$P(A \mid B') = \dfrac{\tfrac{3}{10}}{\tfrac{7}{10}} = \tfrac{3}{7}$

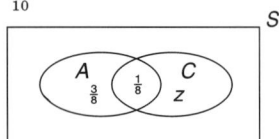

A and C are independent $\Rightarrow P(A \mid C) = P(A)$

$\Rightarrow \dfrac{\tfrac{1}{8}}{z + \tfrac{1}{8}} = \tfrac{1}{2}$

$\Rightarrow z = \tfrac{1}{8}$

$P(C \mid A') = \dfrac{\tfrac{1}{8}}{\tfrac{4}{8}} = \tfrac{1}{4}$

(a) A and B: $P(A \mid B) = \dfrac{\tfrac{2}{10}}{\tfrac{3}{10}} = \tfrac{2}{3} \qquad P(A) = \tfrac{1}{2}$

∴ Not independent
(b) A and C are not mutually exclusive since
$A \cap C \neq 0$.

Exercises

1 There are six blue socks and four red socks in a drawer. When I go for a pair the light is fused, so the room is in darkness. What is the probability that if I choose two socks I will not end up with a pair?

A $\tfrac{4}{15}$ **B** $\tfrac{12}{25}$ **C** $\tfrac{8}{15}$ **D** $\tfrac{29}{50}$ **E** none of these

2 I need to replace the electric element in my kettle. In the nearest town there are 4 shops each of which have a probability of 0.75 of having the required element in stock. I try the shops in turn until I find the element (if one is available). Find
(a) the probability that I visit four shops
(b) the expected number of shops visited.

3 Nine beads, 5 red and 4 white, are taken at random and threaded on a straight rod.
(a) What is the probability that no bead is next to one of the same colour?
(b) What is the probability that the last four beads include 2 red and 2 white beads?

4 Three apple trees and three pear trees are planted in a row in a random order. Find the probability that
(a) the three apple trees are together
(b) the apple and pear trees are planted alternately.
What would be the corresponding probabilities if the trees are planted in a circle?

5 Explain what is meant statistically by the statement that two events, A and B, are independent.

Let A be the event that the result of tossing a coin several times contains both heads and tails.

Let B be the event that there is at most one head. Assuming that the probabilities of obtaining a head or tail are equal, determine whether the events A and B are independent if the coin is tossed
(a) three times
(b) four times.

6 A boat hiring company at the local boating lake has two types of craft: 20 Bluebirds and 35 Herons. The customer has to take the next boat available when hiring. The boats are all distinguishable by their numbers.

A regular customer, John, carefully notes the numbers of the boats which he uses, and finds that he has used 15 different Bluebirds and 20 different Herons. Each boat is equally likely to be the next in line. John hires two boats at the same time (one is for a friend).

If event X is 'John has not hired either boat before', and event Y is 'both boats are Herons', determine
(a) P(X)
(b) P(Y)
(c) P($X \mid Y$)
(d) whether the events are mutually exclusive.

7 A golfer observes that, when playing a particular hole at his local course, he hits a straight drive on 80% of the occasions when the weather is not windy but only on 30% of the occasions when the weather is windy. Local records suggest that the weather is windy on 55% of all days.
  (i) Show that the probability that, on a randomly chosen day, the golfer will hit a straight drive at the hole is 0.525.
  (ii) Given that he fails to hit a straight drive at the hole, calculate the probability that the weather is windy.   (JMB, 1991)

8 A boy found that when he played his sister at snooker he had a probability of 0.4 of winning any one game and 0.6 of losing. He plays three games against his sister.
  (a) Draw a tree diagram showing all possible results and their probabilities assuming that each game is independent.
  (b) What is the probability of the boy winning two of the games and losing the other? (SEB, 1994)

9 The probability that a married man watches a certain television show is 0.4 and the probability that a married woman watches the show is 0.5. The probability that a man watches the show, given that his wife does, is 0.7.

  Calculate
  (i) the probability that a married couple watch the show
  (ii) the probability that a wife watches the show given that her husband does
  (iii) the probability that at least one person of a married couple will watch the show.

10 Julie and Zubair play a board game. The player who starts each game has an advantage. When Julie starts, the probability of her winning that game is 0.6. When Zubair starts, the probability of his winning that game is 0.7. All games end in a win for one player or the other; draws are not possible.

  The players agree to play a series of games which will continue until one or the other has won 3 games. Apart from the first game, each game is started by the winner of the previous game.
  (a) What is the probability of Julie winning the series 3–0 when
    (i) Julie starts the first game
    (ii) Zubair starts the first game
    (iii) the player starting the first game is decided by tossing a fair coin?

(b) Show that when Julie starts the first game the probability that Zubair wins the series 3–1 is 0.1848.   (AEB, 1994)

11 A child has a bag containing 12 sweets of which 3 are yellow, 5 are green and 4 are red. When the child wants to eat one of the sweets, a random selection is made from the bag and the chosen sweet is then eaten before the next random selection is made.
  (a) Find the probability that the child does not select a yellow sweet in the first 2 selections.
  (b) Find the probability that there is at least one yellow sweet in the first 2 selections.
  (c) Find the probability that the fourth sweet selected is yellow, given that the first two sweets selected were red ones.
                (L Specimen Paper, 1996)

12 Two guns fire at a target. For each shot from gun A the probability of hitting the target is $\frac{2}{5}$. For each shot from gun B, the probability of hitting the target is $\frac{1}{4}$.
  (a) Gun A fires three shots at the target. Find the probability that there is exactly one hit.
  (b) Gun A now fires another three shots at the target and gun B fires two shots. Find the probability of
    (i) no hits
    (ii) at least one hit
    (iii) exactly one hit.
  Given that there was exactly one hit, find the probability that this was gun B.   (O&C, 1993)

13 (a) A straight rod is 12 cm long. The rod is cut into three smaller rods so that the length of each smaller rod is a whole number of centimetres. Show that there are 12 distinct sets of three rods which can be formed in this way.

  If one of these sets is chosen at random, find the probability that the three rods so chosen form the sides of a triangle.
  (b) A bag contains 12 identical £1 coins. The coins are all taken from the bag and distributed between 3 students A, B and C, so that each receives at least one coin. Show that this distribution can be made in 55 different ways.

  One of these 55 ways is chosen at random. The event X is 'A has more than 5 coins' and the event Y is 'B has precisely 1 coin'. Find P(X) and P(Y) and show that P($X \cap Y$) = $\frac{1}{11}$. Hence determine the value of P($X \cup Y$).

  Find the probability that A has more than 5 coins, given that B has precisely 1 coin.
                (AEB, 1990)

14 Three events A, B, C are such that $P(B) = \frac{1}{4}$, $P(C) = \frac{1}{3}$, $P(A \cup B) = \frac{3}{5}$ and $P(A \cup C) = \frac{2}{3}$. The events A, B are independent and the events B, C are mutually exclusive.
(i) Find $P(A)$.
(ii) Determine whether or not A, C are independent.
(iii) Show that $P(B \cup C \mid A) = \frac{15}{28}$. (WJEC, 1994)

15 (a) Two fair dice are thrown, and events A, B and C are defined as follows.
A: the sum of the two scores is odd
B: at least one of the two scores is greater than 4
C: the two scores are equal.
Find, showing your reasons clearly in each case, which two of these three events are
(i) mutually exclusive
(ii) independent.
Find also $P(C \mid B)$, making your method clear.
(b) Two players A and B regularly play each other at chess. When A has the first move in a game the probability of A winning that game is 0.4 and the probability of B winning that game is 0.2. When B has the first move in a game, the probability of B winning that game is 0.3 and the probability of A winning that game is 0.2. Any game of chess that is not won by either player ends in a draw.
(i) Given that A and B toss a fair coin to decide who has the first move in a game, find the probability of the game ending in a draw.
(ii) To make their games more enjoyable, A and B decide to change the procedure for deciding who has the first move in a game. As a result of their new procedure, the probability of A having the first move in any game is p. Find the value of p which gives A and B equal chances of winning each game. (UCLES, 1993)

Brief Solutions

1

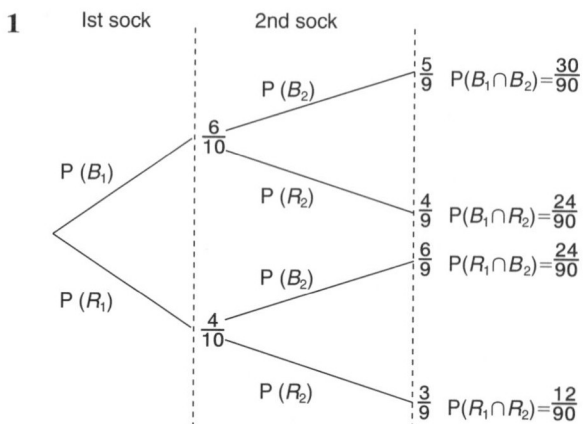

Probability of odd socks
$= P(B_1 \cap R_2) + P(R_1 \cap B_2)$
$= (\frac{6}{10})(\frac{4}{9}) + (\frac{4}{10})(\frac{6}{9}) = (\frac{8}{15})$ <u>Answer **C**</u>

2 (a) I visit 4 shops if the first 3 do not have an element.
$$P(\text{visit 4 shops}) = (\tfrac{1}{4})^3 = \tfrac{1}{64}$$

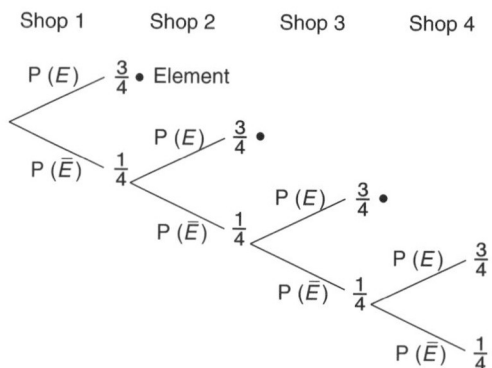

(b) $P(\text{only 1 visited}) = \frac{3}{4}$
$P(\text{only 2 visited}) = (\frac{1}{4})(\frac{3}{4}) = \frac{3}{16}$
$P(\text{only 3 visited}) = (\frac{1}{4})^2(\frac{3}{4}) = \frac{3}{64}$
$P(4 \text{ visited}) = (\frac{1}{4})^3 = \frac{1}{64}$
Expected number
$$= 1(\tfrac{3}{4}) + 2(\tfrac{3}{16}) + 3(\tfrac{3}{64}) + 4(\tfrac{1}{64})$$
$$= \tfrac{85}{64}$$

3 5 red and 4 white can be arranged in
$$\frac{9!}{4!5!} = 126 \text{ ways.}$$
Only one is RWRWRWRWR, therefore
(a) P(no bead is next to one of the same colour) $= \frac{1}{126}$.
(b) Probability that any group of 4 selected are RRWW $= (\frac{5}{9})(\frac{4}{8})(\frac{4}{7})(\frac{3}{6}) = \frac{5}{63}$.
Number of arrangements of 2 red and 2 white
$$= \frac{4!}{2!2!} = 6 \implies \text{probability is } \tfrac{10}{21}.$$

4 (a) $P(AAAPPP) = (\frac{3}{6})(\frac{2}{5})(\frac{1}{4})(1) = \frac{1}{20}$.
Arrangements of $(AAA), P, P, P = 4$.
Therefore probability of 3 apple trees together $= \frac{1}{20} \times 4 = \frac{1}{5}$.
(b) $P(APAPAP) + P(PAPAPA)$
$= (\frac{3}{6})(\frac{3}{5})(\frac{2}{4})(\frac{2}{3})(\frac{1}{2})(\frac{1}{1})(2) = \frac{1}{10}$.
In a circle fix one tree (A) and arrange the others. $2A$ and $3P$ can be arranged in
$$\frac{5!}{2!3!} = 10 \text{ ways.}$$
$A, A, (PPP)$ can be arranged in 3 ways.
(a) P(apples together) $= \frac{3}{10}$
(b) P(trees planted alternately) $= \frac{1}{10}$

5 See Fact Sheet.
(a) $P(A) = \frac{3}{4}$, $P(B) = \frac{1}{2}$.
$$P(A \mid B) = \frac{P(1H, 2T)}{P(B)} = \frac{3}{4} = P(A)$$
thus A and B are independent.

(b) $P(A) = \frac{7}{8}$, $P(B) = \frac{5}{16}$.

$$P(A \mid B) = \frac{P(1H, 3T)}{P(B)} = \frac{4}{5} \neq P(A)$$

thus not independent.

6 (a) P(has not hired 1st boat) $= \frac{20}{55}$
P(has not hired 2nd boat) $= \frac{19}{54}$
$\Rightarrow$ $P(X) = (\frac{20}{55})(\frac{19}{54}) = \frac{38}{297}$

(b) P(1st boat is a Heron) $= \frac{35}{55}$
P(2nd boat is a Heron) $= \frac{34}{54}$
$\Rightarrow$ $P(Y) = (\frac{35}{55})(\frac{34}{54}) = \frac{119}{297}$

(c) $P(X \mid Y) = \dfrac{P(X \cap Y)}{P(Y)}$,

where $P(X \cap Y) = (\frac{15}{55})(\frac{14}{54}) = \frac{7}{99}$.

Therefore $P(X \mid Y) = \dfrac{7/99}{119/297} = \dfrac{3}{17}$.

(d) $P(X \cap Y) \neq 0$ $\Rightarrow$ events are not mutually exclusive.

7 (i) Probability of a straight drive is
$0.8(0.45) + 0.3(0.55) = 0.525$.

(ii) He fails to hit a straight drive with probability
$1 - 0.525 = 0.475$

Probability that it was windy

$= \dfrac{(0.7)(0.55)}{0.475} = 0.811$.

8 (a)

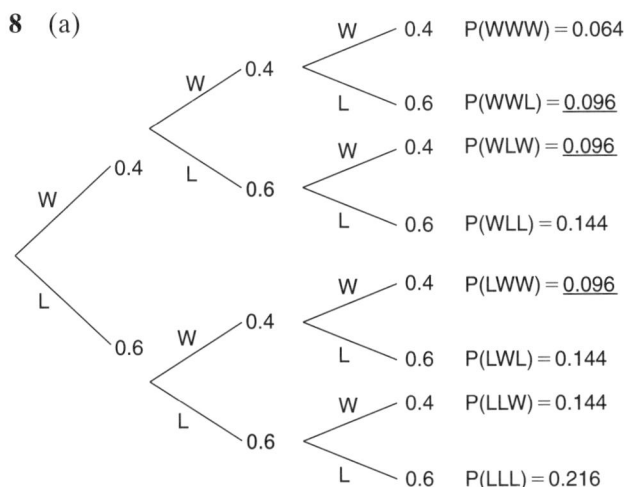

W = 'boy wins'
L = 'boy loses'.

(b) P(boy wins 2 and loses 1) $= 3 \times 0.096 = 0.288$

9 Let M be the event 'man watches the show'.
Let W be the event 'woman watches the show'.

(i) $P(W) = 0.5$ $\qquad P(M) = 0.4$

$$P(M \mid W) = 0.7 = \frac{P(M \cap W)}{P(W)}$$

Hence $P(M \cap W) = 0.7 \times 0.5 = 0.35$.

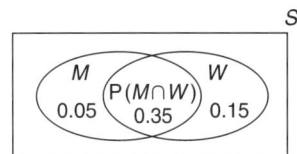

(ii) $P(W \mid M) = \dfrac{0.35}{0.4} = 0.875$

(iii) P(at least one person watches) $= 0.55$

10 (a) (i) Probability that Julie wins 3–0 is $(0.6)^3$
$= 0.216$

(ii) Probability that Julie wins 3–0 is
$0.3 \times 0.6 \times 0.6 = 0.108$.

(iii) Julie wins with probability
$0.5(0.6 \times 0.6 \times 0.6)$
$+ 0.5(0.3 \times 0.6 \times 0.6) = 0.162$.

(b) The probability that Zubair wins 3–1 if Julie starts is

$P(JZZZ) + P(ZJZZ) + P(ZZJZ)$
$= 0.6 \times 0.4 \times 0.7 \times 0.7 + 0.4 \times 0.3 \times 0.4 \times 0.7$
$\quad + 0.4 \times 0.7 \times 0.3 \times 0.4$
$= 0.1848$

11 (a) 1st: P(not yellow) $= \frac{9}{12}$;
2nd: P(not yellow) $= \frac{8}{11}$
$\Rightarrow$ $P(\sim Y, \sim Y) = \frac{9}{12} \times \frac{8}{11} = \frac{6}{11}$

(b) P(at least 1 yellow) $= 1 - \frac{6}{11} = \frac{5}{11}$

(c) After 2 reds taken, there remain 10 sweets:
3Y, 5G, 2R
P(4th sweet is Y) $= \frac{3}{10}$ (it would be the same for the 5th, 6th, . . . sweet)

12 (a) P(A has 1 hit) $= (\frac{2}{5} \times \frac{3}{5} \times \frac{3}{5}) \times 3 = \frac{54}{125} = 0.432$

(b) (i) P(no hits) $= (\frac{3}{5})^3(\frac{3}{4})^2 = 0.1215$
(ii) P(at least one hit) $= 1 - 0.1215 = 0.8785$
(iii) P(exactly one hit)
$= (\frac{2}{5} \times \frac{3}{5} \times \frac{3}{5}) \times 3 \times (\frac{3}{4} \times \frac{3}{4})$
$\quad + (\frac{3}{5})^3 \times (\frac{1}{4} \times \frac{3}{4}) \times 2$
$= 0.324$
P($B \mid$ 1 hit) $= \frac{1}{4}$

13 (a) The 12 distinct sets are:

$(1, 1, 10)$ $\quad (1, 2, 9)$ $\quad (1, 3, 8)$ $\quad (1, 4, 7)$ $\quad (1, 5, 6)$
$(2, 2, 8)$ $\quad (2, 3, 7)$ $\quad (2, 4, 6)$ $\quad (2, 5, 5)$ $\quad (3, 3, 6)$
$(3, 4, 5)$ $\quad (4, 4, 4)$

$(2, 5, 5)$, $(3, 4, 5)$ and $(4, 4, 4)$ could form triangles: Probability $= \frac{1}{4}$.

(b) *Coins*

(4, 4, 4)   1 way

(2, 2, 8), (1, 1, 10), (3, 3, 6), (2, 5, 5)   3 ways each

Remaining 7: 6 ways each

$\Rightarrow$ 1 + 12 + 42 = 55 ways

There are 6 sets with 3 different numbers of coins where one is > 5.

There are 3 sets with 2 different numbers of coins where one is > 5.

No. of ways for X is

$2 \times 6 + 3 = 15 \Rightarrow P(X) = \frac{15}{55}$ or $\frac{3}{11}$

No. of ways for Y is 10 $\Rightarrow P(Y) = \frac{10}{55} = \frac{2}{11}$

$P(X \cap Y) = \frac{5}{55} = \frac{1}{11}$: (10, 1, 1), (9, 1, 2), (8, 1, 3)

(7, 1, 4), (6, 1, 5)

$P(X \cup Y) = \frac{3}{11} + \frac{2}{11} - \frac{1}{11} = \frac{4}{11}$

$P(X \mid Y) = \frac{1}{2}$

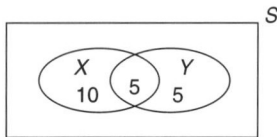

14

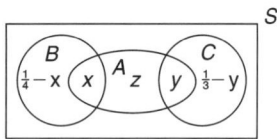

(i)

B and C are mutually exclusive – no intersection

$P(A \cup B) = \frac{1}{4} + y + z = \frac{3}{5} \Rightarrow y + z = \frac{7}{20}$ (1)

$P(A \cup C) = \frac{1}{3} + x + z = \frac{2}{3} \Rightarrow x + z = \frac{1}{3}$ (2)

A and B are independent:

$P(A \cap B) = P(A)P(B) \Rightarrow x = \frac{1}{4}(x + y + z)$

$\Rightarrow 3x = y + z$ (3)

From equations 1, 2 and 3, $x = \frac{7}{60}, y = \frac{8}{60}, z = \frac{13}{60}$.

$P(A) = \frac{7}{15}$

(ii) $P(A \cap C) = y = \frac{8}{60}$

$P(A) \, P(C) = \frac{7}{15} \times \frac{1}{3} = \frac{7}{45}$

Since $P(A \cap C) \neq P(A)P(C)$

A and C are not independent.

(iii) $P(B \cup C \mid A) = \dfrac{x + y}{x + y + z} = \dfrac{15}{28}$

15 (a) P(Sum of scores is odd) $= \frac{1}{2} = P(A)$

P(At least one score > 4) $= \frac{20}{36} = P(B)$

P(The two scores are equal) $= \frac{1}{6} = P(C)$

(i) A and C are mutually exclusive

Since $= A$ all odd, C all even

(ii) $P(A \mid B) = \frac{1}{2}, P(A) = \frac{1}{2}$

$\Rightarrow$ A and B are independent

$P(C \mid B) = \frac{2}{20} = \frac{1}{10}$ since

$P(C \mid B) = \dfrac{P(C \cap B)}{P(B)}$

(b) (i) Probability of a draw =

P(A starts) $\cdot$ P(draw) +

P(B starts) $\cdot$ P(draw) $= \frac{1}{2}(0.4 + 0.5) = 0.45$

(ii) Probability of A winning

$= p(0.4) + (1 - p)(0.2) = 0.2 + 0.2p$

Probability of B winning

$= p(0.2) + (1 - p)(0.3) = 0.3 - 0.1p$

These are equal if $0.3p = 0.1$, i.e. $p = \frac{1}{3}$

16

AS Level			A Level			Topic	Date attempted	Date completed	Self-assessment
CORE	MODULAR	TRADITIONAL	CORE	MODULAR	TRADITIONAL				
	✓	✓		✓	✓	**Expectation**			
	✓	✓		✓	✓	**Variance**			
	✓	✓		✓	✓	**Binomial distribution**			
	✓	✓		✓	✓	**Poisson distribution**			

Fact Sheet

In general, for an independent random variable (r.v.) X:

- Expected value of X or the Mean

$$E(X) = \sum_{\text{all } x} xP(X = x) = \mu$$

- Expected value of X^2

$$E(X^2) = \sum_{\text{all } x} x^2P(X = x)$$

- Variance of X

$$\text{Var } (X) = E(X^2) - (E(X))^2 = E(X^2) - \mu^2$$
$$= E(X - \mu)^2$$

- Expected value of f(x)

$$E(f(x)) = \sum_{\text{all } x} f(x)P(X = x)$$

Special Discrete Probability Distributions

Binomial Distribution

The binomial distribution is used when a trial has only two outcomes – success or failure.

If the probability of success is p and of failure is q ($=1 - p$) then in n trials the probability of r successes is

$$^nC_r p^r q^{n-r} \quad \text{where} \quad ^nC_r = \frac{n!}{r!(n - r)!}$$

This is the term in p^r from the expansion of $(p + q)^n$.

Notation: $X \sim B(n, p)$ or $\text{Bin}(n, p)$

$$E(X) = np \qquad \text{Var } (X) = npq$$

Poisson Distribution

The Poisson distribution is used when counting events which occur randomly in time and space. For example, accidents on a particular road or printing errors on one page.
λ = mean number of occurrences.
The probability density function (p.d.f.) is

$$P(X = x) = e^{-\lambda} \frac{\lambda^x}{x!} \quad \text{for } x = 0, 1, 2, \ldots$$

Notation: $\text{Po}(\lambda)$, λ is the parameter of the distribution

$$E(X) = \lambda \qquad \text{Var } (X) = \lambda$$

Poisson as an Approximation to a Binomial

If n is large ($n > 50$) and np is small ($p < 0.1$) then the Poisson distribution $\text{Po}(np)$ is a good approximation to $B(n, p)$.

Normal as an Approximation to a Binomial and Poisson

For large n:

$$B(n, p) \sim N(np, npq) \qquad n > 30, \quad 5 < np < n - 5$$
$$\text{Po}(\lambda) \sim N(\lambda, \lambda) \qquad \lambda > 25$$

For the Normal distribution, see Chapter 17.

Geometric Distribution

The p.d.f. is

$$P(X = x) = (1 - p)^{r-1}p$$

This occurs when X is the number of trials required before a success occurs.

Worked Examples

1 The discrete variable X has the probability distribution shown in the following table.

r	-1	0	1
$P(X = r)$	$\frac{1}{4}$	$\frac{1}{2}$	$\frac{1}{4}$

(i) Write down the expectation of X, and find the variance of X.

A second random variable Y has the same distribution as X, and the two random variables are independent.

(ii) List all possible values of $X + Y$. Show that $P(X + Y = -2) = \frac{1}{16}$, and find $P(X + Y = 0)$.

(iii) Show in a table the complete probability distribution for $X + Y$.

(iv) Verify that, for these random variables, Var $(X + Y) =$ Var $(X) +$ Var (Y).

(O&C MEI, 1993)

- (i)

r	-1	0	1
$P(X = r)$	$\frac{1}{4}$	$\frac{1}{2}$	$\frac{1}{4}$
$xP(x)$	$-\frac{1}{4}$	0	$\frac{1}{4}$
$x^2P(x)$	$\frac{1}{4}$	0	$\frac{1}{4}$

$\sum xP(x) = 0$ $\qquad$ $E(X) = \sum xP(x)$

$\sum x^2P(x) = \frac{1}{2}$ $\qquad$ Var $(X) = \sum x^2P(x) - (E(X))^2$

$E(X) = 0$ $\qquad$ Var $(X) = \frac{1}{2} - (0)^2 = \frac{1}{2}$

(ii) Possible values of $X + Y$ are $-2, -1, 0, 1, 2$.
Let $X + Y = Z$.
$P(X + Y = -2)$
$= P(X = -1)P(Y = -1) = \frac{1}{16}$
$P(X + Y = 0)$
$= P(X = -1)P(Y = 1) \times 2$
$\quad + P(X = 0)P(Y = 0)$
$= \frac{2}{16} + \frac{1}{4} = \frac{3}{8}$

(iii)

r	-2	-1	0	1	2
$P(X + Y = r)$	$\frac{1}{16}$	$\frac{1}{4}$	$\frac{3}{8}$	$\frac{1}{4}$	$\frac{1}{16}$
$rP(X + Y = r)$	$-\frac{1}{8}$	$-\frac{1}{4}$	0	$\frac{1}{4}$	$\frac{1}{8}$
$r^2P(X + Y = r)$	$\frac{1}{4}$	$\frac{1}{4}$	0	$\frac{1}{4}$	$\frac{1}{4}$

(iv) $\sum zP(z) = 0$
$\sum z^2P(z) = 1$

Variance $(X + Y) = \sum z^2P(z) - (\sum zP(z))^2$
$= 1$
Var $(X) = \frac{1}{2}$ $\quad$ Var $(Y) = \frac{1}{2}$
$\Rightarrow$ $\quad$ Var $(X + Y) = 1 =$ Var $(X) +$ Var (Y)

If there were more than 3 elements in the distribution of X it would be worth drawing up a table for X and Y to find the probabilities of X + Y

Y \ X	-1	0	1
-1	-2	-1	0
0	-1	0	1
1	0	1	2

$P(Y)$ \ $P(X)$	$\frac{1}{4}$	$\frac{1}{2}$	$\frac{1}{4}$
$\frac{1}{4}$	$\frac{1}{16}$	$\frac{1}{8}$	$\frac{1}{16}$
$\frac{1}{2}$	$\frac{1}{8}$	$\frac{1}{4}$	$\frac{1}{8}$
$\frac{1}{4}$	$\frac{1}{16}$	$\frac{1}{8}$	$\frac{1}{16}$

From the diagonal corresponding to the zeros these give $P(X + Y = 0) = \frac{1}{16} + \frac{1}{4} + \frac{1}{16} = \frac{3}{8}$

2 The discrete random variable X has the distribution given in the following table:

x	0	1	2
$P(X = x)$	$0.4 - 0.5\alpha$	α	$0.6 - 0.5\alpha$

Given that the standard deviation of $X = 0.8$, find the value of α.

(WJEC, 1993)

-

x	0	1	2
$P(X = x)$	$0.4 - 0.5\alpha$	α	$0.6 - 0.5\alpha$
$xP(X = x)$	0	α	$1.2 - \alpha$
$x^2P(X = x)$	0	α	$2.4 - 2\alpha$

$\sum xP(X = x) = 1.2$
$\sum x^2P(X = x) = 2.4 - \alpha$

Variance $(X) = \sum x^2P(X = x) - (\sum xP(X = x))^2$
$= (2.4 - \alpha) - 1.44$
$= 0.96 - \alpha$

But standard deviation of $X = 0.8$
$\therefore 0.96 - \alpha = (0.8)^2$ $\quad \Rightarrow \quad \alpha = 0.32$

3 A group of students is conducting a probability experiment using regular tetrahedral dice. The four faces of each die are numbered 1, 2, 3, 4.

(a) One group of students is using two dice and is recording the sum of the outcomes on the two dice.

  (i) Use a suitable diagram to display the anticipated outcome space. Hence, or otherwise, write down the probability distribution for the sum of the outcomes on two tetrahedral dice.

  (ii) Use your probability distribution to show that the mean of the distribution is 5 and its variance is 2.5.

(b) Another group of students decides to roll only one die and to double the score obtained. They claim that they will get the same results. Explain, briefly, why they are in error and demonstrate that the variance of their distribution is different. (OLE, 1993)

• (a) (i)

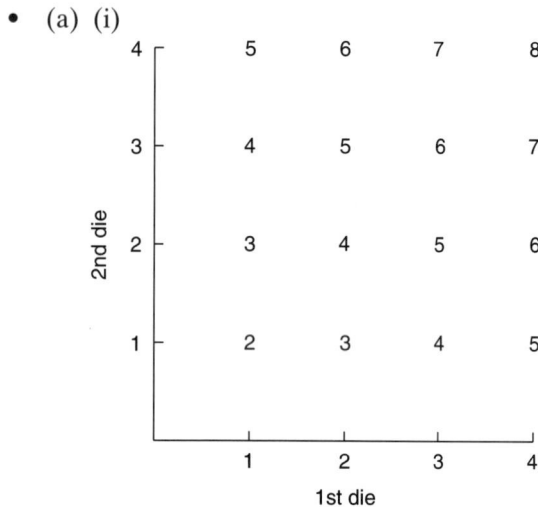

Score x	2	3	4	5	6	7	8
$P(X = x)$	$\frac{1}{16}$	$\frac{2}{16}$	$\frac{3}{16}$	$\frac{4}{16}$	$\frac{3}{16}$	$\frac{2}{16}$	$\frac{1}{16}$
$xP(X = x)$	$\frac{2}{16}$	$\frac{6}{16}$	$\frac{12}{16}$	$\frac{20}{16}$	$\frac{18}{16}$	$\frac{14}{16}$	$\frac{8}{16}$
$x^2P(X = x)$	$\frac{4}{16}$	$\frac{18}{16}$	$\frac{48}{16}$	$\frac{100}{16}$	$\frac{108}{16}$	$\frac{98}{16}$	$\frac{64}{16}$

$\Sigma xP(X = x) = \frac{80}{16}$

$\Sigma x^2P(X = x) = \frac{440}{16}$

(ii) Mean of the distribution is $\frac{80}{16} = 5$

  Variance $= E(x^2) - (E(x))^2 = 27.5 - 25 = 2.5$

(b) The results can only be 2, 4, 6, 8, all with equal probability of $\frac{1}{4}$, mean 5.

  $$\text{Variance} = \frac{4 + 16 + 36 + 64}{4} - (5)^2$$

  $$= 5$$

4 Whenever a particular gymnast performs a certain routine, the probability that she will do so

faultlessly is 0.7. Find, to 3 d.p., the probabilities that

(i) she will perform the routine faultlessly on four occasions out of six

(ii) she will perform the routine faultlessly on at least 14 occasions out of 20.

Use a suitable approximate method to evaluate, to 3 d.p., the probability that she will perform the routine faultlessly on more than 130 occasions out of 200.

Comment briefly on the assumption implied in the first sentence of this question. (JMB, 1991)

• P(success) = 0.7   P(failure) = 0.3

(i) Binomial distribution B(6, 0.7).

  Probability of 4 successes $= {}^6C_4(0.7)^4(0.3)^2$

  $= 0.324$

(ii) Probability of at least 14 successes from 20 attempts

  $= P(14, 15, 16, 17, 18, 19, 20 \text{ successes})$

  $= {}^{20}C_{14}(0.7)^{14}(0.3)^6 + {}^{20}C_{15}(0.7)^{15}(0.3)^5$
  $+ {}^{20}C_{16}(0.7)^{16}(0.3)^4 + {}^{20}C_{17}(0.7)^{17}(0.3)^3$
  $+ {}^{20}C_{18}(0.7)^{18}(0.3)^2 + {}^{20}C_{19}(0.7)^{19}(0.3)^1$
  $+ (0.7)^{20}$

  $= 0.608$ Use ${}^{20}C_r(0.7)^r(0.3)^{20-r}$ $\boxed{8.3}$

For 200 attempts use a Normal approximation N(np, npq) (see Chapter 17)

N(140, 42) $\sigma = \sqrt{42}$

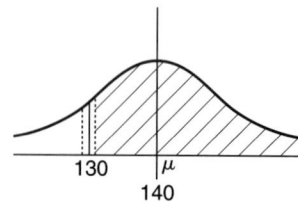

Use a continuity correction P($X > 130$)

$\Rightarrow$ P($X > 130.5$)

  $$\text{Probability} = 1 - Q\left(\frac{9.5}{\sqrt{42}}\right)$$

  $$= 1 - Q(1.4659)$$

  $$= 1 - 0.0713$$

  $$= 0.9287$$

It is assumed that these events are independent – unlikely

5 In the game of 'roll-a-coin', a large board is ruled with parallel equidistant lines. Coins are then rolled on to the board with the object of making them land so that they come to rest with no part touching or cutting a line. In the accompanying diagram five coins have been rolled and two have been successful.

(a) By considering where the centre of a successful coin needs to fall, or otherwise, present a careful argument to show that if the spacing

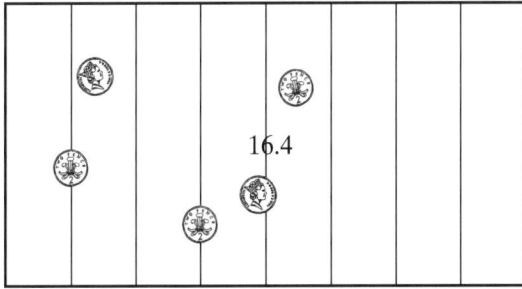

16.4

between the lines is 4 cm and the coins are 2.5 cm in diameter then the probability of a coin being successful is 0.375. (Assume when a coin comes to rest that its centre is a distance from a line which is uniformly distributed.)

(b) Under what conditions might it be appropriate to model the number of successful coins by a binomial distribution? State the parameters of the distribution which might serve to model the situation described above.

(c) Calculate the probability of having:
(i) four successful coins
(ii) five successful coins
giving your answers to four decimal places.

(d) A 'roll-a-coin' game is set up at a fête and a prize of £10 is offered for a stake of 10 pence to anyone who gets five successful coins. How much will the operator of the game lose or win per game in the long run? You are advised to justify your response with appropriate calculation.

(e) Why might the binomial model not be appropriate in the long run?   (OLE, 1992)

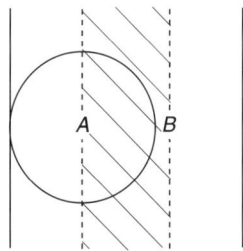

(a) For any coin to succeed, the centre of the coin must lie between lines 1.25 cm from the lines on the board.

Probability of success $= \dfrac{(4 - 2.5)}{4} = 0.375$

(b) As a binomial distribution the conditions to be satisfied are:
(i) only 2 possible outcomes – success, failure
(ii) the probability remains constant.
Parameters: $p = 0.375$ ($q = 0.625$) $n = 5$
$X \sim \text{Bin}(5, 0.375)$

(c) Probability:
$P(X = 4) = 5p^4q = 0.0618$ $^5C_4 = 5$
$P(X = 5) = p^5 = 0.0074$ $^5C_1 = 1$

(d) The probability of 5 coins being successful is

0.0074. This should be reached over a long period.
∴ Probable winnings in one attempt = 7.4p. In the long run a player would lose 2.6p per attempt.
Operator would win 2.6p/game in the long run.

Alternatively
$E(X) = 0.0074 \times £10 + 0.9926 \times £0$
$= £0.074$ or 7.4p

Stake = 10p ⇒ expected loss of 2.6p per game to the player.

(e) The player would probably gain more accuracy with practice and therefore the probability of winning would not remain constant.

6 The number of plants of a particular species found in squares of equal area on a large moor follows a Poisson distribution with mean 3.8.

(a) What is the probability that the number of plants of this species in a particular square is
(i) 5 or fewer
(ii) exactly 4
(iii) 8 or more?

(b) The number of plants of a different species in each square follows a Poisson distribution with mean 27. Using a suitable approximation find the probability that a particular square contains thirty or more plants of this different species.
                                            (AEB, 1994)

• (a) Let X be the random variable (r.v.) 'number of plants' $X \sim \text{Po}(3.8)$

(i) $P(X \leqslant 5) = e^{-3.8}\left(1 + 3.8 + \dfrac{(3.8)^2}{2} + \dfrac{(3.8)^3}{3!}\right.$

$\left. + \dfrac{(3.8)^4}{4!} + \dfrac{(3.8)^5}{5!}\right)$

$= 0.815\,55$

$= 0.816$ to 3 s.f.

(ii) $P(X = 4) = e^{-3.8}\dfrac{(3.8)^4}{4!} = 0.194$ to 3 s.f.

(iii) $P(X \geqslant 8) = 1 - P(X < 8)$
$= 1 - 0.816 - P(X = 6) - P(X = 7)$
$= 0.0401$ to 3 s.f.

(b) Let Y be the r.v. 'number of plants of 2nd species'
$Y \sim \text{Po}(27)$
Use a Normal approximation N(27, 27), $\sigma = \sqrt{27}$ with a continuity correction.

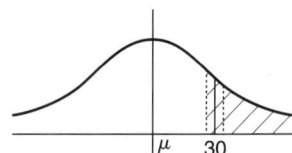

$$P(X \geqslant 30) = Q\left(\frac{29.5 - 27}{\sqrt{27}}\right)$$

$$= Q(0.4811)$$

$$= 0.3152$$

| 10.2 |

Therefore the probability of 30 or more plants in a square is 0.3152.

7 In the manufacture of a particular curtain material small faults occur at random at an average of 0.85 per 10 m².

(a) Find the probability that in a randomly selected 40 m² area of this material there are at most 2 faults.

This curtain material is going to be used in 10 of the rooms of a small block of furnished flats. Each room will require 40 m² of the material.

(b) Find the probability that for the first room to be furnished the material will contain at least one fault.

(c) Find the probability that in exactly half of these 10 rooms the material will contain exactly 3 faults.

The hooks on which these curtains are to hang are produced by a company which claims that only 2% of the hooks it produces are defective.

The owner of the block of flats buys 500 of the hooks which have been selected at random from the production.

(d) Using a suitable approximation find the probability that this sample contains between 8 and 12 defective hooks, inclusive. (L, 1993)

- For 40 m², the expected number of faults is 3.4, i.e. $X \sim Po(3.4)$.

(a) Probability that there are 0, 1 or 2 faults is

$$e^{-3.4}\left(1 + 3.4 + \frac{(3.4)^2}{2}\right) = 0.3397$$

(b) Probability that the first room contains at least 1 fault is

$$1 - P(\text{no faults}) = 0.9666$$

(c) *Consider one room before looking at 10*
Probability in 1 room that there are exactly 3 faults is

$$e^{-3.4}\frac{(3.4)^3}{3!} = 0.2186$$

Each 'room' has 2 outcomes – (3 faults, not 3 faults)
For 10 rooms, $p = 0.2186$, $q = 0.7814$, $n = 10$
Binomial expansion
Probability that exactly 5 rooms have 3 faults is

$$^{10}C_5 p^5 q^5 = 0.0367$$

(d) $n = 500$ $p = 0.02$ $q = 0.98$ $np = 10$
For large n and $np > 5$, use a Normal approximation $X \sim N(np, npq)$:

$$X \sim N(10, 9.8) \qquad \sigma = 3.130$$

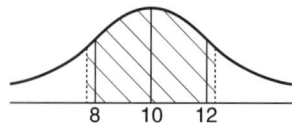

Since the distribution is discrete use the continuity correction:

$$P(8 \leqslant x \leqslant 12) \to P(7.5 < x < 12.5)$$

$$= 1 - 2Q\left(\frac{2.5}{3.130}\right)$$

$$= 1 - 0.4244$$

$$= 0.5756$$

| 10.2 |

8 The discrete random variables X and Y are jointly distributed as shown in the following table.

			x	
		0	1	2
y	0	$\frac{1}{5}$	$\frac{1}{10}$	$\frac{11}{30}$
	1	$\frac{1}{20}$	$\frac{3}{20}$	$\frac{2}{15}$

By first finding the distribution of $T = X + Y$, or otherwise, evaluate Var $(X + Y)$. (WJEC, 1993)

- For X:

$$P(X = 0) = \tfrac{1}{5} + \tfrac{1}{20} = \tfrac{1}{4}$$
$$P(X = 1) = \tfrac{1}{10} + \tfrac{3}{20} = \tfrac{1}{4}$$
$$P(X = 2) = \tfrac{11}{30} + \tfrac{2}{15} = \tfrac{1}{2}$$

For Y:

$$P(Y = 0) = \tfrac{1}{5} + \tfrac{1}{10} + \tfrac{11}{30} = \tfrac{2}{3}$$
$$P(Y = 1) = \tfrac{1}{20} + \tfrac{3}{20} + \tfrac{2}{15} = \tfrac{1}{3}$$

Don't take a short cut and assume the final probability. The final value allows you to check that $P(X) = 1$ and $P(Y) = 1$

Let $T = X + Y$:

$$P(T = 0) = P(X = 0)P(Y = 0) = \tfrac{1}{6}$$
$$P(T = 1) = P(X = 0)P(Y = 1)$$
$$\qquad\qquad + P(X = 1)P(Y = 0) = \tfrac{1}{4}$$
$$P(T = 2) = P(X = 1)P(Y = 1)$$
$$\qquad\qquad + P(X = 2)P(Y = 0) = \tfrac{5}{12}$$
$$P(T = 3) = P(X = 2)P(Y = 1) = \tfrac{1}{6}$$

	t	0	1	2	3
$P(T = t)$		$\frac{1}{6}$	$\frac{1}{4}$	$\frac{5}{12}$	$\frac{1}{6}$
$t\,P(T = t)$		0	$\frac{1}{4}$	$\frac{5}{6}$	$\frac{1}{2}$
$t^2\,P(T = t)$		0	$\frac{1}{4}$	$\frac{5}{3}$	$\frac{3}{2}$

$\Sigma t P(T = t) = \frac{19}{12} \Rightarrow E(t) = \frac{19}{12}$

$\Sigma t^2 P(T = t) = \frac{41}{12}$

$\text{Var } (T) = E(t^2) - (E(t))^2 = \frac{41}{12} - (\frac{19}{12})^2 = 0.9097$

Alternatively, using the data for X and Y:

x	0	1	2
$P(X = x)$	$\frac{1}{4}$	$\frac{1}{4}$	$\frac{1}{2}$
$xP(X = x)$	0	$\frac{1}{4}$	1
$x^2 P(X = x)$	0	$\frac{1}{4}$	2

$\Sigma x P(X = x) = \frac{5}{4}$

$\Sigma x^2 P(X = x) = \frac{9}{4}$

y	0	1
$P(Y = y)$	$\frac{2}{3}$	$\frac{1}{3}$
$yP(Y = y)$	0	$\frac{1}{3}$
$y^2 P(Y = y)$	0	$\frac{1}{3}$

$\Sigma y P(Y = y) = \frac{1}{3}$

$\Sigma y^2 P(Y = y) = \frac{1}{3}$

$\text{Var } X = \frac{9}{4} - (\frac{5}{4})^2 = \frac{11}{16}$

$\text{Var } Y = \frac{1}{3} - (\frac{1}{3})^2 = \frac{2}{9}$

Combining:

$\text{Var } (X + Y) = \text{Var } (X) + \text{Var } (Y) = \frac{131}{144} = 0.9097$

9 The manager of a car showroom monitored the numbers of cars sold during two successive five-day periods. During the first five days the numbers of cars sold per day had mean 1.8 and variance 0.56. During the next five days the number of cars sold per day had mean 2.8 and variance 1.76. Find the mean and variance of the number of cars sold per day during the full ten days. (JMB, 1991)

• 1st period: $\Sigma x_1 = 9$ $\text{Var } (x_1) = 0.56$

2nd period: $\Sigma x_2 = 14$ $\text{Var } (x_2) = 1.76$

Combining: number of cars sold is $9 + 14 = 23$

Therefore the mean over 10 days is 2.3 cars.

$$\text{Variance} = \frac{\Sigma x^2}{n} - \left(\frac{\Sigma x}{n}\right)^2$$

For 1st period: $\left(\dfrac{\Sigma x_1}{5}\right)^2 = 3.24$

$\dfrac{\Sigma x_1{}^2}{5} = 3.24 + 0.56 = 3.8$

For 2nd period: $\left(\dfrac{\Sigma x_2}{5}\right)^2 = 7.84$

$\dfrac{\Sigma x_2{}^2}{5} = 7.84 + 1.76 = 9.6$

Combining:

$$\left(\frac{\Sigma(x)}{10}\right)^2 = 2.3^2 = 5.29$$

$\Sigma x^2 = \Sigma x_1^2 + \Sigma x_2^2$

$\Sigma x^2 = 5(3.8 + 9.6) = 67$

$\therefore \dfrac{\Sigma x^2}{10} = 6.7.$

Variance $= 6.7 - 5.29$

$= 1.41$

10 The probability that a patient suffering from a particular disease will be cured following treatment is 0.97.

(a) If the treatment is given to 10 patients find, correct to 3 d.p., the probability that exactly 8 of them will be cured.

(b) If the treatment is given to 24 patients calculate, correct to 3 d.p., the probability that at least 22 of them will be cured.

(c) If the treatment is given to 64 patients use a Poisson approximation to calculate the probability that at least 62 of them will be cured; give your answer correct to 3 d.p.

The treatment is to be given to 5 patients at Hospital A and to 3 patients at Hospital B.

(d) Calculate, correct to 3 d.p., the probability that of the 5 patients treated at Hospital A the first, second, fourth and fifth to be treated will be cured but the third will not be cured.

(e) Given that 6 of the 8 patients treated at the two hospitals were cured, calculate the conditional probability that 4 of the cured patients were treated at Hospital A.   (WJEC, 1993)

• $p = 0.97, q = 0.03$

i.e. probability of a cure is 0.97

(a) Probability of 8 successes from 10 is

$^{10}C_8 p^8 q^2 = 0.0317$

$= 0.032$ to 3 d.p.

$\boxed{8.3}$

i.e. the probability that exactly 8 from 10 will be cured is 0.032.

(b) P(22, 23 or 24 successes)

$= p^{24} + 24p^{23}q + 276p^{22}q^2$

$= 0.4814 + 0.3573 + 0.1271$

$= 0.966$

(c) From 64 patients, expected number of failures

$= 0.03 \times 64 = 1.92 = \lambda$

Probability of 0, 1 or 2 failures is

$$e^{-1.92}\left(1 + 1.92 + \frac{(1.92)^2}{2}\right) = 0.698$$

Therefore the probability of at least 62 successes is 0.698.

(d) From 5 patients the probability that the 1st, 2nd, 4th and 5th are cured but not the 3rd is

$0.97 \times 0.97 \times 0.03 \times 0.97 \times 0.97 = 0.0266$

(e) If 6 out of 8 were cured then it was

  5 at A and 1 at B
 or 4 at A and 2 at B
 or 3 at A and 3 at B

$P(5_A) \times P(1_B) = (0.97)^5 \times 3(0.97)(0.03)^2$
$= 0.002\ 25$
$P(4_A) \times P(2_B) = 5(0.97)^4(0.03) \times 3(0.97)^2(0.03)$
$= 0.011\ 25$
$P(3_A) \times P(3_B) = 10(0.97)^3(0.03)^2 \times (0.97)^3$
$= 0.007\ 496$

Probability that 4 of the cured patients were treated at A is

$\dfrac{0.011\ 25}{0.020\ 996} = 0.536$

Since each of the 3 different ways include $(0.97)^6(0.03)^2$ the only calculation which is needed is the multiples of these terms:

$\dfrac{15}{3 + 15 + 10} = \dfrac{15}{28} = 0.536$

11 The output of a factory is produced by three machines A, B and C. A produces 50% of the output, B produces 30% and C produces 20%. Of the output of machine A, 5% are defective; 10% of the output of B are defective and 15% of the output of C are defective.
(i) Find the probability that in a random sample of 10 articles produced by A
 (a) none is defective
 (b) more than one are defective.
(ii) Show that the probability that an article chosen at random from the whole output is defective is 0.085.
(iii) An article chosen at random is found to be defective. What is the probability that it came from machine B?
(iv) Four articles are chosen at random. What is the probability that two were produced by A, one by B and one by C?

• A: 0.5 of output, 5% defective
 B: 0.3 of output, 10% defective
 C: 0.2 of output, 15% defective
(i) (a) Probability of 0 defective from 10 produced by A
  $= (0.95)^{10}$ (Binomial distribution)
  $= 0.5987$ (4 d.p.)

 (b) Probability of more than 1 defective
  $= 1 - P(0\text{ defective}) - P(1\text{ defective})$
  $= 1 - (0.95)^{10} - 10(0.95)^9(0.05)$

$= 1 - 0.9139$
$= 0.0861$ (4 d.p.)

(ii) Probability of a defective article is

   A B C
$0.5 \times 0.05 + 0.3 \times 0.1 + 0.2 \times 0.15 = 0.085$

(iii) Probability that a defective article came from B is

$\dfrac{P(\text{defective from }B)}{P(\text{defective})} = \dfrac{0.03}{0.085} = 0.353$

(iv) Probability that 2 from A, 1 from B, 1 from C

$= P(AABC) \times \dfrac{4!}{2!}$
$= (0.5)^2(0.3)(0.2) \times 12$
$= 0.18$

12 The table below summarizes the number of breakdowns on a stretch of motorway on 30 randomly selected days.

Number of breakdowns (x)	3	4	5	6	7	8	9	10	11
Number of days (f)	3	4	2	5	3	6	3	1	3

(a) Calculate unbiased estimates of the mean and the variance of the breakdowns.

Thirty more days were randomly sampled and this sample had a mean of 7.5 breakdowns and

$\dfrac{\sqrt{\Sigma(x - \bar{x})^2}}{30} = 2.5.$

(b) Treating the 60 results as a single sample, obtain further unbiased estimates of the population mean and the population variance.
(c) State, giving a reason, which of these two sets of estimates you would prefer to use.
(L Specimen Paper, 1996)

• (a) Total number of breakdowns

$\Sigma fx = 3 \times 3 + 4 \times 4 + 5 \times 2 + 6 \times 5 + 7 \times 3$
$+ 8 \times 6 + 9 \times 3 + 10 \times 1 + 11 \times 3$
$= 204$

$\Sigma fx^2 = 3(3^2) + 4(4^2) + 2(5^2) + 5(6^2) + 3(7^2)$
$+ 6(8^2) + 3(9^2) + 1(10^2) + 3(11^2)$
$= 1558$

Use the statistics mode on your calculator

Unbiased mean $\bar{\mu}$ (or $\bar{x}$) $= \dfrac{204}{30} = 6.8$

Unbiased variance

$$\hat{s}^2 = \frac{1}{n-1}\left(\Sigma fx^2 - \frac{(\Sigma fx)^2}{n}\right) \qquad \boxed{10.1}$$

$$= \frac{1}{29}\left(1558 - \frac{(204)^2}{30}\right) = 5.89$$

(b) Combining the two samples:

$$\text{Mean} = \frac{6.8 + 7.5}{2} = 7.15 \quad \left(\Sigma fx = 60 \times 7.15\right)$$

If the samples were of different sizes m and n then the mean would be

$$\frac{m\bar{x}_1 + n\bar{x}_2}{m + n}$$

For the second sample:

$$\frac{\sqrt{\Sigma(x-\bar{x})^2}}{30} = 2.5 \quad \Sigma(x-\bar{x})^2 = 30 \times 2.5^2$$
$$= 187.50$$

$$\therefore \Sigma fx^2 = 187.50 + 30 \times \bar{x}^2 = 1875$$

For the combined sample

$$\Sigma fx^2 = 1875 + 1558 = 3433$$

Unbiased variance $= \frac{1}{59}(3433 - 60 \times (7.15)^2)$
$$= 6.197 \text{ or } 6.20 \text{ to 3 s.f.}$$

(c) The larger the sample the better the estimate. Hence use the combined groups for the estimate.

Exercises

1 The random variable X has the binomial distribution B(10, 0.35). Find P($X \leqslant 4$).
   The random variable Y has the Poisson distribution with mean 3.5. Find P($2 < Y \leqslant 5$). (L, 1993)

2 A writer who writes articles for a magazine finds that his proposed articles sometimes need to be revised before they are accepted for publication. The writer finds that the number of days, X, spent in revising a randomly chosen article can be modelled by the following discrete probability distribution.

Number of days, x	0	1	2	4
P($X = x$)	0.8	0.1	0.05	0.05

Calculate E(X) and Var (X).

The writer prepares a series of 15 articles for the magazine. Find the expected value of the total time required for revisions to these articles.

The writer regards articles that need no revisions (i.e. for which $X = 0$) or which need only minor revisions ($X = 1$) as 'successful' articles, and those requiring major revisions ($X = 2$) or complete replacement ($X = 4$) as 'failures'. Assuming independence, find the probability that there will be fewer than 3 'failures' in the 15 articles in the series.

The writer produces 50 articles. Use an appropriate Poisson distribution to find the probability that at least 2 of these 50 articles will need to be completely replaced. (UCLES, 1993)

3 A shop sells a particular make of radio at a rate of 4 per week on average. The number sold in a week has a Poisson distribution.
   (a) Find the probability that the shop sells at least 2 in a week.
   (b) Find the smallest number that can be in stock at the beginning of a week in order to have at least a 99% chance of being able to meet all demands in a week.
   (L Specimen Paper, 1996)

4 The discrete random variable, X, has the following probability distribution.

X	-2	-1	0	1	2
P($X = x$)	0.2	0.1	0.4	0.2	0.1

(i) Find the mean and variance of X.
(ii) If Z is the sum of ten independent observations from this distribution, find the mean and variance of Z.
(iii) If $Y = 2X_1 + 3X_2$ where X_1 and X_2 are two independent observations from the distribution, calculate the mean and variance of Y.

5 An unbiased cubical die has the number 1 on 3 faces, the number 2 on 2 faces and the number 3 on 1 face. The die is rolled twice and X is the total score. Find the probability distribution of X.

Show that E(X) = $\frac{10}{3}$ and Var (X) = $\frac{10}{9}$.

A sample of 25 observations of X is taken. Find, to 3 decimal places, the probability that the mean of this sample is less than $\frac{11}{3}$.

6 Manufactured articles are packed in boxes each containing 200 articles, and on average $1\frac{1}{2}$% of all articles manufactured are defective. A box which contains 4 or more defective articles is substandard. Using a suitable approximation, show that the probability that a randomly chosen box will be substandard is 0.353, correct to three decimal places.

A lorry-load consists of 16 boxes, randomly chosen.

Find the probability that a lorry-load will include at most 2 boxes which are substandard, giving three decimal places in your answer.

A warehouse holds 100 lorry-loads. Show that, correct to two decimal places, the probability that exactly one of the lorry-loads in the warehouse will include at most 2 substandard boxes is 0.06.

(UCLES, 1987)

7

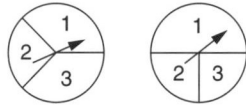

A circular card is divided into 3 sectors scoring 1, 2, 3 and having angles 135°, 90°, 135°, respectively. On a second circular card, sectors scoring 1, 2, 3 have angles 180°, 90°, 90° respectively (see diagram). Each card has a pointer pivoted at its centre. After being set in motion, the pointers come to rest independently in random positions. Find the probability that
(i) the score on each card is 1,
(ii) the score on at least one of the cards is 3.
The random variable X is the larger of the two scores if they are different, and their common value if they are the same. Show that $P(X = 2) = \frac{9}{32}$.
Show that $E(X) = \frac{75}{32}$ and find Var (X).

(UCLES, 1990)

8 (i) The number of times a vending machine breaks down in a year may be described by a random variable, X, which follows a Poisson distribution. If the expected value of X is 10, calculate the probability that:
   (a) 8 breakdowns occur in one year
   (b) fewer than 18 breakdowns occur in 2 years.
(ii) The time required to repair the vending machine is normally distributed with mean 125 minutes and standard deviation 5 minutes. If the machine is out of order for more than 130 minutes the extensive cleaning required and food loss amounts to a cost of £400.

By adding more maintenance personnel the mean of the service time is reduced to 120 minutes; the standard deviation remains at 5 minutes.
   (a) Calculate the difference in expected costs before and after the addition of maintenance personnel.
   (b) State ONE assumption you have made in your calculations. (NICCEA, 1992)

9 An octahedral die has eight faces, numbered 1 to 8. The random variable X is the score obtained when the die is thrown. The bias of the die is such that
$$P(X = r) = c \quad \text{for } r = 1, 2, 3, 4, 5$$
$$P(X = r) = d \quad \text{for } r = 6, 7, 8$$
$$P(X < 6) = P(X \geq 6).$$

(i) Find the values of c and d, show that $E(X) = 5$ and find the variance of X.
(ii) The die is thrown twice. Calculate the probability that the sum of the two scores is 10.
(iii) The random variable Y is the sum of the scores when this die is thrown 48 times. Find the mean and variance Y. Assuming that Y has a normal distribution with this mean and variance, find the probability that Y lies between 220 and 260 inclusive. (O&C, 1991)

10 On average a coastguard station receives one distress call every two days. A 'bad' week is a week in which 5 or more distress calls are received. Show that the probability that a week is a bad week is 0.275, correct to three significant figures.

Find the probability that, in 8 randomly chosen weeks, at least 2 are bad weeks.

Find the probability that, in 80 randomly chosen weeks, at least 30 are bad weeks. (UCLES, 1988)

11 A small business has 12 employees. Their weekly wages, £x, are summarized by
$$\Sigma x = 2501 \quad \Sigma x^2 = 525\,266.8$$

(i) Calculate the mean and standard deviation of the employees' weekly wages.

A second business has 17 employees. Their weekly wages, £y, have a mean of £273.20 and a standard deviation of £23.16.

(ii) Find Σy and show that $\Sigma y^2 = 1\,277\,969$.
(iii) Now consider all 29 employees as a single group. Find the mean and standard deviation of their weekly wages. (O&C MEI, 1993)

Brief Solutions

1 $X \sim B(10, 0.35)$, i.e. $(p + q)^{10}$ where $p = 0.35$, $q = 0.65$

$\boxed{8.3}$

$P(X \leq 4) = q^{10} + 10q^9p + 45q^8p^2 + 120q^7p^3 + 210q^6p^4$
$= 0.751$

Use $^{10}C_r(0.35)^r(0.65)^{10-r}$

$Y \sim Po(3.5)$

$$P(Y = 3, 4 \text{ or } 5) = e^{-3.5}\left(\frac{(3.5)^3}{6} + \frac{(3.5)^4}{24} + \frac{(3.5)^5}{120}\right)$$

$$= 0.537$$

2 $E(X) = 1(0.1) + 2(0.05) + 4(0.05) = 0.4$
$E(X^2) = 1(0.1) + 4(0.05) + 16(0.05) = 1.1$
Var $X = 1.1 - (0.4)^2 = 0.94$

$\boxed{10.1}$

Expected revision time for 15 articles = 15×0.4
= 6 days
P(0, 1 or 2 failures) $= p^{15} + 15p^{14}q + 105p^{13}q^2$
where $p = 0.8 + 0.1 = 0.9$ and $q = 0.1$,
P(0, 1 or 2 failures) $= 0.8159$
50 articles: $\lambda = np = 2.5$

$\quad$ P(at least 2 need replacing)
$\qquad = 1 - $ P(0 or 1 replacement)
$\qquad = 1 - e^{-2.5}(1 + 2.5)$
$\qquad = 0.7127$

3 Poisson distribution $\quad \lambda = 4$

(a) P(at least 2) $= 1 - $ P(0 or 1) $= 1 - e^{-4}(1 + 4)$
$\qquad\qquad\qquad = 0.9084$

(b) $e^{-4}\left(1 + 4 + \dfrac{4^2}{2} + \dfrac{4^3}{3!} + \ldots\right) > 0.99$

Since several terms are required, use the

recurrence formula $P_{n+1} = P_n \dfrac{\lambda}{n+1}$ *and the memory*

$1 + 4 + \dfrac{4^2}{2!} + \ldots + \dfrac{4^n}{n!} > 54.052 \ldots \quad \Rightarrow$

$n = 9$ radios in stock

4 (i) $E(X) = -2(0.2) - 1(0.1) + 1(0.2) + 2(0.1)$
$\qquad\qquad = -0.1$ (Mean) $\qquad \boxed{10.1}$
$\quad$ Var $(X) = 4(0.2) + 1(0.1) + 1(0.2)$
$\qquad\qquad\qquad + 4(0.1) - (-0.1)^2 = 1.49$
(ii) $Z = X_1 + X_2 + \ldots + X_{10}$
$\quad$ Mean $= 10(-0.1) = -1$
$\quad$ Variance $= 10(1.49) = 14.9$
(iii) Mean $2X_1 = -0.2$
$\quad$ Variance $2X_1 = 4(1.49) = 5.96$
$\quad$ Mean $3X_2 = -0.3$
$\quad$ Variance $3X_2 = 9(1.49) = 13.41$
$\quad$ Mean $(2X_1 + 3X_2) = -0.5$
$\quad$ Variance $(2X_1 + 3X_2) = 5.96 + 13.41 = 19.37$

5 $P(1, 1) = \frac{1}{2} \times \frac{1}{2} = \frac{1}{4}$ $\quad$ $P(1, 2) + P(2, 1) = 2(\frac{1}{2} \times \frac{1}{3}) = \frac{1}{3}$
$P(1, 3) + P(3, 1) + P(2, 2) = 2(\frac{1}{2} \times \frac{1}{6}) + (\frac{1}{3})^2 = \frac{5}{18}$
$P(2, 3) + P(3, 2) = \frac{1}{9}$ $\quad$ $P(3, 3) = \frac{1}{36}$
$E(X) = 2(\frac{1}{4}) + 3(\frac{1}{3}) + 4(\frac{5}{18}) + 5(\frac{1}{9}) + 6(\frac{1}{36}) = \frac{10}{3}$
$E(X^2) = 1 + 3 + \frac{40}{9} + \frac{25}{9} + 1 = \frac{110}{9}$
Var $X = \frac{110}{9} - \frac{100}{9} = \frac{10}{9}$

*Hint: for distribution means, see Central Limit Theorem
in Chapter 17*

$n = 25 \quad \bar{X} \sim N\left(\bar{X}, \dfrac{\sigma^2}{n}\right) \sim N(\frac{10}{3}, \frac{1}{25} \cdot \frac{10}{9}) = \sim N(\frac{10}{3}, \frac{2}{45}).$

$\sigma = \frac{1}{3}\sqrt{\frac{2}{5}} = 0.2108$

$\quad P(\bar{X} < \frac{11}{3}) = 1 - Q\left(\dfrac{\frac{1}{3}}{0.2108}\right)$

$\qquad\qquad\quad = 1 - Q(1.5811)$
$\qquad\qquad\quad = 0.943$

---

6 $n = 200$ $\quad$ X is r.v. 'no. of defectives per sample of
200' $\quad$ $X \sim B(200, 0.015)$
Use a Poisson distribution: $X \sim Po(np) = Po(3)$.
Let S be the event 'substandard box':
$P(S) = P(X \geq 4) = 1 - P(X < 4) = 0.353$ (3 d.p.)
Let Y be the number of boxes which are
substandard.
$Y \sim B(16, 0.353)$
$P(Y \leq 2) = {}^{16}C_0(0.647)^{16} + {}^{16}C_1(0.647)^{15}(0.353)$
$\qquad\qquad + {}^{16}C_2(0.647)^{14}(0.353)^2$
$\qquad\qquad = 0.043$
For 100 lorry-loads
P(1 lorry-load includes at least 2 substandard boxes)
$= {}^{100}C_1(0.043)^1(0.957)^{99}$
$= 0.06$ (2 d.p.)

7 Y is r.v. 'score on 1st card'
Z is r.v. 'score on 2nd card'
$P(Y = 1) = \frac{3}{8}$ $\qquad P(Y = 2) = \frac{1}{4}$ $\qquad P(Y = 3) = \frac{3}{8}$
$P(Z = 1) = \frac{1}{2}$ $\qquad P(Z = 2) = \frac{1}{4}$ $\qquad P(Z = 3) = \frac{1}{4}$
(i) $P(Y = 1, Z = 1) = \frac{3}{8} \cdot \frac{1}{2} = \frac{3}{16}$
(ii) $P(Y = 3, Z = $ any$) + P(Y = 1$ or $2, Z = 3)$
$\qquad = \frac{3}{8} + \frac{5}{8} \cdot \frac{1}{4} = \frac{17}{32}$

$$Z$$

		1	2	3
	1	$\frac{3}{16}$	$\frac{3}{32}$	$\frac{3}{32}$
Y	2	$\frac{1}{8}$	$\frac{1}{16}$	$\frac{1}{16}$
	3	$\frac{3}{16}$	$\frac{3}{32}$	$\frac{3}{32}$

$P(X = 1) = \frac{3}{16}$ $\quad$ $P(X = 2) = \frac{1}{8} + \frac{1}{16} + \frac{3}{32} = \frac{9}{32}$
$P(X = 3) = \frac{17}{32}$
$E(X) = 1(\frac{3}{16}) + 2(\frac{9}{32}) + 3(\frac{17}{32}) = \frac{75}{32}$
$E(X^2) = 1(\frac{3}{16}) + 4(\frac{9}{32}) + 9(\frac{17}{32}) = \frac{195}{32}$
$\Rightarrow$ Var $(X) = \frac{195}{32} - (\frac{75}{32})^2 = 0.6006$ or $\frac{615}{1024}$

8 (i) (a) Probability of 8 breakdowns $= e^{-10}\left(\dfrac{10^8}{8!}\right)$
$\qquad\qquad\qquad\qquad\qquad\qquad = 0.1126$

$\qquad$ (b) For 2 years $\lambda = 20$. Use a Normal
$\qquad\qquad$ approximation $\quad$ *with continuity correction*
$\qquad\qquad \mu = 20$ $\quad \sigma^2 = 20$ $\quad \sigma = \sqrt{20}$

$\qquad\qquad$ P(less than 18 breakdowns)
$\qquad\qquad\quad = Q\left(\dfrac{17.5 - 20}{\sqrt{20}}\right) = Q(-0.5590)$
$\qquad\qquad\quad = 0.2881$

$\qquad$ (ii) (a) P(cost of £400) $=$ P(time > 130 min)
$\qquad\qquad\qquad\qquad\qquad\qquad = Q(\frac{5}{5}) = 0.1587$
$\qquad\qquad\qquad$ *No continuity correction since time is
$\qquad\qquad\qquad$ continuous*

Expected cost = £400 × 0.1587 = £63.48
After more maintenance personnel
P(cost of £400) = $Q(2)$ = 0.0228
Expected cost = £400 × 0.0228 = £9.12
Difference = £54.36

(b) It is assumed that there is no extra cost for maintenance personnel.

9 (i) Total probability = $5c + 3d = 1$ and $5c = 3d$
$\Rightarrow \quad c = \frac{1}{10}, d = \frac{1}{6}$
$E(X) = \frac{1}{10}(1 + 2 + 3 + 4 + 5) + \frac{1}{6}(6 + 7 + 8)$
$\quad = 5$
$E(X^2) = \frac{1}{10}(1 + 4 + 9 + 16 + 25)$
$\quad\quad\quad + \frac{1}{6}(36 + 49 + 64) = 30\frac{1}{3}$
Var $(X) = 30\frac{1}{3} - 25 = 5\frac{1}{3}$

(ii) P(score of 10 from 2 throws)
$= P((2, 8), (3, 7), (4, 6), (5, 5), (6, 4), (7, 3), (8, 2))$
$= 6cd + c^2 = 0.11$

(iii) $E(Y) = 240$
Var $(Y) = 48 \times 5\frac{1}{3} = 256 \quad \Rightarrow \quad \sigma = 16$
$P(220 \leqslant Y \leqslant 260) = 1 - 2\left(Q\left(\dfrac{20.5}{16}\right)\right) = 0.8$

10 Use a Poisson distribution.
Average number of calls per week = 3.5 $(= \lambda)$.

Probability of 'a bad week' = $P(X \geqslant 5)$
$= 1 - P(X \leqslant 4)$
$= 0.275$ to 3 s.f.

Let Y be the r.v. 'number of bad weeks'.
$Y \sim B(8, 0.275)$

$P(Y \geqslant 2) = 1 - P(Y = 0) - P(Y = 1)$
$\quad\quad\quad\quad = 0.692$ to 3 s.f.

For 80 weeks use Normal approximation:
$Y \sim N(22, 15.95)$ *with a continuity correction*

$P(Y \geqslant 30) = P(Y > 29.5) = Q\left(\dfrac{29.5 - 22}{\sqrt{15.95}}\right)$
$= Q(1.878)$

$P(Y \geqslant 30) = 0.0302$

11 (i) $\bar{x} = 208.42, \quad s_x^2 = \dfrac{\Sigma x^2}{n} - \bar{x}^2 \quad \Rightarrow \quad s_x = £18.26$

(ii) $\Sigma y = 4644.40, \quad \Sigma y^2 = n(s_y^2 + \bar{y}^2) = 1\,277\,969$
(iii) $\Sigma z = 7145.40, \quad \Sigma z^2 = 1\,803\,235.80$
Mean $\bar{z} = £246.39 \quad s_z = £38.3$

17

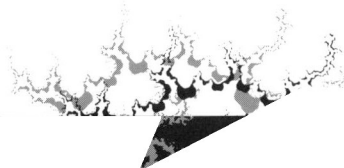

AS Level			A Level				Topic	Date attempted	Date completed	Self-assessment
CORE	MODULAR	TRADITIONAL	CORE	MODULAR	TRADITIONAL					
	✓	✓		✓	✓		Continuous random variables			
	✓	✓		✓	✓		Normal distribution			
	✓	✓		✓	✓		Normal approximation to Binomial			
	✓	✓		✓	✓		Normal approximation to Poisson			

Fact Sheet

Continuous Random Variables (continuous r.v.)

A continuous r.v. is specified by its **probability density function** (p.d.f.) $f(x)$.

(i) $\int_{\text{all } x} f(x)\, dx = 1$

   If $f(x)$ is valid over the range $a \leq x \leq b$ then

   $$\int_a^b f(x)\, dx = 1$$

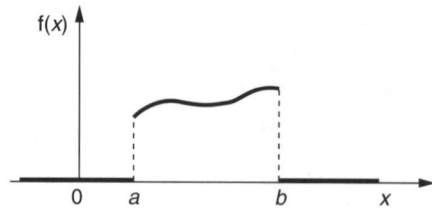

(ii) Probability that $x_1 \leq X \leq x_2$, i.e. $P(x_1 \leq X \leq x_2)$ is

   $$\int_{x_1}^{x_2} f(x)\, dx$$

(iii) Expectation of X is $E(X)$ where

   $$E(X) = \int_{\text{all } x} x f(x)\, dx = \mu$$

(iv) Expectation of X^2 is $E(X^2)$ where

   $$E(X^2) = \int_{\text{all } x} x^2 f(x)\, dx$$

(v) Variance of X is $\text{Var}(X) = E(X^2) - (E(X))^2$ or

   $$\text{Var}(X) = E(X^2) - \mu^2 = \sigma^2$$

(vi) Standard deviation $\sigma = \sqrt{\text{Var}(X)}$

Cumulative Distribution Function

$$F(x) = \int_a^x f(x)\, dx = P(X < x)$$

$$F(b) = 1$$

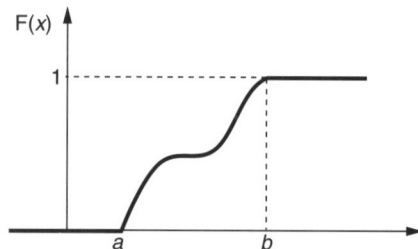

Rectangular Distribution

$$f(x) = \frac{1}{b-a} \text{ for } a \leq x \leq b$$

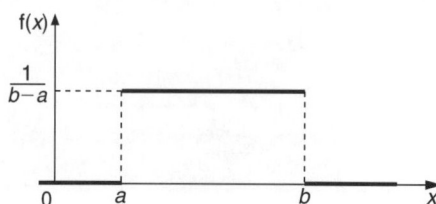

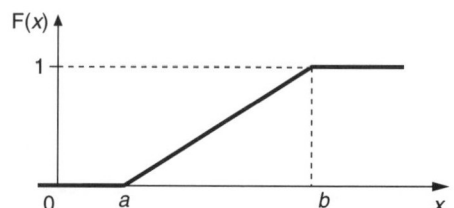

Normal Distribution

Notation $X \sim N(\mu, \sigma^2)$ where μ is the mean and σ^2 is the variance.
Distribution is symmetrical about the mean.
Normal distribution tables usually give the probability of a 'tail', using values of z where

$$z = \frac{X - \mu}{\sigma} \quad \Rightarrow \quad Z \sim N(0, 1)$$

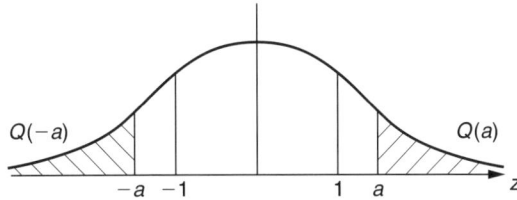

From the tables $Q(z)$ is $P(z > a)$ or $P(z < -a)$.
For a symmetric 90% of the distribution $z = \pm 1.645$
For a symmetric 95% of the distribution $z = \pm 1.96$
For a symmetric 99% of the distribution $z = \pm 2.575$

Sums or Differences of Normally Distributed Variables

$E(X_1 + X_2 + \ldots + X_n) = E(X_1) + E(X_2) + \ldots + E(X_n)$
$E(X_1 - X_2 \ldots) = E(X_1) - E(X_2) \ldots$

Variance

$$\text{Var } (X_1 + X_2 + \ldots + X_n) = \text{Var } (X_1) + \text{Var } (X_2) + \ldots + \text{Var } (X_n)$$
$$\text{Var } (X_1 - X_2 \ldots) = \text{Var } (X_1) + \text{Var } (X_2) + \ldots$$

$\Rightarrow$ if $X_1 \sim N(\mu_1, \sigma_1^2)$ and $X_2 \sim N(\mu_2, \sigma_2^2)$ then

$$X_1 + X_2 \sim N(\mu_1 + \mu_2, \sigma_1^2 + \sigma_2^2)$$
$$X_1 - X_2 \sim N(\mu_1 - \mu_2, \sigma_1^2 + \sigma_2^2)$$

Multiples of a Random Variable

$$X \sim N(\mu, \sigma^2) \qquad kX \sim N(k\mu, k\sigma^2)$$

Central Limit Theorem

- *Sample, mean and variance*

  A sample size n from $N(\mu, \sigma^2)$: $\quad \bar{X} \sim N\left(\mu, \frac{\sigma^2}{n}\right)$

- *Standard error of the mean* $= \dfrac{\sigma}{\sqrt{n}}$

These hold for *large* samples from *any* distribution.

This is the Central Limit Theorem which states, more formally:

If X is a random variable with mean μ and variance σ^2, then the distribution of $\bar{X}$ approximates to a Normal distribution with

mean μ and variance $\dfrac{\sigma^2}{n}$, as $n \to \infty$ (see also Chapter 18).

Normal Approximations to Discrete Distributions

For *large n* the Binomial distribution $B(n, p) \sim N(np, npq)$ providing (i) $n > 30$ and (ii) $5 < np < n - 5$.

The Poisson distribution $Po(\lambda) \sim N(\lambda, \lambda)$ providing $\lambda > 25$ (see also Chapter 16).

Worked Examples

1 The probability that a randomly chosen flight from Stanston Airport is delayed by more than x hours is $\frac{1}{100}(x - 10)^2$, for $x \in \mathbb{R}, 0 \leq x \leq 10$. No flights leave early and none is delayed for more than 10 hours. The delay, in hours, for a randomly chosen flight is denoted by X.
  (i)   Find the median, m, of X, correct to three s.f.
  (ii) Find the cumulative distribution function, F, of X and sketch the graph of F.
  (iii) Find the probability density function, f, of X and sketch the graph of f.
  (iv) Show that $E(X) = \frac{10}{3}$.
A random sample of 2 flights is taken. Find the probability that both flights are delayed by more than m hours, where m is the median of X.
(UCLES, 1990)

• F(x) = 1 − probability delayed more than x hours
  $F(x) = 1 - \frac{1}{100}(x - 10)^2$
  (i)   For the median m, $F(x) = \frac{1}{2}$.
     That is

$$\frac{(x - 10)^2}{100} = \frac{1}{2}$$

| 2.1a and 2.2a |

$$m - 10 = -\sqrt{50} \text{ since } x \leq 10$$
$$m = 10 - 7.07$$
$$ = 2.93$$

Hence the median delay is 2.93 hours.

(ii) $F(x) = \dfrac{100 - (x - 10)^2}{100} = \dfrac{20x - x^2}{100}$ $(0 \leq x \leq 10)$

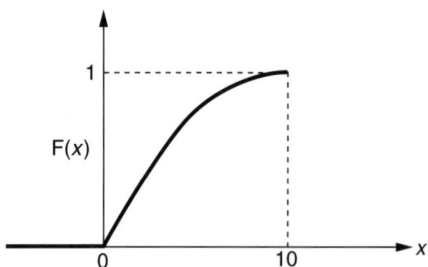

(iii) $f(x) = \dfrac{d}{dx}(F(x)) = \dfrac{10 - x}{50}$ $0 \leq x \leq 10$

$$= 0 \quad \text{all other } x$$

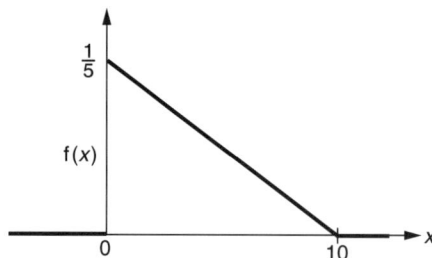

(iv) $E(X) = \dfrac{1}{50}\displaystyle\int_0^{10} x(10 - x)\,dx$

$$= \frac{1}{50}\left[5x^2 - \frac{x^3}{3}\right]_0^{10}$$

| 6.1 and 6.2 |

$$= \frac{10}{3}$$

$P(X_1 > m) = \frac{1}{2}$ $\Rightarrow$
P(both flights are delayed by more than m hours) = $\frac{1}{4}$

2 The heights of the male students at a large college are Normally distributed with mean 180 cm and standard deviation 4 cm. Independently, the heights of the female students at the college are Normally distributed with mean 170 cm and standard deviation 5 cm.
  (a) If one male student and one female student are chosen at random, evaluate, correct to 3 decimal places, the probability that
     (i)   both will be taller than 175 cm
     (ii) the female student will be taller than the male student.
  (b) If two male students and three female students are chosen at random evaluate, correct to 3 decimal places, the probability that
     (i)   the mean of the heights of the three female

students will be greater than the mean of the heights of the two male students

(ii) the sum of the heights of all five students will be greater than 850 cm. (WJEC, 1992)

- For Normal distribution 'think on a diagram'

Let the height of a male student be

$X \sim N(180, 16), \sigma = 4$ [10.2]

Let the height of a female student be

$Y \sim N(170, 25), \sigma = 5$

(a) (i)

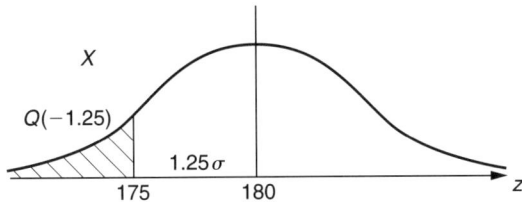

Probability that $X > 175 = 1 - Q(-\frac{5}{4})$
$$= 1 - 0.1056$$
$$= 0.8944$$

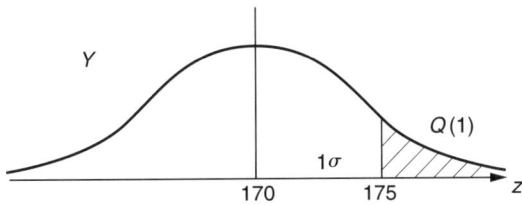

Probability that $Y > 175 = Q(1) = 0.1587$

The probability that both will be taller than 175 cm is

$0.8944 \times 0.1587 = 0.1419$
$$= 0.142 \text{ (3 d.p.)}$$

(ii) Let the height of female $-$ height of male be $Z \sim N(-10, 41), \sigma = \sqrt{41}$.

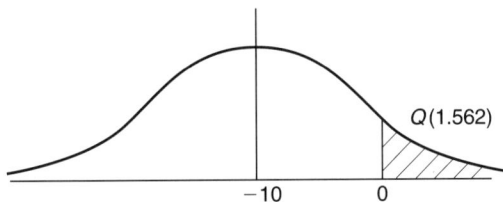

The probability that $Z > 0$ is

$Q\left(\dfrac{10}{\sqrt{41}}\right) = Q(1.562) = 0.0592$

The probability that the female is taller than the male is 0.059 (3 d.p.).

(b) (i) Let the sum of heights of 2 male students be

$X_1 \sim N(360, 32)$, then $\frac{1}{2}X_1 \sim N(180, 8)$

Let the sum of heights of 3 female students be

$Y_1 \sim N(510, 75)$ then $\frac{1}{3}Y_1 \sim N(170, \frac{25}{3})$

Let $W = \frac{1}{3}Y_1 - \frac{1}{2}X_1 \sim N(-10, \frac{49}{3}), \sigma = 4.042$

$$P(W > 0) = Q\left(\dfrac{10}{4.04}\right) = Q(2.474)$$
$$= 0.0067$$

Hence the probability that the mean height of 3 women is greater than the mean height of 2 men is 0.0067.

(ii) Let $T = X_1 + Y_1 \sim N(870, 107), \sigma = 10.34$

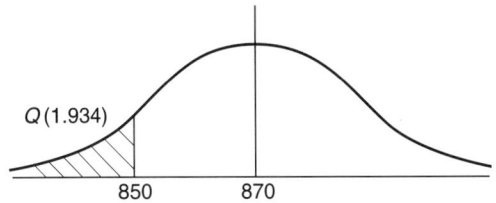

$$P(T > 850) = 1 - Q\left(\dfrac{20}{10.34}\right)$$
$$= 1 - Q(1.934)$$
$$= 1 - 0.0266$$
$$= 0.973$$

Hence the probability that the sum of the heights of the students will exceed 850 cm is 0.973.

3 The following grouped frequency distribution summarizes the time, to the nearest minute, spent waiting by a sample of patients in a doctor's surgery.

Waiting Time (to nearest minute)	Number of Patients
3 or less	6
4–6	15
7–8	27
9	49
10	52
11–12	29
13–15	13
16 or more	9

The mean of the times was 9.63 minutes and the standard deviation was 3.03 minutes.

(a) Using interpolation, estimate the median and semi-interquartile range of these data.

For a normal distribution the ratio of the interquartile range to the standard deviation would be approximately 0.67.

(b) Calculate the corresponding value for the above data. Comment on your result.

For a normal distribution, 90% of times would be expected to lie in the interval
    (mean $\pm$ 1.645 standard deviations)
(c) Find the theoretical limits for these data.
(d) Using appropriate percentiles, estimate comparable limits. Comment on your result.
                                    (L, 1992)

- (a) Total number of patients = 200

    LQ is $\frac{1}{4}$ (201) value $= 8.5 + \dfrac{2.25}{49}$ min

                       $= 8.546$ min (8.55 min)

    Median is $\frac{1}{2}$ (201) value $= 9.5 + \dfrac{3.5}{52}$ min

                       $= 9.567$ min (9.57 min)

    UQ is $\frac{3}{4}$ (201) value $= 10.5 + \dfrac{1.75}{29} \times 2$ min

                       $= 10.621$ min (10.62 min)

    IQR is 2.07 min

- (b) SD = 3.03 (given)

    $\dfrac{\text{IQR}}{\text{SD}} = 0.683$

    This is a close approximation to a Normal distribution.

- (c) Theoretically 90% of the times lie in the interval $(9.63 \pm 1.645(3.03))$
    i.e. 90% of times lie in the interval (4.65, 14.61)

- (d) For the given distribution
    the 5th percentile is $3.5 + \frac{4}{15} \times 3 = 4.3$ min
    the 95th percentile is $12.5 + \frac{12}{13} \times 3 = 15.3$ min

    90% of times lie in the interval (4.3, 15.3)
    i.e. the distribution has a long tail.

4 A manufacturer makes electronic components and the price for which he can sell the components depends on the purity of the material used in their manufacture. The purity, measured in arbitrary units, has a mean of 20 and standard deviation 2, and is normally distributed. Calculate the probability that if a component is selected at random from a large batch
(i) the purity lies less than two standard deviations from the mean
(ii) the purity is measured as 19.8 correct to one d.p.
For components with purity in the range 19 to 21 the manufacturer can make a profit of £20 and on those with a purity above 21 he can make a profit of £30. Components with purity in the range 18 to 19 are sold at the cost price of £100 per component and those with purity below 18 make a loss of £100 per component.

(iii) Calculate the mean profit per component to the nearest £1.
If the manufacturer wishes to raise the mean purity of the components it entails increasing the cost of manufacture of each component. The manufacturer decides to increase the mean purity to 21 units, the standard deviation remaining at 2 units. The manufacturer does not change the price of components to any customer.
(iv) Calculate, to the nearest £1, the increase in cost per component, if it is found that the manufacturer's mean profit per component remains the same.   (NICCEA, 1993)

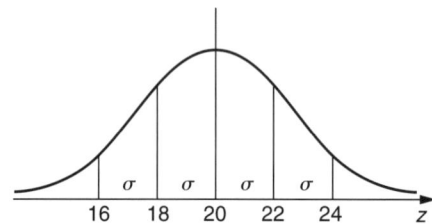

(i) Probability that component is less than 2 SD from mean is $1 - 2Q(2) = 0.9544$

(ii) Purity $19.75 < p < 19.85 = Q\left(\dfrac{0.15}{2}\right) - Q\left(\dfrac{0.25}{2}\right)$

                              $= 0.4701 - 0.4502$
                              $= 0.02$

(iii) Probability of purity $\leqslant 18 = Q(1)$
                                 $= 0.1587$
    Probability of $18 \leqslant$ purity $\leqslant 19 = Q(\frac{1}{2}) - Q(1)$
                                   $= 0.1498$
    Probability of $19 \leqslant$ purity $\leqslant 21 = 1 - 2Q(\frac{1}{2})$
                                   $= 1 - 2(0.3085)$
                                   $= 0.383$
    Probability of purity $> 21 = 0.3085$

    Mean profit $= £(0.3085(30) + 0.383(20)$
                            $- 0.1587(100))$
                $= £1.045$ or £1 per component

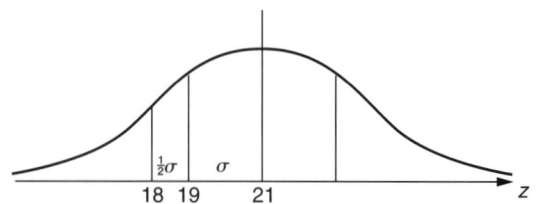

(iv) Probability of purity $> 21 = 0.5$
    Probability of $19 < $ purity $ < 21$
           $= \frac{1}{2} - Q(1) = 0.5 - 0.1587 = 0.3413$
    Probability of $18 < $ purity $ < 19$
           $= Q(1) - Q(1.5) = 0.0919$
    Probability of purity $< 18 = 0.0668$
    Profit over original costs
           $= 0.5(30) + 0.3413(20) - 0.0668(100)$
           $= £15.146$ or £15 to nearest pound
    If the profit remains the same the increase in cost per component is £14.

5 The continuous random variable X has probability density function f given by

$$f(x) = ax^2 + bx \quad \text{for } 0 < x < 1$$
$$f(x) = 0 \qquad \text{otherwise}$$

where a, b are constants.
(i) Show, by integrating f(x), that $2a + 3b = 6$.
(ii) Given that the mean value of X is 0.7, find another equation linking a and b. Hence show that $a = b = 1.2$.
(iii) Find the cumulative distribution function of X.
(iv) The random variable Y is defined by

$$Y = \frac{X}{X + 1}$$

Find the cumulative distribution function of Y.
Deduce that, for $0 < y < \frac{1}{2}$, the probability density function of Y is given by

$$g(y) = \frac{6y}{5(1 - y)^4} \qquad \text{(WJEC, 1994)}$$

• $f(x) = ax^2 + bx \quad \text{for } 0 < x < 1$
  $f(x) = 0 \quad \text{otherwise}$

(i) $\displaystyle\int_0^1 f(x)\,dx = 1 \Rightarrow \left[\frac{ax^3}{3} + \frac{bx^2}{2}\right]_0^1 = 1$

$\Rightarrow \dfrac{a}{3} + \dfrac{b}{2} = 1$

$\Rightarrow 2a + 3b = 6 \qquad (1)$

(ii) $E(X) = \displaystyle\int_0^1 xf(x)\,dx = \left[\frac{ax^4}{4} + \frac{bx^3}{3}\right]_0^1$

$\Rightarrow \dfrac{a}{4} + \dfrac{b}{3} = 0.7$

$\Rightarrow 3a + 4b = 8.4 \qquad (2)$

Solving equations 1 and 2
$\Rightarrow b = 1.2, \ a = 1.2$
Hence $a = b = 1.2$.

(iii) $F(x) = \displaystyle\int_0^x f(x)\,dx = \begin{cases} 0 & x < 0 \\ 0.4x^3 + 0.6x^2 & 0 \le x \le 1 \\ 1 & x \ge 1 \end{cases}$

(iv) $Y = \dfrac{X}{X+1} \Rightarrow X = \dfrac{Y}{1 - Y}$

when $x = 0, y = 0; \quad x = 1, y = \frac{1}{2}$

$$G(y) = \begin{cases} 0 & y < 0 \\ 0.4\left(\dfrac{y}{1-y}\right)^3 + 0.6\left(\dfrac{y}{1-y}\right)^2 & 0 \le y \le \frac{1}{2} \\ 1 & y \le \frac{1}{2} \end{cases}$$

For $0 \le y \le \frac{1}{2}$:

$$g(y) = \frac{d}{dy}(G(y)) = 3(0.4)\left(\frac{y}{1-y}\right)^2 \cdot \frac{1}{(1-y)^2}$$

$$+ 2(0.6)\left(\frac{y}{1-y}\right) \cdot \frac{1}{(1-y)^2}$$

$$= \frac{1.2y^2 + 1.2(y - y^2)}{(1-y)^4}$$

$$= \frac{1.2y}{(1-y)^4} = \frac{6y}{5(1-y)^4}$$

6 The continuous random variable X has probability density function

$$f(x) = \begin{cases} \dfrac{1+x}{6} & 1 \le x \le 3 \\ 0 & \text{otherwise} \end{cases}$$

(a) Sketch the probability density function of X.
(b) Calculate the mean of X.
(c) Specify fully the cumulative distribution function of X.
(d) Find m such that $P(X \le m) = \frac{1}{2}$. (L, 1993)

• $f(x) = \begin{cases} \dfrac{1+x}{6} & 1 \le x \le 3 \\ 0 & \text{otherwise} \end{cases}$

(a)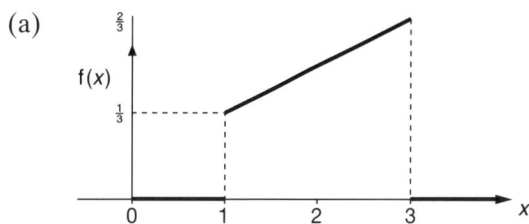

(b) Mean value of $X = \displaystyle\int_1^3 xf(x)\,dx$

$= \frac{1}{6}\left[\frac{x^2}{2} + \frac{x^3}{3}\right]_1^3$

$= \frac{1}{6}(\frac{9}{2} + 9 - \frac{1}{2} - \frac{1}{3})$

$= 2\frac{1}{9}$

(c) Cumulative distribution function $F(x)$ is

$\displaystyle\int_1^x f(x)\,dx = \left[\frac{(1+x)^2}{12}\right]_1^x = \frac{(1+x)^2}{12} - \frac{1}{3}$

Don't forget the lower limit

Hence

$$F(x) = \begin{cases} 0 & x < 1 \\ \dfrac{(1+x)^2}{12} - \dfrac{1}{3} & 1 \le x \le 3 \\ 1 & x > 3 \end{cases}$$

(d) $P(X \leqslant m) = \frac{1}{2}$ when

$$\frac{(1 + m)^2}{12} - \frac{1}{3} = \frac{1}{2}$$

$$\Rightarrow \quad (1 + m)^2 = 10$$

$$\Rightarrow \quad m = \sqrt{10} - 1 = 2.16$$

7 A rectangle of area A square metres has a perimeter of 20 metres and a side of length X metres where X is uniformly distributed between 0 and 2. Show that the probability density function of A is

$$\frac{1}{4\sqrt{25 - A}} \quad 0 \leqslant A \leqslant 16$$

Find the mean and variance of A. (JMB)

• Let the p.d.f. of A be f(a).
Perimeter 20 m ∴ sides are x and $10 - x$

$$\Rightarrow \quad a = x(10 - x)$$
$$a = 10x - x^2$$
$$-a + 25 = x^2 - 10x + 25 = (x - 5)^2$$

Rearrange to make x the subject

$$x - 5 = \pm\sqrt{25 - a}$$

Differentiate:

$$dx = \mp \frac{da}{2\sqrt{25 - a}}$$

Since X is uniformly distributed, the p.d.f. of X is $\frac{1}{2}$, therefore

$$\int_0^2 \tfrac{1}{2} \, dx = 1$$

Change to terms in a:
$x = 0, a = 0$;
$x = 2, a = 16$
Therefore

$$\int_0^{16} \frac{da}{4\sqrt{25 - a}} = 1 \quad \text{p.d.f. must be positive}$$
$$\Rightarrow \quad \text{no minus sign}$$

Hence the p.d.f. of A is $\dfrac{1}{4\sqrt{25 - A}} \quad 0 \leqslant A \leqslant 16$

$$E(A) = \int_0^{16} \frac{a}{4\sqrt{25 - a}} \, da$$

Let $z^2 = 25 - a$, $2z\,dz = -da$
When $a = 0, z = 5$; $a = 16, z = 3$

$$E(A) = \int_3^5 \frac{25 - z^2}{4z} \cdot 2z\,dz = \frac{1}{2}\left[25z - \frac{z^3}{3}\right]_3^5$$

$$= 8\tfrac{2}{3}$$

$$E(A^2) = \frac{1}{2}\int_3^5 (25 - z^2)^2 dz$$

$$= \frac{1}{2}\left[625z - \frac{50}{3}z^3 + \frac{z^5}{5}\right]_3^5$$

$$= 96\tfrac{8}{15}$$

Variance $(A) = 96\tfrac{8}{15} - (8\tfrac{2}{3})^2 = 21\tfrac{19}{45}$ or 21.42

8 Two red balls and two white balls are placed in a bag. Balls are drawn one by one, at random and without replacement. The random variable X is the number of white balls drawn before the first red ball is drawn.
(i) Show that $P(X = 1) = \frac{1}{3}$, and find the rest of the probability distribution of X.
(ii) Find $E(X)$ and show that Var $(X) = \frac{5}{9}$.
(iii) The sample mean for 80 independent observations of X is denoted by $\bar{X}$. Using a suitable approximation, find $P(\bar{X} > 0.75)$.
(UCLES, 1991)

• (i) $P(X = 0) = P(\text{1st ball is red}) = \frac{1}{2}$
$P(X = 1) = P(\text{1st ball is W, 2nd ball is R})$
$\qquad = \frac{1}{2} \times \frac{2}{3} = \frac{1}{3}$
$P(X = 2) = P(\text{1st ball is W, 2nd ball is W})$
$\qquad = \frac{1}{2} \times \frac{1}{3} = \frac{1}{6}$

$X = x$	0	1	2
$P(X = x)$	$\frac{1}{2}$	$\frac{1}{3}$	$\frac{1}{6}$
$xP(X = x)$	0	$\frac{1}{3}$	$\frac{1}{3}$
$x^2P(X = x)$	0	$\frac{1}{3}$	$\frac{2}{3}$

$$\Sigma xP(x) = \tfrac{2}{3}$$
$$\Sigma x^2P(x) = 1$$

(ii) $E(X) = \frac{2}{3}$
Var $(X) = E(X^2) - (E(X))^2$
$\qquad = 1 - \frac{4}{9}$
$\qquad = \frac{5}{9}$

(iii) For large n, $\bar{X} \sim N\left(\mu, \dfrac{\sigma^2}{n}\right)$; see Chapter 18

$$\bar{X} \sim N(\tfrac{2}{3}, \tfrac{5}{720}), \quad \sigma = \tfrac{1}{12}$$

$$P(\bar{X} > 0.75) = Q\left(\frac{0.75 - \frac{2}{3}}{\frac{1}{12}}\right) = Q(1)$$

$$= 0.1587$$

Exercises

1 Mass produced pipes have internal diameters that are Normally distributed with mean 10 cm and standard deviation 0.4 cm.
(i) Calculate, correct to 3 decimal places, the probability that a randomly chosen pipe will have an internal diameter greater than 10.3 cm.

(ii) Find the value of d, correct to two decimal places, if 95% of the pipes have internal diameters greater than d cm. (WJEC, 1992)

2 The random variable X is normally distributed with mean μ and variance σ^2.
Given that $P(X > 58.37) = 0.02$
and $P(X < 40.85) = 0.01$
find μ and σ. (L, 1992)

3 The continuous random variable X has probability density function f given by

$f(x) = kx(4 - x^2)$ $0 \leqslant x \leqslant 2$
$f(x) = 0$ otherwise

(i) Show that $k = \frac{1}{4}$.
(ii) Calculate the mean value of X. (WJEC, 1994)

4 The random variable X has probability density function

$$f(x) = \begin{cases} 3x^k & 0 \leqslant x \leqslant 1 \\ 0 & \text{otherwise} \end{cases}$$

where k is a positive integer.
Find
(a) the value of k
(b) the mode of X
(c) the mean of X
(d) the value, x, such that $P(X \leqslant x) = 0.5$
(L Specimen Paper)

5 The time at which a commuter arrives at her bus stop each morning is a random variable with a normal probability distribution having a mean of 07 40 (twenty minutes to eight) and a standard deviation of 2 minutes. Find the probability that she arrives after 07 45 (quarter to eight). On one day in twenty she arrives before time t. Find t to the nearest 0.1 of a minute.

The time at which her usual bus arrives at this bus stop is also a random variable with a normal distribution having a mean μ and a standard deviation σ. The commuter finds that, if she arrives at the bus stop at 07 40, the probability that she will catch this bus is 0.9 while, if she does not arrive until 07 45, the probability that she will catch it is only 0.2. Write down two equations for μ and σ of the form

$A\mu + B\sigma = C$

where A, B, C are constants to be determined.

Hence find μ and σ.

Find the probability that the commuter both catches her bus and waits less than 5 minutes at the bus stop. (O&C, 1991)

6 On any day, the amount of time, measured in hours, that Mr Goggle spends watching television is a continuous random variable T, with cumulative distribution function given by

$$F(t) = \begin{cases} 0 & (t \leqslant 0) \\ 1 - k(15 - t)^2 & (0 \leqslant t \leqslant 15) \\ 1 & (t \geqslant 15) \end{cases}$$

where k is a constant.
(i) Show that $k = \frac{1}{225}$ and find $P(5 \leqslant T \leqslant 10)$.
(ii) Show that, for $0 \leqslant t \leqslant 15$, the probability density function of T is given by

$f(t) = \frac{2}{15} - \frac{2}{225} t$

(iii) Find the median of T.
(iv) Find Var (T). (UCLES, 1989)

7 The continuous random variable X has probability density function f, where

$f(x) = \dfrac{4x^3}{\alpha^4}$ $0 \leqslant x \leqslant \alpha$

$f(x) = 0$ otherwise

(i) Find the mean and the variance of X in terms of α. Deduce the values of the mean and the variance of

$Y = \dfrac{7\alpha - 5X}{\alpha}$

(ii) If $\bar{X}$ is the mean of a random sample of 24 observations of X show that

$T = \frac{5}{4} \bar{X}$

is an unbiased estimator of α. Express the standard error of T in terms of α. (WJEC, 1992)

8 The random variable X has probability density function

$$f(x) = \begin{cases} 10cx^2 & 0 \leqslant x < 0.6 \\ 9c(1 - x) & 0.6 \leqslant x \leqslant 1.0 \\ 0 & \text{otherwise} \end{cases}$$

where c is a constant.
(a) Find the value of c and sketch the graph of the probability density function.
The mode of a random variable X is the value of x for which the probability density function is a maximum.
(b) Write down the mode of X.
(c) Find the probability that X is less than 0.4. (L, 1992)

9 The continuous random variable U has a uniform distribution on $0 < u < 1$. The random variable X is defined as follows:

$X = 2U$ when $U \leqslant \frac{3}{4}$
$X = 4U$ when $U > \frac{3}{4}$

(i) Give a reason why X cannot take values between $\frac{3}{2}$ and 3, and write down the values of $P(0 < X \leqslant \frac{3}{2})$ and $P(3 < X < 4)$.

(ii) Sketch the complete graph of the probability density function of X.

(iii) Find the lower quartile q of X, i.e. the value of q such that $P(X < q) = \frac{1}{4}$.

(iv) Three independent observations are taken of X. Find the probability that they all exceed q.

(v) Show that $E(X) = \frac{23}{16}$ and find $E(X^2)$.

(UCLES, 1992)

10 The continuous random variable X has probability density function f where

$$f(x) = \begin{cases} (a + bx)/x^{1/2} & 0 < x < 4 \\ 0 & \text{elsewhere} \end{cases}$$

Given that $P(X < 1) = 0.44$, show that $a = 0.21$ and find the value of b.

Find the mean and variance of X.

A computer engineer has 4 hours in which to locate a computer fault before a senior engineer takes over the search. On 75% of occasions, she succeeds in locating the fault. When this happens the time, in hours, that she takes to locate the fault is a random variable with the same probability distribution as X. A is the event that the engineer does not find the fault in the first hour and S is the event that the senior engineer is needed. Find $P(S)$. Explain briefly why $P(S \mid A) = P(S)/P(A)$.

Find the probability that the senior engineer will be needed if the first engineer fails to locate the fault in the first hour. (O&C, 1991)

11 The continuous random variable X has a uniform (rectangular) distribution on the interval $[10, 20]$, i.e. the probability density function f is given by

$$f(x) = \begin{cases} \frac{1}{10} & 10 \leqslant x \leqslant 20 \\ 0 & \text{otherwise} \end{cases}$$

Write down the value of $E(X)$, and show by integration that $\text{Var}(X) = \frac{25}{3}$.

The cumulative distribution function of X is F. Express $F(x)$ in terms of x, for $10 \leqslant x \leqslant 20$. Sketch the form of the graph of F for all values of x.

The random variable Y is defined by $\dfrac{1}{X^2}$. Show that $E(Y) = \frac{1}{200}$ and find $\text{Var}(Y)$. (UCLES, 1993)

Brief Solutions

1 X is the r.v. 'diameter of a pipe'
      $X \sim N(10, 0.16)$, $\sigma = 0.4$

(i)   Probability of $X > 10.3 = Q\left(\dfrac{0.3}{0.4}\right) = 0.227$

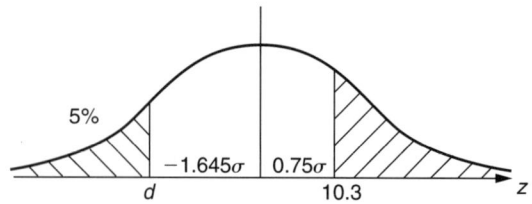

(ii) 95% of pipes have diameter greater than d
$\Rightarrow P(X < d) = 0.05$
But $Q(1.645) = 0.05$
$\therefore d = 10 - 1.645(0.4) = 9.34$ cm

2

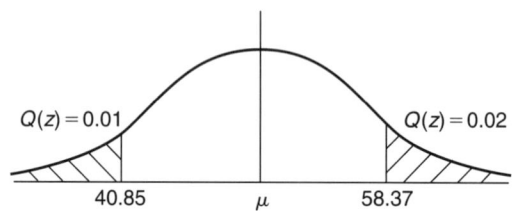

$P(X > 58.37) = 0.02 \quad \Rightarrow \quad z = 2.054$

$$\Rightarrow \quad \frac{58.37 - \mu}{\sigma} = 2.054 \qquad\qquad (1)$$

$P(X < 40.85) = 0.01 \quad \Rightarrow \quad z = -2.327$

$$\Rightarrow \quad \frac{\mu - 40.85}{\sigma} = 2.327 \qquad\qquad (2)$$

2.1b and 2.2b

Adding equations 1 and 2: $\quad 17.52 = 4.381\sigma$
$\Rightarrow \quad \sigma = 4$, $\mu = 50.16$

3 (i) $\displaystyle\int_0^2 kx(4 - x^2)\,dx = 1 \quad \Rightarrow \quad k\left[2x^2 - \frac{x^4}{4}\right]_0^2 = 1$

$$\Rightarrow \quad k = \tfrac{1}{4} \qquad \boxed{6.2}$$

(ii) Mean value of $X = \dfrac{1}{4}\displaystyle\int_0^2 x(4x - x^3)\,dx$

$$= \frac{1}{4}\left[\frac{4x^3}{3} - \frac{x^5}{5}\right]_0^2 = \frac{16}{15}$$

4 (a) $\displaystyle\int_0^1 f(x)\,dx = 1 \quad \Rightarrow \quad \left[\frac{3x^{k+1}}{k+1}\right]_0^1 = 1$

$$\Rightarrow \quad \frac{3}{k+1} = 1 \quad \Rightarrow \quad k = 2$$

$f(x) = 3x^2$

(b) Mode of $X = 1$ Highest value of f(x) occurs at the mode

(c) Mean of $X = \displaystyle\int_0^1 xf(x)\,dx = \left[\frac{3x^4}{4}\right]_0^1 = \frac{3}{4}$

(d) $P(X \leqslant x) = 0.5 \implies \int_0^{x_1} 3x^2 \, dx = \frac{1}{2}$

$x_1 = \sqrt[3]{\frac{1}{2}} = 0.794$

5 Event X: 'time of arrival of the commuter'
$X \sim N(07\,40, 4), \quad \sigma_X = 2$
Event Y: 'time of arrival of the bus'
$Y \sim N(\mu, \sigma^2), \quad \sigma_Y = \sigma$

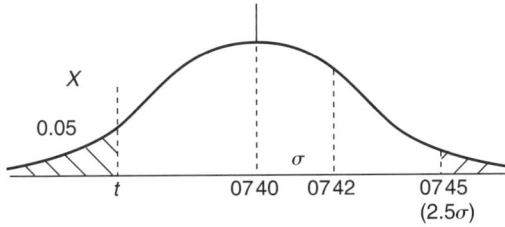

(i) $P(X > 07\,45) = Q(2.5) = 0.0062$
(ii) $P(X < t) = 0.05 = Q(-1.645)$
$\quad t = 07\,40 - 1.645\sigma = 07\,36.7$

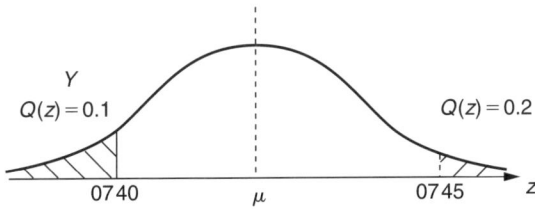

(iii) $P(Y < 07\,40) = 0.1 \implies \mu - 1.281\sigma = 07\,40$
$\quad P(Y > 07\,45) = 0.2 \implies \mu + 0.842\sigma = 07\,45$
$\quad \implies \sigma = 2.355, \mu = 07\,43$
(iv) Event Z = time of arrival of bus − time of
$\quad$ arrival of commuter $(Y - X)$
$\quad Z \sim N(3, 9.55), \quad \sigma_Z = 3.090$

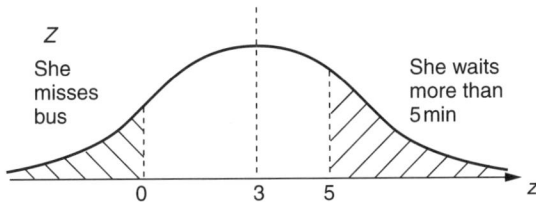

Probability that she catches bus and waits less than
5 min is $P(0 < Z < 5)$

$= 1 - Q\left(\frac{3-0}{3.090}\right) - Q\left(\frac{5-3}{3.090}\right)$

$= 0.5754$

6 (i) $t = 0, F(0) = 0$
$\quad \therefore k \cdot 225 = 1 \implies k = \frac{1}{225}$
$\quad P(5 \leqslant T \leqslant 10) = F(10) - F(5) = \frac{1}{3}$

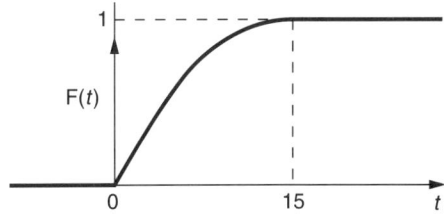

(ii) $f(t) = \frac{d}{dt}(F(t)) = \frac{2}{225}(15 - t) = \frac{2}{15} - \frac{2}{225}t$

(iii) Median of T is t_1 where

$F(t_1) = \frac{1}{2} \implies t_1 = 15 - \frac{15}{\sqrt{2}}$ or 4.393

(iv) $E(T) = \int_0^{15} tf(t) \, dt = 5$

$E(T^2) = \int_0^{15} t^2 f(t) \, dt = 37.5$ $\boxed{6.2}$

$\text{Var}(T) = 37.5 - (5)^2 = 12.5$

7 (i) Mean of $X = \int_0^\alpha \frac{x \cdot 4x^3 \, dx}{\alpha^4} = \left[\frac{4x^5}{5\alpha^4}\right]_0^\alpha = \frac{4\alpha}{5}$

Variance of $X = \int_0^\alpha x^2 \cdot \frac{4x^3}{\alpha^4} \, dx - \left(\frac{4\alpha}{5}\right)^2 = \frac{2\alpha^2}{75}$

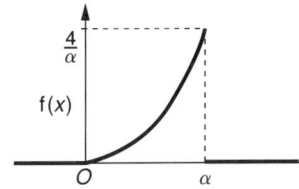

$Y = \frac{7\alpha - 5X}{\alpha} = 7 - \frac{5}{\alpha}X$

Mean of $Y = 7 - \frac{5}{\alpha}(\bar{X}) = 3$

Var of $Y = \frac{25}{\alpha^2}$ Var of $X = \frac{2}{3}$

Adding a constant to the variable does not change
the variance

(ii) $\bar{X}$ = mean of $(X_1 + X_2 + X_3 + \ldots + X_{24})$

$E(\bar{X}) = \frac{X_1 + X_2 + \ldots + X_{24}}{24}$

$T = \frac{5}{4}\left(\frac{X_1 + X_2 + \ldots + X_{24}}{24}\right)$

$= \frac{5}{96}(E(X_1) + E(X_2) + \ldots + E(X_{24}))$

$= \frac{5}{96} \cdot 24 \cdot \frac{4}{5}\alpha = \alpha$
Hence T is an unbiased estimator of α.

$$\text{Standard error} = \frac{\sigma}{\sqrt{n}} = \sqrt{\frac{2\,\alpha^2}{75 \times 24}} = \frac{\alpha}{30}$$

See Chapter 18

8 (a) $\displaystyle\int_{\text{all } x} f(x)\,dx = 1$

$$\Rightarrow \left[\frac{10cx^3}{3}\right]_0^{0.6} + \left[\frac{-9c}{2}(1-x)^2\right]_{0.6}^1 = 1$$

$$\Rightarrow \tfrac{36}{25}c = 1 \Rightarrow c = \tfrac{25}{36} = 0.694$$

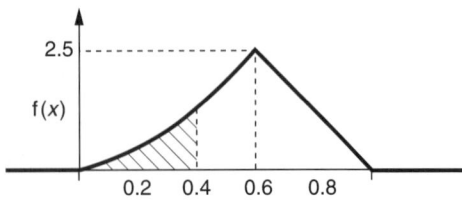

(b) Mode of $X = 0.6$

(c) $P(X) < 0.4 \Rightarrow \left[\dfrac{250}{3 \times 36}x^3\right]_0^{0.4} = 0.148$

9 (i) $0 < U \leqslant \tfrac{3}{4} \Rightarrow 0 < X \leqslant \tfrac{3}{2}$
$\tfrac{3}{4} < U < 1 \Rightarrow 3 < X < 4$
$\therefore X$ cannot take values between $\tfrac{3}{2}$ and 3

$$P(0 < X \leqslant \tfrac{3}{2}) = \tfrac{3}{4} \quad P(3 < X < 4) = \tfrac{1}{4}$$

(ii)

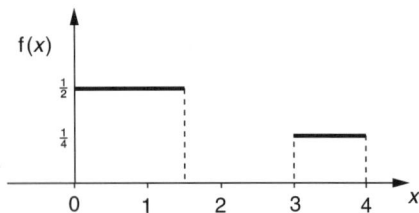

(iii) $P(X < q) = \tfrac{1}{4} \Rightarrow q = \tfrac{1}{2}$ Lower quartile
(iv) Probability that (3 observations all exceed q)
$= (\tfrac{3}{4})^3 = \tfrac{27}{64}$

(v) $E(X) = \displaystyle\int_0^{1.5} \tfrac{1}{2}x\,dx + \int_3^4 \tfrac{1}{4}x\,dx = \tfrac{23}{16}$ ☐ 6.2

$$E(X^2) = \int_0^{1.5} \tfrac{1}{2}x^2\,dx + \int_3^4 \tfrac{1}{4}x^2\,dx = \tfrac{175}{48}$$

10 $f(x) = \begin{cases} (a + bx)/x^{1/2} & 0 < x < 4 \\ 0 & \text{elsewhere} \end{cases}$

Remember $\displaystyle\int_{\text{all } x} f(x)\,dx = 1$

$$\int_0^4 (ax^{-1/2} + bx^{1/2})\,dx = 4a + \tfrac{16}{3}b = 1$$

$$\int_0^1 (ax^{-1/2} + bx^{1/2})\,dx = 0.44 \Rightarrow 2a + \frac{2b}{3} = 0.44$$

$$\Rightarrow b = 0.03, a = 0.21 \text{ as required}$$

$$\text{Mean} = \int_0^4 (ax^{1/2} + bx^{3/2})\,dx = 1.504 = E(X)$$

$$\text{Var}(X) = \int_0^4 (ax^{3/2} + bx^{5/2})\,dx - (1.504)^2 = 1.523$$

$P(\bar{A}) = 0.44 \times 0.75 = 0.33$
$P(A) = 0.67$
$P(S) = 0.25$

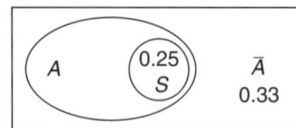

Since S is part of A, $P(S \cap A) = P(S)$.

$$P(S \mid A) = \frac{P(S \cap A)}{P(A)} = \frac{P(S)}{P(A)} = 0.373$$

This is the probability that the senior engineer will be needed if the first engineer did not find the fault in the first hour.

11 $f(x) = \tfrac{1}{10}$ for $10 \leqslant x \leqslant 20$

$$E(X) = 15 \ (\text{by inspection or } \int_{10}^{20} \frac{x}{10}\,dx = 15)$$

$$\text{Var } X = \int_{10}^{20} \frac{x^2}{10}\,dx - (15)^2 = \tfrac{25}{3}$$

$$F(x) = \int_{10}^x \tfrac{1}{10}\,dx = \tfrac{1}{10}(x - 10) \quad 10 \leqslant x \leqslant 20$$

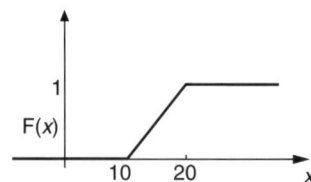

$$Y = \frac{1}{X^2} \Rightarrow X = Y^{-1/2} \Rightarrow dx = -\tfrac{1}{2}y^{-3/2}dy$$

$x = 10, y = \tfrac{1}{100}; \quad x = 20, y = \tfrac{1}{400}$

$$E(Y) = \int_{1/100}^{1/400} y \cdot f(y)\,dy = \int_{1/100}^{1/400} \tfrac{1}{10}y \cdot (-\tfrac{1}{2}y^{-3/2})\,dy$$

$$= \int_{1/400}^{1/100} \tfrac{1}{20}y^{-1/2}\,dy = \frac{1}{200}$$

$$E(Y^2) = \int_{1/400}^{1/100} \tfrac{1}{20}y^{1/2}\,dy = \frac{7}{240\,000}$$

$$\text{Var}(Y) = \frac{7}{240\,000} - \frac{1}{40\,000} = \frac{1}{240\,000}$$

18

Confidence intervals/ samples/hypothesis testing

	AS Level			A Level		Topic	Date attempted	Date completed	Self-assessment
CORE	MODULAR	TRADITIONAL	CORE	MODULAR	TRADITIONAL				
	✓	✓		✓	✓	**Sample mean and variances**			
	✓	✓		✓	✓	**Confidence intervals**			
	✓	✓		✓	✓	**Hypothesis testing**			

Fact Sheet

Samples $n \geqslant 30$

Take many samples, size n, from a population and find their means. The **Central Limit Theorem** proves that these means form a Normal distribution with mean μ (that of the population) and standard deviation $\dfrac{\sigma}{\sqrt{n}}$ (called the **standard error**):

$$\bar{X} \sim N\left(\mu, \frac{\sigma^2}{n}\right)$$

For small samples ($n < 30$) see Chapter 19.

Confidence Intervals (CI) for the Population Mean

A random sample, mean $\bar{x}$, variance s^2, is taken from a normal population with mean μ and variance σ^2.

- If σ^2 is known, the CI for μ is $\bar{x} \pm z\dfrac{\sigma}{\sqrt{n}}$ (where z is taken from the Normal tables).

  90% CI: $\left(\bar{x} - 1.645\dfrac{\sigma}{\sqrt{n}}, \bar{x} + 1.645\dfrac{\sigma}{\sqrt{n}}\right)$

  95% CI: $\left(\bar{x} - 1.96\dfrac{\sigma}{\sqrt{n}}, \bar{x} + 1.96\dfrac{\sigma}{\sqrt{n}}\right)$

  99% CI: $\left(\bar{x} - 2.575\dfrac{\sigma}{\sqrt{n}}, \bar{x} + 2.575\dfrac{\sigma}{\sqrt{n}}\right)$

- If σ^2 is unknown, the CI for μ is

  $$\bar{x} \pm z\frac{\hat{s}}{\sqrt{n}} \text{ where } \hat{s}^2 = \frac{n}{n-1}s^2$$

Confidence Intervals for Proportions

If p_s is the proportion of successes in a sample size n then the CI for the population proportion p is given by

$$p = p_s \pm z\sqrt{\frac{p_s(1 - p_s)}{n}} \quad \text{or} \quad p_s \pm z\sqrt{\frac{p_s q_s}{n}}$$

Hypothesis Testing of the Mean

- If σ^2 is known, use $N\left(\mu, \dfrac{\sigma^2}{n}\right)$ and $z = \dfrac{\bar{x} - \mu}{\sigma/\sqrt{n}}$.

- If σ^2 is unknown but n is large (>30), use $N\left(\mu, \dfrac{\hat{s}^2}{n}\right)$ where $\hat{s}^2 = \dfrac{n}{n-1}s^2$

  and $z = \dfrac{\bar{x} - \mu}{\hat{s}/\sqrt{n}} = \dfrac{\bar{x} - \mu}{s/\sqrt{n-1}} \approx \dfrac{\bar{x} - \mu}{s/\sqrt{n}}$

For small samples ($n < 30$) see Chapter 19.

- Null hypothesis $\quad H_0 : \bar{x} = \mu.$
- Alternative hypothesis $\quad H_1 : \bar{x} \neq \mu \quad$ (two tailed test)

$$\left. \begin{array}{l} \text{or } H_1 : \bar{x} > \mu \\ \text{or } H_1 : \bar{x} < \mu \end{array} \right\} \text{ (both are one tailed tests)}$$

At the 5% level:

two tailed: $z_{\text{critical}} = 1.96$ and -1.96

one tailed: $z_{\text{critical}} = 1.645$ or -1.645

Worked Examples

1 There are many complaints from passengers about the late running of trains in Ruritania. The Ruritanian Railway Board claims that the proportion of express trains that are 'delayed' (i.e. at least 5 minutes late) is 43%. The Railway Passengers' Association conducts a study of a random sample of 200 trains and finds that 92 of these trains are delayed. Test, at the 8% level, whether the Railway Board is understating the proportion of trains that are delayed.

The Association also notes the time t, in seconds, by which each train in the sample is late (no trains are early). The results are summarized by $\sum(t - 300) = 2012$, $\sum(t - 300)^2 = 525\,262$. Find a symmetric 95% confidence interval for the mean time by which a train is late. (UCLES, 1988)

- For $n = 200$, the proportion of the sample delayed is

$$p_s = \tfrac{92}{200} = 0.46$$

Hypothesis test $\quad H_0 : p = 0.43$
$\qquad\qquad\qquad\quad H_1 : p > 0.43$ – a one tailed test

Only a one tailed test since delays are being tested

At the 8% level $z_{\text{critical}} = 1.406$.

$$p_s \sim N\left(p, \frac{pq}{n}\right)$$

$$\sim N\left(0.43, \frac{0.43 \times 0.57}{200}\right)$$

$$\sim N(0.43, 1.2255 \times 10^{-3})$$

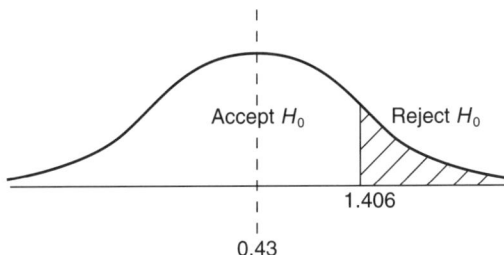

Test statistic $z = \dfrac{0.46 - 0.43}{\sqrt{1.2255 \times 10^{-3}}} = 0.857$

Since $0.857 < 1.406$ there is no reason to reject H_0 at the 8% level.

If t is the time, in seconds, by which each train is late then

$$\sum(t - 300) = 2012 \quad \sum(t - 300)^2 = 525\,262$$
$$n = 200$$

The sample mean is

$$\bar{t} = \frac{\sum(t - 300)}{200} + 300$$

$$= \frac{2012}{200} + 300$$

$$= 310.06$$

The sample variance is

$$s^2 = \frac{\sum(t - 300)^2}{200} - \left(\frac{\sum(t - 300)}{200}\right)^2$$

$$= \frac{525\,262}{200} - (10.06)^2$$

$$= 2525.106$$

95% CI of the population mean $= \bar{t} \pm 1.96 \dfrac{\sigma}{\sqrt{n}}$

$$\sigma \approx \hat{s} = \sqrt{\frac{ns^2}{(n-1)}} = 50.3765$$

$$\text{CI} = 310.06 \pm \frac{1.96(50.3765)}{\sqrt{200}} = 310.06 \pm 6.98$$

CI is (303, 317) for the mean time a train is late (seconds).

2 The length of string in the balls of string made by a particular manufacturer has mean μ m and variance 27.4 m². The manufacturer claims that $\mu = 300$. A random sample of 100 balls of string is taken and the sample mean is found to be 299.2 m. Test whether this provides significant evidence, at the 3% level, that the manufacturer's claim overstates the value of μ.

A piece of string is taken at random from each of the 100 balls of string and the breaking stress x, in suitable units, is measured. The results are summarized by $\sum x = 530$ and $\sum x^2 = 3007$. Find a symmetric 96% confidence interval for the mean breaking stress of the string. (UCLES, 1986)

• Mean μ, variance $\sigma^2 = 27.4$.
Sample size $n = 100$, sample mean $\bar{x} = 299.2$ m.
For a sample size 100:

$$\bar{X} \sim N\left(\mu, \frac{27.4}{100}\right) \text{ i.e. } \bar{X} \sim N(\mu, 0.274)$$

$H_0: \mu = 300$
$H_1: \mu < 300$ (one tailed test)
At a 3% level, $z_{\text{critical}} = -1.881$.

Test statistic $z = \dfrac{299.2 - 300}{\sqrt{0.274}} = -1.528$

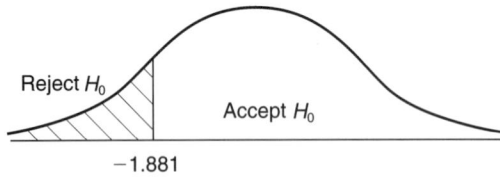

Reject H_0 Accept H_0

-1.881

Since $1.528 < 1.881$, accept H_0.
The manufacturer does not overstate the value of μ.

For the breaking stress x:

$n = 100 \quad \Sigma x = 530 \quad \Sigma x^2 = 3007$

$\bar{x} = 5.3 \quad \text{Var }(x): s^2 = \dfrac{3007}{100} - (5.3)^2 = 1.98$

For a 96% CI of the population mean of the breaking stress:

$$\text{CI} = \bar{x} \pm 2.054 \frac{\sigma}{\sqrt{n}}$$

$$= \bar{x} \pm 2.054 \frac{s}{\sqrt{99}} \quad \left(\frac{\sigma}{\sqrt{n}} \approx \frac{s}{\sqrt{n-1}}\right)$$

$$= 5.3 \pm 2.054 \times 0.1414$$

$$= 5.3 \pm 0.290$$

That is, CI is $(5.01, 5.59)$ for the mean breaking stress.

3 In a random sample of 30 pupils from a large school, 21 pupils received less than £2 per week in pocket money. Treating the sample as a large sample, find a 90% confidence interval for the proportion of pupils in the school receiving less than £2 per week in pocket money.

The amounts (£x) of pocket money received by the 30 pupils in the sample are summarized by $\Sigma x = 54.25$, $\Sigma x^2 = 106.24$. On a previous occasion, the average pocket money per week was found to be £1.67. Test, at the 10% level, whether the average pocket money has increased.

• Proportion with less than £2 per week $= 0.7 = p_s$.

$n = 30 \quad p_s = 0.7 \quad q_s = 0.3$

$$90\% \text{ CI} = p_s \pm 1.645 \sqrt{\frac{p_s q_s}{n}}$$

$$= 0.7 \pm 1.645 \sqrt{\frac{0.21}{30}}$$

$$= (0.5624, 0.8376)$$

$\Sigma x = 54.25 \quad \Sigma x^2 = 106.24 \quad n = 30$

$\bar{x} = \dfrac{54.25}{30} = 1.808 \qquad \text{mean of sample}$

$s^2 = \dfrac{106.24}{30} - (1.808)^2 = 0.2725 \quad \text{variance of sample}$

For the population, $\hat{s}^2 = \dfrac{30 s^2}{29} = 0.282$.

Distribution $\bar{X} \sim N\left(1.67, \dfrac{0.282}{30}\right)$

$\hat{s}/\sqrt{n} = 0.0969$

$H_0: \mu = 1.67$
$H_1: \mu > 1.67$ one tailed test

At 10% level the critical value of z is 1.282.

Test statistic $z = 1.808 - \dfrac{1.67}{0.0969} = 1.424$

Since $1.424 > 1.282$ this is significant at the 10% level. That is, there is evidence that the pocket money has increased.

4 (a) In a survey concerning ownership of television sets in a certain city, 985 randomly chosen households were investigated. It was found that there was no television set in 77 households, one set in 621 households, and two or more sets in 287 households. Calculate a 95% confidence interval for the proportion of households in the city with at least one television set.

(b) A normal distribution has unknown mean μ and known variance σ^2. A random sample of n observations from the distribution has sample mean $\bar{x}$. The null hypothesis $\mu = \mu_0$ is being tested. Find, in terms of μ_0, σ and n, the set of values of $\bar{x}$ for which $\mu = \mu_0$ is rejected in favour of $\mu \neq \mu_0$ at the 1% level of significance.

Find also, in terms of $\bar{x}$, σ and n, the set of values of μ_0 for which $\mu = \mu_0$ is rejected in favour of $\mu < \mu_0$ at the 5% level of significance.
(UCLES, 1990)

• (a) $n = 985$:   No television set 77
One television set 621
Two or more television sets 287
Probability of at least one set is

$$p = \frac{908}{985} = 0.9218 \quad q = 0.0782$$

Taking a Normal approximation to a binomial,

$$\text{proportion} \sim N\left(p, \frac{pq}{985}\right)$$

i.e. proportion $\sim N(0.9218, 7.318 \times 10^{-5})$
SD $= 0.008\,55$

95% confidence interval is
$0.9218 \pm 0.008\,55 \times 1.96$
i.e. $(0.9050, 0.9386)$

(b) Sample mean $\bar{x}$
Null hypothesis $\quad H_0: \mu = \mu_0$
$\qquad\qquad\qquad H_1: \mu \neq \mu_0 \quad$ two tailed test

At the 1% level of significance each tail is $\frac{1}{2}\%$.
Critical value of z are ± 2.576.
H_0 is rejected if

$$\mu_0 - 2.576\,\frac{\sigma}{\sqrt{n}} > \bar{x} > \mu_0 + 2.576\,\frac{\sigma}{\sqrt{n}}$$

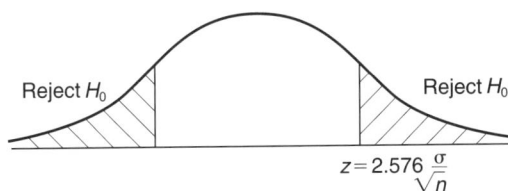

If the alternative hypothesis is $\mu < \mu_0$ then it is a one tailed test where the tail is 5%.

Critical value of z is -1.645.
H_0 is rejected if $\bar{x} < \mu_0 - 1.645\,\dfrac{\sigma}{\sqrt{n}}$.

5 A food processor produces large batches of jars of jam. In each batch the gross weight of a jar is known to be normally distributed with standard deviation 7.5 g.

The gross weights, in grams, of a random sample from a particular batch were:

517, 481, 504, 482, 503, 497, 512, 487, 497, 503, 509

(a) Calculate a 90% confidence interval for the mean gross weight of this batch.
(b) The manufacturer claims that the mean gross weight of a jar in a batch is at least 502 g. Test this claim at the 5% significance level.
(c) Explain why, if the manufacturer had claimed that the mean gross weight was at least 496 g, no further calculations would be necessary to test this claim. (AEB, 1994)

• For the sample

517, 481, 504, 482, 503, 497, 512, 487, 497, 503, 509

$n = 11, \Sigma x = 5492, \bar{x} = 499.3$.
Population standard deviation $\sigma = 7.5$.

(a) A symmetric 90% confidence interval is

$$499.3 \pm 1.645\,\frac{(7.5)}{\sqrt{11}}.$$

That is, 90% CI is $(495.6, 503.0)$.

(b) Hypothesis $\quad H_0: \mu \geqslant 502$.
$\qquad\qquad\quad H_1: \mu < 502 \quad$ (one tailed test)
$\sigma = 7.5, n = 11$.

Rejection criterion:

Reject H_0 if $z < -1.645$

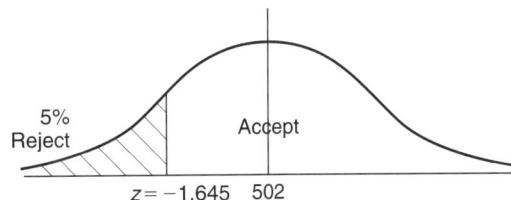

$$z = \frac{499.3 - 502}{7.5/\sqrt{11}} \quad = -\frac{2.7}{2.26}$$

$$= -1.194$$

As $1.194 < 1.645$, accept H_0, i.e. the manufacturer's claim is correct.

(c) If $H_0: \mu \geqslant 496$ g there is no need for a test since the sample mean is greater than 496.

6 (a) A large parent population is normally distributed with mean μ and variance σ^2. A large number of samples of size n are drawn and their means, m, calculated. Write down, in terms of μ, σ and n, the mean and variance of m.

Define the standard error of the mean and, if n is large, state clearly what the central limit theorem tells us about the distribution of the sample means.

(b) Bars of chocolate are produced with a mean mass of 110 g and standard deviation 10 g.

Calculate the minimum sample size necessary to ensure that there is at most a 5% chance of the mean mass being less than 108 g.

Assume the sample size is large.

(c) A random sample of 100 bars of chocolate is drawn from a population consisting of a large number of bars of chocolate.

The following table gives the number of bars of chocolate of each particular mass in the sample:

Mass (g)	108	109	110	111	112
No. of Bars	1	14	72	10	3

Calculate the mean and standard deviation of the sample. Estimate the standard error of the mean, and, assuming a normal population, use it to determine a 99% confidence interval for the mean mass of bars of chocolate in the parent population. (NICCEA, 1993)

• (a) The mean and variance of the means of samples, size n is

$$m = \mu, \quad \text{Var}(m) = \frac{\sigma^2}{n}$$

The standard error of the mean is the standard deviation of a sampling distribution of the mean, i.e. $\dfrac{\sigma}{\sqrt{n}}$.

If n is large the distribution of the sample mean is approximately

normal: $m \sim N\left(\mu, \dfrac{\sigma^2}{n}\right)$.

(b) For the bars of chocolate: $X \sim N(110, 100)$, $\sigma = 10$

For a sample of size n:

$$\bar{X} \sim N\left(110, \frac{100}{n}\right) \quad SD = \frac{10}{\sqrt{n}}$$

(This is the standard error of the mean.)

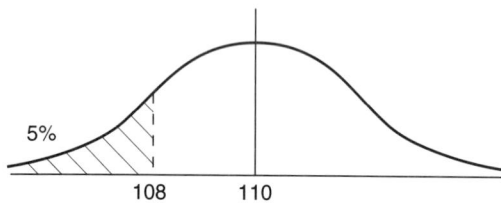

For a tail of 5%, $z = -1.645$

i.e. $z_{\text{crit}} = -1.645 = \dfrac{108 - 110}{10/\sqrt{n}}$

$\Rightarrow \quad 16.45 = 2\sqrt{n}$
$\Rightarrow \quad n = 67.6$

Hence the minimum sample size is 68.

(c) For a large sample of 100 bars

Mass (g)	108	109	110	111	112
No. of Bars	1	14	72	10	3

Mean of the sample is

$$\bar{x} = \frac{\Sigma f(x)}{n} = 110 \text{ g}$$

Variance $s^2 = \dfrac{\Sigma f(x - \bar{x})^2}{n} = 0.4$

Standard deviation $s = \sqrt{0.4} = 0.632$.

Standard error of the mean $= \dfrac{s}{\sqrt{n}} = 0.0632$.

99% CI $= 110 \pm 2.575(0.0632)$
$= 110 \pm 0.163$
$= (109.837, 110.163)$

7 (a) A random sample of 250 adult men undergoing a routine medical inspection had their heights (x cm) measured to the nearest centimetre, and the following data was obtained: $\Sigma x = 43\,205$, $\Sigma x^2 = 7\,469\,107$. Calculate an unbiased estimate of the population variance. Calculate also a symmetric 99% confidence interval for the population mean.

(b) In an investigation into ownership of calculators, 200 randomly chosen school students were interviewed, and 143 of them owned a calculator. Using the evidence of this sample, test, at the 5% level of significance, the hypothesis that the proportion of school students owning a calculator is 75% against the alternative hypothesis that the proportion is less than 75%. (UCLES, 1986)

• (a) $n = 250$, $\Sigma x = 43\,205$, $\Sigma x^2 = 7\,469\,107$

Sample mean $\bar{x} = \dfrac{43\,205}{250} = 172.82$

Sample variance $s^2 = \dfrac{\Sigma x^2}{n} - \bar{x}^2 = 9.6756$

An unbiased estimate of population variance is

$$\hat{s}^2 = \frac{n}{n-1}s^2 = 9.7145 \approx \sigma^2$$

99% confidence interval for the population mean μ is given by

$$\bar{x} \pm 2.575\frac{\sigma}{\sqrt{n}} = 172.82 \pm 0.507$$

$$= (172.31, 173.33)$$

(b) $n = 200$, $p_s = \frac{143}{200} = 0.715$
H_0: proportion owning a calculator is 75%
H_1: proportion owning a calculator <75%
This is a one tailed test.
The critical value for z at 5% level is -1.645.

$X \sim \text{Bin}(200, 0.75)$ $n = 200, p = 0.75, q = 0.25$
$\sim N(150, 37.5)$ $\mu = np, \sigma^2 = npq$

Test statistic is

$$\frac{\bar{x} - \mu}{\sigma} = \frac{143 - 150}{\sqrt{37.5}}$$

$$= -1.143$$

Since $1.096 < 1.645$ accept H_0, i.e. there is evidence at the 5% level that 75% of students own a calculator.

8 Salt is packed in bags which the manufacturer claims contain 25 kg each, on average. A random sample of 80 bags is examined and the mass, x kg, of the contents of each bag is determined. It is found that $\sum(x - 25) = 27.2$ and $\sum(x - 25)^2 = 85.1$, and that exactly 16 bags each contain less than 24.5 kg.
(a) Find a 90% confidence interval for the population proportion of bags containing less than 24.5 kg.
(b) Estimate the population mean and variance of the mass of the contents of a bag.
(c) Test, at the 10% level, whether the manufacturer is understating the average mass of the contents of a bag. (UCLES, 1990)

• (a) Proportion containing less than 24.5 kg is 0.2; $n = 80$.

$$90\% \text{ CI} = 0.2 \pm 1.64 \left(\sqrt{\frac{0.2 \times 0.8}{80}} \right)$$

$$= 0.2 \pm 0.0736$$

The 90% CI is $(0.1264, 0.2736)$.

(b) $\sum(x - 25) = 27.2 \Rightarrow \bar{x} = 25 + \dfrac{27.2}{80} = 25.34$.

$$\sum(x - 25)^2 = 85.1 \Rightarrow s^2 = \frac{85.1}{80} - (0.34)^2$$

$$= 0.948\,15$$

$$\hat{s}^2 = \frac{ns^2}{n - 1} = 0.9602$$

(c) $H_0: \mu = 25$
$H_1: \mu > 25$ *This is a one tailed test*
At 10% level the critical value of $z = 1.282$.

$$\text{Test statistic} = \frac{25.34 - 25}{\sqrt{\dfrac{0.9602}{80}}} = 3.10$$

Since this is much greater than 1.282, reject H_0, i.e. the manufacturer is understating the average mass (strong evidence).

Exercises

1 When a 'Thumbnail' drawing pin is dropped on to the floor, the probability that it lands 'point up' is p.
(i) A teacher drops a Thumbnail drawing pin 900 times and observes that it lands 'point up' 315 times. Test, at the 1% level, the hypothesis that $p = 0.4$ against the alternative $p < 0.4$.
(ii) A student drops a Thumbnail drawing pin 600 times and observes that it lands 'point up' 251 times. Using the student's results, find a symmetric 95% confidence interval for p. As part of a statistics investigation, 1500 students carry out similar experiments and they each calculate (correctly) their own symmetric 95% confidence interval for p. Find the expected number of these intervals that do not contain the true value of p. (UCLES, 1991)

2 In an investigation into the total distance travelled by cars currently in use, 500 randomly chosen cars were stopped and the distances they had travelled were noted. The sample mean was found to be 50 724 km, and the standard deviation of the distances in the sample was 13 112 km. Calculate a 90% confidence interval for the population mean distance travelled.

In a proposed investigation into standards of safety on the road, it is planned to stop 1000 randomly chosen vehicles and examine them for potentially dangerous defects. The purpose of the investigation is to test the null hypothesis that the proportion of vehicles with potentially dangerous defects on the road is 10%, against the alternative hypothesis that the proportion is more than this. Using a 5% significance level, calculate the least number of vehicles with potentially dangerous defects in the sample which would lead to rejection of the null hypothesis.

Explain briefly what could be concluded if the investigation yielded 107 vehicles with potentially dangerous defects. (UCLES, 1989)

3 In a large population of chickens, the distribution of the mass of a chicken has mean μ kg and standard deviation σ kg. A random sample of 100 chickens is taken from the population. The mean mass for the sample is denoted by $\bar{X}$. State the approximate distribution of $\bar{X}$, giving its mean and standard deviation.

The sample values are summarized by $\sum x = 189.1$ and $\sum x^2 = 401.74$, where x kg is the mass of a chicken. Find unbiased estimates of μ and σ^2, and an approximate symmetric 90% confidence interval for μ.

Given that, in fact, $\sigma = 0.71$, test, at the 1% level of significance, the null hypothesis $\mu = 1.75$ against the alternative $\mu > 1.75$, stating whether you are using a one tail or a two tail test and stating your conclusion clearly. (UCLES, 1992)

4 At an early stage in analysing the marks scored by the large number of candidates in an examination paper, the Examining Board takes a random sample of 250 candidates and finds that the marks, x, of these candidates give $\sum x = 11\,872$ and

$\Sigma x^2 = 646\ 193$. Calculate a 90% confidence interval for the population mean mark (μ) for this paper.

Using the figures obtained in this sample, the null hypothesis $\mu = 49.5$ is tested against the alternative hypothesis $\mu < 49.5$ at the $\alpha\%$ significance level. Determine the set of values of α for which the null hypothesis is rejected in favour of the alternative hypothesis.

It is subsequently found that the population mean and standard deviation for the paper are 45.292 and 18.761 respectively. Find the probability of a random sample of size 250 giving a sample mean at least as high as the one found in the sample above.

(UCLES, 1988)

5 A machine is regulated to dispense liquid into cartons in such a way that the amount of liquid dispensed on each occasion is normally distributed with a standard deviation of 20 ml.

Find 99% confidence limits for the mean amount of liquid dispensed if a random sample of 40 cartons has an average content of 266 ml. (L, 1993)

6 A car insurance company receives a large number of claims. A random sample of 120 claims was examined and it was found that the mean amount claimed was £900. Assuming that the population standard deviation is known to be £310 calculate a 95% confidence interval for the mean value of all such claims. (AEB Specimen Paper, 1996)

7 An electronic device is advertised as being able to retain information stored in it 'for 70 to 90 hours' after power has been switched off. In experiments carried out to test this claim, the retention time in hours, X, was measured on 250 occasions, and the data obtained is summarized by $\Sigma(x - 76) = 683$ and $\Sigma(x - 76)^2 = 26\ 132$. The population mean and variance of X are denoted by μ and σ^2 respectively.
(i) Show that, correct to one decimal place, an unbiased estimate of σ^2 is 97.5.
(ii) Test the hypothesis that $\mu = 80$ against the alternative hypothesis that $\mu < 80$, using a 5% significance level.
(iii) Calculate a symmetric 95% confidence interval for μ. (UCLES, 1987)

8 A plant produces steel sheets whose weights are known to be normally distributed with a standard deviation of 2.4 kg. A random sample of 36 sheets had a mean weight of 31.4 kg. Find 99% confidence limits for the population mean. (L, 1993)

9 The diameters in mm of 80 rods selected at random from a large consignment are summarized in the table below.

Diameter (centre of interval) (mm)	11	12	13	14	15
Number of rods		3	22	31	20	4

(i) Estimate the mean and variance of the diameter of the rods in the consignment.
(ii) Find symmetrical two-sided 97.5% confidence limits for the mean of the diameter of the rods in the consignment, correct to two decimal places.
(iii) The cross-section of each rod is a circle with diameter as measured. Estimate the mean cross-section area in mm^2 for the 80 rods in this sample. (O&C, 1992)

Brief Solutions

1 (i) $n = 900, p_s = 0.35$
$H_0: p = 0.4$; $H_1: p < 0.4$ (one tailed)

$$p_s \sim \text{N}\left(0.4, \frac{0.4 \times 0.6}{900}\right) \quad \Rightarrow \quad (0.4, 2.667 \times 10^{-4})$$

Test statistic $z = \left|\dfrac{0.35 - 0.4}{\sqrt{2.667 \times 10^{-4}}}\right| = 3.062$

> 2.326 (z_{crit} at 1% level)
Reject H_0, accept H_1. $p < 0.4$

(ii) $p_s = 0.4183$

$$\text{CI} = 0.4183 \pm 1.96 \sqrt{\frac{0.4183 \times (1 - 0.4183)}{600}}$$

$= (0.3789, 0.4578)$

For an individual,
P(interval does not contain p) $= 0.05$
Expected number of intervals:
$1500 \times 0.05 = 75$

2 $n = 500$, $\bar{x} = 50\ 724$ km, $s^2 = (13\ 112)^2$

90% CI $= 50\ 724 \pm 1.645 \times \dfrac{13\ 112}{\sqrt{500}}$

$= (49\ 760, 51\ 690)$

$n = 1000$, $H_0: p = 0.1$, $H_1: p > 0.1$ (one tailed)
If H_0 is true, $X \sim \text{Bin}(1000, 0.1)$
$\Rightarrow$ $X \sim \text{N}(100, 90)$, SD $= 9.487$

At 5% level $\dfrac{X - 100}{9.487} = 1.645$ $\Rightarrow$ $X = 115.6$

i.e. the least number of defective cars needed to reject H_0 is 116.
If 107 cars had defects, accept H_0.

or $p \sim N\left(0.1, \dfrac{0.09}{1000}\right)$, SD = 0.009 487

$\dfrac{p - 0.1}{0.009\,487} = 1.645$ or $p = 0.1156$

3 Population mean μ kg,   SD σ kg.

$\bar{X}$ is a Normal distribution with mean μ kg,

sample SD $\dfrac{\sigma}{\sqrt{n}}$ or $\dfrac{\sigma}{10}$ kg.

Sample mean = 1.891,   sample variance = 0.4415.
Unbiased estimates $\mu = 1.891, \hat{s} = 0.668$

90% CI for μ is $1.891 \pm 1.645\left(\dfrac{0.668}{10}\right)$,

i.e. (1.78, 2.00).

$H_0: \mu = 1.75, H_1: \mu > 1.75$ 1% one tailed test

Sample size 100: SD = $\dfrac{0.71}{10}$

Test statistic $z = \dfrac{1.891 - 1.75}{0.071} = 1.986$

$z_{\text{crit}} = 2.325$

Accept H_0.

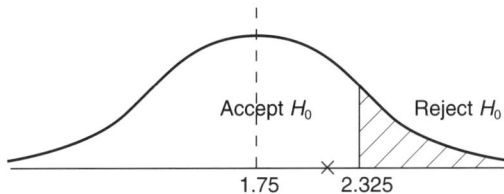

4 $\bar{x} = 47.488,$ $s^2 = 329.66,$ $\hat{s}^2 = \dfrac{ns^2}{n-1} = 330.99$

90% CI for μ: $47.488 \pm \dfrac{1.645}{\sqrt{250}}\sigma$, i.e. (45.60, 49.38)

$H_0: \mu = 49.5,$ $H_1: \mu < 49.5$
$X \sim N(49.5, 1.3240)$, SD = 1.151

Test statistic $z = \dfrac{47.49 - 49.5}{1.151} = -1.747$

$\Rightarrow$ $\alpha = 0.0403$ $\Rightarrow$ $\alpha = 4.03\%$.

To reject H_0, $\alpha > 4.03$.

$\mu = 45.292,$ $\sigma = 18.761,$ $\bar{x} - \mu = 2.196$

$z = \dfrac{2.196}{\sigma/\sqrt{n}} = 1.8507$ $\Rightarrow$ probability of 0.0321.

5 Mean $\bar{x} = 266,$ $n = 40$

99% CI is $\bar{x} \pm 2.576\dfrac{\sigma}{\sqrt{n}} = 266 \pm 2.576 \times \dfrac{20}{\sqrt{40}}$

i.e. (257.9, 274.1) ml.

6 $n = 120, \bar{X} = £900, \sigma = £310$

95% CI is $\bar{x} \pm 1.96\dfrac{\sigma}{\sqrt{n}} = 900 \pm \dfrac{1.96(310)}{\sqrt{120}}$

$= (844.5, 955.5)$

7 (i) $E(X) = 76 + E(X - 76) = 76 + \dfrac{683}{250}$

$= 78.732 = \hat{\mu}$

$\text{Var}\,(X) = \text{Var}\,(X - 36)$

$= \left(\dfrac{26\,132}{250} - \left(\dfrac{683}{250}\right)^2\right)\dfrac{250}{249}$

$= 97.5 = \hat{\sigma}^2$

(ii) $H_0: \mu = 80, H_1: \mu < 80$ – one tailed

$\bar{X} = 78.7$ $\bar{X} \sim N(80, 0.39)$

$z = \dfrac{78.7 - 80}{\sqrt{0.39}} = -2.08$

At 5% significance level,
$z_{\text{crit}} = -1.645$ $\Rightarrow$ Reject H_0, accept $\mu < 80$

(iii) 95% CI for μ: $78.7 \pm \dfrac{1.96\sqrt{97.5}}{\sqrt{250}}$,

i.e. (77.5, 79.9)

8 $\sigma = 2.4$ kg, sample $\bar{x} = 31.4$ kg, $n = 36$

99% CI is $\bar{x} \pm 2.576\dfrac{\sigma}{\sqrt{n}} = (30.37, 32.43)$

9 (i)   Sample: mean $\bar{x} = 13$, variance $0.875, n = 80$

(ii)   97.5% CI is $13 \pm 2.24\sqrt{\dfrac{0.875}{80}}$

$= (12.77, 13.23)$

(iii) Since areas involve d^2, start again to find $\bar{d}^{\,2}$
$\bar{d}^{\,2} = 169.875$.

Mean cross-sectional area $= \dfrac{\pi \bar{d}^{\,2}}{4} = 133.4$ mm^2

19

Miscellaneous statistics topics

AS Level			A Level			Topic	Date attempted	Date completed	Self-assessment
CORE	MODULAR	TRADITIONAL	CORE	MODULAR	TRADITIONAL				
				✓	✓	χ^2 test			
				✓	✓	*t* test			
				✓	✓	Regression			
				✓	✓	Correlation			
				✓	✓	Spearman's rank correlation coefficient			

Fact Sheet

χ^2 Test

Compares observed frequencies with expected frequencies.

The test statistic is $\chi^2_{\text{calc}} = \sum\limits_{i=1}^{n} \dfrac{(O_i - E_i)^2}{E_i}$

χ^2 is a fraction of ν, the number of degrees of freedom (number of classes − number of restrictions) and can be tested at different levels – usually 5% or 1%.

If $\nu = 1$ then χ^2 is not accurate and a correction should be made – **Yates' continuity correction**:

$$\chi^2_{\text{calc}} = \sum\limits_{i=1}^{n} \dfrac{(|\,O_i - E_i\,| - 0.5)^2}{E_i}$$

Small Samples ($n < 30$) – the 't' Test

If the population standard deviation σ is not known and if the sample is small (say $n < 30$), use the 't' test for confidence intervals for μ and for hypothesis testing. (For $20 < n < 30$, 't' and Normal are comparable.)

The statistic is $t = \dfrac{\bar{x} - \mu}{\hat{s}/\sqrt{n}}$

The t distribution is tabulated in terms of the number of degrees of freedom ν. When considering one set of data, $\nu = n - 1$ where $n =$ number in the sample.

(i) Confidence Intervals

The CI for the population μ based on a small sample is

$$\bar{x} \pm t \dfrac{\hat{s}}{\sqrt{n}} \text{ i.e. replace } \sigma \text{ by } \hat{s} \text{ and } z \text{ by } t.$$

(ii) Hypothesis Testing

$$t = \dfrac{\bar{x} - \mu}{\hat{s}/\sqrt{n}} \text{ compare with } t_{\text{crit}} \text{ from tables at the appropriate significance level.}$$

(iii) Equality of Two Population Means

To compare the means μ_A and μ_B of populations A and B the hypothesis test is
$H_0: \mu_A = \mu_B$ or $\mu_A - \mu_B = 0$
$H_1: \mu_A \neq \mu_B$ or $\mu_A - \mu_B \neq 0$ (two tailed)
(If testing for $\mu_A > \mu_B$ it becomes one tailed.)

The statistic is

$$t = \dfrac{\bar{X}_A - \bar{X}_B}{\hat{s}_{\text{p}} \sqrt{\dfrac{1}{n_A} + \dfrac{1}{n_B}}}$$

where

$$\hat{s}_{\text{p}}^2 = \text{unbiased estimate of population variance}$$

$$= \dfrac{n_A s_A^2 + n_B s_B^2}{n_A + n_B - 2}$$

There are now two sets of data so, when using the t tables:

$$\nu = n_A + n_B - 2$$

Regression Lines (the least squares regression lines)

(i) y on x
The line is $y = ax + b$ where

$$\sum y = a\sum x + nb$$
$$\sum xy = a\sum x^2 + b\sum x$$

or $y - \bar{y} = \dfrac{s_{xy}}{s_x}(x - \bar{x})$

where $s_{xy} = \dfrac{\sum xy}{n} - \bar{x}\bar{y}$ and $s_x = \dfrac{\sum x^2}{n} - \bar{x}^2 = $ variance of x

(ii) x on y
The line is $x = cy + d$ where

$$\sum x = c\sum y + nd$$
$$\sum xy = c\sum y^2 + d\sum y$$

or $x - \bar{x} = \dfrac{s_{xy}}{s_{y^2}}(y - \bar{y})$

where $s_y = \dfrac{\sum y^2}{n} - \bar{y}^2$

Both regression lines pass through $(\bar{x}, \bar{y})$.

Correlation

• The **product moment correlation coefficient** r quantifies how near the bivariate data lies to a straight line.

$$r = \frac{\sum xy - n\bar{x}\bar{y}}{\sqrt{(\sum x^2 - n\bar{x}^2)(\sum y^2 - n\bar{y}^2)}} = \frac{s_{xy}}{\sqrt{s_x s_y}}$$

If r is close to -1 there is a good negative correlation.
If r is close to $+1$ there is a good positive correlation.
If $-0.4 < r < 0.4$ there is poor correlation.
Beware of interpreting the value of r without regard to the variables. Comparing shoe size with reading ability would give a high value for r – older children tend to have bigger feet

• **Spearman's rank correlation coefficient**

$$\rho = 1 - \frac{6\sum d^2}{n(n^2 - 1)}$$

where d are the differences in the ranks of corresponding elements in the rankings ($x_1 - y_1, x_2 - y_2$, etc.).

Worked Examples

1 A research worker studying the ages of adults and the number of credit cards they possess obtained the results shown below.

	Number of cards possessed	
Age	**≤3**	**>3**
<30	74	20
≥30	50	35

Use the χ^2 statistic and a significance test at the 5% level to decide whether or not there is an association between age and number of credit cards possessed. (L, 1991)

•

	Number of cards possessed		
Age	**≤3**	**>3**	
<30	74	20	**94**
≥30	50	35	**85**
	124	**55**	**179**

$$\text{Expected value} = \frac{(\text{row total})(\text{column total})}{\text{grand total}}$$

For the element in the top left-hand box:

$$\frac{94 \times 124}{179} = 65.12$$

Age	**≤3**	**>3**	
<30	65.12	28.88	**94**
≥30	58.88	26.12	**85**
	124	**55**	**179**

In a 2 × 2 table there is one degree of freedom. Use Yates' continuity correction $\dfrac{(|O-E|-0.5)^2}{E}$

O	74	50	20	35
E	65.12	58.88	28.88	26.12
$\dfrac{(|O-E|-0.5)^2}{E}$	1.078	1.193	2.432	2.689

$$\sum \frac{(|O-E|-0.5)^2}{E} = 7.392$$

$\chi^2_{\text{TEST}} = 7.39$
$\chi^2_{5\%}(1) = 3.84$
$\Rightarrow \quad \chi_{\text{TEST}}{}^2 > \chi_{5\%}{}^2(1)$
There is a connection between the number of credit cards and age.

2 (a) Daily records of calls for assistance to a fire-fighting unit yielded the following frequency table.

Number of calls (x)	Number of days (f)
0	31
1	53
2	51
3	33
4	20
5	8
6	2
7	2
8 or more	0

(i) Calculate the mean number of calls per day.
(ii) Assuming that the situation can be modelled by a Poisson distribution, use the mean calculated in part (i) to estimate the expected frequencies.
(iii) Carry out a χ^2 goodness-of-fit test and state what can be inferred concerning the occurrence of calls to the unit.

(b) State conditions under which the Poisson distribution may be used to approximate the binomial distribution.

Breakdowns in an electricity supply system occur on average once in every 50 days. Use the Poisson approximation to the binomial distribution to estimate the probability that three or more breakdowns occur in a year. (SEB, 1993)

• (a) (i) Mean no. of calls:

$$\frac{0 + 53 + 102 + 99 + 80 + 40 + 12 + 14}{200}$$

$$= \frac{400}{200} = 2 \text{ calls/day}$$

(ii) Using a Poisson distribution (see Chapter 16) with $\lambda = 2$.

x	Po(x) $\times$ 200
0	27.1
1	54.1
2	54.1
3	36.1
4	18.0
5	7.2
6	2.4
7	0.7
8 or more	0.3

(iii) Consider

$$\sum_{i=1}^{8} \frac{(O_i - E_i)^2}{E_i}$$

$$= 0.561 + 0.022 + 0.178 + 0.266 + 0.222$$
$$+ 0.089 + 0.067 + 2.414 + 0.3$$
$$= 4.119$$

Number of classes $= 9$
Number of instructions $= 2$ (totals the same, means the same).
Hence $v = 7 \Rightarrow \chi^2_{5\%} = 14.1$
Since $4.119 < 14.1$, we infer that the calls follow a Poisson distribution with the same mean.

(b) A Poisson distribution can approximate to a binomial if

(i) n is large
(ii) p is <0.1

For electrical supply
P(breakdown in one day) $= \frac{1}{50}$.

When $n = 365$, $\lambda = \frac{365}{50} = 7.3$

Probability of 0, 1 or 2 breakdowns in one year is

$$e^{-\lambda}\left(1 + \lambda + \frac{\lambda^2}{2}\right) = 0.024$$

∴ Probability of 3 or more breakdowns in one year is 0.9764.

3 The accountant of a company monitors the number of items produced per month by the company, together with the total cost of production. The following table shows the data collected for a random sample of 12 months.

Number of items (x) (1000s)	Production cost (y) (£1000)
21	40
39	58
48	67
24	45
72	89
75	96
15	37
35	53
62	83
81	102
12	35
56	75

(a) Plot these data on a scatter diagram. Explain why this diagram would support the fitting of a regression equation of y on x.
(b) Find an equation for the regression line of y on x in the form $y = ax + b$.
(Use $\sum x^2 = 30\,786$; $\sum xy = 41\,444$.)
The selling price of each item produced is £2.20.
(c) Find the level of output at which total income and total costs are equal. Interpret this value.

(L, 1993)

• (a)

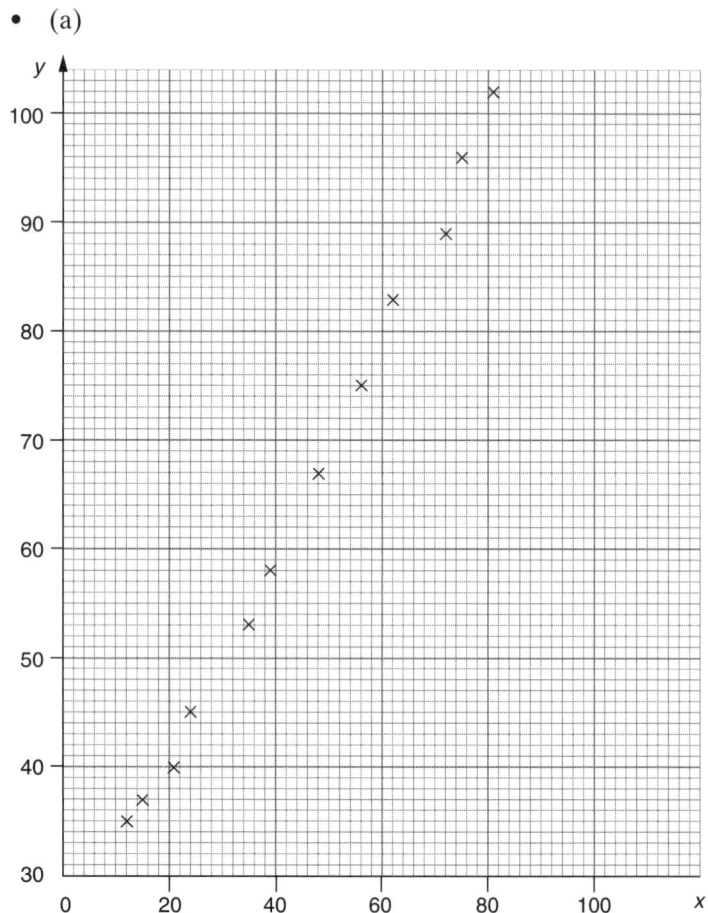

Since all the points appear to lie near a straight line there is a linear correlation and would be suitable for a regression line.

(b) Regression line of y on x: $y = ax + b$
$$\sum y = a\sum x + nb \qquad (1)$$
$$\sum xy = a\sum x^2 + b\sum x \qquad (2)$$
$$\sum x = 540$$
$$\sum y = 780$$
In equation 1, $780 = 540a + 12b$
$$\Rightarrow 65 = 45a + b \qquad (3)$$
In equation 2, $41\,444 = 30\,786a + 540b$
$$\Rightarrow 76.75 = 57.01a + b \qquad (4)$$
From equations 3 and 4,

$$11.75 = 12.01a \Rightarrow a = 0.9784 = 0.98$$
$$b = 20.99$$

Equation of the regression line is
$y = 0.98x + 20.99$.

10.3

(c) When $y = 2.2x$ then $x = 17.205$. This is the breakeven point.

If the number of items sold exceeds 17205 the company makes a profit.

4 A chemist measured the speed, y, of an enzymatic reaction at 12 different concentrations, x, of the substrate and eleven of the results are given below:

x	$\frac{1}{2}$	$\frac{1}{3}$	$\frac{1}{4}$	$\frac{1}{6}$	$\frac{1}{7}$	$\frac{1}{8}$
y	0.204	0.218	0.189	0.172	0.142	0.149

x	$\frac{1}{9}$	$\frac{1}{10}$	$\frac{1}{11}$	$\frac{1}{12}$	$\frac{1}{13}$
y	0.111	0.125	0.123	0.112	0.096

The chemist thought that the model relating y and x could be of the form

$$y = a + \frac{b}{x}$$

(a) Plot a scatter diagram of y against $\frac{1}{x}$.

(b) Find the equation of the regression line in the above form, giving the coefficients to three significant figures.

$$\left(\text{You may use } \sum \left[\frac{1}{x}\right]^2 = 793 \text{ and } \sum \left[\frac{y}{x}\right] = 11.23.\right)$$

(c) Find, to 2 significant figures, the sum of squares of residuals for your equation.
(You may use $S_{yy} = 0.016\ 84$.)

Originally the data included an observation $(\frac{1}{5}, 0.090)$, and the initial model used by the chemist was a linear regression model for

$$y = \frac{1}{x}, \text{ for all observations.}$$

(d) Plot this point on your scatter diagram and explain why you think this value has been omitted.

The sum of squares of the residuals of the equation which included the observation $(\frac{1}{5}, 0.090)$ is 0.0082.

(e) Compare this residual sum of squares with the value calculated in (c) and comment whether the difference is consistent with your answer to (d). (L Specimen Paper, 1996)

• (a)

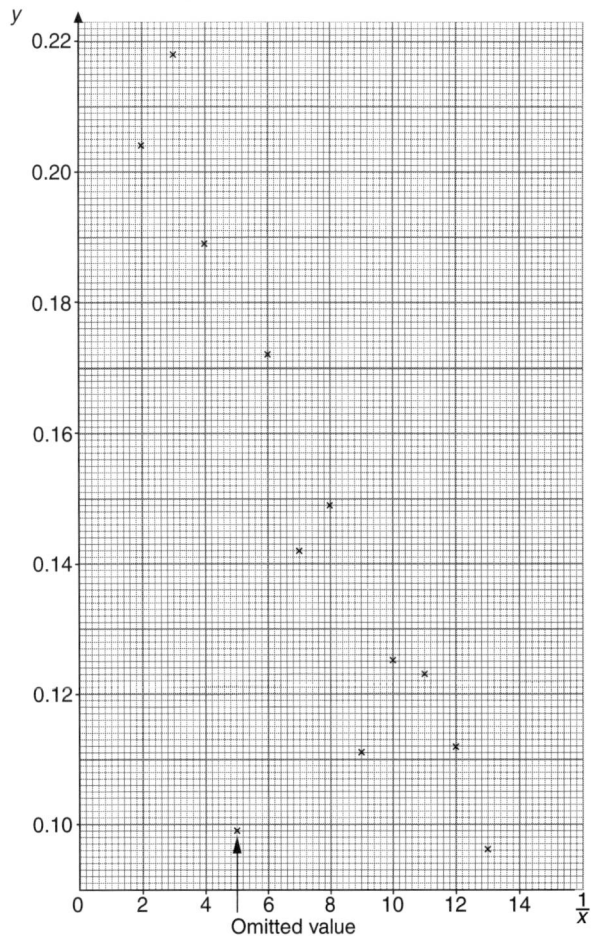

Omitted value

(b) $y = a + b\left(\dfrac{1}{x}\right)$

$$\sum \frac{1}{x} = 85, \quad \sum y = 1.641,$$

$$\sum \left(\frac{1}{x}\right)^2 = 793, \quad \sum \left(y\frac{1}{x}\right) = 11.23$$

$$b = \frac{n\sum y\dfrac{1}{x} - \sum\dfrac{1}{x}\sum y}{n\sum\left(\dfrac{1}{x}\right)^2 - \left(\sum\dfrac{1}{x}\right)^2} \qquad \boxed{\text{See your formula book}}$$

$$= \frac{11 \times 11.23 - 85 \times 1.641}{11 \times 793 - 85^2}$$

$$= \frac{-15.955}{1498}$$

$$= -0.010\ 651$$

$$a = \frac{1.641}{11} - \frac{85}{11}(b) = 0.231\ 48 \ldots$$

This uses the mean point

Regression line is $y = 0.231 - \dfrac{0.011}{x}$

$\boxed{10.3}$

(c) Sum of squares of residuals

$$= S_{yy} - b\left(\Sigma\left(y\dfrac{1}{x}\right) - \dfrac{\Sigma\dfrac{1}{x}\Sigma y}{n}\right) \boxed{\text{See your formulae book}}$$

$$= 0.016\,84 + 0.010\,65\left(11.23 - \dfrac{85 \times 1.641}{11}\right)$$

$$= 0.001\,393 \text{ or } 0.0014$$

(d) This point does not fit in with the others – which is why it has been omitted.

(e) The sum of squares of the residuals is much larger if this point is included.
The lower the sum the better the fit.
This implies that the model fits better without the point $(\tfrac{1}{5}, 0.090)$.

5 A laboratory experiment measuring the crushing stress of ten samples of plastic produced the following results:

578, 572, 579, 568, 576, 570, 572, 596, 584, 572 MPa

Examine the claim that the mean crushing stress is significantly greater than 570. Use a 5% significance level.

• H_0: $\mu = \mu_0 = 570$
H_1: $\mu > \mu_0$ (one tailed test)

Small sample, use t-test $t_0 = \dfrac{\bar{x} - \mu_0}{s/\sqrt{n}}$

$\bar{x} = 576.7 \quad s = 7.8746$
$t_0 = 2.69$

$n = 10$ so $\nu = 10 - 1 = 9$.
From the t tables 5% significance gives $t_{\text{crit}} = 1.833$.
Since $t_0 > t_{\text{crit}}$ we reject H_0 in favour of H_1.

There is evidence to support the claim that the crushing stress is in excess of 570 MPa.

6 A manufacturer claims that the mean elastic modulus of a man-made fibre is 1.2×10^3 units. To test this claim a random sample of eight fibres is taken and the elastic modulus is recorded for each fibre as follows (all in 10^3 units):

1.3 1.2 1.5 1.0 1.7 1.4 1.3 1.2

(i) Calculate the sample mean and an unbiased estimate of the population variance.

(ii) Carry out an appropriate test to determine the validity of the manufacturer's claim.
Use a 5% level of significance.
State any assumption you have made.
(NICCEA, 1993)

• (i) Sample mean $= \dfrac{\Sigma x}{8} = 1.325 = \bar{x}$

Unbiased variance $= \tfrac{1}{7}\Sigma(x_i - \bar{x})^2$
$= \tfrac{1}{7}(0.315)$
$= 0.045$

(ii) H_0: $\mu = 1.2$, H_1: $\mu \neq 1.2$ (2 tailed)
This is a small sample – use the t-test

$t = \dfrac{1.325 - 1.2}{\sqrt{0.045}/\sqrt{8}} = \dfrac{0.125}{0.075} = 1.667$
For the t-test: $8 - 1$ degrees of freedom,
5% level (2 tailed) $t_{\text{crit}} = 2.365$

$1.667 < 2.365 \Rightarrow$ accept the manufacturer's claim.

It is assumed that the distribution of the elastic modulus is Normal.

7 (a) State the effect on the correlation coefficient between two random variables x and y of
(i) changing the units of y
(ii) changing the units of x.

(b) The following table gives data relating to examination marks for six students in A Level Spanish and A Level French.

Spanish (x)	12	10	14	12	11	9
French (y)	18	17	23	20	19	15

Obtain the equation of the line of regression of y on x.

A seventh student has obtained an examination mark of 15 in Spanish. Estimate her examination mark in French. State any assumption you have made.

(c) The equation of the line of regression of y on x may be written as

$y = \alpha + \beta x + \epsilon$

where α and β are the regression coefficients and ϵ is a random error. State THREE assumptions you make concerning ϵ when carrying out a regression analysis.
(NICCEA, 1993)

• (a) (i) and (ii): changing the units of y or x does not change the correlation coefficient.

(b) $\sum x = 68, \sum y = 112, n = 6$

$y = ax + b$ where

$$\sum y = a\sum x + 6b$$
$$\sum xy = a\sum x^2 + b\sum x$$

$\sum xy = 1292, \qquad \sum x^2 = 786.$ Hence

$$112 = 68a + 6b$$
$$\Rightarrow \quad 3808 = 2312a + 204b$$
$$1292 = 786a + 68b$$
$$\Rightarrow \quad 3876 = 2358a + 204b$$
$$\Rightarrow \quad 68 = 46a$$
$$\Rightarrow \quad a = 1.48$$
$$\Rightarrow \quad b = 1.91$$

Regression line is $y = 1.48x + 1.91$ 10.3

Check that $(\bar{x}, \bar{y})$ lies on this line (approximately because of rounding errors)

$$\bar{x} = 11.\dot{3} \quad \bar{y} = 18.\dot{6}$$
$$1.48(11.\dot{3}) + 1.91 = 18.68$$

If $x = 15$ then $y = 24$, i.e. an estimate of the French mark is 24.

Since the mark in Spanish is higher than any of the others the regression line has been extrapolated. This is not generally valid – treat with caution.

(c) $y = \alpha + \beta x + \epsilon$

Assumptions
1 $\sum \epsilon_i = 0$, i.e. $\bar{\epsilon} = 0$
2 $\text{Var}(\epsilon) = 0$
3 Random errors are independent of each other.

8 (a) The masses (x kg) and heights (y cm) of a sample of 10 ten-year-olds are recorded and the following totals calculated:

$\sum x = 315.4$ $\qquad \sum x^2 = 10\,015.52$
$\sum y = 1365$ $\qquad \sum y^2 = 186\,714$
$\qquad\qquad\qquad \sum xy = 43\,165.0$

For a sample of size 10, values of the product moment correlation coefficient greater than 0.549 are significant (one-tailed test at 5% level).

Does this sample lead us to deduce that there is a significant correlation between the height and mass of ten-year-old children?

(b) In a Physics experiment a spring is loaded with various weights and its total length is measured. The weights, W Newtons, and corresponding lengths, d cm, are shown below:

W	10	20	30	40	50	60	70
d	47.9	54.6	61.4	68.3	76.9	83.9	91.1

Assume a linear relationship between W and d.
(i) Obtain the equation of the regression line of d on W.
(ii) Estimate the length of the spring if the weight is 55 Newtons.
(iii) Why could it be considered unnecessary to calculate the regression line of W on d if you were asked to estimate the weight which would cause a length of 70 cm?

(NICCEA, 1994)

• (a) Product moment correlation coefficient is

$$r = \frac{n\sum xy - \sum x\sum y}{\sqrt{n\sum x^2 - (\sum x)^2}\sqrt{n\sum y^2 - (\sum y)^2}} = 0.693$$

This is significant $\Rightarrow$ there is a correlation between height and mass of ten-year-old children.

(b) Regression line of d on W is given by

$$d = aW + b$$

where

$$\sum d = a\sum W + nb \qquad\qquad (1)$$
$$\sum Wd = a\sum W^2 + b\sum W \qquad (2)$$
$$\sum W = 280 \quad \sum W^2 = 14\,000 \quad \sum d = 484.1$$
$$\sum Wd = 21\,401$$

In equation 1: $484.1 = 280a + 7b$
In equation 2: $21\,401 = 14\,000a + 280b$
$\Rightarrow \quad a = 0.7275, \quad b = 40.06$
(i) Equation of the regression line is

$d = 0.7275W + 40.06$ 10.3

(ii) If the weight is 55 Newtons,
$d = 80.07$ cm (80.1 to 3 s.f.)
(iii) $\bar{d} = 69.2, \bar{W} = 40$
The regression lines both pass through $(\bar{W}, \bar{d})$, i.e. (40, 69.2).

Since $d = 70$ is close to this point the d on W line would be sufficiently accurate.

W *appears* to be the controlled variable so d on W would be the line to use.

9 The weights x kg of twenty boys, all of the same age, are summarized as follows:

$\sum x = 684 \quad \sum x^2 = 23\,600$

Find the mean and standard deviation of these measurements.

Assuming that these weights are a random sample from a large normally distributed population, find symmetric 98% confidence limits for the mean of the population.

It is later discovered that the weighing machine used gave erroneous readings, each 0.3 kg larger than it should have been. Give corrected values for the mean and standard deviation found earlier.

(O&C, 1991)

• $\sum x = 684$ $\sum x^2 = 23\,600$ $n = 20$ $\bar{x} = 34.2$ kg

$$\text{Var}(X) = \frac{\sum x^2}{n} - \bar{x}^2 = 10.36 \quad \Rightarrow \quad s = 3.219 \text{ kg}$$

$n = 20$ can be regarded as large or small.
Large – use normal distribution tables
Small – use 't' distribution

(i) *Assume large*
98% CI for μ is

$$\bar{x} \pm \frac{2.327\sqrt{\dfrac{ns^2}{n-1}}}{\sqrt{n}} = \bar{x} \pm \frac{2.327s}{\sqrt{n-1}}$$

i.e. 98% CI is (32.48, 35.92).

(ii) *Assume small*
't' tables with $\nu = 19$: 1% tail

$$\text{CI} = 34.2 \pm \frac{2.539s}{\sqrt{n-1}}$$

$$= 34.2 \pm 1.875$$

i.e. 98% CI is (32.32, 36.08).

The new mean will be 33.9 kg; the standard deviation remains at 3.219 kg.

10 In a ski-jumping contest each competitor made 2 jumps. The orders of merit for the 10 competitors who completed both jumps are shown below.

Ski-jumper	First Jump	Second Jump
A	2	4
B	9	10
C	7	5
D	4	1
E	10	8
F	8	9
G	6	2
H	5	7
I	1	3
J	3	6

(a) Calculate Spearman's rank correlation coefficient for the performances of the ski-jumpers in the two jumps.
(b) Using a 5% level of significance, interpret your result.   (L Specimen Paper)

First jump	Second Jump	Difference d
2	4	2
9	10	1
7	5	2
4	1	3
10	8	2
8	9	1
6	2	4
5	7	2
1	3	2
3	6	3

Spearman's rank correlation coefficient is

$$\rho = 1 - \frac{6\sum d^2}{n(n^2-1)} = 1 - \frac{6 \times 56}{10 \times 99}$$

$$= 0.6606$$

At a 5% level of significance:
Null hypothesis H_0: $\rho = 0$
Alternative hypothesis H_1: $\rho > 0$

From Spearman's tables for a one tailed test the critical value is 0.564.
Reject H_0: There is evidence of correlation between the performances.

Exercises

1 The grades on a statistics examination for a group of students were as follows:

Grade	A	B	C	D	E
Number of students	14	18	32	20	16

Test the hypothesis that the distribution of grades is uniform. Use a 5% level of significance.
(L Specimen Paper, 1996)

2 In an investigation into the effect of personality traits on the incidence of cancer, it was found that a particular trait was possessed by eight patients out of a total of sixteen patients. When a sample of twenty-four patients, who did not suffer from cancer, were examined it was found that only four persons possessed that trait.

Perform an appropriate test to determine whether or not the possession of the trait is independent of having cancer.

Use a 5% level of significance.   (NICCEA, 1993)

3 A calibrated instrument is used over a wide range of values. To assess the operator's ability to read the instrument accurately, the final digit in each of

700 readings was noted. The results are tabulated below.

Final digit	0	1	2	3	4
Frequency	75	63	50	58	73

Final digit	5	6	7	8	9
Frequency	95	96	63	46	81

Use an approximate χ^2 statistic to test whether there is any evidence of bias in the operator's reading of the instrument. Use a 5% significance level and state your null and alternative hypotheses. (L, 1993)

4 (i) The random variable X has a normal distribution with mean 16 and variance 0.64. Find x, such that $P(X < x) = 0.025$.
   (ii) The random variable Y has a χ^2 distribution with 8 degrees of freedom. Find y, such that $P(Y > y) = 0.05$. (L, 1992)

5 Levels of hydrogen sulphide produced in anaerobic fermentation of sewage after 36 hours at 40°C were recorded:

49.1 48.3 50.7 46.2 49.0
50.2 48.7 49.5 51.3 53.8 ppm

Examine the hypothesis that the population mean is significantly higher than 49 ppm.

6 Two purification processes involving the passing of a chemical solution through filters are compared. The following results compare the percentage of impurities remaining. Is there evidence at the 5% level in the mean percentages between the two processes?

Process A	16.85	16.40	17.21	16.35
Process B	17.01	16.98	18.15	18.54

Process A	18.03	17.64	16.96	17.15
Process B	17.25	19.02	18.44	17.65

7 The axial stress in samples of two different struts was measured.

Type 1	72	47	47	52	38
Type 2	64	43	32	24	43

Type 1	61	56	42	34	55
Type 2	53	58	36	44	41

Examine at the 5% level whether the difference in the mean axial stress is significant.

8 The table below gives the annual cigarette consumption per adult in the USA for the years 1977–1983.

Year (t)	Consumption per adult (y)
1977	203
1978	198
1979	193
1980	193
1981	192
1982	187
1983	175

(a) Plot a scatter diagram and comment.
(b) Code the values of t using the equation $x = t - 1980$.
(c) Obtain the equation of the least squares regression line of y on x and plot it on your scatter diagram.
    [You may assume that $\sum x^2 = 28$ and $\sum y = 1341$.]
(d) Use the regression equation to forecast the annual cigarette consumption per adult for both 1985 and 1990.
    Comment on these forecasts. (SEB, 1994)

9 (i) Find the centre of mass of the following four particles: mass 3 kg at the point (1, 1), 3 kg at (2, 3), 1 kg at (3, 5) and 1 kg at (4, 3).
    Find also the centre of mass of four other particles: mass 3 kg at (6, 3), 2 kg at (5, 5), 2 kg at (6, 6) and 1 kg at (8, 5).
   (ii) Sixteen pairs of observations are made of two variables, x and y. The results are summarized below:

Value of x	Value of y	Number of observations
1	1	3
2	3	3
3	5	1
4	3	1
6	3	3
5	5	2
6	6	2
8	5	1

Plot these results on a scatter diagram. Plot on this diagram the positions of the two centres of mass obtained in part (i), and join them with a straight line. Use this line to estimate the value of y when $x = 7$. (O&C, 1993)

10 The following data relate to the percentage unemployment and percentage change in wages over several years.

% Unemployment (x)	% Change in wages (y)
1.6	5.0
2.2	3.2
2.3	2.7
1.7	2.1
1.6	4.1
2.1	2.7
2.6	2.9
1.7	4.6
1.5	3.5
1.6	4.4

(a) Calculate the product moment correlation coefficient between x and y.
(Use $\sum x = 18.9$; $\sum y = 35.2$; $\sum x^2 = 37.01$; $\sum y^2 = 132.22$; $\sum xy = 64.7$.)
It has been suggested that low unemployment and a low rate of wage inflation cannot exist together.
(b) Without further calculation use your correlation coefficient to explain briefly whether or not you think the suggestion is justified. (L, 1993)

11 Draw a diagram showing a few non-collinear points and their regression line of y on x. Mark on your diagram the distances, the sum of whose squares is minimized by the regression line.

Five shells are fired from a gun standing on level ground. After each shell is fired, the angle of elevation of the gun is increased by 1 degree. The rth shell hits the ground at a distance of y_r km from the gun. Given that

$$\sum y_r = 9.6 \quad \sum ry_r = 29$$

find the regression line of y on r in the form $y = \alpha + \beta r$.
A shell is fired at an elevation 2.4 degrees above the initial elevation. Estimate the distance from the gun at which this shell hits the ground. (O&C, 1991)

12 The length of service and the gross annual earnings for 1990 of eleven employees of a large oil company were as follows:

Employee	Length of service (months)	Gross earnings (hundred £)
A	14	121
B	18	117
C	36	124
D	24	118
E	12	104
F	83	60
G	108	74
H	41	52
I	32	54
J	17	47
K	79	69

(a) Calculate Spearman's correlation coefficient for the data, giving your answer to 3 significant figures.
(b) The company statistician found the result of (a) surprising. Explain why she was surprised.
(L, 1991)

Brief Solutions

1 H_0: 'grades are uniform'
$n = 20$ for all grades

$$\chi^2_{calc} = \frac{(6^2 + 2^2 + 12^2 + 4^2)}{20} = 10$$

$\nu = 4$ $\chi^2(5\%) = 9.48$;
$\chi^2_{calc} > \chi^2(5\%) \Rightarrow$ significant
Suggests the grade distribution is not uniform.

2

	Yes	No	
Cancer patients	8	8	**16**
Control group	4	20	**24**
	12	**28**	**40**

$$E(1, 1) = \frac{16 \times 12}{40} \text{ etc}$$

Expected frequencies

4.8	11.2
7.2	16.8

For χ^2 degrees of freedom $= 1 \Rightarrow$ use Yates' correction

$$\chi^2_{calc} = \sum \frac{(|O_i - E_i| - \frac{1}{2})^2}{E_i} = 3.62$$

At 5% level $\nu = 1$ $\chi^2 = 3.84$
Since $\chi^2_{calc} < \chi^2(5\%)$ accept H_0 – the trait is independent of cancer.

3 Expected frequencies are 70, degree of freedom 9.
$\chi^2_{calc} = 38.2$ $\chi^2(5\%) = 16.92$
Reject $H_0 \Rightarrow$ evidence of bias.

4 (i) $X \sim N(16, 0.64)$, $\sigma = 0.8$
P$(X < x) = 0.025 = Q(-1.96)$
Hence $x = 16 - 1.96\sigma = 14.432$
(ii) $Y \sim \chi^2$, 8 degrees of freedom
P$(Y > y) = 0.05$ $\chi^2_{5\%}(8) = 15.51$
$y = 15.51$.

5 $H_0: \mu = \mu_0 = 49$ $H_1: \mu > \mu_0$ (one tailed test)

At 5% level with $n = 10$, $\nu = 9$, $t_{crit} = 1.833$,
$\bar{x} = 49.68$, $s = 1.916$, $\hat{s} = 2.0198$

$$t_0 = \frac{x - \mu_0}{\hat{s}/\sqrt{n}} = 1.0646$$

Since $1.065 < 1.833$, accept H_0.

6 $H_0: \mu_1 - \mu_2 = 0$ $H_1: \mu_1 - \mu_2 \neq 0$ (two tailed test)

Test statistic is

$$t_0 = \frac{\bar{X}_1 - \bar{X}_2}{\hat{s}_p \sqrt{\frac{1}{n_1} + \frac{1}{n_2}}} \quad \text{where } \hat{s}_p^2 = \frac{n_1 s_1^2 + n_2 s_2^2}{n_1 + n_2 - 2}$$

Process A: $n = 8$ $\bar{x}_1 = 17.07$ $s_1^2 = 0.2875$
Process B: $n = 8$ $\bar{x}_2 = 17.88$ $s_2^2 = 0.5173$

$t_0 = -2.39$, $t_{crit} = -2.14$ $(v = 16 - 2 = 14)$
Since $|t| > |t_{crit}|$ reject H_0.

7 $H_0: \mu_1 = \mu_2$ $H_1: \mu_1 \neq \mu_2$ (two tailed test)
$v = 20 - 2 = 18$ $t_{crit} = \pm 2.101$ at 5% level
test statistic $t_0 = 1.226$
$t_0 < t_{crit}$ $\Rightarrow$ accept H_0.

8 (a)

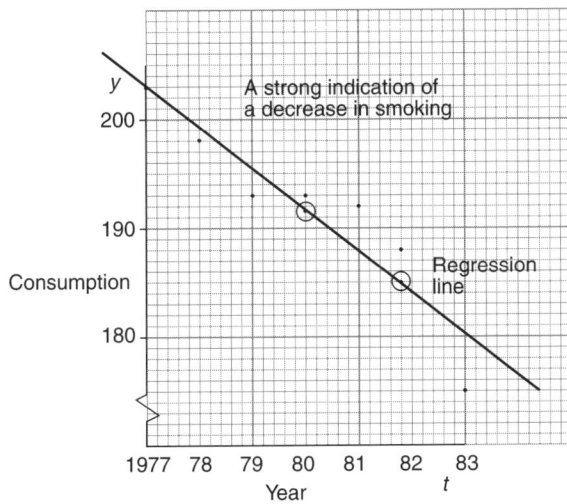

(b) $\quad$ −3 −2 −1 0 1 2 3 x

(c) $\bar{x} = 0$ $\bar{y} = 191.57$ $\sum x^2 = 28$ $\sum y = 1341$
$\sum xy = -107$
Regression line y on x: $a = -3.82$, $b = 191.57$

$y = -3.82x + 191.6$

(d) When $x = 5$, $y = 172.5$ $\boxed{10.3}$
$\quad\quad\quad x = 10$, $y = 153.4$
Since these are *extrapolations* from the regression
line they are not good predictors.

9 (i) Centre of mass $\left(\dfrac{\sum m_i x_i}{\sum m_i}, \dfrac{\sum m_i y_i}{\sum m_i} \right)$

Centre of mass 1: (2, 2.5)
Centre of mass 2: (6, 4.5)

(ii)

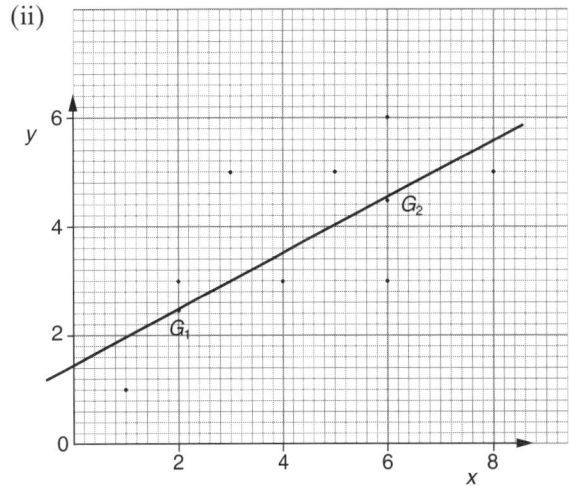

When $x = 7, y = 5$.

10 (a) Product moment correlation coefficient is

$$r = \frac{n\sum x_i y_i - \sum x_i \sum y_i}{\sqrt{(n\sum x_i^2 - (\sum x_i)^2)(n\sum y_i^2 - (\sum y_i)^2)}}$$

There are variations of this – dividing through by n or n^2

$n = 10$ $\Rightarrow$ $r = -0.558$ $\boxed{10.3}$

(b) The statement is supported by a medium
negative correlation coefficient. As
unemployment increases, wages decrease.

11 $\bar{y} = 1.92$ $\bar{r} = 3$ gradient $\dfrac{s_{xy}}{s_{xx}} = 0.02$

Equation of regression line is $y = 0.02r + 1.86$.
When $r = 3.4$, $y = 1.928$ km.

12 (a)

Empl.	Mths.	Pay (£100)	Rankings Mths.	Rankings Pay	d^2
A	14	121	2	10	64
B	18	117	4	8	16
C	36	124	7	11	16
D	24	118	5	9	16
E	12	104	1	7	36
F	83	60	10	4	36
G	108	74	11	6	25
H	41	52	8	2	36
I	32	54	6	3	9
J	17	47	3	1	4
K	79	69	9	5	16

Spearman's rank correlation coefficient is

$$1 - \frac{6\sum d^2}{n(n^2 - 1)} = -0.2455$$

(b) The negative correlation indicates that length
of service does not mean higher salaries.

20

Statics

AS Level			A Level						
CORE	MODULAR	TRADITIONAL	CORE	MODULAR	TRADITIONAL	**Topic**	Date attempted	Date completed	Self-assessment
	✓	✓		✓	✓	**Systems of forces**			
	✓	✓		✓	✓	**Equilibrium**			
	✓	✓		✓	✓	**Friction**			
				✓	✓	**Composite centres of mass (see also Chapter 26)**			

Fact Sheet

Definitions

(i) The component of a force $\mathbf{F}$, having magnitude F, in a direction making θ with $\mathbf{F}$ is $F\cos\theta$. Forces are usually resolved into two perpendicular components F_x and F_y.

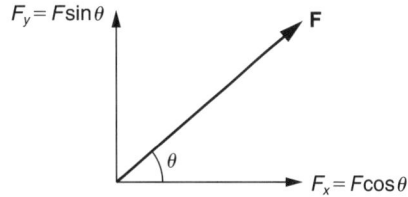

(ii) The moment M of a force $\mathbf{F}$ about a point P is Fd where d is the perpendicular distance of P from the line of action of $\mathbf{F}$.

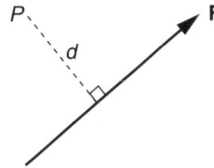

Equilibrium of Three Forces

If three forces are in equilibrium then they must be parallel, or pass through one point. If they pass through one point then they can be represented by the sides of a triangle taken in order.

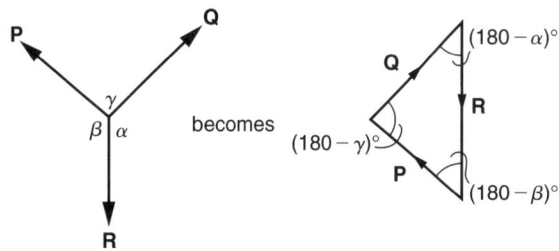

Lami's theorem, $\dfrac{P}{\sin\alpha} = \dfrac{Q}{\sin\beta} = \dfrac{R}{\sin\gamma}$, is equivalent to the sine rule.

Laws of Friction

(i) Friction F opposes the movement of an object across a rough surface.
(ii) Up to a limiting value the magnitude of the frictional forces is just sufficient to prevent motion.
(iii) If μ is the coefficient of friction, and R is the normal reaction, the limiting value of the frictional force is μR; so at all times $F \leqslant \mu R$.

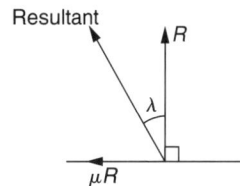

The resultant of R and μR, the limiting value of the frictional force, makes an angle λ with the normal where $\mu = \tan \lambda$. λ is called the angle of friction.

Equilibrium

A system is in equilibrium when the sum of all the forces $= 0$ *and* the sum of moments about any point $= 0$.

When two or more bodies in contact are in equilibrium then
(i) the complete system is in equilibrium, and
(ii) each body is in equilibrium.

Hints

(i)   Good clear diagrams showing the directions of all the forces are essential.
(ii) Questions which state 'on the point of moving', 'limiting equilibrium', 'will just prevent motion' and 'will just move' all require the use of $F = \mu R$.
(iii) A body will slide if the force F required to overcome the other forces is such that $F \geqslant \mu R$.
(iv) A body will topple, rotate, turn or tilt about a point A if the turning moment about A (in the direction of toppling) $\geqslant 0$ before $F = \mu R$.
(v)   Look for the equations which involve the least number of variables first.

Centres of Mass and Gravity

Centre of mass and centre of gravity are different names for the same point.
(i)   If the masses m_i and distances from the y-axis of the centres of mass $\bar{x}_i$ of two or more bodies are known then the distance of the centre of mass $\bar{x}$ of the compound body from the y-axis is given by $\bar{x}\sum m_i = \sum(m_i\bar{x}_i)$, that is:

$$\frac{\text{the moment of}}{\text{the whole}} = \frac{\text{the sum of the moments}}{\text{of the parts}}$$

(ii) If parts of a body are removed then

$$\frac{\text{the moment of}}{\text{the remainder}} = \frac{\text{the moment}}{\text{of the whole}} - \frac{\text{the sum of the moments}}{\text{of the removed parts}}$$

(iii) It is not necessary to use the actual masses of the parts. It is often simpler to use the ratios of the masses, especially with similar figures; for example a cylinder of height h and radius r has mass $\pi r^2 h \rho_1$

a cone of height h and base radius r has mass $\dfrac{\pi}{3} r^2 h \rho_2$

These masses could be represented as $3k\rho_1$ and $k\rho_2$.
(iv) If more than two parts are involved in a compound body it is convenient to tabulate the information of mass, centre of mass and moments:

Body	Mass	Centre of mass	Moments
	m_i	$(\bar{x}_i, \bar{y}_i)$	$(m_i\bar{x}_i, m_i\bar{y}_i)$

(v)   When a composite body is suspended from a point and an angle with the vertical is required, it is usually easier to put a line from the point of suspension to the centre of mass on the original diagram instead of trying to draw a new one, which wastes time and obscures the coordinate system in use.

Worked Examples

1. A packing case of mass 40 kg is to be moved through a vertical distance of 2 m by means of a ramp of length 6 m. The coefficient of friction between the packing case and the ramp is such that, if the packing case was placed on the ramp, it would just begin to slide down the ramp. Draw a diagram showing the forces acting on the packing case and calculate the coefficient of friction.

   Calculate also the least force which would just move the packing case *up* the ramp,
   (a) when the force is applied parallel to the ramp
   (b) when the force is applied in the horizontal
       direction. $(g = 9.8 \text{ m s}^{-2})$
                                                    (SEB, 1994)

- Since the length of the ramp is 6 m:

$$\sin \alpha = \frac{2}{6} = \frac{1}{3}, \quad \cos \alpha = \frac{\sqrt{8}}{3}$$

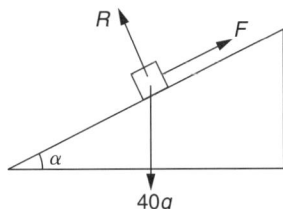

If the case is about to slide down the ramp the frictional force up the ramp is μR and vice versa
Using a triangle of forces

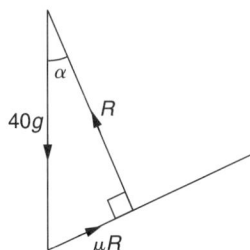

$$\tan \alpha = \frac{\mu R}{R} \quad \Rightarrow \quad \mu = \frac{1}{\sqrt{8}}$$

(a) Let the least force parallel to the ramp $= Q$ N.
    Resolving forces parallel and perpendicular to the slope

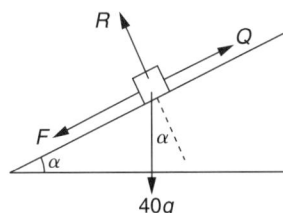

$$R = 40g \cos \alpha = \frac{40\sqrt{8}g}{3}$$

$$F = \mu R = \frac{40g}{3} \quad \text{limiting friction}$$

$$Q = F + 40g \sin \alpha$$

$$= \frac{80g}{3} = 261 \text{ N}$$

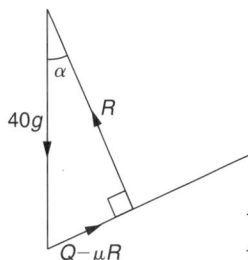

This could have been found by a triangle of forces as above

(b) Let the least horizontal force $= Q_1$

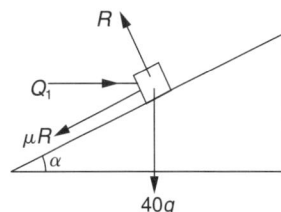

Resolving the forces vertically:

$$R \cos \alpha - \mu R \sin \alpha = 40g$$

$$\Rightarrow \quad R\left(\frac{\sqrt{8}}{3} - \frac{1}{\sqrt{8}} \cdot \frac{1}{3}\right) = 40g$$

$$\Rightarrow \quad R = \frac{40g}{\left(\frac{\sqrt{8}}{3} - \frac{1}{\sqrt{8} \cdot 3}\right)} = 475 \text{ N}$$

Resolving horizontally:

$$Q_1 = R \sin \alpha + \mu R \cos \alpha$$
$$= R(\tfrac{1}{3} + \tfrac{1}{3})$$
$$= 317 \text{ N}$$

Alternative method using the angle of friction and Lami's Theorem

2.

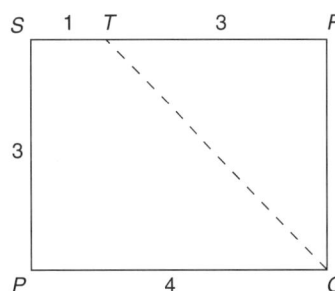

The diagram shows a rectangular sheet $PQRS$ of uniform thin metal with $PQ = 4$ m and $QR = 3$ m. T is a point on RS such that $RT = 3$ m. The sheet is folded about the line QT until R lies on PQ.
(a) Find the distances from PQ and PS of the centre of mass of the folded sheet.
The folded sheet is freely suspended from the point S and hangs in equilibrium.
(b) Calculate the angle of inclination of the edge PS to the vertical.   (AEB, 1994)

- For a triangle, the centre of mass is on a median, $\frac{1}{3}$ distance from the base.

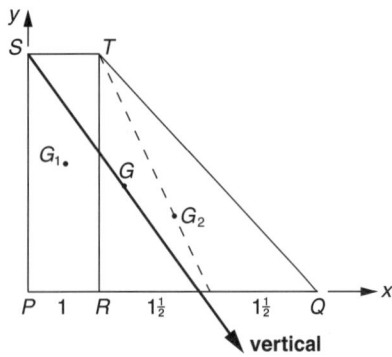

Centre of mass of $PSTR$ (G_1) is at $(\frac{1}{2}, 1\frac{1}{2})$.
Centre of mass of TRQ (G_2) is at $(2, 1)$.

Mass of $PSTR = 3M$
Mass of $TRQ = 9M$
Total mass $= 12M$
where M = mass per unit area.

- (a) Taking moments about PS:

  $$3M \times \tfrac{1}{2} + 9M \times 2 = 12M\bar{x}, \quad \bar{x} = \tfrac{39}{24}$$

  Taking moments about PQ:

  $$3M \times 1\tfrac{1}{2} + 9M \times 1 = 12M\bar{y}, \quad \bar{y} = \tfrac{27}{24}$$

  Distance of G from PQ is $1\frac{1}{8}$ m.
  Distance of G from PS is $1\frac{5}{8}$ m.

- (b) When suspended from S, SG will be vertical.
  Angle of inclination of PS to the vertical is
  $\angle PSG$.

  $$\tan PSG = \frac{1\frac{5}{8}}{1\frac{7}{8}} = \tfrac{13}{15} \quad \Rightarrow \quad \text{angle } PSG = 40.9°$$

  *Do not try to draw another diagram showing SG
  vertical. It is much easier to use the coordinate
  system by drawing the 'vertical' line SG on your
  original diagram*

3 A rigid rectangular frame $ABCD$ is fixed in a
horizontal position. The sides AB, CD are of length
12 m and the sides BC, DA are of length 6 m. A
body P of mass 16 kg hangs a vertical distance of
2 m below the frame, supported by four equal light
inextensible strings, PA, PB, PC, PD.

- (a) (i) Unit vectors $\mathbf{i}$, $\mathbf{j}$, and $\mathbf{k}$ are defined with $\mathbf{i}$, $\mathbf{j}$
    parallel to AB, AD respectively and $\mathbf{k}$
    vertically upwards.
    Show that the force exerted on P by the
    string PA is

    $$\tfrac{1}{7}T(-6\mathbf{i} - 3\mathbf{j} + 2\mathbf{k})$$

    where T is the magnitude of the tension in
    each string.
    Write down similar expressions for the
    forces exerted on P by the strings PB, PC,
    and PD.
  - (ii) Hence calculate the value of T.
- (b) Each of the strings has a breaking tension of
  250 N.
  Determine whether it would be possible for the

body to be supported by the strings PA and PC
alone. $(g = 9.8 \text{ m s}^{-2})$
(OLE)

(a) (i)

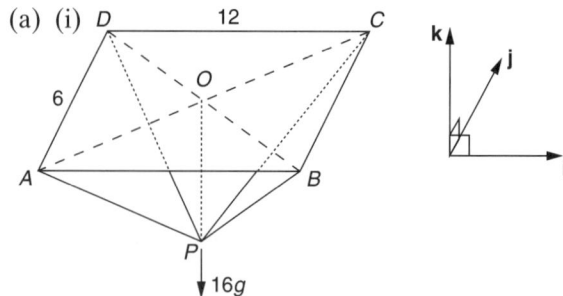

If the mid-point of AC is O with coordinates
$(0, 0, 0)$ then
$\overline{OA} = -6\mathbf{i} - 3\mathbf{j} \quad \overline{OB} = 6\mathbf{i} - 3\mathbf{j} \quad \overline{OC} = 6\mathbf{i} + 3\mathbf{j}$
$\overline{OD} = -6\mathbf{i} + 3\mathbf{j} \quad \overline{OP} = -2\mathbf{k}$

By symmetry the tensions in each string are
equal (T).

Direction of $\overline{PA} = \mathbf{a} - \mathbf{p} = -6\mathbf{i} - 3\mathbf{j} + 2\mathbf{k}$

with magnitude $\sqrt{36 + 9 + 4} = 7$.

Tension in $\overline{PA}$ is $\dfrac{T}{7}(-6\mathbf{i} - 3\mathbf{j} + 2\mathbf{k})$.

Similarly:

in $\overline{PB}$ tension is $\dfrac{T}{7}(6\mathbf{i} - 3\mathbf{j} + 2\mathbf{k})$

in $\overline{PC}$ tension is $\dfrac{T}{7}(6\mathbf{i} + 3\mathbf{j} + 2\mathbf{k})$

in $\overline{PD}$ tension is $\dfrac{T}{7}(-6\mathbf{i} + 3\mathbf{j} + 2\mathbf{k})$

(ii) For equilibrium, Σforces at $P = 0$. That is

$$-16g\mathbf{k} + \dfrac{T}{7}(8\mathbf{k}) = 0 \quad \Rightarrow \quad T = 14g \text{ or } 137.2 \text{ N}$$

(b) If the strings PB and PD were removed the
new tension T_1 would be $28g$ which is greater
than 250 N. Therefore the strings would break.

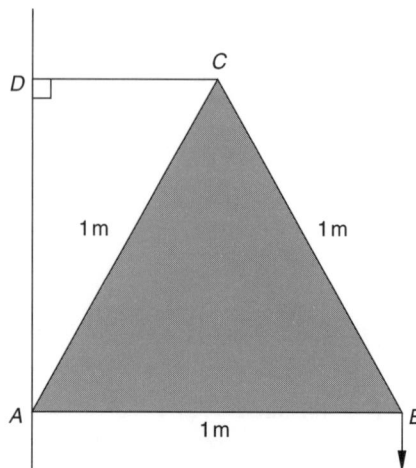

Figure 1

The diagram at the bottom of the previous page shows a light plate ABC in the shape of an equilateral triangle of side 1 m which has a vertical load W attached at the point B. The plate is freely hinged to a vertical wall at A and held with AB horizontal by a light, horizontal string CD. A, B, C and D are in a vertical plane.

(i) Copy Figure 1 and mark in the forces acting on the plate at the points A, B and C.

Give your answers to the following questions in terms of W:

(ii) By taking moments about the point A, or otherwise, find the tension in the string CD.

(iii) Find the horizontal and vertical components of the forces acting on the plate from the hinge at A.

The plate is replaced by three equal light rods AB, BC and CA, freely hinged at A, B and C. The load and string are attached as before. Figure 2 shows the forces in the rods acting at the hinges at A, B and C.

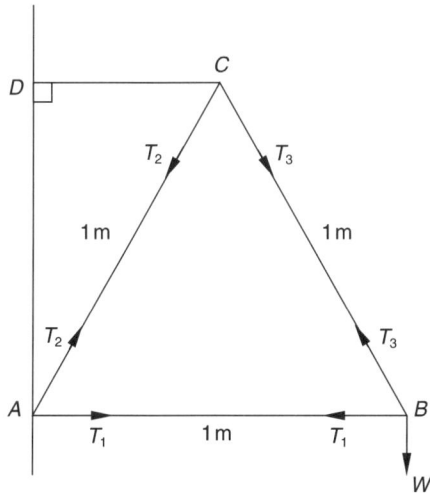

Figure 2

(iv) Copy Figure 2 and mark on it the other forces acting at A and C. Hence find the magnitude of the forces in the rods AB and AC and state whether each of these rods is in tension or compression. (MEI, 1993)

• (i)

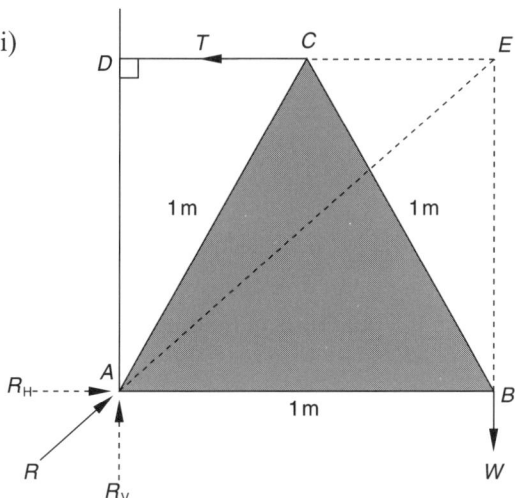

R_H and R_V are the horizontal and vertical components of R.

T is the horizontal tension in the string.

R is the reaction from the hinge on the triangle. By the principle of concurrency its direction will be along AE where E is the point of intersection of the lines of action of W and T.

(ii) Since it is in equilibrium, $\sum M = 0$ about A:

$$T \times AD - W \times AB = 0$$

$$\Rightarrow \quad T = \frac{W}{\cos 30°} = \frac{2W}{\sqrt{3}} = \frac{2\sqrt{3}W}{3}$$

(iii) Since it is in equilibrium, $\sum F = 0$ in any direction so

$$R_H = T = \frac{2\sqrt{3}W}{3}$$

$$R_V = W$$

(iv)

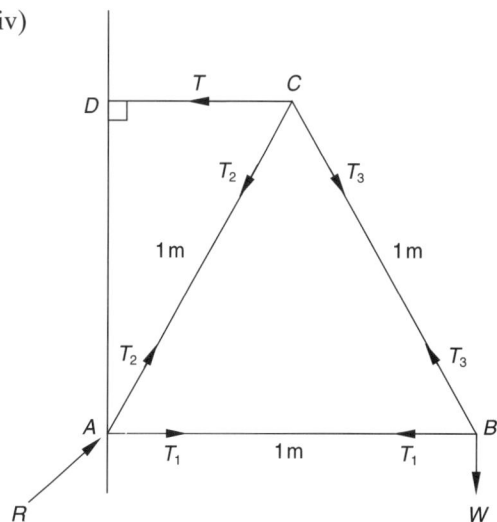

• The forces T and R are the same as in part (i).

Consider the equilibrium of joint A, and use the components of R already found.

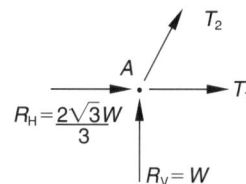

Resolving vertically:

$$-T_2 \cos 30° = W$$

$$T_2 = -\frac{2\sqrt{3}W}{3} = \text{compressive force in } AC$$

Resolving horizontally:

$$\frac{2\sqrt{3}W}{3} + T_2 \cos 60° + T_1 = 0$$

$$\Rightarrow \quad T_1 = \frac{2\sqrt{3}W}{3}(\cos 60° - 1)$$

$$= -\frac{\sqrt{3}W}{3} = \text{compressive force in } AB$$

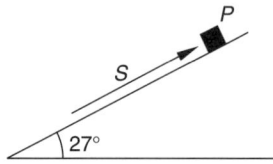

A small parcel P, of mass 1.5 kg, is placed on a rough plane inclined at an angle of 27° to the horizontal. The coefficient of friction between the parcel and the plane is 0.3. A force $\mathbf{S}$, of variable magnitude, is applied to the parcel as shown in the diagram. The line of action of $\mathbf{S}$ is parallel to a line of greatest slope of the inclined plane.

Determine, in N to 1 decimal place, the magnitude of $\mathbf{S}$ when the parcel P is in limiting equilibrium and on the point of moving
(a) down the plane
(b) up the plane.
    $(g = 9.8 \text{ m s}^{-2})$ (L, 1992)

•   (a)   Friction will be limiting $(=\mu R)$ and will act up the plane:

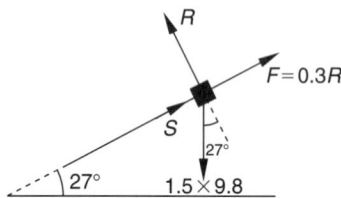

The parcel is in equilibrium so Newton's 1st Law applies (see page 263):

Parallel to slope
$$S + 0.3R - 1.5 \times 9.8 \sin 27° = 0 \qquad (1)$$
Perpendicular to slope
$$R = 1.5 \times 9.8 \cos 27° = 13.098 \qquad (2)$$

Therefore in equation 1 $\Rightarrow$ $S = 2.74$ N

(b)   Same as part (a) except that friction will act down the plane.

So equation 1 is now

$$S - 0.3R - 1.5 \times 9.8 \sin 27° = 0 \qquad (3)$$

and equation 2 is as before so $R = 13.098$.

Therefore equation 3 $\Rightarrow$ $S = 10.6$ N

6

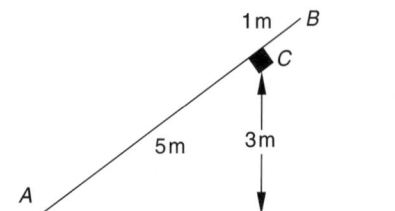

A smooth horizontal rail is fixed at a height of 3 m above a horizontal playground whose surface is rough. A straight uniform pole AB, of mass 20 kg and length 6 m, is placed to rest at a point C on the rail with the end A on the playground. The vertical plane containing the pole is at right angles to the rail. The distance AC is 5 m and the pole rests in limiting equilibrium (see above diagram).

Calculate
(a) the magnitude of the force exerted by the rail on the pole, giving your answer to the nearest N
(b) the coefficient of friction between the pole and the playground, giving your answer to 2 decimal places
(c) the magnitude of the force exerted by the playground on the pole, giving your answer to the nearest N.   (L, 1992)

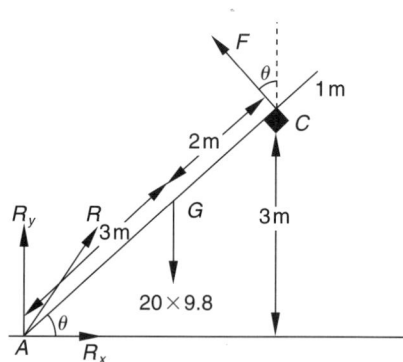

The weight of the pole acts at G, 3 m from either end.

$$\sin \theta = \tfrac{3}{5} \quad \Rightarrow \quad \cos \theta = \tfrac{4}{5}$$

Let R be the force exerted by the playground on the pole. Its horizontal component R_x is the friction force at the ground and its vertical component R_y is the normal reaction force at the ground.

(a) $\sum M = 0$ about A

$$\Rightarrow \quad 5F = 20 \times 9.8 \times 3 \cos \theta$$
$$= 20 \times 9.8 \times 3 \times \tfrac{4}{5}$$
$$\Rightarrow \quad F = 94.08$$
$$= 94 \text{ N to the nearest Newton}$$

(b) $\sum F = 0$
    Horizontally $R_x = F \sin \theta$
$$= 56.448$$
    Vertically $R_y + F \cos \theta - 20 \times 9.8 = 0$
$$\Rightarrow \quad R_y = 120.736$$

Since the pole rests in limiting equilibrium,
$R_x = \mu R_y$

$$\Rightarrow \quad \mu = 0.47 \text{ (2 d.p.)}$$

(c) $R = \sqrt{(R_x^2 + R_y^2)}$
$$= 133 \text{ N to the nearest Newton}$$

7

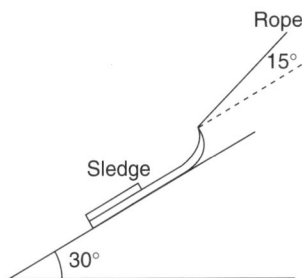

One winter day, Veronica is pulling a sledge up a hill with slope 30° to the horizontal at a steady speed. The weight of the sledge is 40 N. Veronica pulls the sledge with a rope inclined at 15° to the slope of the hill (that is at 45° to the horizontal). The tension in the rope is 24 N.

(i) Draw a force diagram showing the tension in the rope, the weight of the sledge, the normal reaction R of the ground and the frictional force F on the sledge.

(ii) By considering separately the forces parallel to and perpendicular to the slope, find the values of R and F.

(iii) Show that the coefficient of friction is slightly more than 0.1.

Veronica stops and when she pulls the rope to start again it breaks and the sledge begins to slide down the hill. The coefficient of friction is now 0.1.

(iv) Find the new value of the frictional force and the acceleration down the slope.

$$(g = 10 \text{ m s}^{-2})$$
(MEI, 1993)

• (i)

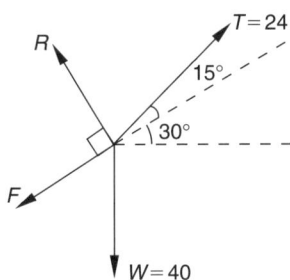

(ii) Since it is moving at constant velocity, Newton's 1st Law applies.

Parallel to slope $\sum F = 0$

$\Rightarrow \quad 24 \cos 15° - 40 \sin 30° - F = 0$
$\Rightarrow \quad F = 3.18 \text{ N}$

Perpendicular to slope $\sum F = 0$

$\Rightarrow \quad 24 \sin 15° + R - 40 \cos 30° = 0$
$\Rightarrow \quad R = 28.4 \text{ N}$

(iii) Since the sledge is sliding, friction will be at its maximum value, $F = \mu R$

$$\therefore \mu = \frac{F}{R} = 0.112$$

which is 'slightly more' than 0.1.

(iv)

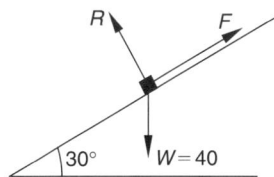

Perpendicular to the slope $\sum F = 0$:

$\Rightarrow \quad R = W \cos 30 \quad \Rightarrow \quad R = 34.6 \text{ N}$

Since the sledge is sliding:

$F = \mu R = 3.46 \text{ N}$

Parallel to the slope $\sum F = $ mass $\times$ acceleration

$$\Rightarrow \quad W \sin 30 - F = \frac{W}{g} \times \text{acceleration}$$

$20 - 3.46 = 4 \times \text{acceleration}$

$\Rightarrow \quad$ acceleration down the slope is 4.13 m s^{-2}

8 A rectangle $ABCD$ has sides $AB = 3a$, $BC = 4a$. Forces of magnitude $7W$, $6W$, $10W$, $13W$ and $15W$ act along the lines BA, BC, DC, DA and AC respectively in the directions indicated by the order of the letters. Find the resultant of this system in magnitude and direction and the distance from A at which its line of action cuts AD.

An extra force P is now added at D so that the system of forces reduces to a couple. Find the value of P and the magnitude of the resulting couple.

• From given dimensions,

$$\tan \alpha = \tfrac{3}{4}, \quad \sin \alpha = \tfrac{3}{5}, \quad \cos \alpha = \tfrac{4}{5}$$

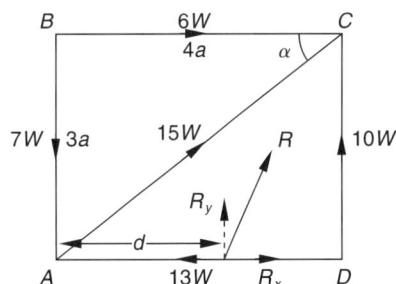

Resolving forces,

parallel to AD, $R_x = 6W + 15W \cos \alpha - 13W$
$\qquad\qquad = 5W$
parallel to AB, $R_y = 10W + 15W \sin \alpha - 7W$
$\qquad\qquad = 12W$

Hence the resultant force is given by

$$R = \sqrt{[(5W)^2 + (12W)^2]} = 13W$$

making an angle $\tan^{-1}(\tfrac{5}{12})$ with AB, i.e. 22.6° with AB, and 67.4° with AD.

Let the line of action of R cut AD at a distance d from A.
Taking moments about A,

$(10W)(4a) - (6W)(3a) = R_y d = (12W)(d)$
$(22W)(a) = (12W)(d)$, so $d = \frac{11}{6}a$

The line of action of the resultant force cuts AD at a distance $\frac{11}{6}a$ from A.

If the resultant of P and R is a couple, then P is an equal and opposite force to R, i.e. P has a magnitude of $13W$.

The magnitude of the resulting couple is its moment about D.
Magnitude $= R_y(4a - d) = 12W \left(\frac{13}{6}\right) = 26Wa$.

Hint: It is usually easier to use the components of R when taking moments, rather than calculating the perpendicular distance from A to the line of action of R.

In this case with the line of action of R cutting AD, R has components

$R_x = 5W$ along AD (with no moment about A)

and

$R_y = 12W$ perpendicular to AD (with moment $12W(d)$ about A).

9 A light rod AB, of length $2l$, is hinged to a vertical wall at A and is supported in a horizontal position, perpendicular to the wall, by an inextensible string joining B to a point C vertically above A so that $\angle ABC = 60°$.
Loads of weight W are hung from the mid-point of AB and from B.

Find the tension in the string and the magnitude and direction of the reaction at the hinge.

• Let the tension be T and the horizontal and vertical components of the reaction at A be R and S respectively.

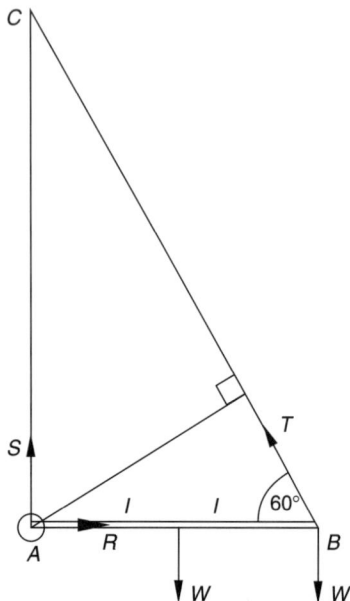

Taking moments about A,

$(T)(2l \sin 60°) = Wl + W(2l)$

So

$$T = \frac{3}{\sqrt{3}}W = \sqrt{3}W$$

Resolving horizontally,

$$R = T \cos 60° = \frac{\sqrt{3}}{2}W$$

Resolving vertically,

$$S + T \sin 60° = 2W$$

$$\Rightarrow \quad S = 2W - \sqrt{3}W\left(\frac{\sqrt{3}}{2}\right) = \frac{W}{2}$$

Thus the reaction at the hinge is $\sqrt{(R^2 + S^2)} = W$ at an angle θ to the rod where

$$\tan \theta = \frac{S}{R} = \frac{1}{\sqrt{3}} \text{ , i.e. at } 30° \text{ to the rod.}$$

10 In this question all answers should be given correct to three significant figures.
ABC is a triangle with $A = 30°$ and $C = 40°$.
(a) Find the magnitude of the forces acting at A in directions $\overline{AB}$ and $\overline{BC}$ which are equivalent to a force of 20 N acting along AC.
(b) A force of 30 N acts along AC and another force $\mathbf{P}$ of magnitude 30 N acts at A. What is the direction of $\mathbf{P}$ if the resultant acts in the direction $\overline{BC}$ and what is the magnitude of the resultant?
(c) A force of 40 N acts along AC and a force $\mathbf{Q}$ along BA. What is the magnitude of $\mathbf{Q}$ if the resultant acts in a direction perpendicular to BC? (SUJB)

• (a) If the resultant of the forces acts along AC then forces must be proportional to the sides of the triangle.

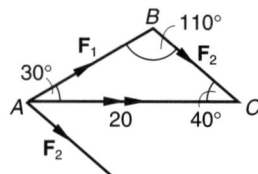

If the forces are $\mathbf{F}_1$ and $\mathbf{F}_2$ along AB and BC respectively, then

$$\frac{F_1}{\sin 40°} = \frac{F_2}{\sin 30°} = \frac{20}{\sin 110°}$$

giving $\quad F_1 = 13.7$ N (to 3 s.f.)
and $\quad F_2 = 10.6$ N (to 3 s.f.).

(b) The resultant of two equal forces bisects the angle between the forces; therefore P acts at $80°$ to AC, on the opposite side of AC from B.

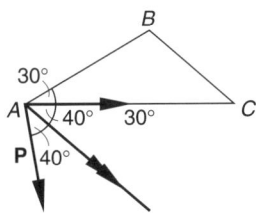

Resolving the forces along the bisector of the angle,

$$R = 2P \cos 40° = 46.0 \text{ N (to 3 s.f.)}$$

(c) If the resultant acts perpendicular to BC, resolving parallel to BC gives:

$$Q \cos 70° = 40 \cos 40°$$
$$\Rightarrow \quad Q = 89.6 \text{ N (to 3 s.f.)}$$

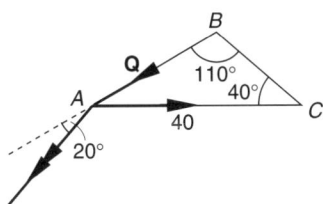

11 A uniform square lamina $ABCD$ of mass M rests in a vertical plane with AB in contact with a horizontal table. The coefficient of friction between the lamina and the table is μ.

A gradually increasing horizontal force P is applied at C in the plane of the lamina in the direction $\overrightarrow{DC}$. Prove that equilibrium is broken by sliding if $\mu < 0.5$, and by the lamina tilting about B if $\mu > 0.5$. Find the value of P which breaks equilibrium in the latter case. (OLE)

• Let the lamina have side $2a$, R denote the normal reaction acting somewhere along the base and F the frictional force.

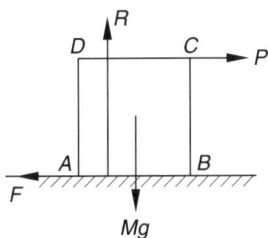

For sliding:
Resolving horizontally, $P \geqslant F = \mu R$.
Resolving vertically, $R = Mg$.
Therefore for sliding $P \geqslant \mu Mg$.

For tilting:
When about to tilt, R acts at B.
Taking moments about B, $P(2a) \geqslant Mga$.
Therefore for tilting $P \geqslant Mg/2$.

If $\mu < \frac{1}{2}$, P attains the sliding condition first.
If $\mu > \frac{1}{2}$, P attains the tilting condition first.

Value of P which breaks equilibrium by tilting is $P = Mg/2$.

12 An L-shaped structure is formed by rigidly joining two uniform rods AB and AC at A so that the angle BAC is a right angle. The rod AB is of mass $3m$ and length $4a$ and the rod AC is of mass $2m$ and length $2a$.

The structure is free to turn in a vertical plane about a smooth horizontal axis through A. A horizontal force is applied at C so that the structure remains in equilibrium with AB and AC equally inclined to the vertical with B and C lower than A. Determine

(a) the magnitude of the horizontal force
(b) the magnitude of the reaction at A
(c) the tangent of the angle that the reaction at A makes with the horizontal.

The applied force at C is now removed and the axis at A roughened, so that it exerts a frictional couple on the structure, which remains in equilibrium with AB making an acute angle θ with the upward vertical and B and C on the same side of the vertical through A.

(d) Find the value of the frictional couple necessary to maintain this equilibrium.
(e) The nature of the frictional couple is such that it will take whatever value is necessary to maintain equilibrium, up to a maximum value of G. Find the least such G to ensure that equilibrium is possible for all values of the acute angle θ. (AEB)

• Let the force applied at C be F, and the reaction at A have horizontal and vertical components X and Y.

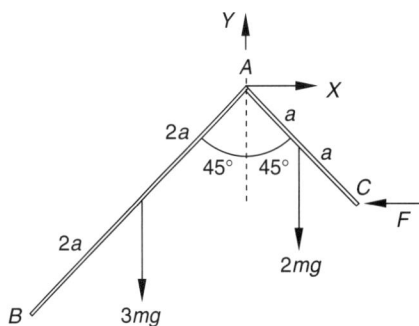

(a) Taking moments about A,

$$3mg(2a \sin 45°) = 2mg(a \sin 45°)$$
$$+ F(2a \cos 45°)$$
$$F = 2mg$$

Magnitude of the horizontal force is $2mg$.

(b) Resolving horizontally and vertically:

$$X = F = 2mg \quad \text{and} \quad Y = 5mg$$

Magnitude of the reaction at A
$$= mg\sqrt{(2^2 + 5^2)} = mg\sqrt{29}.$$

(c) Reaction at A makes an angle with the horizontal of $\tan^{-1}(\frac{5}{2}) = 68.2°$.

(d) Let the frictional couple about A be G.

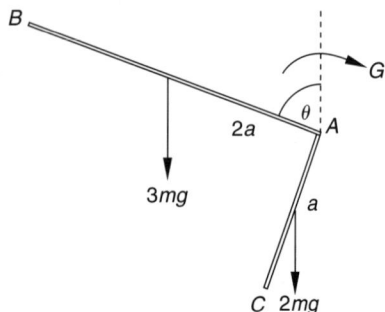

Taking moments about A,

$$3mg(2a \sin \theta) + 2mg(a \cos \theta) = G$$

Frictional couple is
$2mga(3 \sin \theta + \cos \theta)$

$$= 2mga\sqrt{10}\,[\sin(\theta + \alpha)] \text{ where } \tan \alpha = \tfrac{1}{3}.$$

(e) The least value of G to ensure that equilibrium is possible for all θ is $2mga\sqrt{10}$.

13 A uniform circular hoop of weight W hangs over a rough horizontal peg A. The hoop is pulled with a gradually increasing horizontal force P which is applied at the other end B of the diameter through A and acts in the vertical plane of the hoop.

Given that the system is in equilibrium and that the hoop has not slipped when AB is inclined at an angle θ to the downward vertical, find the value of P in terms of W and θ. Show also that the ratio of the frictional force to the normal reaction at the peg is $(\tan \theta)/(2 + \tan^2 \theta)$.

Show that, when the coefficient of friction is 0.5, the hoop never slips, however hard it is pulled. (L)

• The forces acting on the hoop are as shown in the sketch.
If the system is in equilibrium then the turning moment about A is zero.

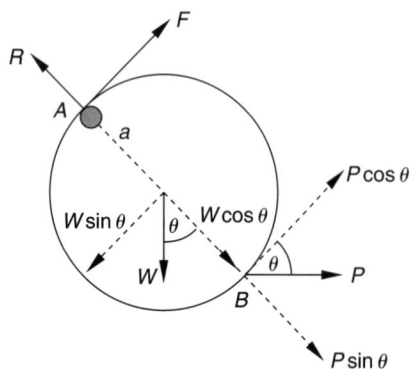

$$W(a \sin \theta) = P(2a \cos \theta) \quad \Rightarrow \quad P = \frac{W}{2} \tan \theta$$

Resolving radially,

$$R = W \cos \theta + P \sin \theta = \frac{W}{2}(2 \cos \theta + \sin \theta \tan \theta)$$

Resolving tangentially,

$$F = W \sin \theta - P \cos \theta = W \sin \theta - \frac{W}{2} \sin \theta$$

$$= \frac{W}{2} \sin \theta$$

$$\frac{F}{R} = \frac{\sin \theta}{(2 \cos \theta + \sin \theta \tan \theta)} = \frac{\tan \theta}{(2 + \tan^2 \theta)}$$

This may be written:

$$\frac{F}{R} = \frac{\sin \theta \cos \theta}{2 \cos^2 \theta + \sin^2 \theta} = \frac{\sin 2\theta}{2(1 + \cos^2 \theta)}$$

Now $\sin 2\theta \le 1$ for all θ and $1 + \cos^2 \theta \ge 1$ for all θ.

Since $\sin 2\theta = 1$ and $1 + \cos^2 \theta = 1$ occur at different values of θ, $\dfrac{F}{R} < \tfrac{1}{2}$ for all θ.

Hence if $\mu = \tfrac{1}{2}$ the hoop never slips.

14 A uniform straight wire of length $14a$ is bent into the shape shown in the diagram. Find the distances of the centre of mass from AB and AD.

If the wire is freely suspended from B and hangs in equilibrium, find the tangent of the angle of inclination of BC to the vertical.

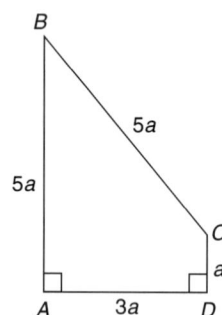

• Let the wire have a mass of $14M$

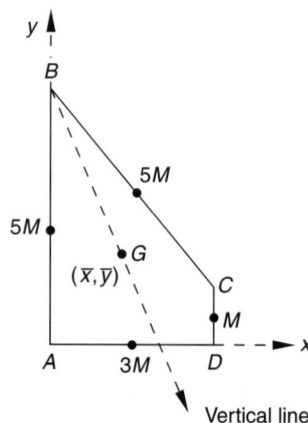

Taking axes of reference along AD (x-axis) and AB (y-axis),

Wire	Mass	Centre of mass	Moments
AD	$3M$	$\left(\dfrac{3a}{2}, 0\right)$	$\left(\dfrac{9Ma}{2}, 0\right)$
DC	M	$\left(3a, \dfrac{a}{2}\right)$	$\left(3Ma, \dfrac{Ma}{2}\right)$
CB	$5M$	$\left(\dfrac{3a}{2}, 3a\right)$	$\left(\dfrac{15Ma}{2}, 15Ma\right)$
AB	$5M$	$\left(0, \dfrac{5a}{2}\right)$	$\left(0, \dfrac{25Ma}{2}\right)$
Total	$14M$	$(\bar{x}, \bar{y})$	$(14M\bar{x}, 14M\bar{y})$

Taking moments about the y-axis,

$$\frac{9Ma}{2} + 3Ma + \frac{15Ma}{2} + 0 = 14M\bar{x}$$

$$\Rightarrow \quad 15Ma = 14M\bar{x}, \quad \bar{x} = \frac{15a}{14}$$

Taking moments about the x-axis,

$$0 + \frac{Ma}{2} + 15Ma + \frac{25Ma}{2} = 14M\bar{y}$$

$$\Rightarrow \quad 28Ma = 14M\bar{y}, \quad \bar{y} = 2a$$

Therefore the distances of the centre of mass from

AB and AD are $\dfrac{15a}{14}$ and $2a$ respectively.

When the wire is suspended at B, the line BG will be vertical (G is the centre of mass). The required angle is

$$\angle CBG = \angle ABC - \angle ABG$$

$$\tan A\hat{B}G = \frac{\bar{x}}{5a - \bar{y}} = \frac{15a/14}{3a} = \frac{5}{14}$$

$$\tan A\hat{B}C = \frac{3a}{4a} = \frac{3}{4}$$

$$\tan C\hat{B}G = \tan(A\hat{B}C - A\hat{B}G) = \frac{3/4 - 5/14}{1 + 15/56}$$

$$= \frac{42 - 20}{56 + 15} = \frac{22}{71}$$

The tangent of the angle of inclination of BC to the vertical is $\frac{22}{71}$.

15 Prove that the centre of mass of a uniform solid right circular cone of height h lies on its axis a distance $\frac{3}{4}h$ from the vertex.

The diagram represents the section of a uniform solid in the shape of a cylinder of radius 10 cm and length 12 cm joined to a cone of height 24 cm.

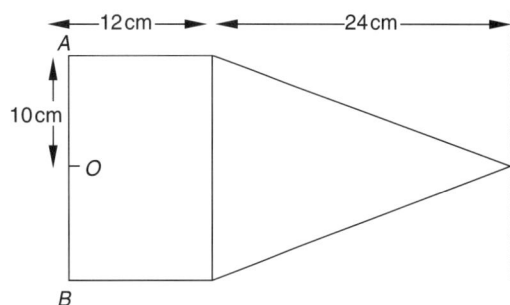

(a) Find the distance from O to the centre of mass of the solid.
(b) If the solid were freely suspended from A to hang in equilibrium under gravity, find (correct to the nearest degree) the angle that AB would make with the vertical.
(c) Show that the solid could rest in equilibrium on a horizontal table with either the curved surface of the cone or the curved surface of the cylinder in contact with the table. (SUJB)

• Bookwork.
Mass of a circular cone $= \frac{1}{3}\pi r^2 h\rho$.

(a) Cylinder radius 10 cm, height 12 cm:

volume $= \pi(100)\,12 = 1200\pi$

Cone radius 10 cm, height 24 cm:

volume $= \frac{1}{3}\pi(100)\,24 = 800\pi$

Mass of cylinder : mass of cone $= 3 : 2$.

Let masses be $3M$ and $2M$ respectively, with centres of mass at distance 6 cm and 18 cm from O, and the centre of mass of the composite solid a distance $\bar{x}$ from O.

Taking moments about AB,

$$3M(6) + 2M(18) = 5M(\bar{x})$$

$$\bar{x} = \frac{54}{5}\ \text{cm} = 10.8\ \text{cm}$$

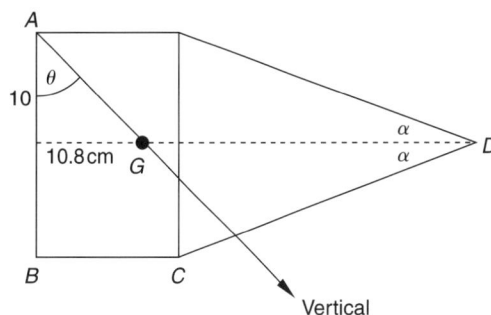

(b) If AB makes an angle θ with the vertical (see bottom of previous page) then

$$\tan\theta = \frac{10.8}{10} = 1.08$$

$$\theta = 47.2°$$

(c) (i) When the cylindrical curved surface is in contact with the table, the line of action of the weight comes within BC, so the solid can rest in equilibrium.

(ii) When the conical curved surface is in contact with the table,

$$GD = 25.2 \text{ cm}; \quad \tan\alpha = \frac{10}{24} = \frac{5}{12}$$

$$\Rightarrow \quad \cos\alpha = \frac{12}{13}.$$

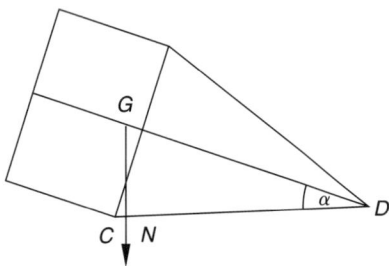

Distance $DN = GD\cos\alpha$
$\qquad\qquad = 25.2\,(\frac{12}{13}) = 23.3$ cm
But $CD^2 = 10^2 + 24^2$, $CD = 26$ cm.
$DN < CD$, so line of action comes within the line CD, and the solid can rest in equilibrium.

Exercises

1 A uniform rod PQ, of mass 6 kg, is smoothly hinged to a vertical wall at P. The rod makes an angle of 60° with the vertical. It is kept in equilibrium by a horizontal force of magnitude T newtons acting at Q. T is approximately equal to:

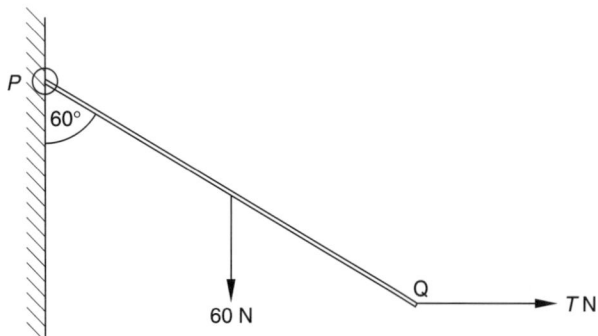

A 17   **B** 34   **C** 52   **D** 104   **E** 208   (L)

2 The diagram shows a uniform rod AB of weight W resting at an angle θ to the horizontal with its lower end in contact with a smooth vertical wall which is perpendicular to the vertical plane containing the rod. The rod passes under a peg C and over a peg D, both pegs being fixed, smooth, horizontal and parallel to the wall. A weight $2W$ is suspended from B and the rod rests in equilibrium. Given that $AC = DB = a$ and $CD = 2a$, find the reactions at C and D and show that $\cos^2\theta \geqslant \frac{9}{10}$.

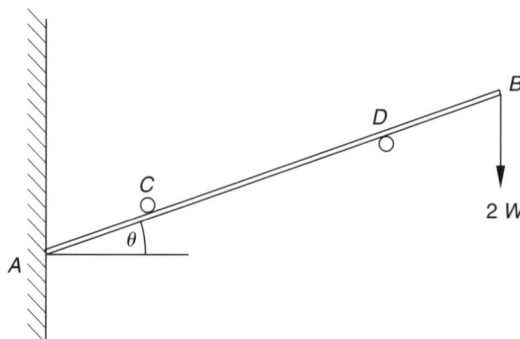

3 A uniform rod AB, of length $2a$ and weight W, is hinged to a vertical wall at A and is supported at an angle θ to the horizontal (B above A) by a string of length $2a$ attached to B and to a point C on the wall vertically above A. A load of weight W is hung from B.

Find the tension in the string and the force exerted by the hinge on the rod, in terms of W and θ.

Show that, if the reaction at the hinge A is perpendicular to the string, then $AC = a\sqrt{6}$.

4 A solid uniform cylindrical piece of metal, of height h and radius r, has a cone shape removed from it as shown in the diagram. The base of the cone is of radius r and its height is h.

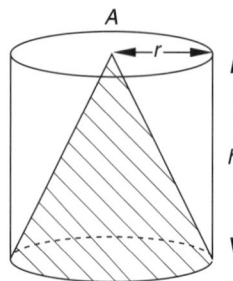

(i) Show that the centre of gravity of the resulting solid is at a distance of $\dfrac{3h}{8}$ from the point A measured along the axis of symmetry.

The solid is placed on an inclined plane with the open end in contact with the surface of the plane.

(ii) Given that $h = 4r$ and the coefficient of friction between the solid and the plane is 0.75, show that, as the inclination of the plane increases from zero degrees the solid will topple before it slides. (NICCEA, 1994)

5 A particle of weight 60 N is attached to two inextensible strings each of length 13 cm. The other ends of the strings are attached to two points A and B on the same horizontal level at a distance of 24 cm apart. Find the tension in the strings when the particle hangs in equilibrium.

The inextensible strings are then replaced by elastic ones, each of natural lengths 13 cm and of the same modulus of elasticity. The particle then hangs in equilibrium 9 cm below the line AB. Find the elastic modulus of the strings. (WJEC, 1994)

6

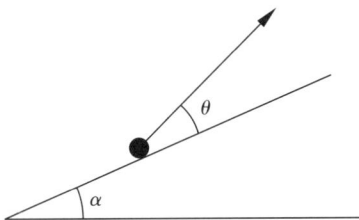

A particle of mass m lies on a smooth plane inclined at angle α to the horizontal. The particle is held in equilibrium by a string which lies in a vertical plane through a line of greatest slope and makes an angle θ with the plane, as shown in the diagram. The tension in the string is of magnitude T, and the force exerted by the plane on the particle is of magnitude R.
(a) Find, in terms of m, g, α and θ, an expression for T.
(b) Show that $R = mg \cos \alpha \, (1 - \tan \alpha \tan \theta)$.
(L, 1993)

7 A uniform ladder of length 5 m and weight 80 N stands on rough level ground and rests in equilibrium against a smooth horizontal rail which is fixed 4 m vertically above the ground. If the inclination of the ladder to the vertical is θ where $\tan \theta \leqslant \frac{3}{4}$, find expressions in terms of θ for the vertical reaction R of the ground, the friction F at the ground and the normal reaction N at the rail.

Given that the ladder does not slip, show that F is a maximum when $\tan \theta = \dfrac{1}{\sqrt{2}}$ and give the maximum value.

The coefficient of friction between the ladder and the ground is $\frac{1}{5}$. How much extra weight should be added at the bottom of the ladder so that the ladder will not slip when $\theta = \arctan (\frac{3}{4})$? (O&C, 1993)

8 (i) An object is in equilibrium under the action of exactly three forces. What must be true about the forces?

(ii)

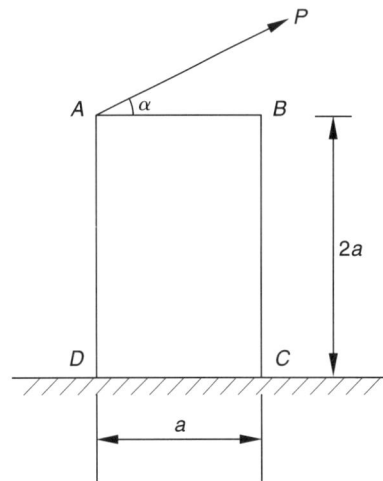

The figure shows a cross-section $ABCD$ of a uniform rectangular box of weight W. The centre of mass of the box lies in the plane $ABCD$ and $AB = a$, $BC = 2a$. The box rests with CD on a rough horizontal floor and is pulled with force P by a rope attached at A. The rope is at an angle α to the horizontal, as shown in the figure, and the box is on the point of rotating about C. Draw a diagram showing all the forces acting on the box.

By taking moments about C, prove that

$P = W/(2 \sin \alpha + 4 \cos \alpha)$

Hence, or otherwise, show that P is a minimum when $\alpha = \arctan (\frac{1}{2})$. State this minimum value of P.
(iii) When $\alpha = \arctan (\frac{1}{2})$ and the box is on the point of rotating about C without first sliding along the floor, show that μ, the coefficient of friction between the box and the floor, must be at least 2/9.

What is the magnitude of the reaction at C?
(iv) If $\mu = 1/5$, what is the minimum value of P necessary to cause the box to rotate about C without sliding, and what is the corresponding value of α? (O&C, 1992)

9 Two uniform cylinders of equal radius rest against each other on a fixed plane inclined at an angle α to

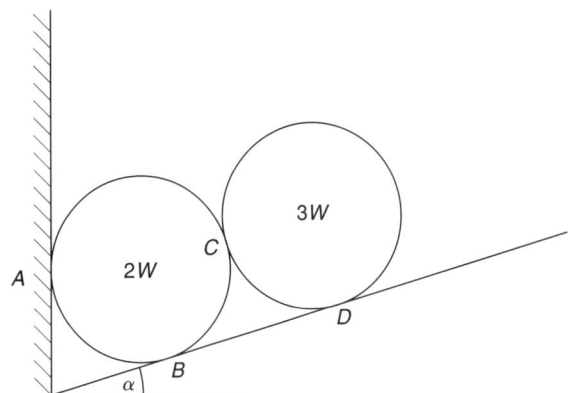

the horizontal, the lower cylinder resting against a fixed vertical plane. The line of intersection of the planes is perpendicular to the section shown. The lower cylinder has a weight $2W$ and the upper cylinder has weight $3W$. Given that all the contacts are smooth, find the magnitudes of the reactions at all points of contact, A, B, C and D.

10 A particle of weight W is placed on a rough plane which is inclined at an angle α to the horizontal. The coefficient of friction between the particle and the plane is μ. Show that if $\mu < \tan \alpha$ then the particle will slide down the plane.

   If $\alpha = 45°$ and $\mu = 0.5$, find the magnitude of the least horizontal force needed to maintain the particle in equilibrium.

11 A particle is placed on the inner surface of a fixed rough hollow sphere of internal radius a. Given that the coefficient of friction between the particle and the sphere is 3/4, show that the particle rests in limiting equilibrium at a depth $4a/5$ below the centre of the sphere.

12 [Take the acceleration due to gravity to be 10 m s^{-2} in this question, and give your answers correct to 3 significant figures where appropriate.]

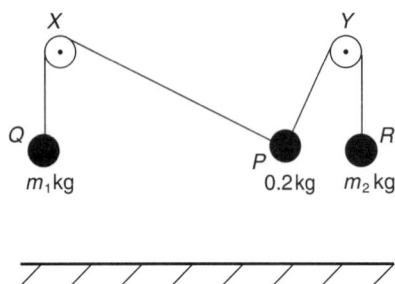

In the diagram, X and Y represent two small smooth pulleys which are fixed at equal heights above a horizontal surface. One light inextensible string passes over pulley X, and another light inextensible string passes over pulley Y. One end of each string is attached to the particle P, and the other ends are attached to particles Q and R. The masses of Q and R are m_1 kg and m_2 kg respectively, and the mass of P is 0.2 kg.
(i)   The system is in equilibrium, with all three particles hanging freely. Given that PX is inclined at $60°$ to the vertical and that angle XPY is $90°$, find m_1 and m_2.
(ii) The system is in equilibrium, with P in contact with the surface, and with Q and R hanging freely. PX is inclined at $30°$ to the vertical and PY is inclined at $45°$ to the vertical. The mass of Q is 0.1 kg.
   (a) Given that the surface is smooth, find m_2 and the magnitude of the reaction of the surface on P.

(b) Given that the coefficient of friction between P and the surface is $\frac{1}{3}$ and that P is about to slip towards the right, find m_2.
(UCLES, 1993)

13 A uniform sphere, radius a, weight W, stands in a vertical plane upon a rough horizontal floor with a point A of its circumference in contact with an equally rough vertical wall. Weights are added to the point A of the sphere until the sphere is on the point of slipping. If the added weight is $3W/2$, find the coefficient of friction.

14 Show, without use of any result quoted in the booklet of formulae, that the centre of gravity of a uniform triangular lamina lies at the point of intersection of the medians.

   A lamina $ABCD$, in the form of a rhombus, consists of two uniform triangular laminae ABC and ADC, joined along AC. Both triangles are equilateral of side $2a$. ABC is of mass $2m$ and ADC is of mass m. Determine the perpendicular distance of the centre of gravity G of the rhombic lamina from AC and from AB.

   The rhombic lamina is suspended by a string attached to A. Determine:
   (a) the tangent of the angle between AC and the vertical
   (b) the tension in the string when a mass is attached at D so that AC is vertical.

15 Show, by integration, that the centre of mass of a uniform, solid, right circular cone of height h is

   at a distance $\dfrac{3h}{4}$ from the vertex.

   Two uniform, solid, right circular cones, each with base radius a, have heights h and $3h$ and densities 3ρ and ρ. These cones are joined together with their bases coinciding to form a spindle.

   Show that the centre of mass is distance $\dfrac{11h}{4}$ from

   the vertex of the larger cone. Find a condition on a and h, if the spindle can rest in equilibrium with either curved surface in contact with a smooth horizontal table.

16 A uniform diving board AB, of length 4 metres and mass 40 kg, is fixed at A to a vertical wall and is maintained in a horizontal position by means of a light strut DC. D is a point on the wall 1 metre below A and C is a point on the board where $AC = 1$ metre. An object of mass 60 kg is placed at end B.
(i)   Draw a neat and clearly labelled diagram of the board showing all the forces acting on it.
(ii) Find the position of the centre of mass of the

60 kg mass and the mass of the board AB combined.

(iii) Using a triangle of forces, or otherwise, find
  (a) the thrust in the strut and
  (b) the magnitude of the reaction at A.

$$(g = 10 \text{ m s}^{-2})$$
(NICCEA, 1994)

Brief Solutions

1 Taking moments about P,
$T(2l) \cos 60° = 60 (l) \sin 60°$.
$T = 30\sqrt{3} \approx 52$.

Answer **C**

2 Moments about A:
$P(3a) = S(a) + W(2a) \cos \theta + 2W(4a) \cos \theta$.
Resolving vertically, $3W + S \cos \theta = P \cos \theta$.

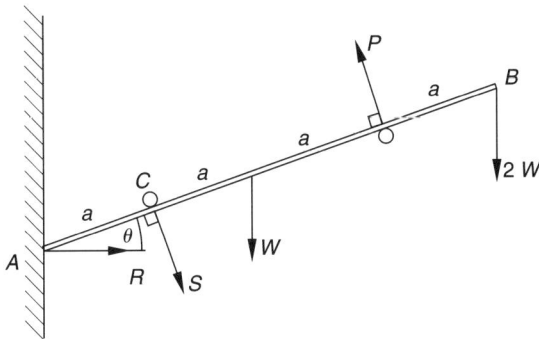

Solving, $S = W \dfrac{(10 \cos^2 \theta - 9)}{2 \cos \theta}$,

$$P = W \frac{(10 \cos^2 \theta - 3)}{2 \cos \theta}.$$

But $P \geqslant 0 \Rightarrow 10 \cos^2 \theta \geqslant 3$;
$\quad S \geqslant 0 \Rightarrow 10 \cos^2 \theta \geqslant 9$, so $\cos^2 \theta \geqslant \frac{9}{10}$.

3 Taking moments about A:
$T(2a \sin 2\theta) = Wa \cos \theta + W(2a \cos \theta)$,

$$T = \frac{3W}{4 \sin \theta}$$

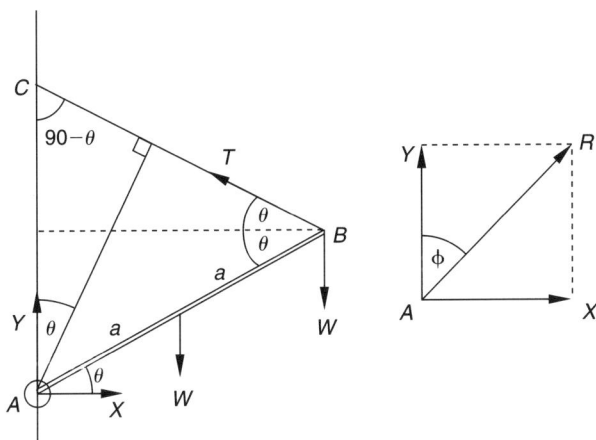

Resolving, $X = T \cos \theta = \dfrac{3W \cot \theta}{4}$,

$$Y = 2W - T \sin \theta = \frac{5}{4} W.$$

Reaction $R = \dfrac{W}{4} \sqrt{(9 \cot^2 \theta + 25)}$

at $\tan^{-1} \dfrac{X}{Y} = \tan^{-1} \left(\dfrac{3 \cot \theta}{5} \right)$

$$= \phi \text{ to the vertical.}$$

If R is perpendicular to BC, $\phi = \theta$, so $\tan^2 \theta = \frac{3}{5}$.
$AC = 2(2a \sin \theta) \Rightarrow AC = a\sqrt{6}$.

4

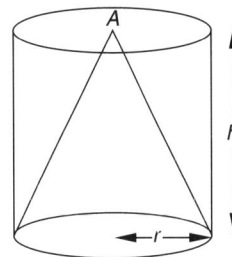

(i) Mass of cylinder $3M$; distance of C of G from base $\dfrac{h}{2}$; mass of cone M;

distance of C of G from base $\dfrac{h}{4}$.

Take moments about the base:

$$3M\frac{h}{2} - M\frac{h}{4} = 2M\bar{h} \Rightarrow \bar{h} = \frac{5h}{8}$$

C of G is $\dfrac{3h}{8}$ from A.

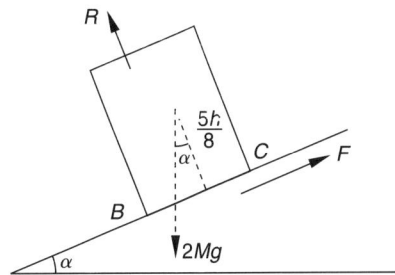

(ii) $R = 2Mg \cos \alpha$, $F = 2Mg \sin \alpha$

$$\frac{F}{R} = \tan \alpha$$

The solid will slide when $\tan \alpha > 0.75$
$\Rightarrow \alpha > 36.9°$ $F > \mu R$
The solid will be about to topple when the weight acts through B, that is

$\tan \alpha = \frac{2}{5} \Rightarrow \alpha = 21.8°$

Thus the solid topples before sliding.

5

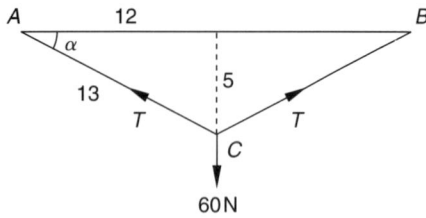

Depth of particle below AB is 5 cm.

$$2T \sin \alpha = 60 \quad \Rightarrow \quad T = 78 \text{ N}$$

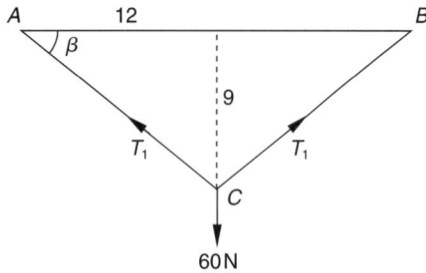

$AC = BC = 15$ cm, extension 2 cm

$$2T_1 \sin \beta = 60 \quad \Rightarrow \quad T_1 = 50 \text{ N}$$

$$\text{Tension} = \frac{\lambda(\text{extension})}{\text{natural length}} = \frac{2\lambda}{13} \quad \Rightarrow \quad \lambda = 325 \text{ N}$$

6

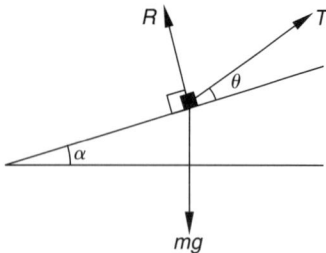

(a) Parallel to the plane

$$T \cos \theta = mg \sin \alpha \quad \Rightarrow \quad T = \frac{mg \sin \alpha}{\cos \theta}$$

(b) Perpendicular to the plane

$$R + T \sin \theta = mg \cos \alpha$$

$$R = mg \cos \alpha - \frac{mg \sin \alpha \sin \theta}{\cos \theta}$$

$$= mg \cos \alpha \left(1 - \frac{\sin \alpha \sin \theta}{\cos \alpha \cos \theta} \right)$$

$$= mg \cos \alpha (1 - \tan \alpha \tan \theta)$$

7

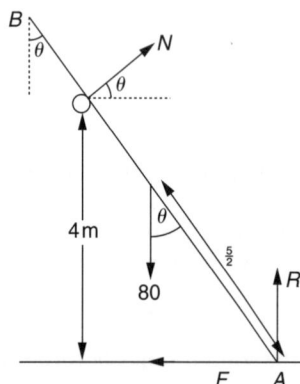

Moments about A:

$$N\frac{4}{\cos \theta} = 80 \frac{5}{2} \sin \theta$$

$$N = 50 \sin \theta \cos \theta \tag{1}$$

Resolving vertically
$$R = 80 - 50 \sin^2 \theta \cos \theta \tag{2}$$
Resolving horizontally
$$F = 50 \sin \theta \cos^2 \theta \tag{3}$$

Since the ladder does not slip, the maximum value of F occurs when

$$\frac{\mathrm{d}F}{\mathrm{d}\theta} = 0 \quad \Rightarrow \quad \text{at } \tan \theta = \frac{1}{\sqrt{2}}$$

$$F_{\max} = 50 \cdot \frac{1}{\sqrt{3}} \cdot \frac{2}{3} = \frac{100\sqrt{3}}{9} \text{ N}$$

When $\tan \theta = \frac{3}{4}$, $F = \frac{96}{5}$ N, $R = \frac{328}{5}$ N.

Since $\mu = \frac{1}{5}$, $R_1 \geqslant 5F$ if the ladder does not slip. But $R_1 = R +$ extra weight. Hence the extra weight required is 30.4 N.

8 (i) Sum of forces = 0; sum of moments about any point = 0.

(ii)

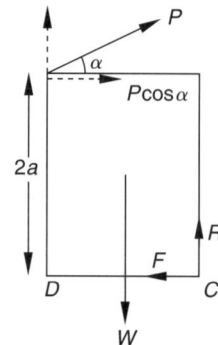

When about to rotate about C the normal contact force acts at C, so take moments about C

$$W \cdot \frac{a}{2} = P \cos \alpha \cdot 2a + P \sin \alpha \cdot a$$

$$\Rightarrow \quad P = \frac{W}{4 \cos \alpha + 2 \sin \alpha}$$

$$= \frac{W}{\sqrt{20} \cos(\alpha - \phi)} \text{ where } \tan \phi = \frac{2}{4}$$

$$P_{\min} = \frac{W}{\sqrt{20}} \text{ or } \frac{W\sqrt{5}}{10}, \quad \alpha = \arctan \frac{1}{2}$$

(iii) Rotates before sliding if $\dfrac{F}{R} < \mu$ when

$$\alpha = \arctan \frac{1}{2}$$

$$F = P \cos \alpha, \quad R = W - P \sin \alpha$$

$$\frac{F}{R} = \frac{P \cos \alpha}{W - P \sin \alpha} = \frac{2}{9}$$

Hence it rotates before sliding if $\mu > \frac{2}{9}$

If the box is about to rotate, $\mu = \frac{2}{9}$.

The Normal reaction at C is $R = \dfrac{9W}{10}$.

The total reaction at C is $\sqrt{R^2 + F^2} = \dfrac{W\sqrt{85}}{10}$

(iv) $\mu = \frac{1}{5}$; box rotates without sliding

$$\Rightarrow \quad \frac{P \cos \alpha}{W - P \sin \alpha} \leqslant \frac{1}{5}$$

$$\Rightarrow \quad P(5 \cos \alpha + \sin \alpha) \leqslant W$$

But $P = \dfrac{W}{4 \cos \alpha + 2 \sin \alpha}$ (from ii) $\Rightarrow$

$$P(5 \cos \alpha + \sin \alpha) \leqslant P(4 \cos \alpha + 2 \sin \alpha)$$

$$\Rightarrow \quad \cos \alpha \leqslant \sin \alpha \quad \Rightarrow \quad \alpha \geqslant \frac{\pi}{4}$$

When $\alpha = \dfrac{\pi}{4}$, $\quad P = \dfrac{W}{3\sqrt{2}}$,

the minimum value of P.

9

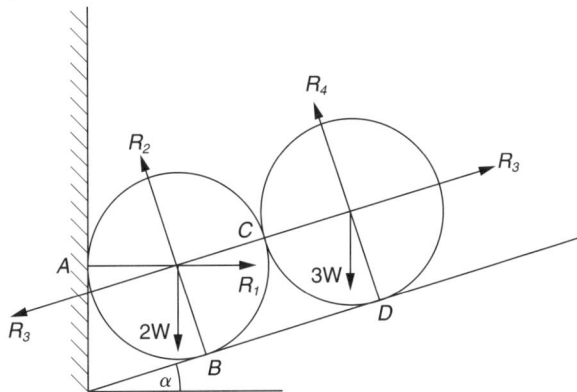

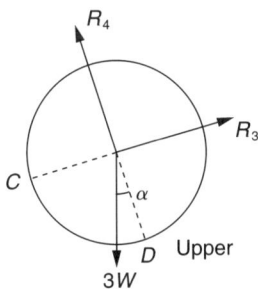

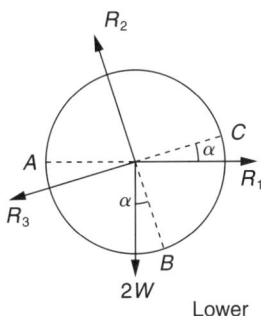

Resolving:

for upper cylinder,

$R_4 = 3W \cos \alpha$, $R_3 = 3W \sin \alpha$;

for lower cylinder,

$2W \sin \alpha + R_3 = R_1 \cos \alpha \quad \Rightarrow \quad R_1 = 5W \tan \alpha$

$2W \cos \alpha + R_1 \sin \alpha = R_2$

$$\Rightarrow \quad R_2 = \frac{W(2 \cos^2 \alpha + 5 \sin^2 \alpha)}{\cos \alpha}$$

10 *For equilibrium:*

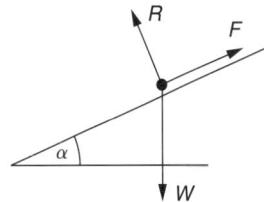

Resolving parallel to plane, $\qquad F = W \sin \alpha$.

Resolving perpendicular to plane, $\quad R = W \cos \alpha$.

So maximum frictional force $= \mu R = \mu W \cos \alpha$.

Particle slides if

$W \sin \alpha > \mu W \cos \alpha \quad \Rightarrow \quad \mu < \tan \alpha$.

If horizontal force P is applied:

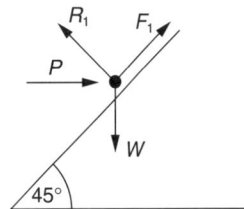

Resolving perpendicular to plane, $R_1 = \dfrac{(W + P)}{\sqrt{2}}$.

Resolving parallel to plane, $\qquad F_1 = \dfrac{(W - P)}{\sqrt{2}}$.

P is minimum when $F_1 = \dfrac{R_1}{2} \quad \Rightarrow \quad P = \dfrac{W}{3}$.

11 Limiting equilibrium $\Rightarrow F = \frac{3}{4}R$.

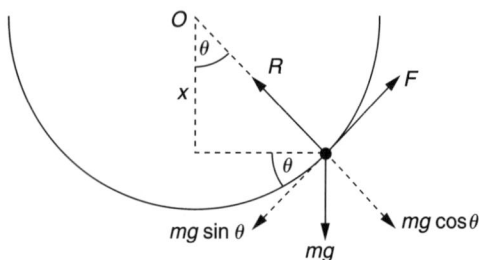

If R makes angle θ with vertical,

$F = mg \sin \theta, \quad R = mg \cos \theta$

$\Rightarrow \quad \tan \theta = \dfrac{F}{R} = \dfrac{3}{4}.$

Depth below centre, $x = a \cos \theta = \frac{4}{5}a$.

12 (i)

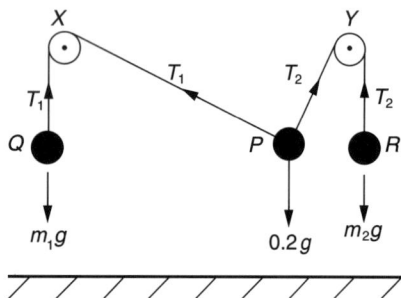

At P, resolving horizontally:
$T_1 \cos 30 = T_2 \sin 30$ (1)
resolving vertically:
$T_1 \sin 30 + T_2 \cos 30 = 0.2g = 2$ (2)
Combining equations 1 and 2
$T_1(\cos^2 30 + \sin^2 30) = 2 \sin 30$

$T_1 = 1, \quad T_2 = \sqrt{3}, \quad m_1 = 0.1, \quad m_2 = \dfrac{\sqrt{3}}{10}$

(ii) (a) See diagram below.

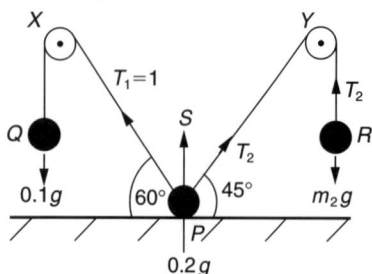

Let reaction at $P = S$ N.
At P, resolving horizontally:

$1 \cos 60° = T_2 \cos 45° \quad \Rightarrow \quad T_2 = \dfrac{\sqrt{2}}{2}$

at R, $T_2 = m_2 g \quad \Rightarrow \quad m_2 = \dfrac{\sqrt{2}}{20} = 0.0707$

At P, resolving vertically:

$S + T_1 \cos 30° + T_2 \cos 45° = 0.2g$
$S = 0.634$ N

(b)

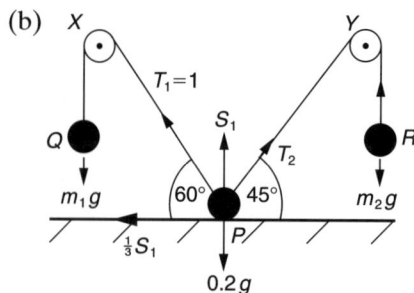

Let reaction at $P = S_1$ N
At P, resolving vertically:
$1 \cos 30° + S_1 + 10m_2 \cos 45° = 2$
Resolving horizontally:
$1 \sin 30° + \frac{1}{3}S_1 - 10m_2 \sin 45° = 0$
Eliminating S_1:

$\dfrac{\sqrt{3}}{2} - \dfrac{3}{2} + 10m_2 \dfrac{\sqrt{2}}{2} + 30m_2 \dfrac{\sqrt{2}}{2} = 2$

$m_2 = \dfrac{7 - \sqrt{3}}{40\sqrt{2}} = 0.0931$

13

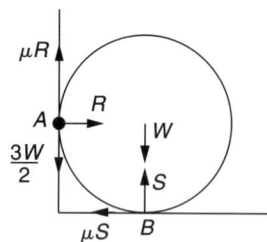

Resolving horizontally: $R = \mu S$

Resolving vertically: $\mu R + S = \frac{5}{2}W$,

giving $S = \dfrac{5W}{2(\mu^2 + 1)}.$

Taking moments about A, $W = S(1 - \mu)$.
Eliminating S gives $2\mu^2 + 5\mu - 3 = 0 \quad \Rightarrow \quad \mu = \frac{1}{2}.$

14

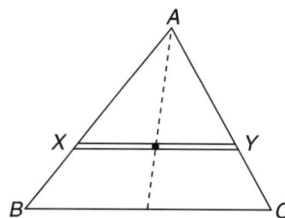

The centre of gravity of the element lies at the midpoint of XY, i.e. on the median from A to BC. This is true for all elements, so c.g. of triangle lies on this median. By symmetry, c.g. lies on all three medians, and is therefore at their intersection.

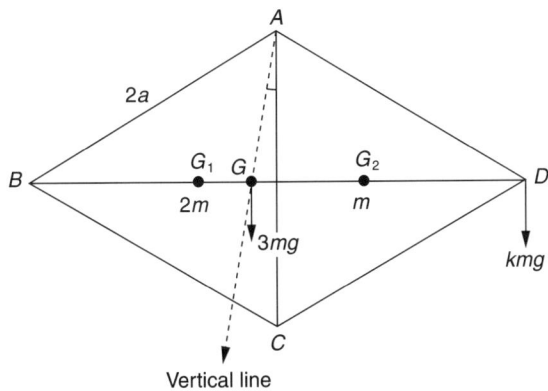

C.G. (G_1) of ABC is $\dfrac{1}{3}(2a\cos 30) = \dfrac{\sqrt{3}}{3}\,a$ from AC.

C.G. (G_2) of ADC is also $\dfrac{\sqrt{3}}{3}\,a$ from AC.

C.G. of mass $2m$ at G_1 and mass m at G_2 is at G, on G_1G_2, distance $\dfrac{\sqrt{3}}{9}\,a$ from AC. $\quad BG = \dfrac{8\sqrt{3}}{9}\,a$.

Perpendicular distance from AB to G is

$$\dfrac{8\sqrt{3}}{9}\,a\sin 30 = \dfrac{4\sqrt{3}}{9}\,a.$$

(a) When suspended from A, AG is vertical, so

$$\tan C\hat{A}G = \dfrac{\sqrt{3}}{9}.$$

(b) Attach mass km at D to make AC vertical. Take moments about the midpoint of AC to get $k = \tfrac{1}{3}$.
Tension $= 3mg + kmg = \tfrac{10}{3}\,mg$.

15 Centre of mass: standard bookwork.
Masses of cones are equal.

Taking moments about O gives $\bar{x} = \dfrac{11h}{4}$ from O.

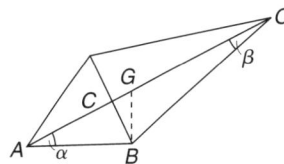

Let the semi-vertical angles of small and large cones be α and β.
When the small cone rests on the surface, equilibrium is possible if $AG\cos\alpha \leqslant AB$;
i.e. if $\tfrac{5}{4}h^2 \leqslant (AB)^2 = a^2 + h^2$
i.e. if $h \leqslant 2a$.
When the large cone rests on the surface, the spindle is always in equilibrium, since c.g. is within the large cone. This implies that the spindle can rest in equilibrium on either surface provided $h \leqslant 2a$.

16 (i)

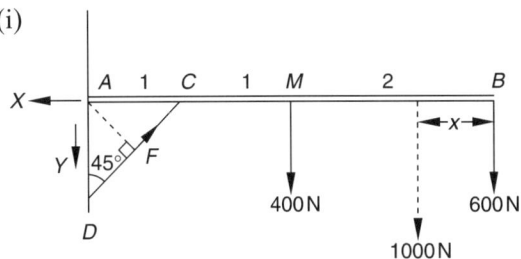

(ii) Moments about B:
$400 \times 2 = 1000x \quad\Rightarrow\quad x = 0.8$
$\Rightarrow$ centre of mass is 0.8 m from B

(iii) Moments about A:
(a) $F\cos 45 = 1000 \times 3.2 \quad\Rightarrow\quad F = 4525$ N
(b) Resolving horizontally:
$X = F\cos 45 = 3200$ N
Resolving vertically:
$Y = F\sin 45 - 1000 = 2200$ N
Magnitude of the force at A is
$\sqrt{X^2 + Y^2} = 3880$ N to 3 s.f.

21

Constant acceleration

$$\int \frac{x}{1 - \frac{v}{c}^2} \, dx$$

	AS Level			A Level		Topic	Date attempted	Date completed	Self-assessment
CORE	MODULAR	TRADITIONAL	CORE	MODULAR	TRADITIONAL				
	✓	✓		✓	✓	**Distance/velocity/acceleration/ time formulae**			
				✓		**Distance/velocity/acceleration/ time graphs**			
	✓	✓		✓	✓	**Equations of motion**			
		✓			✓	**Conservation of momentum**			
		✓		✓	✓	**Impacts**			
	✓	✓		✓	✓	**Connected particles**			

Fact Sheet

Constant-Acceleration Formulae

If a point is moving in a straight line and the distance of the point from O at time t is x (or s) then

$$v = \frac{dx}{dt} \qquad a = \frac{dv}{dt} = \frac{d^2x}{dt^2} = v\frac{dv}{dx}$$

With constant acceleration a,

$$v = \int a\, dt = at + u \qquad\qquad v = u + at \tag{1}$$

$$x = \int v\, dt = ut + \tfrac{1}{2}at^2 + c \qquad s \text{ or } x = ut + \tfrac{1}{2}at^2 \tag{2}$$

From (1) and (2), $s = \dfrac{u+v}{2}t$ and $v^2 = u^2 + 2as$ can be obtained.

Newton's Laws

(i) Every body will remain at rest or continue to move with constant velocity unless an external force is applied to it.
or If a body has an acceleration there is a force acting to cause it.
At constant, or maximum, speed the sum of forces = 0.
(ii) When an external force is applied to a body of constant mass, the force produces an acceleration directly proportional to the force.
or The sum of forces in the direction of motion
= mass × acceleration in that direction ($F = ma$).
(iii) To every action there is an equal and opposite reaction.

Velocity–Time Graphs

If the units of time are the same in velocity and time then:
(i) The gradients of the graphs represent accelerations.
(ii) The area under the graph represents the distance travelled.
It is not necessary to use SI units for these graphs.

Momentum

Momentum is defined as mass × velocity = $m\mathbf{v}$.
It is a vector.

Impulse

Impulse is defined as change in momentum:

$$\mathbf{I} = m\mathbf{v} - m\mathbf{u} \tag{1}$$

where $\mathbf{u}$ is the velocity before impulse $\mathbf{I}$ and $\mathbf{v}$ is the velocity afterwards.

Acceleration

A force $\mathbf{F}$ acting on particle mass $\mathbf{m}$ gives it an acceleration $\mathbf{a} = \dfrac{\mathbf{F}}{m}$.

$$\mathbf{v} = \mathbf{u} + \mathbf{a}t \quad \text{so} \quad \mathbf{v} = \mathbf{u} + \frac{\mathbf{F}}{m}t \quad \text{or} \quad \mathbf{F}t = m\mathbf{v} - m\mathbf{u} \tag{2}$$

Comparison of (1) and (2) gives $\mathbf{I} = \mathbf{F}t$, which is an alternative definition of impulse; usually $\mathbf{F}$ is very large and t, the time during which $\mathbf{F}$ acts, is very small.

Conservation of Momentum

When two particles collide, equal but opposite forces act between them for a short time, giving equal but opposite impulses $\mathbf{I}$ and $-\mathbf{I}$.

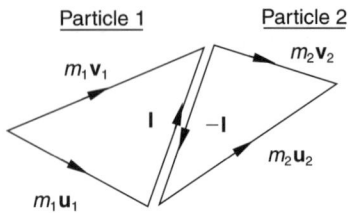

Assuming motion to be unrestricted:

Particle 1   Particle 2

$m_1\mathbf{v}_1$ $m_2\mathbf{v}_2$

$\mathbf{I}$ $-\mathbf{I}$

$m_2\mathbf{u}_2$

$m_1\mathbf{u}_1$

For first particle: $\qquad\qquad m_1\mathbf{v}_1 - m_1\mathbf{u}_1 = \mathbf{I}$
For second particle: $\qquad\quad m_2\mathbf{v}_2 - m_2\mathbf{u}_2 = -\mathbf{I}$
Adding: $\qquad\qquad\qquad m_1\mathbf{v}_1 + m_2\mathbf{v}_2 = m_1\mathbf{u}_1 + m_2\mathbf{u}_2$
This is the conservation of momentum equation.

Newton's Law of Restitution

When two particles (1) and (2) are in direct impact then

$$\frac{\text{speed of separation}}{\text{speed of approach}} = e$$

e is the coefficient of restitution. $0 \leqslant e \leqslant 1$.
If $e = 1$ then the collision is perfectly elastic.
If $e = 0$ then the collision is inelastic and the particles coalesce.

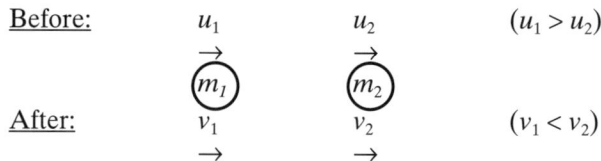

Before: $\qquad\qquad u_1 \qquad\qquad u_2 \qquad\qquad (u_1 > u_2)$
$$\rightarrow \qquad\qquad \rightarrow$$
$$\boxed{m_1} \qquad\quad \boxed{m_2}$$
After: $\qquad\qquad v_1 \qquad\qquad v_2 \qquad\qquad (v_1 < v_2)$
$$\rightarrow \qquad\qquad \rightarrow$$

$$v_2 - v_1 = e(u_1 - u_2)$$

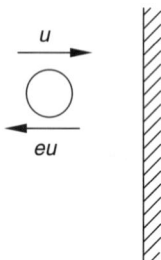

Impact of a Particle Moving Perpendicular to a Fixed Plane

u

eu

Since the plane is fixed the principle of conservation of momentum cannot be applied, but the law of restitution still applies.

Connected Particles

When particles are connected by a taut string the tension is the same throughout its length even if the string passes over a (frictionless) pulley. When the system is moving, the motion of each particle satisfies $\mathbf{F} = m\mathbf{a}$, the particles having accelerations of equal magnitude.

When a system experiences an impulsive change of state (for example acquiring an extra mass which has a different velocity or if the string suddenly becomes taut), then momentum is conserved through the impulsive tension.

Worked Examples

1   The brakes of a train, which is travelling at 108 km h^{-1}, are applied as the train passes point A. The brakes produce a constant retardation of magnitude $3f$ m s^{-2} until the speed of the train is reduced to 36 km h^{-1}. The train travels at this speed for a distance and is then uniformly accelerated at f m s^{-2} until it again reaches a speed of 108 km h^{-1} as it passes point B. The time taken by the train in travelling from A to B, a distance of 4 km, is 4 minutes.

Sketch the speed–time graph for this motion and hence calculate
(a) the value of f
(b) the distance travelled at 36 km h^{-1}. (L)

• 108 km h^{-1} = 30 m s^{-1} 36 km h^{-1} = 10 m s^{-1}.
If train takes t_1 seconds to decelerate then it takes $3t_1$ seconds to accelerate. Therefore it is at constant speed 10 m s^{-1} for $240 - 4t_1$ seconds.

Distance travelled = area under graph
$$= 20t_1 + 10(240 - 4t_1) + 60t_1$$
$$= 4000 \text{ m}$$
$$\Rightarrow \quad 40t_1 = 1600, \quad t_1 = 40.$$

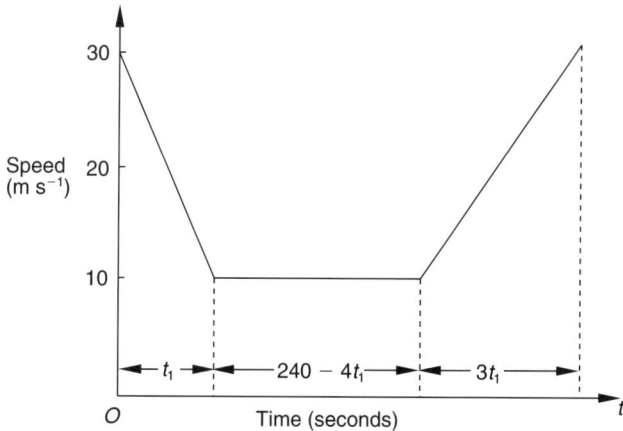

Therefore train takes 40 seconds to decelerate from 30 m s^{-1} to 10 m s^{-1}, a deceleration of 0.5 m s^{-2}.

Hence $f = \frac{1}{6}$ m s^{-2}.

Distance travelled at 36 km h^{-1} = 10(240 − 160)

$\qquad\qquad\qquad\qquad\qquad\qquad = 800$ m.

2 A car, of mass M kilograms, is pulling a trailer, of mass λM kilograms, along a straight horizontal road. The tow-bar connecting the car and the trailer is horizontal and of negligible mass. The resistive forces acting on the car and trailer are constant and of magnitude 300 N and 200 N respectively. At the instant when the car has an acceleration of magnitude 0.3 m s^{-2}, the tractive force has magnitude 2000 N.

Show that $M(\lambda + 1) = 5000$

Given that the tension in the tow-bar is 500 N at this same instant, find the value of M and the value of λ. (L, 1992)

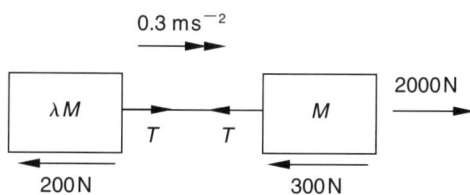

ΣForces acting horizontally = 1500 N
$\Sigma F = \Sigma$mass $\times$ accel. *Newton's Law*
$1500 = M(1 + \lambda) \times 0.3$

$\therefore M(\lambda + 1) = \dfrac{1500}{0.3} = 5000$ $\qquad$ (1)

If the tension in the tow-bar is 500 N

$\quad \Sigma$Forces acting horizontally on the trailer
$\quad = 500 - 200$
$\quad = 300$ N

$300 = \lambda M(0.3) \quad \Rightarrow \quad \lambda M = 1000$ $\qquad$ (2)
From equations 1 and 2: $\quad M = 4000, \quad \lambda = \frac{1}{4}$

3 Two identical small smooth spheres A and B are at rest on a smooth horizontal table which has a

smooth plane vertical rim at one end. The centres of the spheres are in a line perpendicular to the rim of the table with B nearer to the rim than A. The coefficient of restitution between A and B is $\frac{1}{2}$, and that between B and the rim is e ($0 < e < 1$). Sphere A is projected directly towards B with speed u. Show that the speed of B after impact is

$\dfrac{3u}{4}$ and find the speed of A. Sphere B then strikes

the rim and rebounds to hit A again. Find the speeds of A and B after this impact and deduce that B will then be moving towards the rim whatever the value of e whilst A will be moving towards the rim provided that $e < \frac{1}{9}$. (AEB, 1988)

• Let the spheres have mass m.

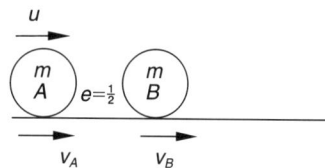

Conservation of linear momentum
1st collision: $\qquad mu = mv_A + mv_B$ $\qquad$ (1)
Law of restitution $\quad \frac{1}{2}u = v_B - v_A$ $\qquad$ (2)

Equation 1 + equation 2:

$\dfrac{3u}{2} = 2v_B \quad \Rightarrow \quad v_B = \dfrac{3u}{4}, \quad v_A = \dfrac{u}{4}$

Hence speeds of A and B are $\dfrac{u}{4}$ and $\dfrac{3u}{4}$.

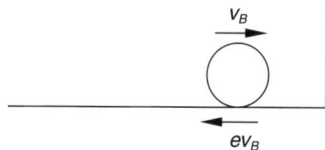

2nd collision: B with rim

Direction reversed. Speed of $B = -\frac{3}{4}ue$

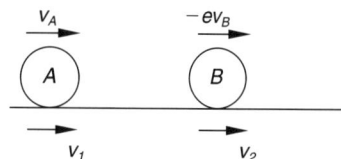

3rd collision:

Conservation of linear momentum

$m\left(\dfrac{u}{4}\right) - m\left(\dfrac{3}{4}ue\right) = mv_1 + mv_2$ $\qquad$ (3)

Law of restitution

$\dfrac{1}{2}\left(\dfrac{u}{4} + \dfrac{3}{4}ue\right) = v_2 - v_1$ $\qquad$ (4)

Equation 3 + equation 4: $\dfrac{3u}{8} - \dfrac{3}{8}ue = 2v_2$

$$v_2 = \dfrac{3u}{16}(1 - e), \qquad v_1 = \dfrac{u}{16}(1 - 9e)$$

The speed of B (v_2) is always positive, i.e. moving towards the rim.
The speed of A (v_1) is positive if $1 > 9e$, i.e. $e < \frac{1}{9}$.

4 A small ring is pulled along a line AB on a horizontal surface by two forces. The two forces of magnitude P N and Q N are applied to the ring in a horizontal plane and make angles α and β with AB as shown in the diagram.

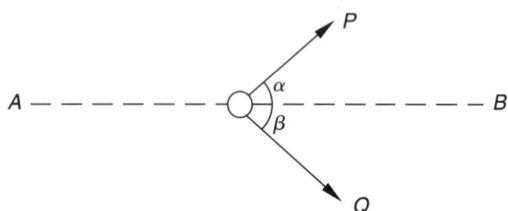

(a) When $P = 120$ N, $Q = 60\sqrt{2}$ N, $\alpha = 30°$ and $\beta = 45°$ their resultant force has magnitude R N.
  (i)   Sketch a vector triangle for the two forces and their resultant.
  (ii) Calculate R.
  (iii) Explain why R must be less than $120 + 60\sqrt{2}$ N.
  (iv) Explain why, if three forces of magnitudes 20 N, 30 N and 30 N are used, regardless of their directions, they could not move the ring if the total resistance force acting had magnitude 90 N.
(b) When $P = 30$ N, $Q = 40$ N, $\alpha = 45°$ and $\beta = 60°$,
  (i)   find the component of the resultant in the direction of AB
  (ii) explain why the ring cannot move along AB
  (iii) find the magnitude and direction of the resultant force.   (OLE, 1993)

• (a) When $\alpha = 30°$ and $\beta = 45°$
    $P = 120$ N, $Q = 60\sqrt{2}$ N
    (i)

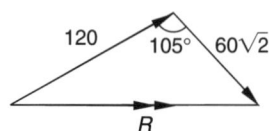

  (ii) $R^2 = (120)^2 + (60\sqrt{2})^2$
        $\quad - 2 \cdot 120 \cdot 60\sqrt{2} \cos 105°$
        $R = 164$ N
  (iii) The maximum value of R occurs when forces P and Q are in the same direction. In this case $R = 120 + 60\sqrt{2}$, otherwise $R \cdot 120 \cdot 60\sqrt{2}$.

(iv) Forces of 20 N, 30 N and 30 N would have a maximum resultant of 80 N. This would not overcome the fictional force of 90 N.

(b)

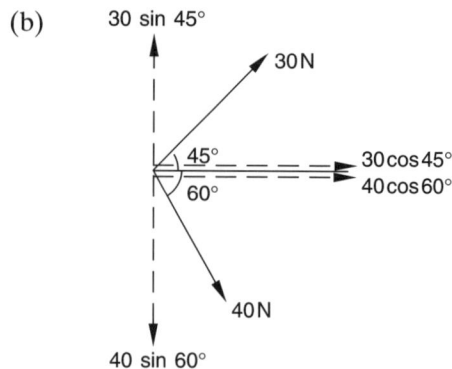

(i)   The component along AB is
      $(30 \cos 45° + 40 \cos 60°)$ N $= 41.2$ N

(ii) The component perpendicular to AB is
      $(40 \sin 60° - 30 \sin 45°)$ N $= 13.4$ N

      Hence the particle cannot move along AB since the resultant force makes an angle of $18°$ with AB.

(iii)

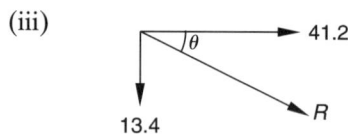

The resultant force has magnitude
$\sqrt{41.2^2 + 13.4^2} = 43.3$ N

The direction of R is $\tan^{-1}\left(\dfrac{13.4}{41.2}\right)$ with AB

(nearer to Q) i.e. $\theta = 18.0°$.

5

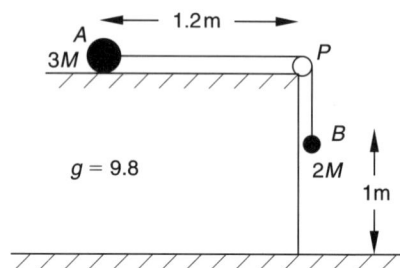

A particle A of mass $3M$ lies on a rough horizontal table. The particle is attached to a light inextensible string which passes over a small smooth pulley P fixed at the edge of the table. To the other end of the string is attached a particle B of mass $2M$, which hangs freely. AP is perpendicular to the edge of the table, and A, P and B are in the same vertical plane.

The system is released from rest with the string taut, when A is 1.2 m from the edge of the table and B is 1 m above the floor, as shown in the figure. The

particle B reaches the floor after 2 s, and does not rebound.
(a) Find the acceleration of A during the first two seconds.
(b) Find, to 2 decimal places, the coefficient of friction between A and the table.
(c) Determine, by calculation, whether A reaches P. $(g = 9.8 \text{ m s}^{-2})$
(L, 1993)

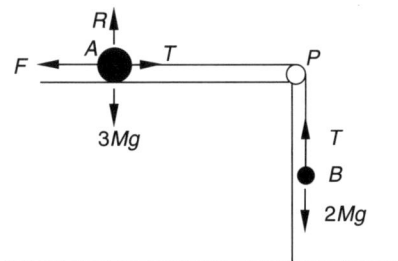

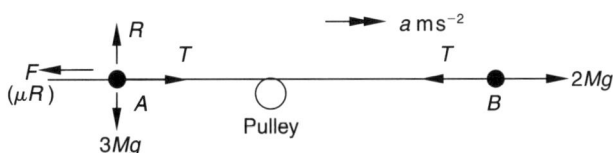

'Straighten' the string and look at the forces in the direction of the string
For the whole system

$$2Mg - F = 5Ma \qquad (1)$$

where a is the mutual acceleration.
For B: $s = 1, \quad t = 2, \quad u = 0$.
(a) Using $s = ut + \frac{1}{2}at^2$

$$1 = 2a \quad \Rightarrow \quad \text{mutual acceleration } a = \tfrac{1}{2} \text{ m s}^{-2} \quad (2)$$

(b) Substituting in equation 1: $F = M(2g - \tfrac{5}{2})$ (3)
Since A is moving, $F = \mu R = 3Mg\mu$
$\Rightarrow \quad 3g\mu = 2g - \tfrac{5}{2} \quad \Rightarrow \quad \mu = 0.58$ (2 d.p.)
(c) When B strikes the ground

$$v^2 = 2as \quad \Rightarrow \quad v = 1 \text{ m s}^{-1}$$

When B is on the ground $T = 0$.
Accelerating force acting on A is $-3Mg\mu$

$$\Rightarrow \quad \text{acceleration is } -\mu g$$

Using $v^2 = u^2 + 2as$, A comes to rest when

$$0 = 1 - 2\mu g s_1 \quad \Rightarrow \quad s_1 = \frac{1}{2\mu g} = 0.088 \text{ m}.$$

Since A was 0.2 m from P it does not reach P.

6

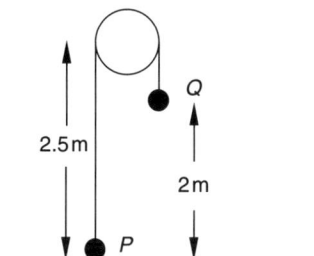

Two particles P and Q, of masses 0.3 kg and 0.7 kg respectively, are attached to the ends of a light inextensible string. The string passes over a small, smooth pulley fixed at a height of 2.5 m above horizontal ground. Initially both parts of the string are taut and vertical, with P resting on the ground and Q held at a height of 2 m above the ground. At time $t = 0$, Q is released from rest and the system then moves freely under gravity with the string taut. Find the tension in the string.

Show that, when $t = 0.5$ s, Q has fallen a distance of 0.5 m.

At the instant when $t = 0.5$ s, part of Q becomes detached, leaving a particle Q' of mass 0.2 kg attached to the string and with unchanged velocity. Find
(i) the minimum height above the ground reached by Q'
(ii) the speed of P just before it hits the ground.
When P hits the ground, it rebounds vertically with a speed of 2 m s^{-1}. Find the impulse exerted on the ground by P. $(g = 10 \text{ m s}^{-2})$
(UCLES, 1990)

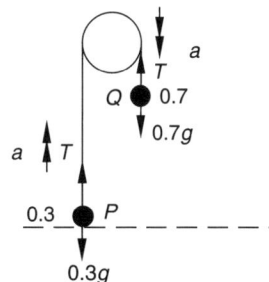

Let the acceleration be a m s^{-2} and the tension in the string T.
For P: $T - 0.3g = 0.3a$ (1)
For Q: $0.7g - T = 0.7a$ (2)
Adding equations 1 and 2, putting $g = 10$:

$$4 = 1a \quad \Rightarrow \quad \text{acceleration} \quad a = 4 \text{ m s}^{-2}$$
$$T = 4.2 \text{ N}$$

Using $s = ut + \frac{1}{2}at^2$, $u = 0$, $a = 4$.
When $t = 0.5$,
distance fallen by $Q = \frac{1}{2}(4)(\frac{1}{2})^2 = 0.5$ m.

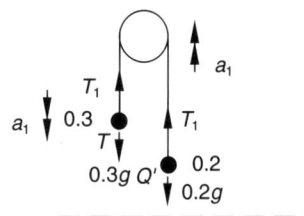

Velocity of $Q = at = 2$ m s^{-1}.
When part of the particle Q falls off the acceleration is now a_1 m s^{-2} *upwards*.
For P: $0.3g - T_1 = 0.3a_1$ (3)
For Q': $T_1 - 0.2g = 0.2a_1$ (4)
Adding equations 3 and 4:

$$0.1g = 0.5a_1 \quad \Rightarrow \quad a_1 = 0.2g \text{ or } 2 \text{ m s}^{-2}$$

Q' has a velocity 2 m s^{-1} downwards and acceleration of 2 m s^{-2} upwards.

At the lowest point $v = 0$:

$\Rightarrow$ Using $v^2 = u^2 + 2as$ $\quad 0 = 4 - 2(2)s_1$

$\quad s_1 = 1$ m

(i) Total distance fallen by Q' is 1.5 m $\Rightarrow$ Q' is 0.5 m above the ground (minimum height).

(ii) P is 1.5 m above the ground.
When P strikes the ground
$(v_P)^2 = 2 \cdot 2 \cdot (1.5) = 6$

$\Rightarrow$ speed of P is $\sqrt{6}$ m s^{-1} = 2.45 m s^{-1}

Impulse exerted by the ground
$= $ change in momentum of P
$= 0.3\,(\sqrt{6} + 2)$ N s
$= 1.33$ N s

7

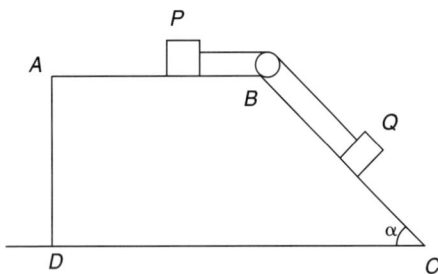

$g = 9.8$ m s^{-2}

The diagram shows a vertical section $ABCD$ of a block of wood fixed on a horizontal plane. AB is horizontal and BC is inclined at an angle α to the horizontal where $\sin \alpha = \frac{4}{5}$. Particles P and Q, of mass m and $5m$ respectively, are placed on AB and BC and joined together by a light inextensible string passing over a smooth pulley at B. The particles are then released from rest. Find, assuming that P does not reach B and Q does not reach C, the acceleration of the particles and the tension in the string when

(a) AB and BC are smooth,

(b) AB and BC are both equally rough, the coefficient of friction being $\frac{1}{3}$.

Find, for the second case, the loss of energy due to friction after both particles have moved a distance d. (WJEC, 1993)

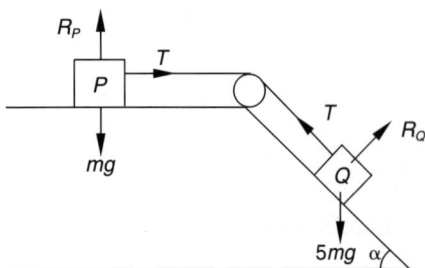

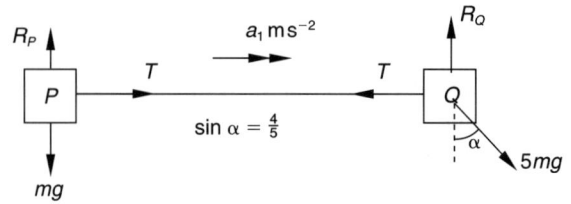

(a) For the whole system, with acceleration a_1 m s^{-2}:

$$6ma_1 = 5mg \sin \alpha \quad \Rightarrow \quad a_1 = \frac{5}{6} \cdot \frac{4}{5} \cdot g = \frac{2g}{3}$$

For P: $\quad T = m \cdot \dfrac{2g}{3}$

Hence acceleration is 6.53 m s^{-2}; tension $= 6.53m$ N

(b) With friction $\mu = \frac{1}{3}$:

$$F_P = \frac{mg}{3}, \quad F_Q = \frac{5mg \cos \alpha}{3} = mg$$

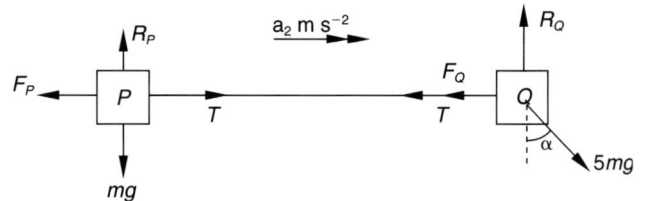

For the whole system with acceleration a_2 m s^{-2}

$$6ma_2 = 5mg \sin \alpha - \frac{mg}{3} - mg = \frac{8mg}{3}$$

$$\Rightarrow \quad a_2 = \frac{4g}{9}$$

For P: $T - \dfrac{mg}{3} = \dfrac{4mg}{9} \quad \Rightarrow \quad T = \dfrac{7mg}{9}$

Hence acceleration is 4.36 m s^{-2} and tension $T = 7.62m$ N.

When the particles have moved a distance d loss of energy due to friction is

$$(F_P + F_Q)d = \tfrac{4}{3}mgd$$

8

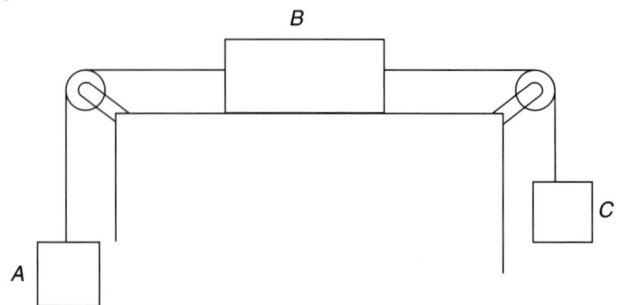

The figure shows a block B of mass 5 kg lying on a smooth table. It is connected to blocks A of mass 6 kg and C of mass 3 kg, which are hanging over the edges of the table, by light inextensible strings running over smooth pulleys. Initially the system is held at rest. After being released, the tension (in newtons) in the string joining A and B is P and the tension (in newtons) in the string joining B and C is Q. If the magnitude of the acceleration of each block is a m s^{-2}, write down three equations connecting a with P or Q (or both) and hence calculate a, P and Q.

Calculate the time taken by A to descend 60 cm from rest, assuming that B remains on the table and C remains hanging during this time.

[Take g, the acceleration due to gravity, as 9.8 m s^{-2}.] (O&C, 1992)

•

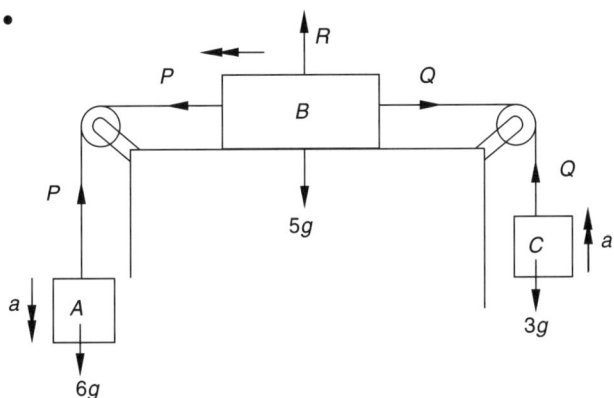

For A: $6g - P = 6a$ (1)
$\quad$ B: $P - Q = 5a$ (2)
$\quad$ C: $Q - 3g = 3a$ (3)

Adding equations 1, 2 and 3:

$$3g = 14a \quad \Rightarrow \quad a = \frac{3g}{14} = 2.1 \text{ m s}^{-2}$$

As an alternative method:

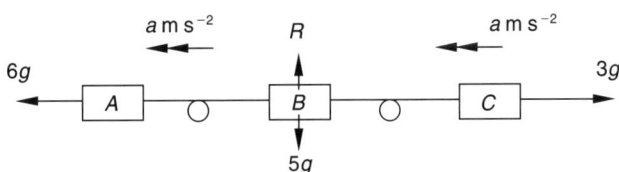

From the diagram

$$6g - 3g = 14a \quad \Rightarrow \quad a = \frac{3g}{14} = 2.1 \text{ m s}^{-2}$$

From equation 1:

$$P = 6(9.8 - 2.1) = 46.2 \text{ N}$$

From equation 3:

$$Q = 3(9.8 + 2.1) = 35.7 \text{ N}$$

With acceleration 2.1 m s^{-2}, the time to fall distance 0.6 m is

Using $s = ut + \frac{1}{2}at^2$
$0.6 = \frac{1}{2}(2.1)t^2 \quad \Rightarrow \quad t = 0.756 \text{ s}$

9 Two small identical uniform perfectly elastic smooth spheres A and B are moving directly towards each other with speeds u and v respectively. Show that after collision the direction of motion of both spheres is reversed and that the speeds of A and B, immediately after collision, are v and u respectively.

Sphere A is projected vertically upwards with speed 29.4 m s^{-1} from a point O and some time later sphere B is also projected vertically upwards from O with the same speed. Given that the spheres collide when they are both moving with speed 4.9 m s^{-1} and that the acceleration due to gravity is 9.8 m s^{-2}, find the time T taken for sphere B to reach the point of collision and the time that elapsed between the projection of the two spheres.

Given that O is on a perfectly elastic horizontal plane, determine the velocities of the spheres at a time t ($t > T$) after their first collision and before their second collision. Hence find the time that elapses before the spheres next collide. (AEB, 1990)

• For perfectly elastic spheres $e = 1$

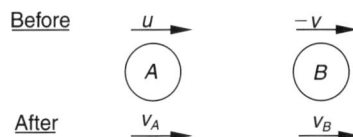

Law of restitution:

$$u + v = v_B - v_A \quad (1)$$

Conservation of linear momentum:

$$mu - mv = mv_B + mv_A \quad (2)$$

From equations 1 and 2:

$$2u = 2v_B \quad \Rightarrow \quad v_B = u, \quad v_A = -v$$

Hence directions are reversed and the speeds of A and B are v and u respectively.

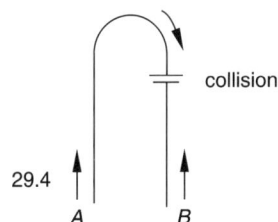

Since both balls were projected with the same velocity then they will have the same speed at the same height:

$$v_A = -4.9 \text{ m s}^{-1} \quad \text{and} \quad v_B = 4.9 \text{ m s}^{-1}$$

$v_A = 29.4 - 9.8t_A = -4.9 \quad \Rightarrow \quad t_A = 3.5 \text{ s}$
$v_B = 29.4 - 9.8t_B = 4.9 \quad \Rightarrow \quad t_B = 2.5 \text{ s}$

B takes 2.5 seconds to reach the point of collision,
$\Rightarrow \quad T = 2.5 \text{ s}$.
1 second elapsed between the projections.

After collision the velocities are reversed: A has velocity 4.9 m s^{-1}; B has velocity -4.9 m s^{-1}.

At time t seconds after the projection of B

$$\text{vel } A = 4.9 - (t - T)9.8 = 4.9(1 - 2t + 2T)$$

B takes 2.5 s to reach O, so

$$\begin{aligned} \text{vel } B &= 29.4 - (t - T - 2.5)9.8 \\ &= 4.9(11 - 2t + 2T) \end{aligned}$$

Particles collide:

$$\begin{aligned} 1 - 2t + 2T &= -11 + 2t - 2T \\ 12 = 4t - 4T &\Rightarrow \quad 3 + T = t \\ \Rightarrow \quad t &= 5.5 \text{ s} \end{aligned}$$

i.e. 3 seconds after the first collision.

10 Two particles of masses $4m$ and $6m$ respectively are attached one to each end of a light inextensible string. The string passes over a small smooth pulley and the particles are released from rest with the string vertical and taut. Find, in terms of m and g, the tension in the string during the subsequent motion. (WJEC, 1994)

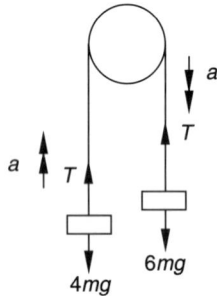

Let the tension in the string be T and the mutual acceleration a.

For mass $6m$: $6mg - T = 6ma$ (1)
For mass $4m$: $T - 4mg = 4ma$ (2)
Adding equations 1 and 2: $2mg = 10ma$.

Hence acceleration $= \dfrac{g}{5}$.

Tension $T = 4m(a + g) = \dfrac{24\,mg}{5}$.

11 A train starts from rest and moves with constant acceleration $\frac{1}{3}$ m s^{-2} for 2 minutes. For the next 4 minutes the train moves with zero acceleration, after which a uniform retardation of 2 m s^{-2} brings it to rest. Find the total distance travelled by the train from starting to stopping. (AEB, 1994)

- Initial velocity $= 0$ m s^{-1},
  acceleration $= \frac{1}{3}$ m s^{-2} for 120 s.

$$v = \tfrac{1}{3} \times 120 \text{ m s}^{-1} = 40 \text{ m s}^{-1}$$

Uniform retardation 2 m s^{-2} $\therefore$ Loses 40 m s^{-1} in 20 s.

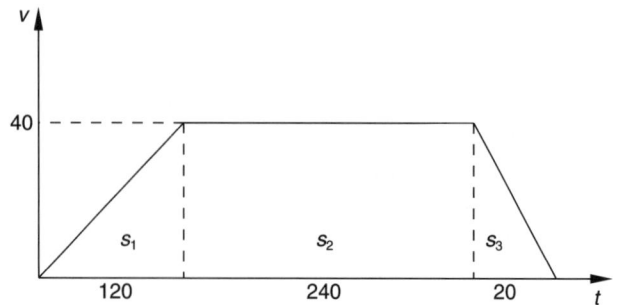

Distance travelled by the train is

$$\frac{40}{2}(380 + 240) \text{ metres} = 12.4 \text{ km}$$

Alternatively

While accelerating: $s_1 = \frac{1}{2}(\frac{1}{3})(120)^2 = 2400$ m
At constant speed: $s_2 = 40 \times 240 = 9600$ m
While retarding: $s_3 = \frac{1}{2}(2)(20)^2 = 400$ m

Total distance $= 12400$ m or 12.4 km.

12 A pile of mass 1000 kg is driven into the ground by a piledriver of mass 5000 kg, which is dropped vertically on to the pile from a height of 4 metres. The piledriver falls freely on to the pile, remaining in contact with the pile after the impact, and drives the pile 25 centimetres into the ground.

Calculate
(i) the speed with which the pile starts to penetrate the ground
(ii) the resistance of the ground to penetration, assuming this resistance to be constant.
[Take $g = 9.8$ m s^{-2}] (SEB, 1994)

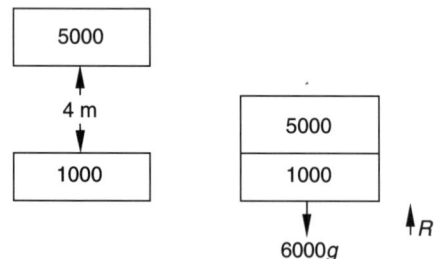

As the piledriver falls 4 m it gains velocity v where

$$v^2 = 2g(4) \quad \Rightarrow \quad v = \sqrt{8g}$$

By conservation of momentum the combined velocity v_1 immediately after impact is given by

$$6000\, v_1 = 5000\, v \quad \Rightarrow \quad v_1 = \tfrac{5}{6}\sqrt{8g}$$

If the resistance is R N then

$$\text{acceleration} = \frac{-(R - 6000g)}{6000}$$

Distance travelled is 0.25 m before coming to rest.

Using $v^2 = u^2 + 2as$

$$\frac{25}{36} \cdot 8g = 2 \times \frac{(R - 6000g)}{6000} \times (0.25)$$

$$R - 6000g = 12\,000 \times \frac{50g}{9} \quad \Rightarrow \quad R = 712\,000 \text{ N}$$

13 A circular groove, of radius 0.5 m, has been cut in a horizontal surface; the points A and B are at opposite ends of a diameter of the groove. Two small smooth spheres P, Q, with equal radii and masses 0.01 kg and 0.02 kg respectively, are constrained to move round the groove. Initially P and Q are at rest at A and B respectively. The sphere P is projected from A, along the groove, so that its speed immediately before collision with Q is u m s^{-1}. The coefficient of restitution between the spheres is $\frac{1}{8}$. Find, in terms of u, the speeds of P and Q immediately after collision.

(a) Assuming that the groove is smooth and that $u = 12$, find the time that elapses before the spheres next collide.

(b) Assume now that the groove is rough and that P and Q are at rest at A and B respectively. The sphere P is projected from A with speed 12 m s^{-1} and the speed of Q immediately after collision is 3 m s^{-1}. Find

    (i) the impulse acting on P during its collision with Q

    (ii) the work done by friction as P travelled from A to B. (WJEC, 1994)

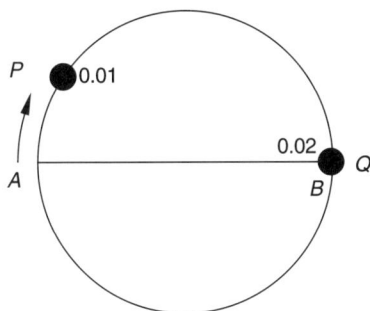

Just before collision, P has speed u, Q has speed 0. Just after collision, P has speed v_P, Q has speed v_Q. By conservation of linear momentum:

$$0.01u = 0.01v_P + 0.02v_Q \quad \Rightarrow \quad u = v_P + 2v_Q \quad (1)$$

Law of restitution: $v_Q - v_P = eu$ (2)

Adding equations 1 and 2:

$$3v_Q = u(e + 1)$$

$$\Rightarrow \quad v_Q = \frac{u}{3}(e + 1), \quad v_P = \frac{u}{3}(1 - 2e)$$

But $e = \frac{1}{8} \quad \Rightarrow \quad v_Q = \frac{3u}{8}, \quad v_P = \frac{u}{4}$

(a) If $u = 12$, $v_P = 3$ m s^{-1}, $v_Q = 4.5$ m s^{-1}. Relative velocity is 1.5 m s^{-1}, circumference of the circle = π m.

The particles collide again after $\dfrac{\pi}{1.5}$ s or 2.09 s.

(b) (i) Let P have a speed of v immediately before collision. Then the speed of Q after collision is

$$\frac{3v}{8} = 3 \text{ m s}^{-1}$$

Hence $v = 8$ m s^{-1}.

Impulse $= 0.02 \times 8 = 0.16$ N s

(ii) For P, loss of kinetic energy
$$= \tfrac{1}{2}(0.01)(12^2 - 8^2)$$
$$= 0.4 \text{ N m}$$
Work done by friction = loss of kinetic energy
$$= 0.4 \text{ N m or } 0.4 \text{ J}$$

Exercises

1 Three forces $\mathbf{F}_1$, $\mathbf{F}_2$ and $\mathbf{F}_3$ act on a particle and

$$\mathbf{F}_1 = (-3\mathbf{i} + 7\mathbf{j}) \text{ N} \quad \mathbf{F}_2 = (\mathbf{i} - \mathbf{j}) \text{ N}$$
$$\mathbf{F}_3 = (p\mathbf{i} + q\mathbf{j}) \text{ N}$$

(a) Given that this particle is in equilibrium, determine the value of p and the value of q.

The resultant of the forces $\mathbf{F}_1$ and $\mathbf{F}_2$ is $\mathbf{R}$.

(b) Calculate, in N, the magnitude of $\mathbf{R}$.

(c) Calculate, to the nearest degree, the angle between the line of action of $\mathbf{R}$ and the vector $\mathbf{j}$. (L, 1992)

2 Two identical uniform smooth spheres, A and B, each of mass m, moving in opposite directions with speeds u and $3u$ respectively, collide directly. Given that the sphere B is brought to rest, find the coefficient of restitution between the spheres. Find also, in terms of m and u,

(a) the magnitude of the impulse experienced by A

(b) the kinetic energy lost in the collision. (AEB, 1994)

3 Points P_1 and P_2 start together at the origin and move along the x-axis so that their respective displacements x_1 and x_2 in metres at time t seconds are given by

$$x_1 = t, \quad x_2 = t^2.$$

1   P_2 begins to move ahead of P_1.
2   P_1 and P_2 will coincide again at some later time.
3   P_1 and P_2 have equal velocities at time $t = 0.5$.

A 1, 2, 3 are correct   **B** only 1 and 2 are correct
C only 2 and 3 are correct   **D** only 1 is correct
E only 3 is correct   (L)

4 A small smooth sphere P of mass $6m$ moving on a smooth plane in a straight line with constant speed $8u$ collides directly with a small smooth sphere Q of the same radius but of mass $4m$ and moving in the same direction with speed $6u$. The direction of motion of the sphere Q is unchanged by the collision and immediately after collision it moves with speed $8u$. Find

(i) the speed of P immediately after collision
(ii) the coefficient of restitution.   (WJEC, 1993)

5 Given that s, defined by $s = ut + \frac{1}{2}at^2$, where u and a are constants, represents the displacement of a particle at time t show, by differentiation, that u is the velocity at time $t = 0$ and that the acceleration is equal to a.

A train starting from rest is uniformly accelerated during the first minute of its journey when it covers 600 m. It then runs at a constant speed until it is brought to rest in a distance of 1 km by applying a constant retardation.
(a) Find the maximum speed attained by the train.
(b) Determine the magnitude of the retardation.
(c) Given that the total journey time is 5 minutes determine the distance covered at constant speed.
(d) Given that the magnitude of the retardation, instead of being constant, is directly proportional to the speed and the train comes to rest from the same constant speed in a distance of 500 m, find the magnitude of the retardation in m s^{-2} when the train's speed is 10 m s^{-1}.   (AEB)

6 [In this question take the value of g to be 10 m s^{-2}.]

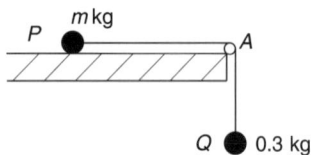

The diagram shows a particle P of mass m kg on a rough horizontal table. The coefficient of friction between P and the table is $\frac{1}{4}$. The particle P is attached to one end of a light inextensible string which passes over a small smooth pulley A fixed at the edge of the table. A particle Q of mass 0.3 kg is attached to the other end of the string. The system is released from rest with the string taut and Q hanging freely. Initially P is at a distance 0.7 m from the pulley. When P reaches the pulley its speed is 2 m s^{-1}. Find the value of m.

Find also the magnitude and direction of the resultant force exerted on the pulley by the string.

The system is returned to its initial position and is maintained in equilibrium in that position by a horizontal force of magnitude X newtons acting on the particle P in the direction of AP. Find the set of possible values of X.   (UCLES, 1991)

7 Two particles of masses 2 kg and 3 kg are connected by a light inelastic string passing over a smooth fixed peg. The particles start from rest and after 1 second are in the same horizontal line. Show that the magnitude of the velocity of the particles is

then 2 m s^{-1}, if the acceleration due to gravity is taken to be 10 m s^{-2}.

If the string then breaks describe the initial motion of each particle. State with reasons whether you would expect the particles to be in the same horizontal line again during this motion.   (OLE, 1988)

8 Three small bodies, A of mass m, B of mass $2m$, C of mass $3m$, are such that A is connected to B and B to C by two equal light inextensible strings. The bodies lie together at rest with A, B, and C in that order in a straight line on a smooth horizontal surface. Body C is given a speed u along the surface away from B and A. The strings do not impede the motion of the particles when slack.
(a) Find the speed with which all three bodies begin to move together.
(b) The motion of A, B, and C is in a line perpendicular to a fixed plane barrier and the body C is next in a perfectly elastic collision with the barrier. The coefficient of restitution between the bodies is $\frac{1}{2}$.
    (i) Show that after C's collision with B its speed is $\frac{1}{10}u$ and determine the speeds of A and B.
    (ii) Show that the total momentum of the bodies after C's collision with B is zero and say where the momentum of the bodies has been lost.
(c) What happens next? Give reasons. (OLE, 1993)

Brief Solutions

1 (a) $\mathbf{F}_1 + \mathbf{F}_2 + \mathbf{F}_3 = ((-3 + 1 + p)\mathbf{i} + (7 - 1 + q)\mathbf{j})$
$= 0$ when $p = 2, q = -6$
(b) $\mathbf{F}_1 + \mathbf{F}_2 = (-2\mathbf{i} + 6\mathbf{j})\,\text{N} = \mathbf{R}$
$|\mathbf{R}| = \sqrt{2^2 + 6^2} = 2\sqrt{10}\,\text{N}$
(c) If the angle between $\mathbf{R}$ and $\mathbf{j}$ is θ then
$\mathbf{R} \cdot \mathbf{j} = 2\sqrt{10} \cos \theta$

$$\begin{pmatrix} -2 \\ 6 \end{pmatrix} \cdot \begin{pmatrix} 0 \\ 1 \end{pmatrix} = 6$$

$$\therefore \cos \theta = \frac{6}{2\sqrt{10}}$$

$\Rightarrow \quad \theta = 18°$ to the nearest degree

2

Law of restitution:

$$e(3u + u) = v_A \qquad (1)$$

Law of conservation of linear momentum:

$$3mu - mu = mv_A \qquad (2)$$

From equation 2: $v_A = 2u \Rightarrow e = \frac{1}{2}$

(a) The impulse experienced by A is $3mu$
(b) Loss of kinetic energy
$= \frac{1}{2}m(3u)^2 + \frac{1}{2}m(u)^2 - \frac{1}{2}m(2u)^2 = 3mu^2$

3. $x_1 = t$, $x_2 = t^2$.
$x_1 = x_2$ when $t = 0$ or 1   (2 correct)
$\dot{x}_1 = 1$, $\dot{x}_2 = 2t$,
when $t = 0$, $\dot{x}_1 = 1, \dot{x}_2 = 0$ (1 incorrect)
when $t = 0.5$, $\dot{x}_1 = \dot{x}_2 = 1$ (3 correct)

Answer **C**

4. (i)

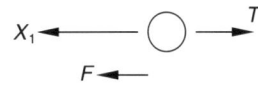

Conservation of linear momentum:

$6m(8u) + 4m(6u) = 6mv_P + 4m(8u)$
$\Rightarrow v_P = \frac{40}{6}u = \frac{20}{3}u$

(ii) Law of restitution:

$e(8u - 6u) = 8u - \frac{20}{3}u \Rightarrow e = \frac{2}{3}$

5. (a) $s = 600$, $u = 0$, $t = 60$, $s = \dfrac{u + v}{2}t$

$\Rightarrow v = 20$. Maximum speed $= 20$ m s^{-1}.
(b) $u = 20$, $s = 1000$, $v = 0$
$\Rightarrow a = 0.2$, retardation is 0.2 m s^{-2}
$v^2 = u^2 - 2as$
(c) $t = 100$ s for retardation $v = u - at$
Time at constant speed $= 300 - (100 + 60)$
    $= 140$ s.
Distance at constant speed $= 20 \times 140$
    $= 2800$ m.

(d) Retardation $= kv \Rightarrow v\dfrac{\mathrm{d}v}{\mathrm{d}s} = -kv$

$\Rightarrow \dfrac{\mathrm{d}v}{\mathrm{d}s} = -k$, $v = -ks + c$.

Initially, $v = 20$, $s = 0 \Rightarrow c = 20$.
Finally, $v = 0$, $s = 500 \Rightarrow k = \frac{1}{25}$.

Thus retardation $= \dfrac{v}{25}$.

When $v = 10$, retardation $= \frac{2}{5}$ m s^{-2}.

6.

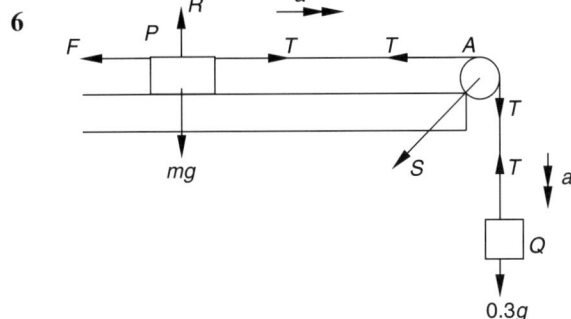

Put all the forces on to the diagram and then straighten the string

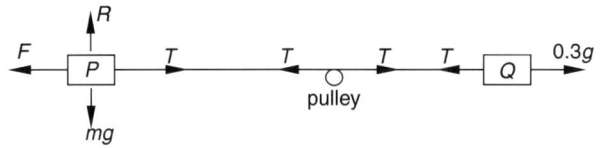

As you can see, all the tension forces cancel out

Whole system

$u = 0$, $v = 2$, $s = 0.7$, $v^2 = u^2 + 2as$
$\Rightarrow a = \frac{20}{7}$ m s^{-2}

Accelerating force

$3 - F = 3 - 2.5m = (m + 0.3)\frac{20}{7} \Rightarrow m = 0.4$

Resultant force on the pulley $= 2T \cos 45 = S$

For Q: $3 - T = 0.3a \Rightarrow T = \frac{15}{7}$

Force on the pulley $= \dfrac{15\sqrt{2}}{7}$ N at $45°$ below AP.

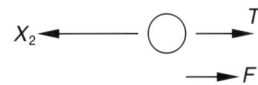

For equilibrium, $T = 0.3g = 3$, $F_{\max} = 1$.
Extreme conditions
Either $X_1 + F = T \Rightarrow X_1 = 2$
  or $F + T = X_2 \Rightarrow X_2 = 4$.

$\therefore 2 \leqslant X \leqslant 4$

7.

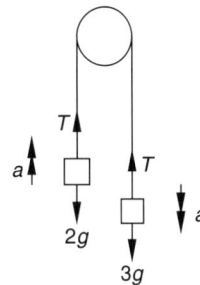

$a = \dfrac{g}{5} = 2$ m s^{-2}, $u = 0$, $t = 1$

$\Rightarrow v = 2$ m s^{-1} after 1 s.

When the string breaks, $T = 0$.
3 kg particle will fall under gravity, with starting velocity 2 m s^{-1}.
2 kg particle will continue upwards for $\frac{1}{5}$ s and *then* fall from rest under gravity.
$\Rightarrow$ They would not be on the same horizontal line again.

8

m 2m 3m
(A)(B)(C)

m 2m   3m
(A)(B)——(C)→ u

(a) Momentum of C is $3mu$.
When strings are taut the total momentum is
$3mu \Rightarrow$ all three are moving with common

speed $\dfrac{u}{2}$.

(b) (i) When C rebounds

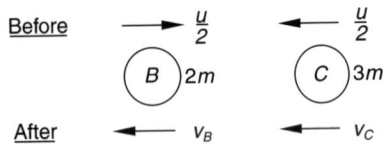

Before   → $\dfrac{u}{2}$ ← $\dfrac{u}{2}$

(B) 2m   (C) 3m

After   ← v_B ← v_C

Law of restitution:

$$\frac{1}{2}\left(\frac{u}{2} + \frac{u}{2}\right) = v_B - v_C \quad \Rightarrow \quad v_B - v_C = \frac{u}{2} \qquad (1)$$

Conservation of linear momentum ($\leftarrow$):

$$3m\frac{u}{2} - 2m\frac{u}{2} = 3mv_C + 2mv_B$$

$$\Rightarrow \quad \frac{u}{2} = 3v_C + 2v_B \qquad (2)$$

From equations 1 and 2:

$$v_C = \frac{-u}{10} \qquad v_B = \frac{2u}{5} \qquad v_A = \frac{u}{2}$$

in the directions shown.

(ii) Total momentum in initial direction is

$$m\frac{u}{2} - 2m\left(\frac{2u}{5}\right) + 3m\left(\frac{u}{10}\right) = 0$$

The loss of momentum is caused by the change
in direction of C striking the wall.

(c) B will strike A and C will move back towards
the wall.

22

Projectiles

AS Level			A Level			Topic	Date attempted	Date completed	Self-assessment
CORE	MODULAR	TRADITIONAL	CORE	MODULAR	TRADITIONAL				
	✓	✓		✓	✓	**2D motion under gravity**			
	✓	✓		✓	✓	**Maximum range and height**			
	✓	✓		✓	✓	**Equation of trajectory**			

Fact Sheet

Vector Approach

Newton's second law $F = ma$ can be generalized to $\mathbf{F} = m\ddot{\mathbf{r}}$ for motion in two or three dimensions.

For motion under gravity,

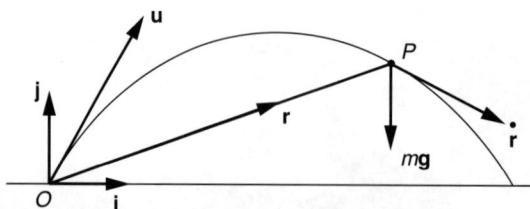

$$\mathbf{F} = m\ddot{\mathbf{r}} = m\mathbf{g} \quad \text{therefore} \quad \ddot{\mathbf{r}} = \mathbf{g}$$

Integrating, $\mathbf{v} = \dot{\mathbf{r}} = \mathbf{u} + t\mathbf{g}$, where $\mathbf{u}$ is the initial velocity.

Integrating, $\mathbf{r} = t\mathbf{u} + \frac{1}{2}t^2\mathbf{g} + \mathbf{r}_0$

where $\mathbf{r}_0$ is the position vector at $t = 0$ (usually zero).

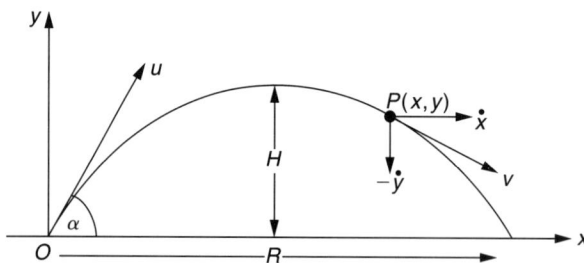

Cartesian Approach

If $\mathbf{i}$ and $\mathbf{j}$ are unit vectors horizontally and vertically upwards respectively, then

$$\mathbf{r} = x\mathbf{i} + y\mathbf{j}$$

and $\ddot{\mathbf{r}} = \mathbf{g}$ may be written $\ddot{x} = 0$, $\ddot{y} = -g$.

If the particle is projected with speed u at an angle α to the horizontal, then

$$\dot{x} = u \cos \alpha \tag{1}$$
$$\dot{y} = u \sin \alpha - gt \tag{2}$$
$$x = (u \cos \alpha)t \tag{3}$$
$$y = (u \sin \alpha)t - \tfrac{1}{2}gt^2 \tag{4}$$

taking $x = 0$, $y = 0$ as the point of projection.

Velocities

From equation 2:

$$\dot{y}^2 = u^2 \sin^2 \alpha - 2ugt \sin \alpha + g^2 t^2$$
$$= (u \sin \alpha)^2 - 2g(ut \sin \alpha - \tfrac{1}{2}gt^2)$$
$$\dot{y}^2 = (u \sin \alpha)^2 - 2gy \tag{5}$$
$$\dot{x}^2 + \dot{y}^2 = u^2 \cos^2 \alpha + u^2 \sin^2 \alpha - 2gy$$
$$\text{or} \qquad v^2 = u^2 - 2gy \tag{6}$$

where v is the speed at any height y.

This can be rewritten as $\frac{1}{2}mu^2 - \frac{1}{2}mv^2 = mgy$, or

loss of kinetic energy = gain in potential energy

This is the conservation of energy equation for a particle.

Range R on the Horizontal Plane

The range R is the horizontal displacement (x) of the particle when its vertical displacement (y) is again zero.
From equation 4, $y = 0$ when $t = 0$ (initially), and when

$$t = \frac{2u \sin \alpha}{g}$$

This is the 'time of flight'.

Substituting in equation 3 gives

$$R = u \cos \alpha \left(\frac{2u \sin \alpha}{g} \right) = \frac{2u^2 \sin \alpha \cos \alpha}{g}$$
(7)

This can be changed to $R = \dfrac{u^2 \sin 2\alpha}{g}$ to find the two possible (complementary) angles of projection for a given range.

Maximum Height H

Using equation 5, at maximum height, $\dot{y} = 0$,

$$u^2 \sin^2 \alpha = 2gH \quad \Rightarrow \quad H = \frac{u^2 \sin^2 \alpha}{2g}$$
(8)

There are several alternative methods for finding H and R.

Equation of the Trajectory

The trajectory is the parabolic path followed by the particle.
Eliminate t from equations 3 and 4.

From equation 3:

$$t = \frac{x}{u \cos \alpha}$$

In equation 4:

$$y = u \sin \alpha \left(\frac{x}{u \cos \alpha} \right) - \frac{gx^2}{2u^2 \cos^2 \alpha}$$

$$= x \tan \alpha - \frac{gx^2}{2u^2} \sec^2 \alpha$$

or $\quad y = x \tan \alpha - \dfrac{gx^2}{2u^2} (1 + \tan^2 \alpha)$
(9)

This may be used in questions where a particle has to pass through, or strike, a given point.

Differentiating equation 9 with respect to x gives

$$\frac{dy}{dx} = \tan \alpha - \frac{gx}{u^2} (1 + \tan^2 \alpha)$$

This gives the direction of motion at a given horizontal displacement.

Worked Examples

1 A cricketer hits a ball so that it first lands on the ground at a point P at a distance 75 m from him. At the highest point of its path the ball reaches a height of 20 m. Assuming that the ball is projected from ground level determine
   (a) the horizontal and vertical components of the velocity of projection
   (b) the tangent of the angle between the horizontal and the direction of motion of the ball 1 s after it has been hit
   (c) the furthest distance from P that a fielder who can run at 8 m s^{-1} can stand, in order that, starting when the ball is hit, he can arrive at P before the ball lands
   (d) the tangent of a different angle of projection which is such that the ball, when projected with the same initial speed, again first lands at P.
   (Let $g = 10$ m s^{-2}).   (AEB, 1982)

• (a) Let the initial components of velocity be v_x and v_y.

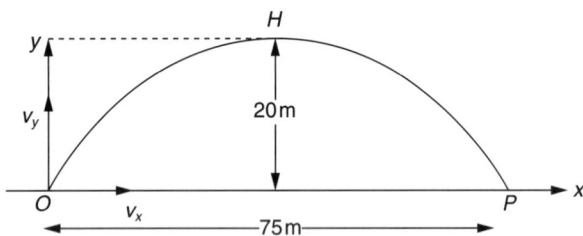

   At the greatest height, the vertical velocity $\dot{y} = 0$.
   Using $\dot{y}^2 = v_y^2 - 2gy$, $v_y^2 = 2g(20)$
   $\Rightarrow \quad v_y = 20$ m s^{-1}.

   Time taken to reach highest point $= \dfrac{v_y}{g} = 2$ s.

   Therefore time of flight $= 4$ s (symmetry).

   $v_x = \dfrac{75}{4} = 18.75$ m s^{-1} (remains constant).

   (b) After one second, $\dot{y} = 20 - g = 10$ m s^{-1}.

   The tangent of the angle between the path and the horizontal after 1 second $= \dfrac{10}{18.75}$ or $\dfrac{8}{15}$.

   (c) Since the time of flight is 4 s, the fielder can be up to 32 m from P.
   (d) For any given range there are (usually) two angles of projection which are complementary.

   For given projection $\tan \alpha = \dfrac{20}{18.75} = \dfrac{16}{15}$.

   Second angle has a tangent of $\dfrac{15}{16}$.

2 Two particles A and B are projected simultaneously under gravity; A from a point O on horizontal ground and B from a point 40 m vertically above O. B is projected horizontally with speed 28 m s^{-1}. If the particles hit the ground simultaneously at the same point, taking g as 9.8 m s^{-2}, calculate,
   (a) the time taken for B to reach the ground and the horizontal distance it has then travelled
   (b) the magnitude and direction of the velocity with which A is projected.
   Show that, just prior to hitting the ground, the directions of motion of A and B differ by about $18\frac{1}{2}°$.   (SUJB)

• Take the horizontal and vertically upward displacements from O as x and y respectively.

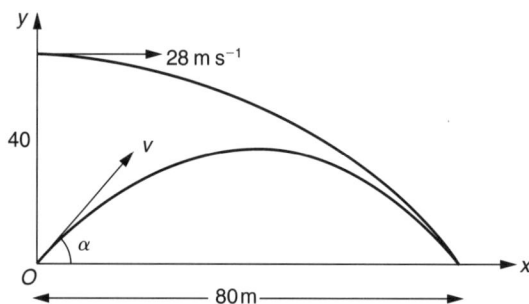

   (a) *For B:* $\dot{x}_B = 28, \quad \dot{y}_B = -gt$
          $x_B = 28t, \quad y_B = 40 - \tfrac{1}{2}gt^2$
   When B strikes the ground $y_B = 0$, so

   $t^2 = \dfrac{40}{4.9}, t = \dfrac{20}{7}$ s (or 2.86 s)

   At this time $x_B = 28t = 80$ m.

   B strikes the ground after 2.86 s having travelled 80 m horizontally.

   (b) *For A:* Let the horizontal and vertically upwards components of the velocity of projection be $v \cos \alpha$ and $v \sin \alpha$. Then

   $x_A = (v \cos \alpha)t, \qquad y_A = (v \sin \alpha)t - \tfrac{1}{2}gt^2$

   Since A and B travel the same distance horizontally in the same time, in order to collide they must have the same horizontal speed, i.e. $v \cos \alpha = 28$.

   When $t = \dfrac{20}{7}, y_A = 0$, so
          $v (\sin \alpha)t = \tfrac{1}{2}gt^2$ (1)

          $v \sin \alpha = \dfrac{9.8}{2}\left(\dfrac{20}{7}\right) = 14$ (2)

   and $\dot{y}_B = -9.8\left(\dfrac{20}{7}\right) = -28$

   Therefore, from equations 1 and 2, $\tan \alpha = \tfrac{1}{2}$,
   $v = \sqrt{(14^2 + 28^2)} = 14\sqrt{5}$ m s^{-2}.

The velocity of projection $= 31.3$ m s^{-1} at $26.6°$ to the ground.

Just before striking the ground, velocity of A makes the same angle as the angle of projection but below the horizontal, and velocity of B

makes an angle $\arctan\left(\dfrac{\dot{y}_B}{\dot{x}_B}\right) = 45°$ below the

horizontal. Therefore the directions of motion of A and B differ by about $18\frac{1}{2}°$.

3 Two boys, John and Paul, play football, kicking a ball on level ground against a wall of height 5 m. John can kick the ball with maximum speed 12 m s^{-1} and Paul with maximum speed 8 m s^{-1}.
   (a) Calculate the heights that the ball would reach if projected vertically with these speeds. Use your answers to decide whether each of the boys could kick the ball over the wall in any position.
   (b) John kicks the ball with maximum speed, at $60°$ to the horizontal from a point 10 m from the wall, so that the ball travels in a plane perpendicular to the wall. Determine the ball's height at the wall to decide whether it will pass over the wall.　　　　($g = 10$ m s^{-2})
   　　　　　　　　　　　　　　　　　　(OLE, 1993)

• (a) *For John:* $u = 12$ m s^{-1}, $v = 0$, $g = 10$ m s^{-2}
   At maximum height

   $2gh = u^2 = 144$
   $h = 7.2$ m

   Since this is higher than the wall, John could kick the ball over the wall.
   For Paul: $u = 8$ m s^{-1}, $v = 0$, $g = 10$ m s^{-2}
   At maximum height

   $2gh = u^2 = 64$
   $h = 3.2$ m

   Paul cannot kick the ball over the wall.

   (b) Horizontal component of velocity is

   $12\cos 60 = 6$ m s^{-1}　*Remains constant*

   Time taken to reach the wall $= \frac{5}{3}$ s.

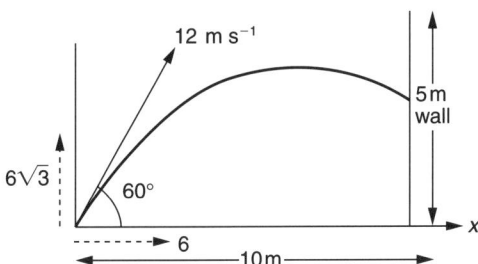

The initial vertical component of velocity $= 6\sqrt{3}$ m s^{-1}

$h = 6\sqrt{3}\,t - \frac{1}{2}gt^2$

When $t = \frac{5}{3}$

$h = 6\sqrt{3} \cdot \frac{5}{3} - 5(\frac{5}{3})^2$
$= 3.43$ m

Hence the ball will not clear the wall.

4 A particle is projected with speed v at an angle of elevation θ above the horizontal, and moves freely under gravity. Prove that the range on a horizontal plane through the point of projection is

$$\frac{v^2 \sin 2\theta}{g}$$

The point O is situated on the ground 10 m in front of a vertical wall. A particle is projected from O, with speed v m s^{-1} at an angle of elevation $30°$ above the horizontal, in the vertical plane through O perpendicular to the wall. The particle hits the wall while still moving upwards. Show that, correct to 3 significant figures, $v > 15.2$.

At the impact with the wall, the vertical component of the particle's velocity is unchanged. Show that the time from the instant of projection until the particle hits the ground is $\frac{1}{10}v$ s.

It is also given that, at the impact with the wall, the horizontal components of the particle's velocity is reversed in direction and halved in magnitude. The particle returns to O without first hitting the ground. Find the value of v, giving your answer correct to 3 s.f.　　　　($g = 10$ m s^{-2})
　　　　　　　　　　　　　　　　　(UCLES, 1993)

•

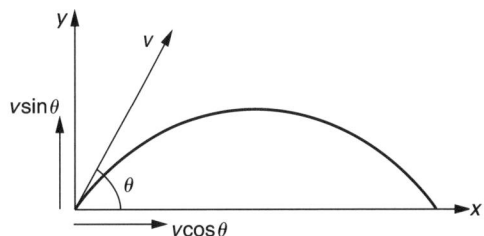

At time t the vertical velocity is $v \sin \theta - gt$.
At the maximum height

$$\text{velocity} = 0 \quad \Rightarrow \quad t = \frac{v \sin \theta}{g}$$

Hence time of flight $= \dfrac{2v \sin \theta}{g}$.

For horizontal motion, velocity remains constant at
$v \cos \theta$

Horizontal range $= v \cos \theta \cdot \dfrac{2v \sin \theta}{g}$

$$= \dfrac{v^2}{g}(2 \sin \theta \cos\theta)$$

$$= \dfrac{v^2 \sin 2\theta}{g}$$

When $\theta = 30°$

initial horizontal velocity $= \dfrac{v\sqrt{3}}{2}$

initial vertical velocity $= \dfrac{v}{2}$

The particle has travelled 10 m horizontally when it hits the wall

Time taken to reach the wall is

$$\dfrac{10}{v\sqrt{3}/2}\,\text{s} = \dfrac{20}{\sqrt{3}v}\,\text{s}$$

At this time the vertical velocity is

$$v \sin 30 - \dfrac{20g}{\sqrt{3}v} = \dfrac{v}{2} - \dfrac{200}{\sqrt{3}v}$$

Since the particle is still moving upwards

$$\dfrac{v}{2} > \dfrac{200}{\sqrt{3}v} \quad \Rightarrow \quad v > 15.2 \text{ m s}^{-1}$$

If the vertical velocity remains unchanged, the time of flight is the time taken for a particle projected

vertically upwards with velocity $\dfrac{v}{2}$ to return to the ground

Time taken is

$$2 \times \dfrac{v/2}{g} = \dfrac{v}{10}\,\text{s}.$$

After rebound, horizontal velocity is $\dfrac{v\sqrt{3}}{4}$ m s^{-1}.

In $\left(\dfrac{v}{10} - \dfrac{20}{\sqrt{3}v}\right)$ s the horizontal distance travelled

is 10 m (since it returns to O)

$$\therefore v = 18.6 \text{ m s}^{-1}$$

5 A shot-putter (illustrated in the diagram) finds that he can project a shot a maximum distance of 20 m. He wishes to determine the shot's projection speed. (The shot will be modelled as a particle travelling in a vacuum.)
(a) For his first attempt he decides to consider the shot to be projected from the level ground.

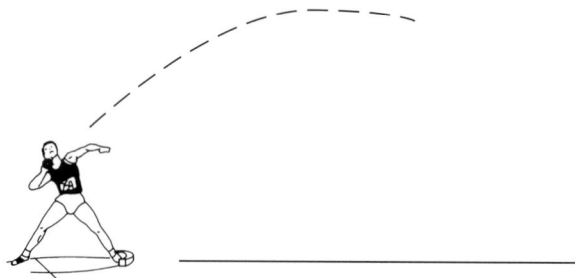

Using this model determine the minimum possible speed of projection V_1.
(b) Not satisfied with this value for the projection speed, he decides to take account of the height from which the shot is projected. He measures this to be 1.5 m.
   (i) Show that the range R m of the shot along the level ground, when projected from a height of 1.5 m, satisfies the equation

$$10R^2 \tan^2 \theta - 2v^2R \tan \theta + 10R^2 - 3v^2 = 0$$

where v m s^{-1} is the speed of projection of the shot and θ is its angle of projection above the horizontal.
   (ii) If the equation, regarded as a quadratic in $\tan \theta$, gives real values of $\tan \theta$, show that

$$R \leqslant \dfrac{v}{10}\sqrt{(30 + v^2)}$$

   (iii) Use the result in (ii) to determine the minimum speed of projection V_2 for the shot in this model.
   (iv) If the shot is projected with speed V_2, would you expect that more than one angle of projection would be possible for the range of 20 m?
   $(g = 10 \text{ m s}^{-2})$ (OLE, 1992)

• (a)

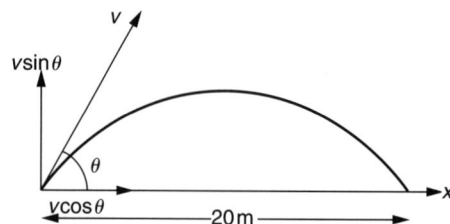

Time taken for range is $\dfrac{2v \sin \theta}{g}$ $(g = 10 \text{ m s}^{-2})$.

Horizontal range $= \dfrac{2v \sin \theta}{g} \cdot v \cos \theta = 20$ m

$$v^2 = \dfrac{20g}{\sin 2\theta}$$

For minimum speed $\sin 2\theta = 1$
$$\Rightarrow \quad v_1 = \sqrt{20g} \text{ m s}^{-1}$$

(b)

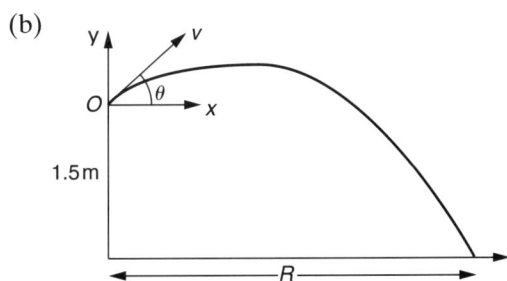

Equation of the trajectory is

$$y = x \tan \theta - \frac{gx^2 \sec^2 \theta}{2v^2}$$

(i) At the point of impact $y = -1.5, x = R$. Therefore

$$-1.5 = R \tan \theta - \frac{gR^2(1 + \tan^2 \theta)}{2v^2}$$

$$-3v^2 = 2v^2 R \tan \theta - gR^2 - gR^2 \tan^2 \theta$$

Putting $g = 10$

$$10R^2 \tan^2 \theta - 2v^2 R \tan \theta + 10R^2 - 3v^2 = 0 \quad (1)$$

(ii) Solving equation 1 for values of $\tan \theta$

$$\tan \theta = \frac{2v^2 R \pm \sqrt{4v^4 R^2 - 40R^2(10R^2 - 3v^2)}}{20R^2}$$

$$= \frac{v^2 \pm \sqrt{v^4 + 30v^2 - 100R^2}}{10R} \quad (2)$$

Since $\tan \theta$ is real

$$v^4 \geq 100R^2 - 30v^2 \qquad `b^2 \geq 4ac'$$

i.e. $R \leq \dfrac{\sqrt{v^2(v^2 + 30)}}{10} = \dfrac{v}{10}\sqrt{v^2 + 30}$

(iii) When $R = 20$

$$200 \leq v\sqrt{v^2 + 30}$$
$$v^4 + 30v^2 - 40\,000 \geq 0$$
$$(v^2 + 15)^2 \geq 40\,225 \quad \text{'Completing the square'}$$
$$v^2 > 185.6 \Rightarrow v \geq 13.6 \text{ m s}^{-1}$$

Minimum speed of projection

$$V_2 = 13.6 \text{ m s}^{-1}$$

(iv) With this speed

$$\tan \theta = \frac{v^2}{10R} \quad \text{(from equation 2)}$$

$$\Rightarrow \theta = 42.8°$$

i.e. there is only one possible angle θ

6

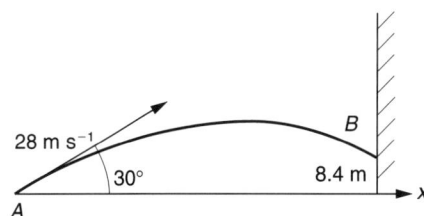

A golf ball is driven from a point A with a velocity which is of magnitude 28 m s^{-1} and at an angle of elevation of 30°. The ball moves freely under gravity. On its downward flight, the ball hits a vertical wall, at a point B, which is 8.4 m above the level of A, as shown in the diagram.

Calculate
(a) the greatest height achieved by the ball above the level of A
(b) the time taken by the ball to reach B from A.
By using the principle of conservation of energy, or otherwise,
(c) find the speed, in m s^{-1} to 1 decimal place, with which the ball strikes the wall.

(Use $g = 9.8$ m s^{-2}) (L, 1992)

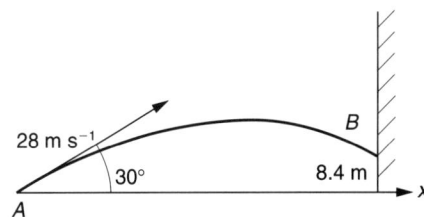

(a) Initially vertical velocity $= 28 \sin 30 = 14$ m s^{-1}. Maximum height occurs when vertical velocity $= 0$.
Using $v^2 = u^2 - 2gh$

height (max) $= \dfrac{(14)^2}{2g} = 10$ m above A

(b) Using $s = ut + \frac{1}{2}at^2, h = 8.4$:

$$h = 14t - \tfrac{1}{2}(9.8)t^2 \Rightarrow 4.9t^2 - 14t + 8.4 = 0$$
$$\Rightarrow 7t^2 - 20t + 12 = 0$$
$$\Rightarrow (7t - 6)(t - 2) = 0$$

Taking the larger value of t, the time to reach B is 2 s.

(c) Gain in gravitational energy $= (9.8 \times 8.4m)$ J. By conservation of energy

$$\tfrac{1}{2}m(28)^2 = \tfrac{1}{2}mv^2 + m(82.32)$$
$$v^2 = 619.36 \Rightarrow v = 24.9$$

The ball strikes the wall with speed 24.9 m s^{-1}

Using $x = (28 \cos 30)T$,
$y = (28 \sin 30)T - 4.9T^2$

7.3

7 A particle P is projected, from a point O on horizontal ground, with speed V at an angle θ above the horizontal, where $\tan\theta = \frac{1}{3}$ The particle passes through the point with coordinates $(3a, \frac{3}{4}a)$ relative to horizontal and vertical axes at O in the plane of motion. Show that $V^2 = 20ga$.

A particle Q is projected from O at the instant when P is moving horizontally. It strikes the ground at the same place and at the same instant as P. Show that the speed of projection of Q is

$$\sqrt{\left(\frac{145ga}{2}\right)}$$ and find the tangent of the angle of

projection.

$(g = 10 \text{ m s}^{-1})$ (UCLES, 1988)

•

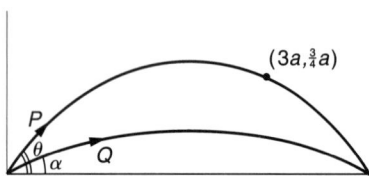

$\tan\theta = \frac{1}{3}$

$\cos\theta = \dfrac{3}{\sqrt{10}}$

$\sin\theta = \dfrac{1}{\sqrt{10}}$

For P
The equation of the trajectory is

$$y = x\tan\theta - \frac{gx^2(1+\tan^2\theta)}{2V^2}$$

At point $(3a, \frac{3}{4}a)$

$$\frac{3}{4}a = 3a\frac{1}{3} - \frac{9a^2g(1+\frac{1}{9})}{2V^2}$$

$$\tfrac{1}{2}V^2 = 10ag$$
$$V^2 = 20ag$$

Range of $P = V\cos\theta \cdot t_R$

Range of $Q = U\cos\alpha \cdot \dfrac{t_R}{2}$

where U is the speed of projection of Q and α is the angle of projection.

Since the ranges are equal

$$2 \cdot \sqrt{20ag} \cdot \frac{3}{\sqrt{10}} = U\cos\alpha = 6\sqrt{2ga} \qquad (1)$$

Time for maximum height of $Q = \frac{1}{2}$ time for maximum height of P

$$\frac{V\sin\theta}{2g} = \frac{U\sin\alpha}{g} \quad \Rightarrow \quad U\sin\alpha = \frac{\sqrt{2ga}}{2} \qquad (2)$$

From equations 1 and 2: $\tan\alpha = \frac{1}{12}$.

$$U^2 = 72ga + \frac{ga}{2} = \frac{145ga}{2}$$

Speed of projection of Q is $\sqrt{\left(\dfrac{145ga}{2}\right)}$ at angle of projection arctan $(\frac{1}{12})$.

8 The muzzle speed of a gun is v and it is desired to hit a small target at a horizontal distance a away and at a height b above the gun. Show that this is impossible if $v^2(v^2 - 2gb) < g^2a^2$, but that, if $v^2(v^2 - 2gb) > g^2a^2$, there are two possible elevations for the gun.

Show that if $v^2 = 2ag$ and $b = \frac{3}{4}a$, there is only one possible elevation, and find the time taken to hit the target.   (OLE)

• Let the angle of elevation be α.
The trajectory equation is

$$y = x\tan\alpha - \frac{gx^2(1+\tan^2\alpha)}{2v^2}$$

For the shell to hit the target at $x = a$, $y = b$,

$$b = a\tan\alpha - \frac{ga^2(1+\tan^2\alpha)}{2v^2}$$

$$\Rightarrow \quad ga^2\tan^2\alpha - 2av^2\tan\alpha + 2v^2b + ga^2 = 0 \quad (1)$$

This is a quadratic equation in $\tan\alpha$, which must have real solutions for the target to be hit.

The target cannot be hit if the discriminant < 0, i.e. if

$$4a^2v^4 - 4(ga^2)(2v^2b + ga^2) < 0$$
$$v^4 - 2gbv^2 - g^2a^2 < 0$$
$$\Rightarrow \quad v^2(v^2 - 2gb) < g^2a^2$$

If the discriminant > 0, i.e. if $v^2(v^2 - 2gb) > g^2a^2$, then equation 1 has two real distinct solutions for $\tan\alpha$ and hence for the elevation.
If $v^2 = 2ga$ and $b = \frac{3}{4}a$, equation 1 becomes

$$ga^2\tan^2\alpha - 4ga^2\tan\alpha + 3ga^2 + ga^2 = 0$$
$$\Rightarrow \quad \tan^2\alpha - 4\tan\alpha + 4 = 0 \quad \Rightarrow \quad \tan\alpha = 2$$

i.e. only one possible elevation, arctan (2).

Horizontal displacement

$$a = (v\cos\alpha)t = \sqrt{(2ga)}\left(\frac{1}{\sqrt{5}}\right)t$$

$$\Rightarrow \quad t = \frac{a\sqrt{5}}{\sqrt{(2ga)}} = \sqrt{\left(\frac{5a}{2g}\right)}$$

Time taken for the shell to reach the target is

$$\sqrt{\left(\frac{5a}{2g}\right)} \text{ s.}$$

9 A particle is projected from a point O with a velocity which has a horizontal component U and a vertically upward component V. Show that R, the range on a horizontal plane through O, and H, the maximum height of the particle above the plane, are given by

$$R = \frac{2UV}{g} \quad \text{and} \quad H = \frac{V^2}{2g}$$

A tennis ball is served horizontally from a point which is 2.5 m vertically above a point A. The ball first strikes the horizontal ground through A at a distance 20 m from A. Show that the ball is served with speed 28 m s^{-1}.

During its flight the ball passes over a net which is a horizontal distance 12 m from A. Find the vertical distance of the ball above the horizontal ground at the instant when it passes over the net. ($g = 9.8$ m s^{-2}) (AEB, 1994)

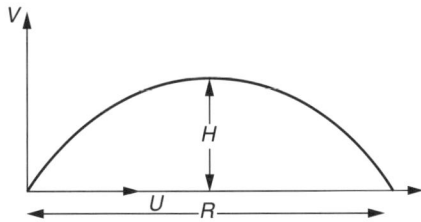

At maximum height, vertical velocity $= 0$

$$\Rightarrow \quad V^2 = 2gH \text{ or } H = \frac{V^2}{2g} \text{ after time } \frac{V}{g}$$

$\Rightarrow$ horizontal range after time $\dfrac{2V}{g}$ is $U\dfrac{2V}{g}$ or

$\dfrac{2UV}{g}$.

$$\therefore R = \frac{2UV}{g} \quad H = \frac{V^2}{2g}$$

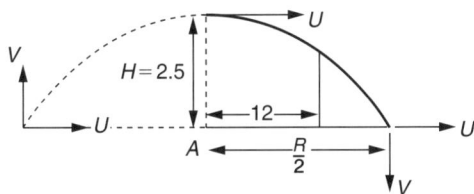

From above, $H = 2.5$, $\quad \dfrac{R}{2} = 20$ m

$$\therefore \frac{UV}{g} = 20, \quad 2.5 = \frac{V^2}{2g}$$

$$\therefore V = 7 \text{ m s}^{-1} \quad U = 28 \text{ m s}^{-1}$$

Hence the ball is served with speed $U = 28$ m s^{-1}.

When the horizontal distance is 12 m from A, time taken $= \frac{12}{28} = \frac{3}{7}$.

Height above the ground is $2.5 - \dfrac{9.8t^2}{2} = 1.6$ m.

10 (a) A particle is projected from a point O with speed u at an angle α above the horizontal.
  (i) Write down expressions for the horizontal and vertical displacements of the particle from O at time t after projection.
  (ii) Deduce that the particle first hits the horizontal plane through O at a distance

$$\frac{u^2 \sin 2\alpha}{g}$$

  (iii) Show that the greatest height reached by the particle above the level of O is

$$\frac{u^2 \sin^2 \alpha}{2g}.$$

(b) A particle P is projected from a point O with speed $\sqrt{12ga}$ at the angle which gives maximum range on the horizontal plane through O. Find the tangent of the angle between the velocity of P and the horizontal

at time $\sqrt{\left(\dfrac{3a}{8g}\right)}$ after projection.

(c) A particle Q is projected with speed $\sqrt{12ga}$ from a point A on the horizontal floor of a

room with a horizontal ceiling at a height $\dfrac{3a}{2}$

above the floor. Assuming that Q must not hit the ceiling, find the maximum value of the distance from A of the point at which Q first hits the floor. ($g = 9.8$ m s^{-2}) (WJEC, 1994)

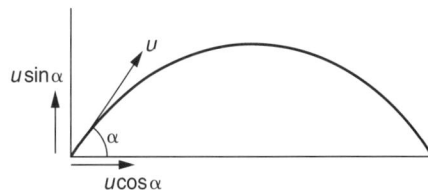

(a) (i) Horizontal displacement $= u \cos \alpha \,.\, t$.

Vertical displacement $= u \sin \alpha \,.\, t - \dfrac{g}{2}t^2$.

(ii) When vertical height $= 0$, $t = 0$ or $\dfrac{2u \sin \alpha}{g}$.

Horizontal range is

$$u \cos \alpha \,.\, t = \frac{2u^2 \sin \alpha \cos \alpha}{g} = \frac{u^2 \sin 2\alpha}{g}$$

(iii) Maximum height occurs when $v = 0$, that is

$$s = \frac{u^2 \sin^2 \alpha}{2g}$$

(b) Angle of projection for maximum range = 45°.

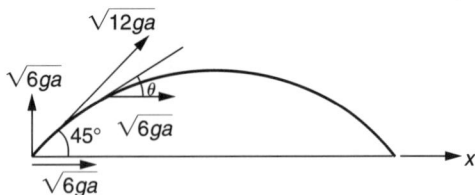

After time $\sqrt{\dfrac{3a}{8g}}$ horizontal velocity = $\sqrt{6ga}$

Vertical velocity = $\sqrt{6ga} - g\sqrt{\dfrac{3a}{8g}}$

$$= \sqrt{6ga} - \tfrac{1}{4}\sqrt{6ga}$$
$$= \tfrac{3}{4}\sqrt{6ga}$$

Direction = $\tan^{-1}(\tfrac{3}{4})$ above the horizontal
$\Rightarrow \quad \tan\theta = \tfrac{3}{4}$.

(c) If the initial speed is $\sqrt{12ga}$, the maximum height reached is $6a\sin^2\alpha$ where α is the angle of projection.

If maximum possible height is $\dfrac{3a}{2}$ then

$\sin^2\alpha = \tfrac{1}{4} \quad \sin\alpha = \tfrac{1}{2} \quad \alpha = 30°$

Maximum distance from A with angle of projection θ is

$$\frac{u^2\sin 2\theta}{g}$$

$\theta \leqslant 30° \quad \Rightarrow \quad 2\theta \leqslant 60°$
Maximum distance from A when Q first hits the

floor is $\dfrac{12ga\sin 60°}{g} = 6\sqrt{3}a$

Exercises

1 A particle is projected with velocity 40 m s⁻¹ at an angle of elevation arctan $\tfrac{3}{4}$. After 2 seconds the particle is moving at an angle θ to the horizontal.

Then $\tan\theta =$

A $\tfrac{4}{3}$ **B** $\tfrac{16}{7}$ **C** $\tfrac{7}{16}$ **D** $\tfrac{3}{4}$ **E** $\tfrac{1}{8}$

2 A tennis player hits a ball at a point O, which is at a height of 2 m above the ground and at a horizontal distance 4 m from the net, the initial speed being in a direction of 45° above the horizontal in a vertical plane perpendicular to the net. The ball just clears the net which is 1 m high.

(a) Taking the horizontal and vertical through O in the plane of motion as axes of x and y respectively, show that the equation of the path of the ball may be written in the form

$$y = x - \frac{5x^2}{16}, \text{ assuming that the only force}$$

acting on the ball is that due to gravity.

Find the initial speed of the ball.

Find, also,
(b) the distance from the net at which the ball strikes the ground
(c) the magnitude and direction of the velocity with which the ball strikes the ground.
All answers should be given correct to two significant figures. $(g = 10 \text{ m s}^{-2})$
(SUJB)

3 [In this question take the value of g to be 10 m s^{-2}.]
A small stone is fired from a catapult. At time $t = 0$, the stone leaves the catapult at the point O, with velocity 30 m s⁻¹ at an angle of 30° above the horizontal. The stone moves freely under gravity and the effect of air-resistance may be ignored. The stone passes through a point A and then through a point B, both points being at a height of 5 m above the level of O. Find the times at which the stone passes through A and B, and find the distance AB, giving three significant figures in your answers.

Find the maximum height above AB reached by the stone.

Find the magnitude and direction of the velocity of the stone as it passes through B. (UCLES, 1991)

4 A point O is vertically above a fixed point A of a horizontal plane. A particle P is projected from O with speed $5V$ at an angle $\cos^{-1}(\tfrac{3}{5})$ above the horizontal and hits the plane at a point B at a

distance $\dfrac{48V^2}{g}$ from A.

(i) Show that the height of O above A is $\dfrac{64V^2}{g}$.

(ii) Find the distance of P from O when it is directly level with it.
A second particle is now projected with speed $24W$ from O at an angle of α above the horizontal and it also hits the plane at B. Find an equation involving V, W, and α.

Given that one value of α is 45° find W in terms of V and show that the other value of α is such that

$$7\tan^2\alpha - 6\tan\alpha - 1 = 0 \qquad \text{(WJEC, 1992)}$$

5 [In this question you should assume $g = 9.8 \text{ m s}^{-2}$.]

At time $t = 0$ a particle is projected from a point O with speed 49 m s⁻¹ and in a direction which makes an acute angle θ with the horizontal plane through O. Find, in terms of θ, an expression for R, the horizontal range of the particle from O.

The particle also reaches a height of 9.8 m above the horizontal plane through O at times t_1 seconds and t_2 seconds. Find, in terms of θ, expressions for t_1 and t_2.

Given that $t_2 - t_1 = \sqrt{17}$ seconds, find θ.

Hence show that $R = \dfrac{245\sqrt{3}}{2}$.   (AEB, 1992)

6

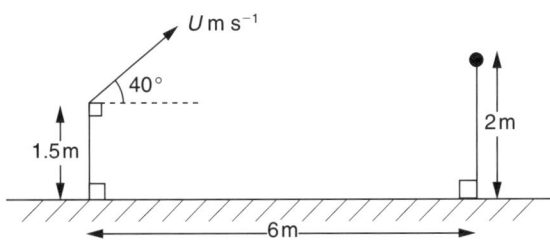

In a fairground game, a child throws a small ball from a height of 1.5 m above level ground, aiming at a small target. The target is on top of a vertical pole of height 2 m, and the horizontal displacement of the child from the pole is 6 m, as shown in the figure. The initial velocity of the ball has magnitude U m s^{-1} and angle of elevation 40°. The ball moves freely under gravity.
(a) For $U = 10$, find the greatest height above the ground reached by the ball, giving your answer in metres to 1 decimal place.
(b) Calculate, to 1 decimal place, the value of U for which the ball hits the target.
    ($g = 9.8$ m s^{-2})
    (L, 1993)

7 O is a point on horizontal ground. D is a point vertically above O. A particle A is projected from O with a speed u m s^{-1} at an angle of elevation α. Simultaneously, a second particle B is projected horizontally from D with a speed v m s^{-1} on the same side of OD as A. Show that, if the particles collide, then $v = u \cos \alpha$. Find a second condition which must also be satisfied.

Given that $u = 51$, $v = 45$, $\tan \alpha = \frac{8}{15}$ and $d = 60$ satisfy these conditions, verify that a collision will occur.

Find (a) the position of the particles on collision, and (b) the speed of A just before the collision.
    ($g = 10$ m s^{-1})

8 Two parallel walls each of height $a/4$ are on level ground and are distant a and $2a$ from a point O on the ground. A ball is projected from O in a vertical plane perpendicular to the walls and just clears both walls. Find the magnitude and direction of the velocity with which it is projected. If P is the highest point of the ball's trajectory, find the angle that OP makes with the horizontal.   (SUJB)

9 A particle A of mass 0.7 kg is moving with speed 3 m s^{-1} on a horizontal smooth table of height 0.9 m above a horizontal floor. Another particle B, of mass M kg, is at rest on the edge of the table top. Particle A strikes particle B, and they coalesce into a single particle C. The particle C then falls from

the table. From the point of leaving the table to the point of hitting the floor, the horizontal displacement of C is 0.6 m.
(a) Show that C takes $\frac{3}{7}$ s to fall to the floor.
(b) Find the value of M.
    ($g = 9.8$ m s^{-2})   (L, 1993)

10 A stone is thrown horizontally from a point 44 metres above a horizontal plane. The stone hits the plane 39 metres from the point vertically below the point of projection.

Find the time of flight and the initial speed of the stone.
    ($g = 9.8$ m s^{-2})   (SEB, 1994)

11 [Take the acceleration due to gravity to be 10 m s^{-2} in this question.]

A particle is projected with speed 100 m s^{-1} at an angle of elevation θ from a point O on a horizontal plane, and it moves freely under gravity. The horizontal and upward vertical displacements of the particle from O at any subsequent time t s are denoted by x m and y m respectively. Express x and y in terms of θ and t, and hence show that

$$y = x \tan \theta - \frac{x^2}{2000}(1 + \tan^2 \theta)$$

Given that the particle passes through the point (800, 160), find the two possible values of $\tan \theta$.

Given instead that the particle passes through the point $(a, 160)$, show that

$$(a \tan \theta - 1000)^2 = 680\,000 - a^2$$

and deduce, or find otherwise, the greatest possible value of a as θ varies.   ($g = 10$ m s^{-2})
    (UCLES, 1987)

12

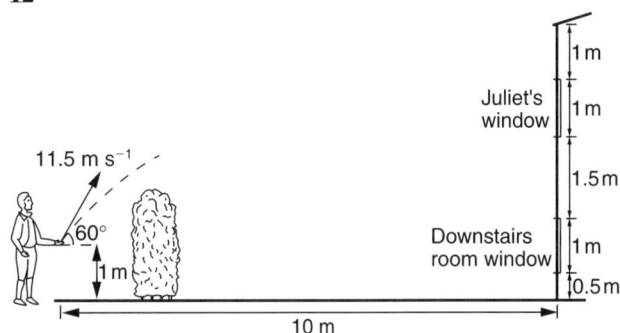

The picture shows Romeo trying to attract Juliet's attention without her nurse, who is in a downstairs room, noticing. He stands 10 m from the house and lobs a small pebble at her bedroom window. Romeo throws the pebble from a height of 1 m with a speed of 11.5 m s^{-1} at an angle of 60° to the horizontal.

(i)   How long does the pebble take to reach the house?

(ii) Does the pebble hit Juliet's window, the wall of the house or the downstairs room window?

(iii) What is the speed of the pebble when it hits the house? $(g = 9.8 \text{ m s}^{-2})$
(O&C, MEI, 1993)

13 A netball player throws the ball from a point b metres below and h metres horizontally from the net with the purpose of the ball falling through the net. The ball is thrown at a speed of v m s^{-1} at an angle of projection of α to the horizontal. Throughout the question you should neglect air resistance and model the ball as a particle. You should further assume that the acceleration g due to gravity is constant and of magnitude $g = 10$ m s^{-2}.

(a) From Newton's second law of motion, show that during its flight the horizontal and upward vertical displacements x and y of the ball from its point of projection satisfy the equation

$$y = x \tan \alpha - \frac{gx^2 \sec^2 \alpha}{2v^2}$$

(b) If the player throws the ball with a speed of 7 m s^{-1}, $b = 1$ and $h = 2$, find the possible angles of projection of the ball for the ball to fall through the net.

Brief Solutions

1 $\dot{x} = 32$, $\dot{y} = 4$, $\tan \theta = \dfrac{\dot{y}}{\dot{x}} = \frac{1}{8}$.   <u>Answer **E**</u>

2 $x = \dfrac{vt}{\sqrt{2}}$ in $y = \dfrac{vt}{\sqrt{2}} - \frac{1}{2}gt^2$ gives $y = x - \dfrac{gx^2}{v^2}$.

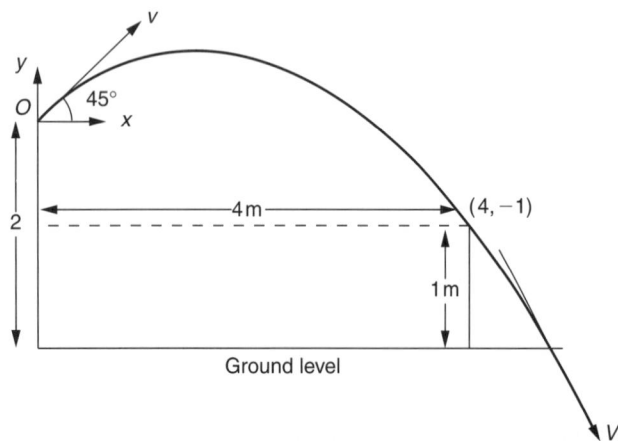

When $x = 4$, $y = -1$ so $v^2 = \dfrac{16g}{5} = 32 \implies$

(a) initial speed $= \sqrt{\left(\dfrac{16g}{5}\right)} = 5.7$ m s^{-1}

and $y = x - \dfrac{5x^2}{16}$.

(b) When $y = -2$,

$$-2 = x - \frac{5x^2}{16} \implies x = 4.6 \text{ m.}$$

Ball strikes the ground 0.6 m beyond the net.

(c) Using $v^2 = u^2 - 2gy$
$V^2 = 32 + 4g = 72$.
Speed when ball strikes ground $= 8.5$ m s^{-1}.

$$\frac{dy}{dx} = 1 - \tfrac{5}{8}x;$$

when $x = 4.6$, $\dfrac{dy}{dx} = -1.875 = \tan(-62°)$.

Direction is $62°$ below the horizontal.

3

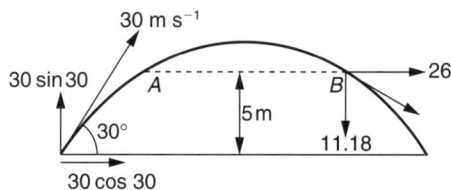

Time taken to reach height of 5 m:

$$5 = 15t - 5t^2 \quad t^2 - 3t + 1 = 0 \quad t = \frac{3 \pm \sqrt{5}}{2}$$

$\implies t = 0.382$ or 2.618 s   2.1a and 2.2a

Horizontal velocity $= 26$ m s^{-1}
Distance $AB = 58.1$ m

Maximum height $= \dfrac{(15)^2}{20} = 11.25$ m

$\implies$ height above $AB = 6.25$ m

At time $t = 2.618$, vertical velocity $= -11.18$.
Direction is $\tan^{-1}(-0.43)$, i.e. $23.3°$ below horizontal.
Velocity $= 28.3$ m s^{-1}.   7.3

4

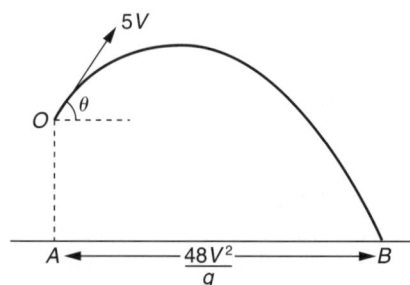

$\cos \theta = \frac{3}{5}$ $\tan \theta = \frac{4}{3}$

Equation of trajectory is

$$y = x \tan \theta - \frac{gx^2 (1 + \tan^2 \theta)}{2u^2}$$

At B

$$y = \frac{48V^2}{g} \cdot \frac{4}{3} - \frac{g}{50V^2} \left(\frac{48V^2}{g}\right)^2 \cdot \frac{25}{9}$$

$$= -64 \frac{V^2}{g}$$

(i) Height of O above A is $\dfrac{64V^2}{g}$.

(ii) Horizontal range $= \dfrac{24V^2}{g}$

For the second particle, the equation of the trajectory gives

$$W^2(24 \tan \alpha + 32) = V^2(1 + \tan^2 \alpha) \qquad (1)$$

If $\alpha = 45°$, $56W^2 = 2V^2 \Rightarrow V^2 = 28W^2$
Substitute into equation 1 to get
$7 \tan^2 \alpha - 6 \tan \alpha - 1 = 0$.

5 $g = 9.8$ Range $= \dfrac{u^2 \sin 2\theta}{g} = 245 \sin 2\theta$

$$9.8 = 49 \sin \theta\, t - 4.9t^2 \Rightarrow t^2 - 10 \sin \theta\, t + 2 = 0$$

$$t_1 + t_2 = 10 \sin \theta, \quad t_1 t_2 = 2, \quad t_2 - t_1 = \sqrt{17}$$

$$(t_2 - t_1)^2 + 4t_1 t_2 = (t_1 + t_2)^2$$
$$\Rightarrow \quad 17 + 8 = (t_1 + t_2)^2$$
$$\Rightarrow \quad t_1 + t_2 = 5, \quad \theta = 30°$$

$$R = 245 \sin 60° = \frac{245\sqrt{3}}{2}$$

6 Consider the point of projection as the origin and then adjust the answer to give the height above the ground.

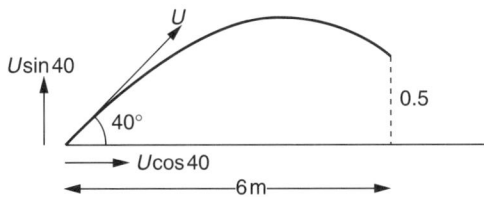

(a) Greatest height $= 1.5 + \dfrac{U^2 \sin^2 40}{2g} = 3.6$ m.

(b) To hit the target:

$$0.5 = 6 \tan 40 - \frac{4.9 \times 36}{U^2 \cos^2 40}$$

$$\Rightarrow \quad U = 8.1 \text{ m s}^{-1}.$$

7 $x_A = (u \cos \alpha)t, \quad y_A = (u \sin \alpha)t - \frac{1}{2}gt^2$.
$x_B = vt, \quad y_B = d - \frac{1}{2}gt^2$.
For collision, $(u \cos \alpha)t = vt$ and
$(u \sin \alpha)t - \frac{1}{2}gt^2 = d - \frac{1}{2}gt^2$

$$\Rightarrow \quad v = u \cos \alpha \qquad (1)$$

and

$$(u \sin \alpha)t = d \qquad (2)$$

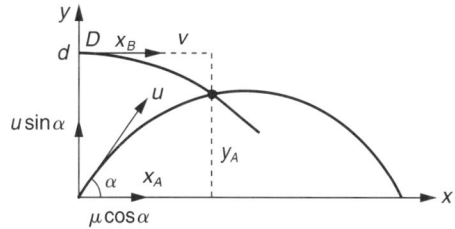

Given values satisfy equations 1 and 2 when $t = 2.5$.
(a) Collide at $x = 112.5$ m, $y = 28.75$ m.
(b) $\dot{x}_A = u \cos \alpha = 45$;
    when $t = 2.5$, $\dot{y}_A = u \sin \alpha - gt = -1$.
    Speed of A at that instant
    $= \sqrt{[(45)^2 + 1^2]} = 45.01$ m s^{-1}.

8 Substitute $\left(a, \dfrac{a}{4}\right)$, and $\left(2a, \dfrac{a}{4}\right)$ into trajectory

equation to get $\tan \alpha = \frac{3}{8}$ and $v = \frac{1}{4}\sqrt{(73ag)}$.

At highest point, $x = \frac{3}{2}a$, $y = \frac{9}{32}a$, so $\tan P\hat{O}X = \frac{3}{16}$.
$P\hat{O}X = 10.6°$.

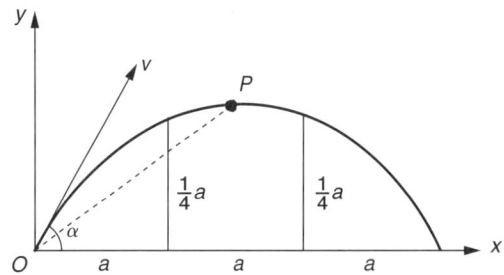

9 Conservation of linear momentum:

$$0.7 \times 3 = (0.7 + M)v \Rightarrow v = \frac{2.1}{M + 0.7} \qquad (1)$$

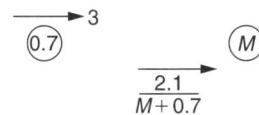

(a) Time taken to fall 0.9 m with initial vertical velocity $= 0$:

$$0.9 = 4.9t^2 \Rightarrow t = \frac{3}{7} \text{ s}$$

(b) Horizontal velocity $= \dfrac{0.6}{3/7} = \dfrac{7}{5}$

$$\Rightarrow \quad M = 0.8 \text{ (from equation 1)}$$

10 Time taken to fall 44 m from rest $= \sqrt{\dfrac{44}{4.9}} = 3$ s.

Horizontal speed $= \dfrac{39}{3}$ m s$^{-1} = 13$ m s^{-1}.

11 Bookwork.
If $x = 800$, $y = 160$ then

$160 = 800 \tan \theta - 320(1 + \tan^2 \theta)$
$\Rightarrow \quad \tan \theta = 1 \text{ or } \frac{3}{2}$.

If $x = a$, $y = 160$ then

$$160 = a \tan \theta - \frac{a^2}{2000}(1 + \tan^2 \theta).$$

Completing the square is a useful method for finding maximum or minimum values

$a^2 \tan^2 \theta - 2000a \tan \theta + (1000)^2$
$\qquad\qquad = -320\,000 - a^2 + (1000)^2$
$(a \tan \theta - 1000)^2 = 680\,000 - a^2$

Since LHS $\geqslant 0$ then RHS $\geqslant 0$, i.e.
$a^2 \leqslant 680\,000 \quad \Rightarrow \quad \text{Max } a = \sqrt{680\,000} = 825 \text{ m}$

12 (i) Horizontal velocity $= 11.5 \cos 60 = 5.75 \text{ m s}^{-1}$.

Time taken for 10 m $= \dfrac{10}{5.75} = 1.74 \text{ s}$.

(ii) Vertical height $= 1 + 11.5 \sin 60(1.74) - 4.9(1.74)^2$
$\qquad\qquad\qquad = 3.5 \text{ m above the ground.}$
Stone hits Juliet's window.

(iii) If speed is v at the window:
$v^2 = 11.5^2 - 2(9.8)(2.5)$
$\Rightarrow \quad v = 9.12 \text{ m s}^{-1}$

7.3

13 (a) Bookwork.
(b)

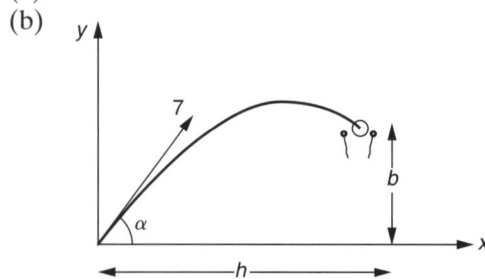

$$y = x \tan \alpha - \frac{gx^2(1 + \tan^2 \alpha)}{2v^2}$$

$v = 7$, $x = 2$, $y = 1$ gives
$20 \tan^2 \alpha - 98 \tan \alpha + 69 = 0$
$\alpha = 76.1° \text{ or } 40.4°$

9.1

$\alpha = 40.4° \quad \Rightarrow \quad$ Ball takes 0.375 s, then vertical velocity is +ve, i.e. upwards – no good.
$\alpha = 76.1° \quad \Rightarrow \quad$ Ball takes 1.19 s, then vertical velocity is −ve, i.e. downwards – a Basket.

23

Work, power and energy

AS Level			A Level			Topic	Date attempted	Date completed	Self-assessment
CORE	MODULAR	TRADITIONAL	CORE	MODULAR	TRADITIONAL				
	✓	✓		✓	✓	**Hooke's Law**			
	✓	✓		✓	✓	**Work**			
	✓	✓		✓	✓	**Energy**			
		✓		✓	✓	**Power**			

Fact Sheet

Hooke's Law

When an elastic string, natural length a, modulus of elasticity λ, is stretched by a distance x, the tension in the string is $T = \lambda \dfrac{x}{a}$.

For a spring in compression, the force is a thrust.

Work Done

The work done by a force $\mathbf{F}$ moving from A to B is

$$\int_A^B \mathbf{F} \cdot d\mathbf{r}$$

For motion in a straight line, work done is

$$\int_A^B F \, dx$$

$F = ma = mv \dfrac{dv}{dx}$, so work done is

$$\int_{v_1}^{v_2} mv \, dv = \tfrac{1}{2}mv_2^2 - \tfrac{1}{2}mv_1^2$$

$$= \text{increase in kinetic energy}$$

For motion under gravity, work done is

$$\int_{h_1}^{h_2} mg \, dx = mgh_2 - mgh_1$$

$$= \text{increase in gravitational potential energy}$$

For stretching an elastic string, work done is

$$\int_{x_1}^{x_2} \frac{\lambda x}{a} \, dx = \frac{\lambda(x_2^2 - x_1^2)}{2a}$$

$$= \text{increase in elastic potential energy}$$

Conservation of Energy

When a particle is oscillating on the end of an elastic string, the sum of
(i)   kinetic energy (k.e.)
(ii) gravitational potential energy (g.p.e.)
(iii) elastic potential energy (e.p.e.),
remains constant.

Power

Power is the rate of working, measured in watts.
For a given speed of v metres per second and tractive force of F newtons, the power developed $= Fv$ watts.

Worked Examples

1 A car of mass 800 kg tows a caravan of mass 480 kg along a straight level road. The tow-bar connecting the car and the caravan is horizontal and of negligible mass. With the car's engine working at a rate of 24 kW, the car and caravan are travelling at a constant speed of 25 m s^{-1}.

(a) Calculate the magnitude of the total resistance, in N, to the motion of the car and the caravan.

The resistance to the motion of the car has magnitude 800λ newtons and the resistance to the motion of the caravan has magnitude 480λ newtons, where λ is a constant.

(b) Find the value of λ.

(c) Find the tension, in N, in the tow-bar. (L, 1993)

• (a) Consider the car and caravan as one body:

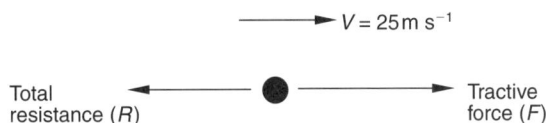

$$\longrightarrow V = 25\,\text{m s}^{-1}$$

Total resistance (R) $\longleftarrow$ ● $\longrightarrow$ Tractive force (F)

Power = tractive force × speed
$24 \times 10^3 = F\,25 \quad \Rightarrow \quad F = 960$

Since speed is constant, Newton's 1st Law applies, therefore $R = 960$ N.

(b) $800\lambda + 480\lambda = 960 \quad \Rightarrow \quad \lambda = 0.75$

(c) Consider just the caravan:

Resistance = 480λ = 360 $\longleftarrow$ ● $\longrightarrow$ Tension in tow-bar (T)

Again by Newton's 1st Law:

$T = 360$ N

2 A sledge of mass M is pulled, by means of a light rope, up a line of greatest slope of a rough plane inclined at an angle α, where $\tan \alpha = \frac{3}{4}$. The rope lies in a vertical plane through the line of greatest slope and is inclined to this line of greatest slope at a constant angle β, where $\tan \beta = \frac{1}{2}$. The coefficient of friction between the sledge and the plane is $\frac{1}{4}$ and the sledge moves at constant speed. Show that the tension in the rope is $\left(\dfrac{16}{45}\sqrt{5}\right)Mg$ and find the work done by this tension in moving the sledge a distance d up the plane.

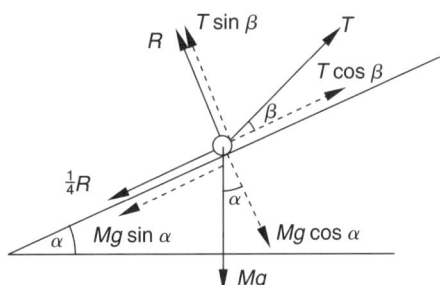

•

$$\tan \alpha = \frac{3}{4} \quad \sin \alpha = \frac{3}{5} \quad \cos \alpha = \frac{4}{5}$$

$$\tan \beta = \frac{1}{2} \quad \sin \beta = \frac{1}{\sqrt{5}} \quad \cos \beta = \frac{2}{\sqrt{5}}$$

$$\mu = \tfrac{1}{4}$$

When the sledge is moving the frictional force = μR
Resolving forces perpendicular to the plane and parallel to the plane

$\nwarrow R + T \sin \beta - Mg \cos \alpha = M \times$ acceleration

$\nearrow T \cos \beta - \tfrac{1}{4}R - Mg \sin \alpha = M \times$ acceleration

Since the sledge has constant speed, acceleration = 0 $\Rightarrow$

$$R + \frac{T}{\sqrt{5}} - Mg \cdot \frac{4}{5} = 0 \tag{1}$$

$$\frac{2T}{\sqrt{5}} - \frac{1}{4}R - Mg \cdot \frac{3}{5} = 0 \tag{2}$$

Equation 1 + 4 × equation 2

$$\Rightarrow \quad \frac{9T}{\sqrt{5}} - \frac{16Mg}{5} = 0$$

$$\Rightarrow \quad T = \frac{16Mg\sqrt{5}}{45} = \frac{16\sqrt{5}Mg}{45} \text{ as required}$$

Work done = force × distance
= component of T up the slope × distance

$$= \frac{16\sqrt{5}Mg}{45} \cdot \frac{2}{\sqrt{5}}d$$

$$= \frac{32Mgd}{45}$$

Since no units are given in the question there are no units (such as N or J) in the answer

3 A car of mass 1.2 tonnes is travelling up a slope of 1 in 150 at a constant speed of 10 m s^{-1}. If the frictional and air resistances are 100 N, calculate the power exerted by the engine. The car descends the same slope working at a rate of 2 kW. What will be its acceleration when its speed is 20 m s^{-1} if the resistances are the same? If the engine is shut off when the speed of the car is 25 m s^{-1} as it descends the slope, how long will it be before the car comes to rest? (SUJB, 1982)

• Take $g = 10$ m s^{-2}.

Ascending:

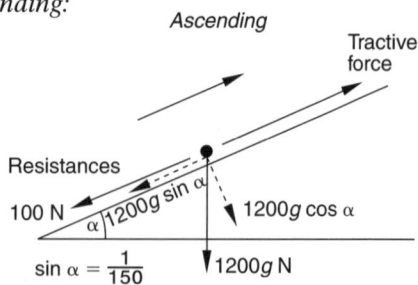

Accelerating force
$$= \text{tractive force} - 100 - 1200g \sin \alpha$$
$$= 0 \text{ when speed is constant.}$$
Therefore tractive force $= 100 + 80 = 180$ N.
Power of the engine $= $ tractive force $\times$ speed
$$= 180 \times 10$$
$$= 1.8 \text{ kW}$$

Descending:

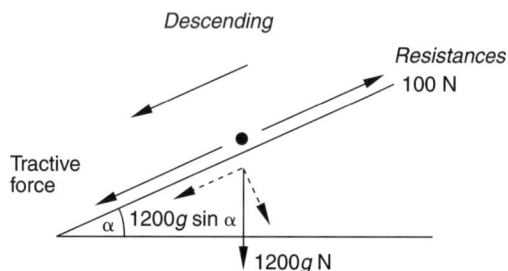

If power $= 2$ kW and speed $= 20$ m s^{-1} then
tractive force $= 100$ N.
Accelerating force $= 1200g \sin \alpha + 100 - 100$
$$= 80 \text{ N}$$

Therefore acceleration $= \dfrac{80}{1200} = \dfrac{1}{15}$ m s^{-2}

When the engine is shut off the accelerating force is
$80 - 100 = -20$ N. Therefore the car has a

retardation of $\dfrac{20}{1200}$ m s$^{-2} = \dfrac{1}{60}$ m s^{-2}.

Therefore the car will take $\dfrac{v}{a}$ s $= 60 \times 25$ s to come

to rest, i.e. 25 minutes.

4 A mass of 6 kg is accelerated vertically upwards
from a position of rest. When it attains a height of
7 m it has a velocity of 8 m s^{-1}. Neglecting air
resistance, find the work done by the force causing
the motion. (Assume $g = 10$ m s^{-2}.)

• Work done by force raising 6 kg through 7 m is

$(6g)(7) = 420$ J

Work done giving 6 kg a velocity of 8 m s^{-1} is

$\frac{1}{2}(6)(8)^2 = 192$ J

Total work done by the force $= 612$ J.

5 Prove, by integration, that the work done in
stretching an elastic string, of natural length l and

modulus of elasticity λ, from a length l to a length

$l + x$ is $\dfrac{\lambda x^2}{2l}$.

A particle of mass m is suspended from a fixed
point O by a light elastic string of natural length l.
When the mass hangs freely at rest the length of the

string is $\dfrac{13l}{12}$. The particle is now held at rest

at O and released. Find the greatest extension of
the string in the subsequent motion.

By considering the energy of the system when the
length of the string is $l + x$ and the velocity of the
particle is v explain why

$\frac{1}{2}mv^2 = mg(l + x) - 6mg\dfrac{x^2}{l}$

Hence show that the kinetic energy of the particle
in this position may be written as

$\dfrac{mg}{l}\{\alpha l^2 - 6(x - \beta l)^2\}$

where α and β are positive constants which must be
found. Hence deduce that the maximum kinetic
energy of the particle during the whole of its
motion occurs when it passes through the
equilibrium position. (AEB, 1990)

• Tension in an elastic string $= \dfrac{\lambda x}{l}$, where x is the

extension. Work done in stretching the string is

$\displaystyle\int_0^x \dfrac{\lambda x}{l}\, \mathrm{d}x = \dfrac{\lambda x^2}{2l}$

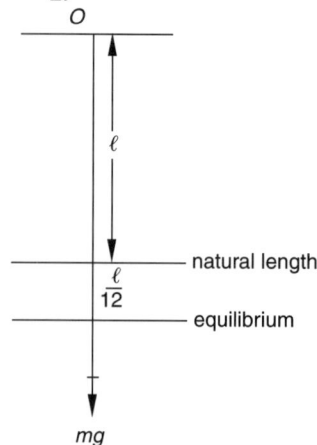

At the point of equilibrium

$T = mg = \dfrac{\lambda(l/12)}{l} \quad \Rightarrow \quad \lambda = 12mg$

At O, g.p.e. $= 0$, k.e. $= 0$, e.p.e. $= 0$.
When the extension is x the particle has velocity v

g.p.e. $= -mg(l + x)$, k.e. $= \frac{1}{2}mv^2$

e.p.e. $= \dfrac{12mgx^2}{2l}$

By conservation of energy

$$\tfrac{1}{2}mv^2 = mg(l + x) - 6mg\,\frac{x^2}{l}$$

At the lowest point $v = 0$

$$\Rightarrow \quad 6x^2 - xl - l^2 = 0$$
$$\Rightarrow \quad (2x - l)(3x + l) = 0$$

$$\therefore x = \frac{l}{2} \quad \left(\text{or } \frac{-l}{3} - \text{rejected} \right)$$

Hence the maximum extension $= \dfrac{l}{2}$.

Kinetic energy: $= \tfrac{1}{2}mv^2 = \dfrac{mg}{l}(l^2 + xl - 6x^2)$

By completing the square

$$= \frac{mg}{l}\left(-6\left(x^2 - \frac{1}{6}xl + \frac{l^2}{144} \right) + \frac{25}{24}l^2 \right)$$

$$= \frac{mg}{l}\left\{ \frac{25}{24}l^2 - 6\left(x - \frac{l}{12} \right)^2 \right\}$$

$$\Rightarrow \quad \alpha = \tfrac{25}{24}, \beta = \tfrac{1}{12}$$

Maximum kinetic energy is $\tfrac{25}{24}mgl$ when

$x = \dfrac{l}{12}$ (equilibrium position).

6   A ball B, of mass 0.125 kg, is attached to one end of a light elastic string OB, the end O being attached to the ceiling. The modulus of elasticity of the string is 52.5 N, and the string has natural length 1.5 m. In equilibrium the ball is at E. Show that the depth of E below O is 1.54 m, correct to 3 significant figures.

The ball is released from rest at O, and does not hit the floor. State an assumption necessary for conservation of energy to apply.

Hence find
(i)   the speed as B passes through E
(ii) the maximum depth of B below O.
(UCLES Specimen Paper, 1994)

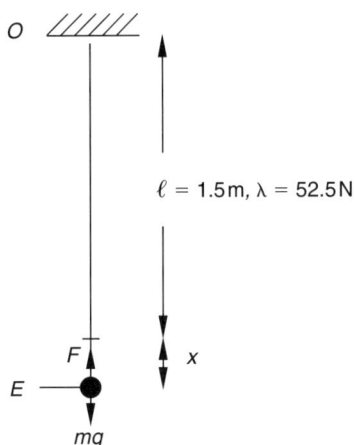

$\ell = 1.5\text{m}, \lambda = 52.5\text{N}$

In equilibrium at E,

$$F = mg \quad \Rightarrow \quad \frac{\lambda x}{l} = mg$$

$$\Rightarrow \quad x = \frac{mgl}{\lambda} = \frac{0.125 \times 9.81 \times 1.5}{52.5} = 0.035$$

$$\Rightarrow \quad OE = l + x = 1.54 \text{ m (3 s.f.)}$$

Assumption: no air resistance.

(i)   Let speed at E be v.
      From O to E,

      gain of k.e. + gain of e.p.e. = loss of g.p.e.

$$\frac{1}{2} \times 0.125v^2 + \frac{1}{2}\frac{52.5(0.035)^2}{1.5}$$

$$= 0.125 \times 9.81 \times 1.535$$

$$\Rightarrow \quad v = 5.46 \text{ m s}^{-1}$$

(ii) Let maximum depth below natural length be y metres.
      From O to lowest point,

      Gain of e.p.e. = loss of g.p.e.

$$\frac{1}{2}\frac{52.5y^2}{1.5} = 0.125 \times 9.81 \times (1.5 + y)$$

$$\Rightarrow \quad 17.5y^2 - 1.226\,25y - 1.839\,375 = 0$$
$$y = 0.361 \text{ (or } -0.291) \quad \boxed{2.1a \text{ and } 2.2a}$$
$$\Rightarrow \quad \text{Maximum depth below } O = 1.5 + y$$
$$= 1.86 \text{ metres}$$

7   A car of mass 1500 kg is travelling along a straight horizontal road at a constant speed of 40 m s^{-1} against a resistance of 900 N. Calculate, in kW, the power being exerted.

Calculate the power required to go down a hill of 1 in 25 (along the slope) at a steady speed of 40 m s^{-1} against the same resistance.

Given that the resistance is proportional to the square of the speed of the car, calculate the acceleration of the car up this hill with the engine working at 18 kW at the instant when the speed is 20 m s^{-1}. (Take $g = 10$ m s^{-2})

•   At constant speed the accelerating force = 0. Therefore the tractive force = resistances.
    ∴ On a level road tractive force = 900 N.
    At 40 m s^{-1} the power being exerted is

$$Fv = 900 \times 40 \text{ W}$$
$$= 36 \text{ kW}$$

Descending

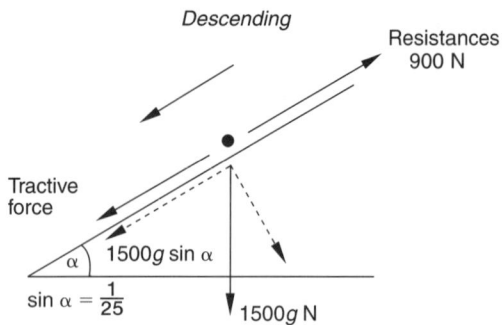

To go down the hill at a steady speed,

tractive force + $mg \sin \alpha$ = 900

$mg \sin \alpha$ = 15 000/25 = 600.
Therefore the tractive force = 300 N.
The power exerted at 40 m s^{-1} = Fv = 300 × 40 W
                                    = 12 kW.

Ascending

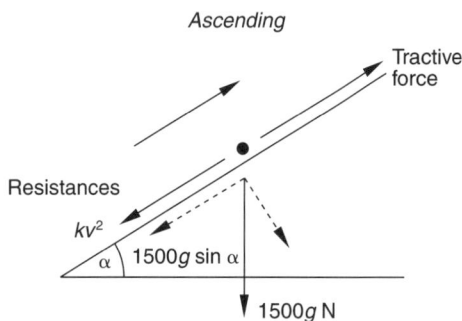

If resistance is proportional to the square of the speed then $R = kv^2$.
When $R = 900$, $v = 40$, so $k = \frac{9}{16}$ (=0.5625).

To go up the hill:
when $v = 20$ m s^{-1} and power = 18 kW,
tractive force is

$$\frac{18 \times 10^3}{20} = 900 \text{ N}$$

Resistance = $0.5625(20)^2$ = 225 N.
Tractive force − ($mg \sin \alpha$ + resistance)
        = accelerating force
⇒   900 − (600 + 225) = 1500a

Acceleration $a = \dfrac{75}{1500} = 0.05$ m s^{-2}.

8 A bus of mass 18 tonnes travels up a slope of gradient $\sin^{-1}\left(\frac{1}{50}\right)$ against a resistance of 0.1 N per kilogram. Find the tractive force required to produce an acceleration of 0.05 m s^{-2} and the power which is developed when the speed is 10 m s^{-1}.

A second bus of mass 25 tonnes experiencing the same resistance and with a maximum power 120 kW follows the first bus up the slope. If the first bus maintains the same power while on the slope, will the distance between the buses decrease when both are travelling at maximum speed?

(Take $g = 10$ m s^{-2})

• *First bus:*

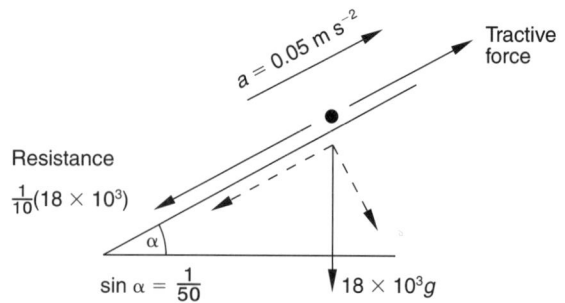

Mass = 18×10^3 kg, resistance $R = 1.8 \times 10^3$ N.
Component of weight down the slope is

$$W \sin \alpha = \frac{180 \times 10^3}{50} = 3.6 \text{ kN}$$

Accelerating force = $18 \times 10^3 \times 0.05$ N
                   = 0.9 kN
Accelerating force
    = tractive force − (R + component of weight)

0.9×10^3 = tractive force − $(1.8 \times 10^3 + 3.6 \times 10^3)$

Hence tractive force = 6.3 kN

When the speed is 10 m s^{-1},
power = $10 \times 6.3 \times 10^3$ W
      = 63 kW
At maximum speed,

tractive force = R + component of weight
              = 1.8 kN + 3.6 kN = 5.4 kN

Power = 63 kW,

$$\Rightarrow \quad \text{maximum speed} = \frac{\text{power}}{\text{tractive force}}$$

$$= \frac{63}{5.4} = \frac{35}{3} \text{ m s}^{-1}$$

$$= 11\tfrac{2}{3} \text{ m s}^{-1}$$

Second bus:
Mass = 25×10^3 kg, resistance $R = 2.5 \times 10^3$ N*.
Component of weight down the slope is

$$W_s = \frac{25 \times 10^4}{50} \text{ N} = 5 \text{ kN}$$

At maximum speed

accelerating force = 0
    tractive force = $R + W_s$
                   = (2.5 + 5) kN = 7.5 kN

But power = tractive force × speed:

$120 \times 10^3 = 7.5 \times 10^3 \times$ speed

Therefore the maximum speed of the bus is 16 m s^{-1}.
Since the second bus travels faster than the first the distance between them will decrease.

* This assumes that the 'same' resistance means 0.1 N/kg.

9 In the dangerous sport of bungee diving an individual attaches one end of an elastic rope to a fixed point on a river bridge. He/she is attached to the other end and jumps over the bridge so as to fall vertically downwards towards the water. The rope should be such that the diver comes to rest just above the surface of the water. In order to find out which particular ropes are suitable experiments are carried out with weights attached to the rope rather than people. In one experiment it was found that when a weight of mass m was attached to a particular rope of natural length a and dropped from a bridge at a height of $3a$ above the water level then the weight just reached the level of the water. Show that the modulus of elasticity of the rope is $\dfrac{3mg}{2}$.

The weight of mass m is then removed and a weight of mass $\dfrac{5m}{2}$ is then attached to this rope and dropped from the same height. Find the speed of the weight just as it reaches the water.

When the weight emerges from the water its speed has been reduced to zero by the resistance of the water. Show by using conservation of energy or otherwise and assuming that the rope does not slacken, that the subsequent speed v of the weight at height h above the water level is given by

$$v^2 = \frac{gh}{5a}(2a - 3h)$$

Describe the subsequent motion of the weight.

(WJEC, 1993)

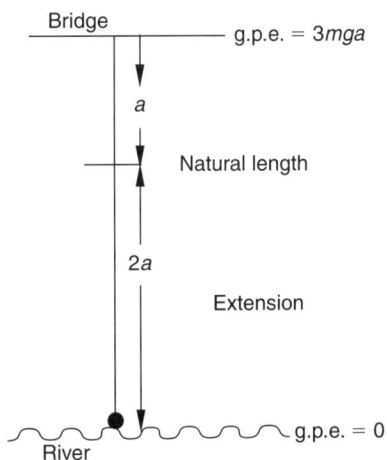

By conservation of energy
At the bridge:

g.p.e. $= 3mga$,   k.e. $=$ e.p.e. $= 0$

At the surface of the river:

g.p.e. $= 0$,   k.e. $= 0$,   e.p.e. $= \dfrac{\lambda(2a)^2}{2a}$

Hence $3mga = \lambda \cdot 2a \quad \Rightarrow \quad \lambda = \dfrac{3mg}{2}$

If mass $= \dfrac{5m}{2}$

At the bridge:

$$\text{g.p.e.} = \frac{15mga}{2}, \qquad \text{k.e.} = \text{e.p.e.} = 0$$

At the water:

$$\text{g.p.e.} = 0, \qquad \text{k.e.} = \frac{1}{2}\left(\frac{5m}{2}\right)v^2$$

$$\text{e.p.e.} = \frac{3mg}{2}(2a) = 3mga$$

By conservation of energy

$$\frac{15mga}{2} = \frac{5m}{4}v^2 + 3mga$$

$$\Rightarrow \quad v^2 = \frac{9ga}{2} \cdot \frac{4}{5} = 3.6ga$$

The speed as the particle reaches the water $= \sqrt{3.6ga}$.

As the particle comes out of the water:

k.e. $= 0$,   e.p.e. $= 3mga$,   g.p.e. $= 0$

At height h:

$$\text{k.e.} = \frac{1}{2}\left(\frac{5m}{2}\right)v^2,$$

$$\text{e.p.e.} = \frac{3mg(2a - h)^2}{2 \cdot 2a},$$

$$\text{g.p.e.} = \frac{5mgh}{2}$$

By conservation of energy

$$3mga = \frac{5m}{4}v^2 + \frac{3mg(2a - h)^2}{4a} + \frac{5mgh}{2}$$

$$12mga^2 = 5mv^2a + 3mg(4a^2 - 4ah + h^2) + 10mgha$$

$$0 = 5v^2a + 3g(-4ah) + 3gh^2 + 10gha$$

$$\Rightarrow \quad 2gha - 3gh^2 = 5v^2a$$

$$v^2 = \frac{gh(2a - 3h)}{5a}$$

The particle will come to rest instantaneously when $h = \dfrac{2a}{3}$ and then fall again coming to rest at the surface of the water.

10 A child's toy consists of a bat, an elastic string of natural length 0.3 m, and a small ball of mass 0.05 kg. One end of the string is attached to the centre of the bat and the other end is attached to the ball. When the bat is held horizontally the ball will hang in equilibrium 0.33 m below the centre of the bat.

(i) Show that the modulus of elasticity of the string is 5 N.

(ii) Referred to the level of the bat, what is the gravitational potential energy and the energy stored in the string, when the string is extended downwards by x m?

The ball is brought to the centre of the bat and the bat then propels the ball vertically downwards at an initial speed of 0.8 m s^{-1}.

(iii) By using the principle of the conservation of mechanical energy, show that, when the ball is travelling at v m s^{-1}, the extension of the string, x m, satisfies

$$5x^2 - 0.3x - 0.0996 + 0.015v^2 = 0$$

(iv) Hence, find the maximum extension of the string.
$(g = 10 \text{ m s}^{-2})$ (NICCEA, 1992)

•
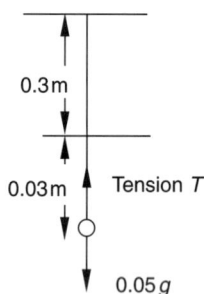

When hanging in equilibrium the tension in the string is equal to the weight of the particle

Extension = 0.03 m   modulus of elasticity = λ

(i) In equilibrium

$$T = \frac{\lambda(0.03)}{0.3} = 0.05g$$

$$\Rightarrow \quad \lambda = \frac{0.5 \times 0.3}{0.03} = 5 \text{ N}$$

(ii)
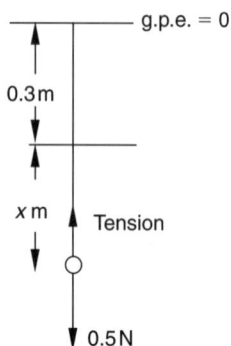

When the string is extended by x m

g.p.e. $= -0.5(0.3 + x)$ J

e.p.e. $= \dfrac{\lambda x^2}{2l} = \dfrac{5x^2}{0.6} = \dfrac{25x^2}{3}$ J

(iii) Initial energy of the system:

g.p.e. = 0,   e.p.e. = 0,   k.e. = $\frac{1}{2}(0.05)(0.8)^2$

i.e. initial energy of the system is 0.016 J.

When the extension of the string is x and speed of particle is v m s^{-1} then
k.e. + g.p.e. + e.p.e = 0.016

$$\tfrac{1}{2}(0.05)v^2 - 0.05 \times 10 \times (0.3 + x) + \frac{25x^2}{3} = 0.016$$

$$0.15v^2 - 3(0.3 + x) + 50x^2 = 0.096$$

$$5x^2 - 0.3x + 0.015v^2 = 0.0996$$

$$5x^2 - 0.3x - 0.0996 + 0.015v^2 = 0$$

(iv) At maximum extension $v = 0$.

Then $5x^2 - 0.3x - 0.0996 = 0$

2.1a and 2.2a

$$x = \frac{0.3 \pm \sqrt{0.09 + 1.992}}{10}$$

$$= \frac{0.3 + 1.4429}{10}$$

Maximum extension = 0.174 m.

11 One end of a light elastic string of natural length l and modulus of elasticity $3mg$ is attached to a fixed point O. To the other end is attached a particle P of mass m. The particle is projected vertically downwards from O with a speed $\sqrt{(3lg)}$. By the principle of conservation of energy, or otherwise, find the speed of P when it is at a depth x $(x > l)$ below O. Show that the greatest depth below O reached by P is $\frac{8}{3}l$.

Find the maximum speed of P.

• Take the potential energy datum as O.

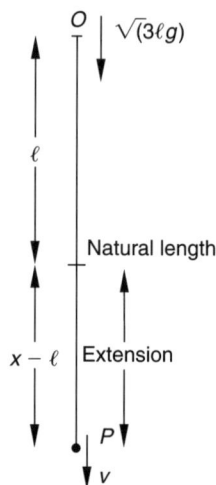

When at O, kinetic energy $= \frac{1}{2}m(3lg) = \dfrac{3mgl}{2}$

potential energy of particle = 0
potential energy in string   = 0

When particle has descended a distance x $(x > l)$, extending the string by $(x - l)$,

k.e. of particle $= \frac{1}{2}mv^2$, g.p.e. $= -mgx$

e.p.e. in string $= \frac{\lambda(\text{ext.})^2}{2l} = \frac{3mg(x - l)^2}{2l}$

By conservation of energy,

$$\frac{3}{2}mgl = \frac{1}{2}mv^2 - mgx + \frac{3mg(x - l)^2}{2l}$$

$$v^2 = 3gl + 2gx - \frac{3g(x - l)^2}{l}$$

$$= \frac{g}{l}(3l^2 + 2xl - 3x^2 + 6xl - 3l^2)$$

$$v^2 = \frac{g}{l}(8xl - 3x^2) \qquad (1)$$

Speed of P when at a depth x $(x > l)$ below O is

$$v = \sqrt{\left(\frac{g}{l}(8xl - 3x^2)\right)}$$

$v = 0$ when $x = 0$ (not applicable), or $x = \frac{8}{3}l$.
Greatest depth below O reached by P is $\frac{8}{3}l$.

Maximum speed is attained when $\dfrac{dv}{dt} = v\dfrac{dv}{dx} = 0$.

Differentiating equation 1,

$$2v\frac{dv}{dx} = \frac{g}{l}(8l - 6x).$$

Maximum speed occurs when $x = \frac{4}{3}l$.

Then $v^2 = \dfrac{g}{l}\left(\dfrac{32l^2}{3} - \dfrac{48l^2}{9}\right) = \dfrac{16gl}{3}$.

Maximum speed of P is $4\sqrt{\left(\dfrac{gl}{3}\right)}$.

Exercises

1 The mass of a car is 800 kg and the total resistance to its motion is constant and equal to a force of 320 N.
  (a) Find, in kW, the rate of working of the engine of the car when it is moving along a level road at a constant speed of 25 m s^{-1}.
  (b) What is the acceleration in m s^{-2}, when the car is moving along a level road at 20 m s^{-1} with the engine working at 11 kW?
  (c) What is the maximum speed attained by the car with the engine working at this rate?

2 A car of mass 10^3 kg is travelling up a slope, inclination $\sin^{-1}(0.05)$, at a constant speed of 25 m s^{-1}. The engine is generating its maximum power of $\frac{5}{4} \times 10^5$ W. Show that, excluding gravity,

the car is experiencing a resistance to motion of 4.5×10^3 N. (Let $g = 10$ m s^{-2}.)

Assuming that the resistance is proportional to the speed of the car determine:
  (a) the maximum speed of the car on the level road,
  (b) the maximum acceleration of the car when it is travelling at 10 m s^{-1} on a level road.

3 A car has an engine capable of developing 15 kW. The maximum speed of the car on a level road is 120 km h^{-1}. Calculate the total resistance in newtons at this speed.

Given that the mass of the car is 1000 kg and that the resistance to motion is proportional to the square of the speed, obtain the rate of working, in kW to two decimal places, of the engine when the car is moving at a constant speed of 40 km h^{-1} up a road of inclination θ, where $\sin\theta = 1/25$. (Take g as 9.8 m s^{-2}.) $\qquad$ (L)

4 A particle of mass m is attached to one end of an elastic string of natural length a, the other end of which is attached to a fixed point O. When the particle hangs freely the string extends a further distance $a/4$. The particle is placed at O and allowed to fall freely. Find the maximum extension of the string in the subsequent motion. $\qquad$ (L)

5 A car of mass 800 kg tows a trailer of mass 200 kg. The constant resistances acting on the car and trailer are R and 200 N respectively.
  (i) If the car is travelling at a constant speed of 54 km h^{-1} and the engine is working at a rate of 9.75 kW, find the tension in the tow-bar and the value of R.
Due to a fault in the tow-bar the maximum tension which the tow-bar can take is 350 N. When the car is travelling at 72 km h^{-1}, find
  (ii) the maximum acceleration if the resistances remain unchanged
  (iii) the maximum power at which the engine can work, if the resistances remain unchanged. $\qquad$ (NICCEA, 1993)

6

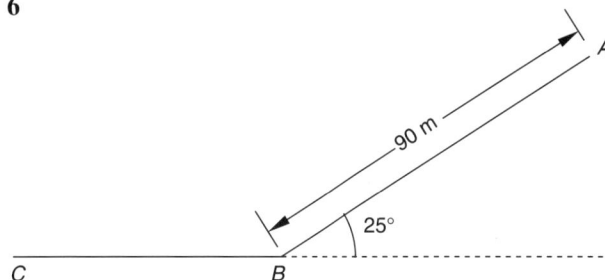

The figure represents a path which consists of a slope AB, 90 m long, inclined at 25° to the horizontal and a horizontal section BC. A boy on a skate-board starts from rest at A and glides down AB before coming to rest between B and C. The magnitude of the resistive forces opposing the motion are constant throughout the journey. The combined mass of the boy and skate-board is 40 kg and the boy reaches B with a speed of 14 m s^{-1}. By using a suitable model for the boy and his skate-board, find

(a) the energy lost, in J, by the boy and the skate-board in going from A to B

(b) the magnitude, in N, of the resistive forces

(c) the distance, in m, the boy travels along BC before coming to rest. $(g = 9.8$ m s$^{-2})$

(L Specimen Paper, 1996)

7

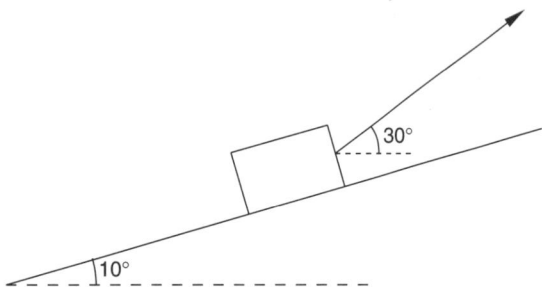

A crate of mass 5 kg is pulled directly up a rough slope, of inclination 10°, by a constant force of magnitude 20 N, acting at an angle of 30° above the horizontal. Find the work done by the force as the crate moves a distance of 3 m up the slope.

Find also the work done against gravity in this displacement. $(g = 9.81$ m s$^{-2})$

(UCLES Specimen Paper, 1994)

8 The mass of a car is 800 kg. Its engine works at a constant rate of 20 kW, and the motion of the car is subject to a constant resistance of magnitude 600 N.

(a) Find, in kJ to 2 significant figures, the kinetic energy of the car when it is travelling at maximum speed on a level road.

(b) Find, in m s^{-2}, the acceleration of the car when it is travelling at 25 m s^{-1} on a level road.

The car now ascends a straight road, inclined at 5° to the horizontal, with the same power output and against the same constant resistance.

(c) Find, in m s^{-1} to 2 significant figures, the maximum speed of the car up the slope.

$(g = 9.8$ m s$^{-2})$

(L, 1993)

9 A uniform rod AB, of length $2l$ and weight W, has its end A in contact with a rough vertical wall and its end B connected by a string BC to a point C vertically above A with $AC = 2l$. The rod is in equilibrium in a vertical plane perpendicular to the wall with the angle $ACB = \alpha$, where $\alpha < 45$. Show that the tension in the string is $W \cos \alpha$ and that the frictional force on the rod at A must act upwards.

Show also that the least possible value of the coefficient of friction at A is $\tan \alpha$. Given that $\cos \alpha = 3/4$ and that the string is an elastic one with modulus of elasticity $3W/2$, find the natural length of the string. (L)

10 A particle is projected with initial speed u up a rough slope, inclination α, where $\tan \alpha = \frac{5}{12}$. The coefficient of friction is $\frac{1}{6}$. By considering work done against external forces, find

(a) the distance moved up the slope before coming instantaneously to rest

(b) the speed of the particle when it returns to its initial position.

Brief Solutions

1 (a) Tractive force = resistance = 320 N.
Power = rate of work = Fv = (320)(25)
$= 8$ kW.

(b) Tractive force $= \dfrac{11 \times 10^3}{20} = 550$ N.

Acceleration $= \dfrac{\text{tractive force} - \text{resistance}}{\text{mass}}$

$= 0.2875$ m s^{-2}

(c) At maximum speed t.f. = resistance = 320 N.

Maximum speed $= \dfrac{\text{power}}{\text{t.f.}} = 34.375$ m s^{-1}.

2 Tractive force $= \dfrac{1.25 \times 10^5}{25}$ N $= 5000$ N.

Tractive force = resistance + component of weight
$\Rightarrow$ resistance $= 5000 - 500 = 4500$ N.

(a) $R = kv = 180v$ from the first part.
On level road t.f. = resistance = $180v$.

Maximum speed $= \dfrac{\text{power}}{\text{t.f.}}$,

i.e. $v = \dfrac{1.25 \times 10^5}{180v}$,

$v = 26.4$ m s^{-1}.

(b) $v = 10$,   resistance $= 1800$,

t.f. $= \dfrac{1.25 \times 10^5}{10}$

$= 12\,500$ N.
Accelerating force $= 12\,500 - 1800$
$= 10\,700$ N.

Maximum acceleration $= \dfrac{10\,700}{\text{mass}} = 10.7$ m s^{-2}.

3 Maximum speed $= \dfrac{100}{3}$ m s^{-1}.

Tractive force $= \dfrac{\text{power}}{\text{speed}} = 450$ N.

At maximum speed,
resistance = tractive force = 450 N.

$R = kv^2 \quad \Rightarrow \quad k = \dfrac{R}{v^2} \quad \Rightarrow \quad k = 0.405$

$\Rightarrow \quad R = 0.405v^2$

At 40 km h^{-1} $(=\frac{100}{9}$ m s$^{-1})$, $\quad R = 50$ N.
Tractive force $= R + mg \sin \theta = 442$ N.

Therefore power $= 442 \times \dfrac{100}{9}$ W $= 4.91$ kW.

4

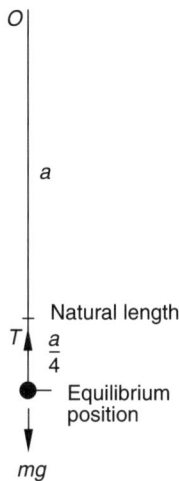

$mg = \dfrac{\lambda(a/4)}{a} \quad \Rightarrow \quad \lambda = 4mg.$

At O, e.p.e. (string) $= 0$, k.e. $= 0$.
When string has extension x,

k.e. $= \frac{1}{2}mv^2$, e.p.e. (string) $= \dfrac{2mgx^2}{a}$,

g.p.e. (particle) $= -mg(a + x)$

Conservation of energy:

$\frac{1}{2}mv^2 + \dfrac{2mgx^2}{a} - mg(a + x) = 0.$

When $v = 0$ (lowest point), this gives $x = a$ or $-a/2$.
Since $x > 0$, $x = a$.
Maximum extension in subsequent motion is a.

5 *Remember to change all units to SI units; km h^{-1} must be changed to m s^{-1}. 54 km h^{-1} = 15 m s^{-1}*

(i)
Whole system

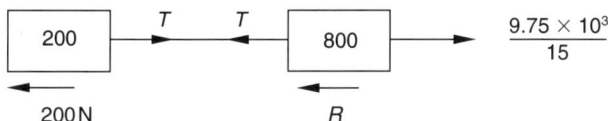

Tractive force $= \dfrac{\text{power}}{\text{velocity}} = 650$ N.

For the whole system (avoids tension):

accelerating force $= 0 \quad \Rightarrow \quad R = 450$ N.

For the trailer:

$\sum F = 0 \quad \Rightarrow \quad$ Tension $T = 200$ N

(ii)

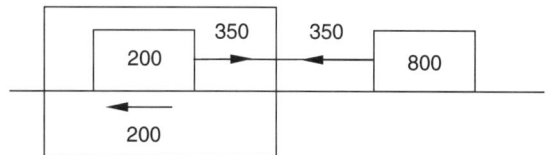

$v = 20$ m s^{-1}

Tension = 350 N
$\Rightarrow$ accelerating force (trailer) = 150 N
$\Rightarrow$ Maximum acceleration = 0.75 m s^{-2}

(iii) For the car:

Accelerating force $= \dfrac{\text{power}}{20} - 450 - 350$ N

$= 800 \times 0.75$ (maximum)

Maximum power $= (450 + 350 + 600)20$
$= 28$ kW

6 Model the boy and skate-board as a single particle of mass 40 kg.
   (a) Loss of energy = loss of g.p.e. − gain in k.e.
   $= 40 \times 9.8 \times 90 \sin 25° - \frac{1}{2} \times 40 \times 14^2$
   $= 10\,990$ J
   (b) Loss of energy
   = work done against resistances = 10 990 J.
   Resistive forces $\times$ distance moved = 10 990
   $\Rightarrow$ resistive forces = 122 N.
   (c) From B to final position, distance x.
   Loss of k.e. = work done against resistance
   $\frac{1}{2} \times 40 \times 14^2 = 122x \quad \Rightarrow \quad x = 32.1$ m

7 Angle between direction of the force and direction of displacement $= 20°$
   Work done by the force $= 20 \times 3 \times \cos 20°$
   $= 56.38$ J
   Work done against gravity = increase in g.p.e.
   $= mg \times 3 \sin 10°$
   $= 25.55$ J

8

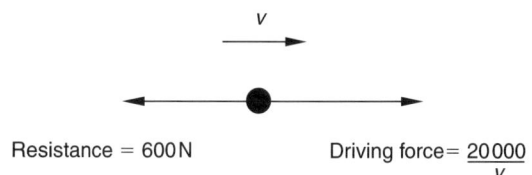

(a) At maximum speed: $\sum$ forces $= 0$

$$\Rightarrow \quad \frac{20\,000}{v} - 600 = 0 \quad \Rightarrow \quad v = \frac{100}{3} \text{ m s}^{-1}$$

$$\text{k.e.} = \frac{1}{2}(800)\left(\frac{100}{3}\right)^2 = 440 \text{ kJ}$$

(b)

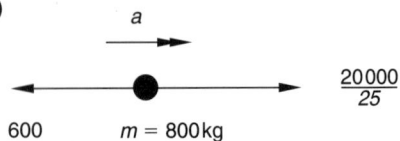

When accelerating: $\sum F = m \times$ acceleration

$$\frac{20\,000}{25} - 600 = 800a \quad \Rightarrow \quad a = 0.25 \text{ m s}^{-2}$$

(c)

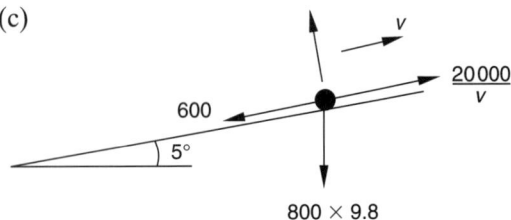

$$\sum \text{forces} = 0$$

$$\frac{20\,000}{v} - 600 - 800 \times 9.8 \sin 5° = 0$$

$$\Rightarrow \quad v = 16 \text{ m s}^{-1}$$

9

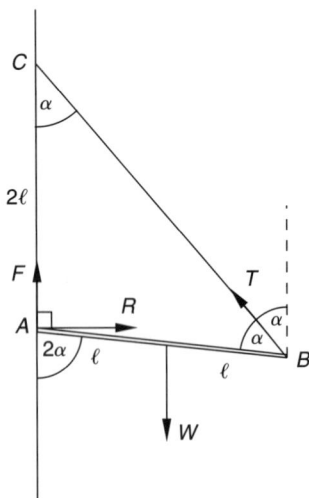

$\angle ACB = \angle ABC = \alpha$.
Taking moments about A,
$T(2l \sin \alpha) = W(l \sin 2\alpha)$.
Thus $T = W \cos \alpha$.
Resolving vertically,
$F + T \cos \alpha = W \quad \Rightarrow \quad F = W \sin^2 \alpha$, which is
positive $\Rightarrow$ F acts upwards.

Resolving horizontally,
$R = T \sin \alpha = W \sin \alpha \cos \alpha \quad \Rightarrow \quad \dfrac{F}{R} = \tan \alpha$.

Hence $\mu \geqslant \tan \alpha$.

$\cos \alpha = \tfrac{3}{4}$, $\quad CB = 4l \cos \alpha = 3l$.
Let the natural length of CB be a, then
extension $= 3l - a$.

$$T = \frac{3W}{4} = \frac{3W}{2a}(3l - a) \quad \Rightarrow \quad a = 6l - 2a \quad \Rightarrow \quad a = 2l.$$

10

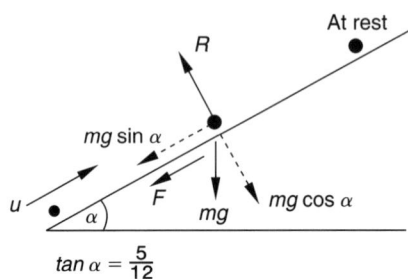

(a) Let total distance moved up the slope $= x$.

$$R = mg \cos \alpha = \frac{12}{13} mg, \quad F = \mu R = \frac{2mg}{13}.$$

Work done against weight and friction

$$= \left(mg \sin \alpha + \frac{2mg}{13}\right)x$$

$$= \frac{7mg}{13} x$$

Therefore $\dfrac{1}{2} mu^2 = \dfrac{7mg}{13} x$, $\quad x = \dfrac{13u^2}{14g}$.

(b)

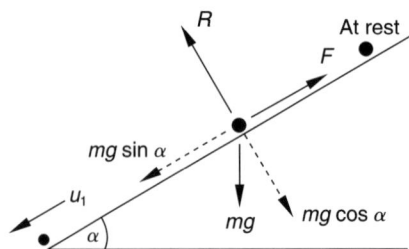

Down slope: gain in k.e. $= \dfrac{5mgx}{13} - \dfrac{2mgx}{13}$

$$\frac{1}{2} mu_1^2 = \frac{3mgx}{13} = \frac{3mu^2}{14}$$

Speed of particle at initial position

$$u_1 = u \sqrt{\left(\frac{3}{7}\right)}.$$

24

Motion in a circle

AS Level			A Level			Topic	Date attempted	Date completed	Self-assessment
CORE	MODULAR	TRADITIONAL	CORE	MODULAR	TRADITIONAL				
		✓		✓	✓	Angular velocity and acceleration			
		✓		✓	✓	Equations of motion			
		✓		✓	✓	Motion in a horizontal circle			
		✓		✓	✓	Motion in a vertical circle			
		✓		✓	✓	Satellites			

Fact Sheet

Angular Velocity

- The angular velocity ω of a particle P moving relative to a point O is defined as $\omega = v/r$, where r is the distance OP and v the velocity of the particle perpendicular to OP. ω is measured in radians per second. In the case of circular motion, r is the radius of the circle (often denoted by a) and v is the speed of the particle, so $\omega = v/a$ or $v = a\omega$.
- If the circular motion is at constant speed (both v and ω constant) and if in a time t the distance travelled is s and the angle moved through is θ then

$$v = \frac{s}{t} \quad \text{and} \quad \omega = \frac{\theta}{t}$$

If one revolution takes time T seconds then the second of these equations becomes

$$\omega = \frac{2\pi}{T} \quad \text{or} \quad T = \frac{2\pi}{\omega}$$

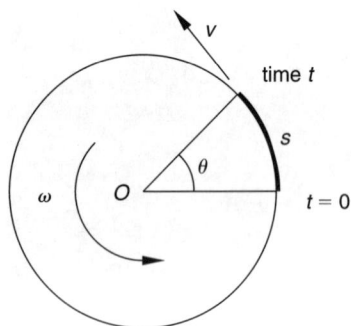

Circular Motion at Constant Speed

(i) *Vector derivation of velocity and acceleration*

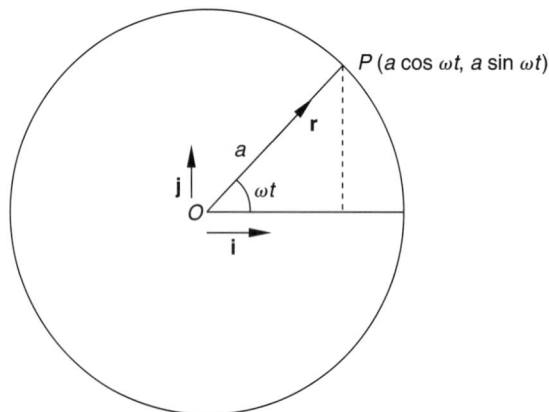

P moves in a circle, centre O, radius a, with constant angular velocity ω.
Position $\overline{OP} = \mathbf{r} = a\cos\omega t\,\mathbf{i} + a\sin\omega t\,\mathbf{j}$, distance $OP = |\mathbf{r}| = a$.
Velocity $\dot{\mathbf{r}} = -a\omega\sin\omega t\,\mathbf{i} + a\omega\cos\omega t\,\mathbf{j}$,
speed $v = |\dot{\mathbf{r}}| = a\omega$, $\qquad \mathbf{r}\cdot\dot{\mathbf{r}} = 0$ so $\mathbf{r}$ is perpendicular to $\dot{\mathbf{r}}$.
Acceleration $\ddot{\mathbf{r}} = -a\omega^2\cos\omega t\,\mathbf{i} - a\omega^2\sin\omega t\,\mathbf{j} = -\omega^2\mathbf{r}$
magnitude of acceleration $= |\ddot{\mathbf{r}}| = a\omega^2 = v^2/a$, direction towards O.

Summary of equations
Velocity $v = a\omega$ directed along the tangent

$$\text{Acceleration} = \frac{v^2}{a} = v\omega = a\omega^2 \text{ towards the centre of the circle}$$

(ii) *Equations of motion*
Since the body performing circular motion has an acceleration, Newton's Second Law is the appropriate equation of motion to use. It can be applied in any direction provided the correct components of all the forces and of acceleration are calculated for the chosen direction. In horizontal circular motion the most direct solution to problems is obtained by applying Newton's Second Law horizontally (towards the centre of the circle) and vertically. The

acceleration towards the centre is v^2/a or $v\omega$ or $a\omega^2$ depending on what is given in the question and what is wanted in the answer, and the component of acceleration vertically is zero. Typically therefore the two equations that are used are as follows.

Resolving horizontally towards the centre of the circle:

$$\sum F = \frac{mv^2}{a}$$

Resolving vertically:

$$\sum F = 0$$

In the solutions that follow, these two 'standard' equations may be referred to as 'resolving horizontally' ('r.h.') and 'resolving vertically' ('r.v.').

(iii) *Satellites*

A satellite is assumed to move in a circular path at constant speed and to be subject only to the force of gravitation from the body it is orbiting. The gravitational force of attraction between any two objects is given by Newton's Law of Gravitation which states:

$$F = \frac{GM_1M_2}{r^2}$$

where
G is the Universal Gravitational Constant ($\approx 6.673 \times 10^{-11}$ N m^2 kg^{-2})
M_1 and M_2 are the masses of the two objects
r is the distance between the centres of mass of the objects

In the case of a satellite mass m orbiting the Earth the downward force on the satellite can be expressed as $F = km/d^2$ where d is the distance from the satellite to the centre of the Earth (= height above surface + radius of Earth) and the constant k is equal to $G \times$ mass of Earth.

When $d = a$ (= the radius of the planet), $mg = km/a^2$ where g is the acceleration due to gravity experienced on the surface of the planet, giving $k = ga^2$.

Geostationary satellites above the Earth have a period of 24 hours to enable them to remain above the same point on the equator.

Circular Motion with Variable Speed

This frequently occurs when motion takes place in a vertical circle, for example when an object is swung in a vertical circle on the end of a piece of string. In this case neither ω nor v will be constant. Newton's Second Law, applied towards the centre of the circle, takes the same form as before (sometimes ω is written as $\dot{\theta}$), but vertically the sum of the forces is no longer zero; the Law of Conservation of Energy is used in place of $\sum F = 0$.

Some Common Situations

In each of the following examples acceleration is labelled f – this would then be written as v^2/a or $a\omega^2$ or $v\omega$ where a is the radius of the circular path.

Note that when solving problems it is often necessary to use simple trigonometry to relate the radius of the circular path to other parameters given in the question. In the first of the following examples, if the length of the string is l then $a = l \sin \theta$.

- *Particle on a string – 'conical pendulum'*

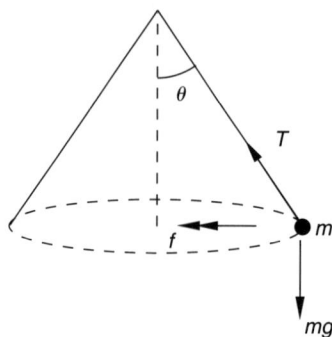

Resolving vertically: $\qquad$ $T \cos \theta = mg$
Resolving horizontally: $\qquad$ $T \sin \theta = mf$

- *Car on a banked track*
  (i) On the point of sliding outwards (at maximum speed)

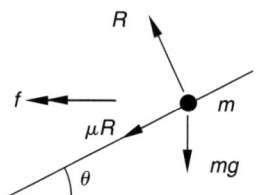

Resolving vertically: $\qquad$ $R \cos \theta - \mu R \sin \theta - mg = 0$
Resolving horizontally: $\qquad$ $R \sin \theta + \mu R \cos \theta = mf$

(ii) On the point of sliding inwards (at minimum speed)

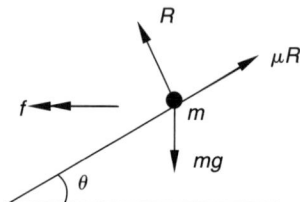

Resolving vertically: $\qquad$ $R \cos \theta + \mu R \sin \theta - mg = 0$
Resolving horizontally: $\qquad$ $R \sin \theta - \mu R \cos \theta = mf$

- *Particle tied to two separate strings*
  Note that the two tensions are not equal and that the angles α and β are determined by the lengths of the two strings.

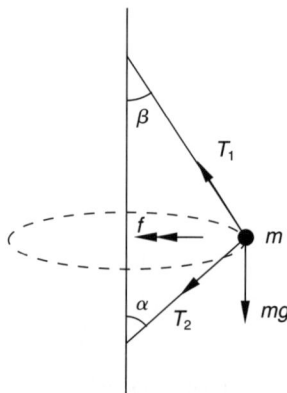

Resolving vertically: $\qquad$ $T_1 \cos \beta - T_2 \cos \alpha - mg = 0$
Resolving horizontally: $\qquad$ $T_1 \sin \beta + T_2 \sin \alpha = mf$

- *Particle threaded on a smooth string*
  Note that the tension is the same in both portions of the string. The angles α and β will depend on the mass and the speed $(\beta < \alpha)$.

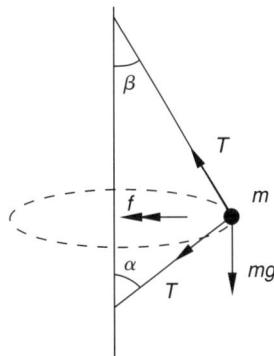

Resolving vertically: $T \cos \beta - T \cos \alpha - mg = 0$
Resolving horizontally: $T \sin \beta + T \sin \alpha = mf$

- *Particle on the inside surface of a smooth bowl*
  Note that

$$\text{radius of circular path} = \text{radius of bowl} \times \sin \theta$$

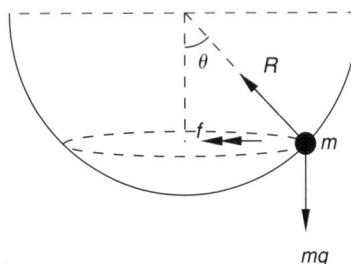

Resolving vertically: $R \cos \theta = mg$
Resolving horizontally: $R \sin \theta = mf$

- *Bead threaded on a smooth wire bent into a circle*
  R_v and R_h are the vertical and horizontal components of the reaction of the wire on the bead.

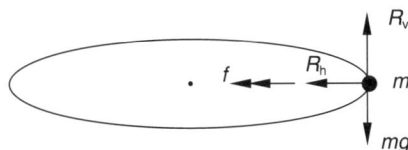

Resolving vertically: $R_v = mg$
Resolving horizontally: $R_h = mf$

Worked Examples

1   A particle P of mass m is attached to one end of a light inextensible string of length d. The other end of the string is fixed at a height h, where $h < d$, vertically above a point O on a smooth horizontal table. The particle describes a circle, centre O, on the surface of the table, with constant angular speed ω.

Find expressions, in terms of m, d, h, ω and g as appropriate, for
(a) the tension in the string,
(b) the magnitude of the force exerted on P by the plane.
(c) Hence show that

$$\omega \leqslant \sqrt{\left(\frac{g}{h}\right)}$$

The angular speed of P is now increased to $\sqrt{\left(\dfrac{3g}{h}\right)}$, and P now describes a horizontal circle, centre Q, *above* the surface of the table, with this constant angular speed.

(d) Determine, in terms of h, the distance OQ.

(L, 1993)

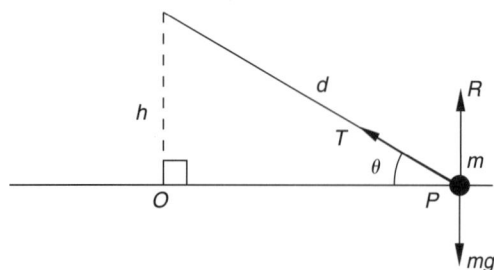

$OP = \sqrt{(d^2 - h^2)}$

(a)   r.h.: $T \cos \theta = m(OP)\omega^2$

$$T\left(\frac{OP}{d}\right) = m(OP)\omega^2$$

$$T = md\omega^2$$

(b)   r.v.: $T \sin \theta + R - mg = 0$

$$R = mg - md\omega^2\left(\frac{h}{d}\right)$$

$$= m(g - h\omega^2)$$

(c)   Particle is in contact with the plane,
∴ $R \geqslant 0$ so $m(g - h\omega^2) \geqslant 0$
But $m > 0$ ∴ $g - h\omega^2 \geqslant 0$

$$\omega^2 \leqslant \frac{g}{h}$$

$$\omega \leqslant \sqrt{\frac{g}{h}}$$

(d)

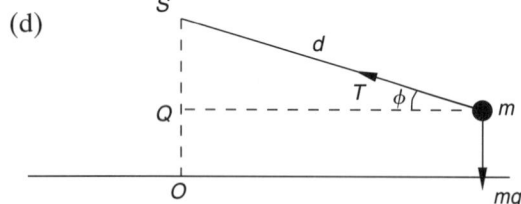

r.h.: $T \cos \phi = m(d \cos \phi)\dfrac{3g}{h}$

$$T = \frac{3mgd}{h}$$

r.v.: $T \sin \phi = mg$

$$\frac{3mgd}{h}\frac{SQ}{d} = mg$$

$$SQ = \frac{h}{3}$$

$$\therefore OQ = \frac{2h}{3}$$

2   The diagram shows a small bead B of mass m which is free to slide on a smooth, light, inextensible string of length a. One end of the string is attached to a fixed point O on a smooth horizontal table and the other end to a point A vertically above O, where $OA = \lambda a$ ($\lambda < 1$).

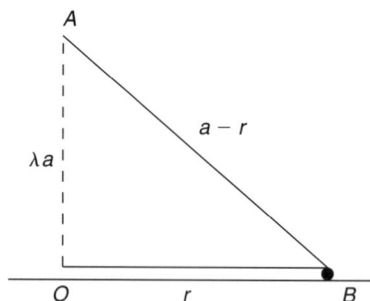

The bead remains in contact with the table and describes on it a circle of radius r with constant angular speed ω, the string being taut. Show that $r = \frac{1}{2}a(1 - \lambda^2)$.

Prove that the magnitude of the tension in the string is $\frac{1}{4}ma\omega^2(1 - \lambda^4)$.

Find, in terms of m, a, ω, λ and g, an expression for the reaction between the bead and the table.

Deduce that P, the period of the circular motion, satisfies

$$P > \pi \sqrt{\left[\frac{2a\lambda(1 - \lambda^2)}{g}\right]}$$

(AEB, 1993)

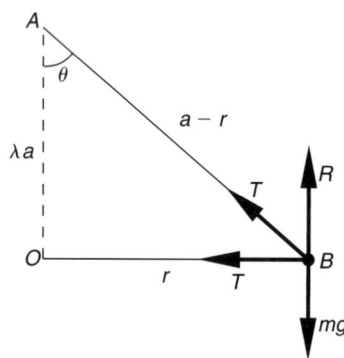

mass = m
angular speed = ω

By Pythagoras' Theorem:

$$(\lambda a)^2 + r^2 = (a - r)^2$$
$$\lambda^2 a^2 + r^2 = a^2 - 2ar + r^2$$
$$2ar = a^2 - \lambda^2 a^2$$
$$= a^2(1 - \lambda^2)$$
$$2r = a(1 - \lambda^2)$$
$$r = \tfrac{1}{2}a(1 - \lambda^2)$$

So $AB = a - r = a - \tfrac{1}{2}a(1 - \lambda^2) = \tfrac{1}{2}a(1 + \lambda^2)$

r.h.: $T + T \sin \theta = mr\omega^2$
$$T(1 + \sin \theta) = mr\omega^2$$

$$T\left(1 + \frac{r}{a-r}\right) = mr\omega^2$$

$$T\left(\frac{a}{a-r}\right) = mr\omega^2$$

$$T\left(\frac{a}{\frac{1}{2}a(1+\lambda^2)}\right) = m \times \frac{1}{2}a(1-\lambda^2)\omega^2$$

$$T = m \times \frac{1}{2}a(1-\lambda^2)\omega^2 \times \frac{1}{2}(1+\lambda^2)$$
$$= \frac{1}{4}ma\omega^2(1-\lambda^2)(1+\lambda^2)$$
$$= \frac{1}{4}ma\omega^2(1-\lambda^4)$$

r.v.: $\qquad T\cos\theta + R - mg = 0$

$$R = mg - \frac{1}{4}ma\omega^2(1-\lambda^2)(1+\lambda^2)\frac{2\lambda a}{a(1+\lambda^2)}$$

$$= m\left(g - \frac{\lambda}{2}a\omega^2(1-\lambda^2)\right)$$

For the motion to take place as described the particle must remain in contact with the plane, i.e. $R > 0$. Therefore

$$g - \frac{\lambda a\omega^2}{2}(1-\lambda^2) > 0$$

$$\omega^2 < \frac{2g}{a\lambda(1-\lambda^2)}$$

But $P = \dfrac{2\pi}{\omega}$, therefore

$$\frac{(2\pi)^2}{P^2} < \frac{2g}{a\lambda(1-\lambda^2)}$$

$$\Rightarrow \quad P^2 > \frac{2\pi^2 a\lambda(1-\lambda^2)}{g}$$

Inverting reverses the inequality

$$\Rightarrow \quad P > \pi\sqrt{\frac{2a\lambda(1-\lambda^2)}{g}}$$

3 A particle P, of mass M, moves on the smooth inner surface of a fixed hollow spherical bowl, centre O and inner radius r, describing a horizontal circle at constant speed. The centre C of this circle is at a depth $\frac{1}{3}r$ vertically below O. Determine
(a) the magnitude of the force exerted by the surface of the sphere on P
(b) the speed of P. $\qquad$ (L, 1992)

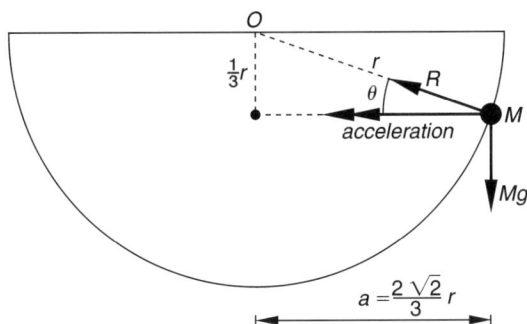

$$a = \frac{2\sqrt{2}}{3}r$$

$$\sin\theta = \frac{1}{3}$$

$$\Rightarrow \quad \cos\theta = \frac{2\sqrt{2}}{3}$$

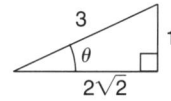

(a) r.v.: $\qquad R\sin\theta = Mg \qquad\qquad (1)$

r.h.: $\qquad R\cos\theta = M\dfrac{v^2}{a} \qquad\quad (2)$

Equation 1 $\quad\Rightarrow\quad R = 3Mg$

(b) $\therefore$ equation 2 $\quad\Rightarrow\quad v^2 = \dfrac{3Mga\cos\theta}{M}$

$$= 3g\left(\frac{2\sqrt{2}r}{3}\right)\left(\frac{2\sqrt{2}}{3}\right)$$

$$= \frac{8gr}{3}$$

$$\therefore v = 2\sqrt{\frac{2gr}{3}}$$

Be careful not to confuse the radius of the circular path, a, with the radius of the bowl, r

4 [Assume $g = 10 \text{ m s}^{-2}$.]

A circular track of radius r is banked at an angle α to the horizontal. A motorcyclist travels around the track at speed V without slipping.

The coefficient of friction between the tyres and the track is μ.

(i) Show that the least value of V is given by

$$V^2 = rg\frac{(\sin\alpha - \mu\cos\alpha)}{(\cos\alpha + \mu\sin\alpha)}$$

A fairground show called 'The Wall of Death' consists of a large cylindrical cage mounted with the axis vertical. A motorcyclist starts off on the floor of the cage and, as his speed increases, steers the motorcycle up onto the walls of the cage until he is riding around on the vertical inside surface of the cage.

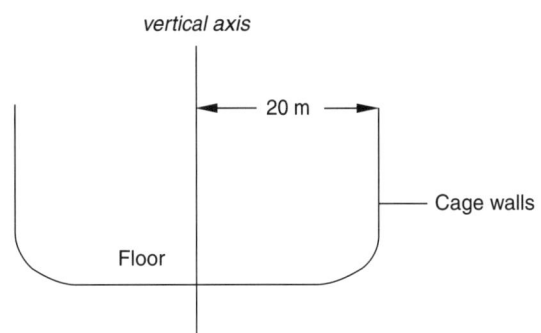

The radius of the cage is 20 metres and the coefficient of friction between the tyres and the walls of the cage is 0.8.

(ii) Using the result of (i), or otherwise, find the least speed he must maintain if he is to stay on the walls of the cage. (NICCEA, 1994)

• (i) *When travelling at the least value of V, the friction force will be maximum ($= \mu R$) and it will act up the banked track*

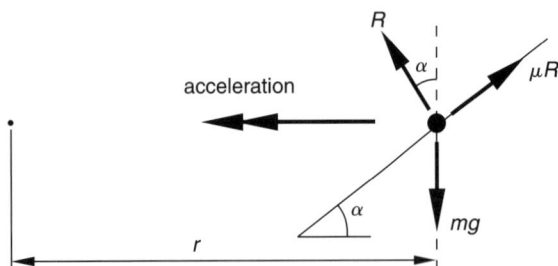

r.h.: $R \sin \alpha - \mu R \cos \alpha = m \dfrac{V^2}{r}$ (1)

r.v.: $R \cos \alpha + \mu R \sin \alpha = mg$ (2)

Equation 1 ÷ equation 2 gives:

$$\frac{R(\sin \alpha - \mu \cos \alpha)}{R(\cos \alpha + \mu \sin \alpha)} = \frac{V^2}{rg}$$

$$\Rightarrow V^2 = rg \frac{(\sin \alpha - \mu \cos \alpha)}{(\cos \alpha + \mu \sin \alpha)}$$

(ii) Using the result of part (i), with $r = 20$, $\mu = 0.8$, $\alpha = 90°$ and $g = 10$

$$V^2 = \frac{200}{0.8} = 250$$

$$V = 5\sqrt{10} \text{ m s}^{-1}$$

5 One end of a light inextensible string AB, of length $2a + b$, is attached to a point A distance $\dfrac{a}{2}$ vertically above a point C on a horizontal table. A particle of mass m is attached to the point P of the string where $AP = a$. End B of the string is threaded down through a small hole in the table at C and a mass $\dfrac{m}{2}$ is then attached at B.

The mass at B is too large to pass through the hole in the table which has a thickness b.

The particle at P is made to rotate about AC in a horizontal circle with constant angular velocity ω so that the particle at B is in contact with the underside of the table.

Find the tensions in both parts of the string and show that $a\omega^2 \geqslant 5g$.

Find the least tension the string has to sustain for this motion to be possible.

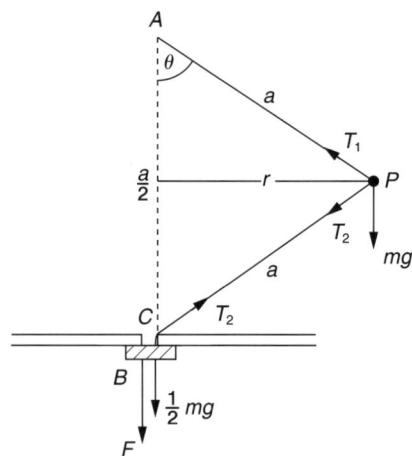

At P, r.v.: $(T_1 - T_2) \cos \theta = mg$

$\Rightarrow T_1 - T_2 = 4mg$ (1)

r.h.: $(T_1 + T_2) \sin \theta = m\omega^2 a \sin \theta$

$\Rightarrow T_1 + T_2 = m\omega^2 a$ (2)

Adding equations 1 and 2: $2T_1 = m(a\omega^2 + 4g)$

$$T_1 = \frac{m}{2}(a\omega^2 + 4g)$$ (3)

$$T_2 = \frac{m}{2}(a\omega^2 - 4g)$$ (4)

At B, force exerted by the table on particle $= F$.

r.v.: $F + \dfrac{mg}{2} = T_2$ (At B the string is vertical)

$$F = \frac{m}{2}(a\omega^2 - 5g)$$

$F \geqslant 0$ since particle is in contact with table, so

$a\omega^2 \geqslant 5g$

$$T_1 \geqslant \frac{m}{2}(5g + 4g) = \frac{9}{2}mg$$

$$T_2 \geqslant \frac{m}{2}(5g - 4g) = \frac{mg}{2}$$

$\Rightarrow$ least tension that the string must sustain $= \frac{9}{2}mg$

6 [Assume $g = 10$ m s^{-2}.]

A particle is attached to the end A of a light inextensible string OA of length 1 m. The end O is fixed at a distance of 1 m above the horizontal ground. The particle describes a horizontal circle at a height of 0.5 m above the ground.

(a) Calculate the steady speed of the particle.

(b) The string is then released at O.

(i) Describe in words the initial direction of the subsequent motion of the particle.

(ii) Calculate the horizontal displacement of the particle from its position when the string is released to where it first strikes the ground. (OLE, 1992)

- (a)

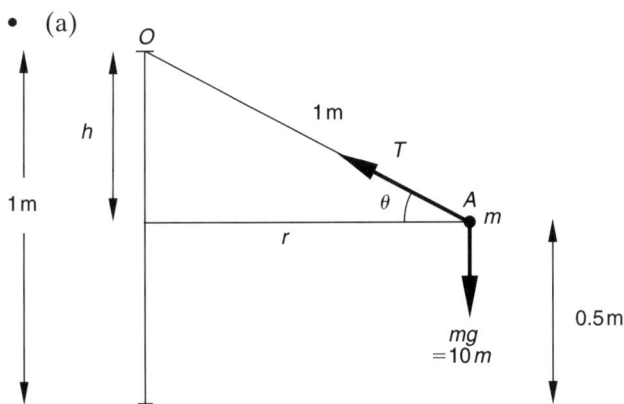

$h = 1 - 0.5 = 0.5$ m

$\sin \theta = 0.5 \quad \Rightarrow \quad \theta = 30° \quad \Rightarrow \quad r = \dfrac{\sqrt{3}}{2}$

r.h.: $T \cos \theta = \dfrac{mv^2}{r}$ (1)

r.v.: $T \sin \theta = mg$ (2)

Equation 2 ÷ equation 1 gives

$\tan \theta = \dfrac{gr}{v^2}$

$v^2 = \dfrac{gr}{\tan \theta}$

$= \dfrac{10 \times \dfrac{\sqrt{3}}{2}}{\dfrac{1}{\sqrt{3}}}$

$= 15$

$v = \sqrt{15}$ m s^{-1} or   3.87 m s^{-1}

(b) (i)   Initially the particle's velocity is tangential to the circle (so is horizontal).

(ii)

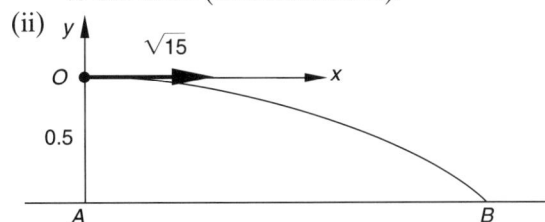

$a_x = 0, a_y = -10$

$u_x = \sqrt{15}, u_y = 0$

Using $s = ut + \frac{1}{2}at^2$

in the y-direction: $-0.5 = \frac{1}{2}(-10)t^2$

$t^2 = 0.1$

in the x-direction: $AB = \sqrt{15}\,\sqrt{0.1}$

horizontal displacement $= \sqrt{\frac{3}{2}}$ m

7   [Assume $g = 9.8$ m s^{-2}.]

At a fairground the 'chair o' planes' are small chairs attached by chains 2.5 metres long to the ends of horizontal arms 1.5 metres long, radiating from a central pillar about which the arms rotate. To simplify the model, ignore the weight of the chains

and consider a stage in the motion when the horizontal arms are rotating uniformly with the chains making an angle of 30° to the vertical, in the vertical plane containing the horizontal arm, as shown in the diagram.

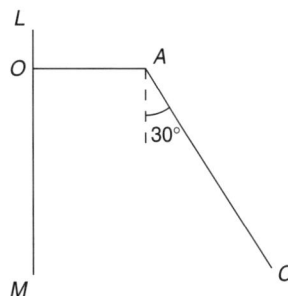

LM represents the central pillar, OA a horizontal arm and AC the chain attaching a chair to this arm. Assume that the chain AC is in the vertical plane containing the arm OA.

For the particular stage of the motion described above in this simplified model, calculate the speed of the chairs and the time for one complete revolution.

(SEB, 1993)

- Let the mass of a chair be m kg

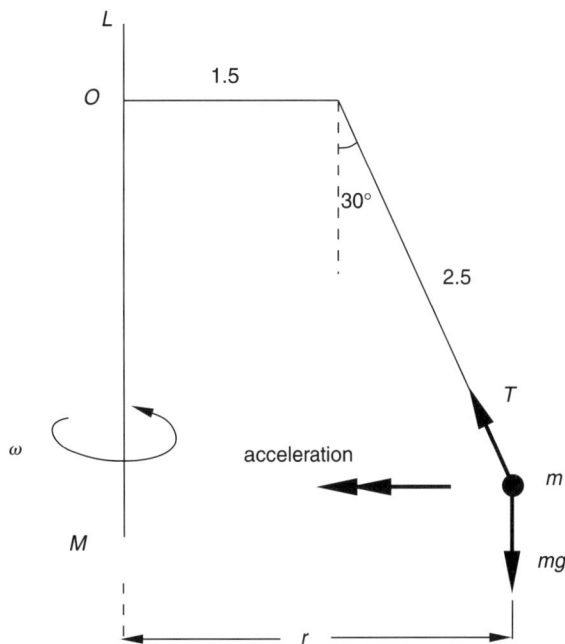

r.h.: $T \sin 30° = mr\omega^2$

$= m(1.5 + 2.5 \sin 30°)\omega^2$

$T = 5.5m\omega^2$ (1)

r.v.: $T \cos 30° = mg$

$T = 9.8 \times \dfrac{2}{\sqrt{3}}m$

$= 11.316m$ (2)

Equations 1 and 2 give

$$5.5\omega^2 = 11.316$$
$$\omega = 1.434 \text{ rad s}^{-1}$$

$$v = r\omega \quad \Rightarrow \quad v = (1.5 + 2.5 \sin 30°) \times 1.434$$
$$= 3.94 \text{ m s}^{-1}$$

Time for one revolution $= \dfrac{2\pi}{\omega} = 4.38$ s

Alternatively, in equation 1 use v^2/r instead of $r\omega^2$, then equations 1 and 2 give v directly. Then find ω using $\omega = v/r$ and hence the time as above

8 A particle P of mass m is placed at the highest point on the outside of a fixed smooth hollow sphere of radius a and centre O. The particle P is just disturbed from rest. Assuming that P remains in contact with the sphere, show that the reaction of the sphere on P is

$$mg(3 \cos \theta - 2)$$

where θ is the angle between the upward vertical and the radius OP.

Write down the value of $\cos \theta$ at the point where P leaves the sphere.

A particle Q, on the inside of the sphere, is projected horizontally from the lowest point of the sphere with speed u. Find u, in terms of a and g, so that both P and Q leave the surface of the sphere at the same height above O. (WJEC, 1994)

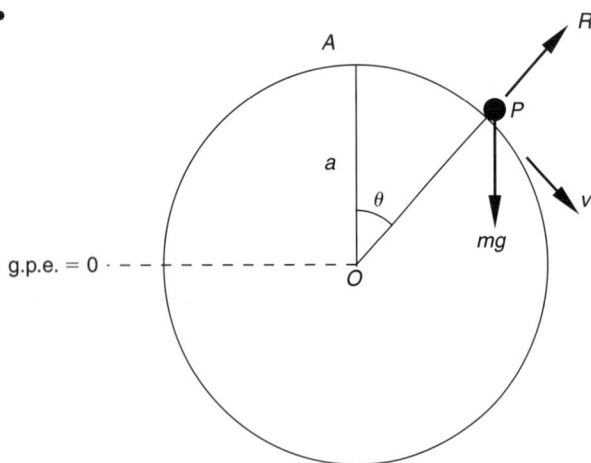

Newton's 2nd Law in direction PO:

$$mg \cos \theta - R = m \frac{v^2}{a} \tag{1}$$

Conservation of energy from A to P:

$$mga(1 - \cos \theta) = \tfrac{1}{2}mv^2 \tag{2}$$

Gravitational potential datum has been taken to be at A; any choice would do since it is the change of potential energy that really matters

From equation 2: $\quad mv^2 = 2mga(1 - \cos \theta)$
Substitute this into equation 1:

$$R = mg \cos \theta - 2mg(1 - \cos \theta)$$
$$= mg \cos \theta - 2mg + 2mg \cos \theta$$
$$= mg(3 \cos \theta - 2)$$

It leaves the sphere when $R = 0$ so $\cos \theta = \tfrac{2}{3}$.

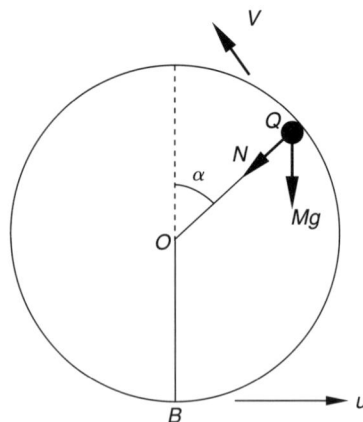

Let the particle Q have mass M and consider it in the position shown while still in contact with the sphere.

Newton's 2nd Law in direction QO:

$$N + Mg \cos \alpha = \frac{MV^2}{a} \tag{1}$$

Conservation of energy from B to Q:

$$\tfrac{1}{2}Mu^2 = \tfrac{1}{2}MV^2 + Mga(1 + \cos \alpha) \tag{2}$$

At the point where it leaves contact, $N = 0$ and $\cos \alpha = \tfrac{2}{3}$ (from the question and from the answer to the first part), so, from equation 1:

$$V^2 = \frac{2ga}{3}$$

Substitute into equation 2:

$$u^2 = \frac{2ga}{3} + 2ga(1 + \tfrac{2}{3})$$

$$= 4ga$$
$$\therefore u = 2\sqrt{ga}$$

9 [Assume $g = 9.8$ m s^{-2}.]

A 'big wheel' at a fairground is rotating in a vertical circle at a constant rate of 1 revolution per minute.
(i) Calculate the angular speed of the wheel in radians per second.
(ii) Calculate the linear speed of a point 10 m from the axis of rotation.
A child of mass 40 kg is sitting 10 m from the axis of rotation on a seat on the wheel. The situation may be modelled by assuming that there is negligible distance between the seat and the child's centre of mass and that the seat remains horizontal at all times. When the radius joining the seat to the axis is inclined at θ to the upward vertical, the reaction R

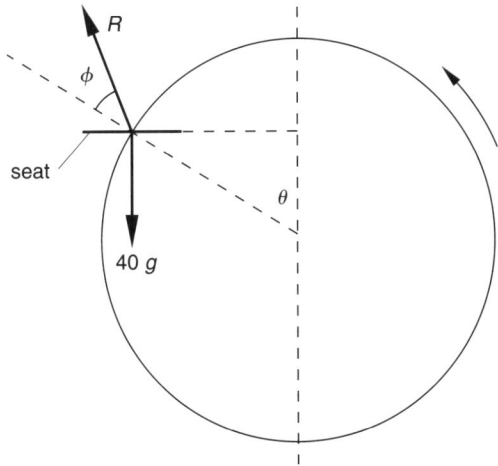

of the seat on the child is inclined at ϕ to that radius, as shown in the diagram.

(iii) By resolving the forces acting on the child in a direction perpendicular to the radius explain why

$$R \sin \phi = 40g \sin \theta$$

(iv) Write down the equation of motion of the child for the radial direction.

(v) Given that $\theta = \dfrac{\pi}{6}$, find R and ϕ.

(O&C MEI, 1993)

• (i) 1 r.p.m. $= \dfrac{2\pi}{60} = \dfrac{\pi}{30}$ rad s^{-1}

(ii) $v = r\omega = 10\left(\dfrac{\pi}{30}\right) = \dfrac{\pi}{3}$ m s^{-1}

(iii)

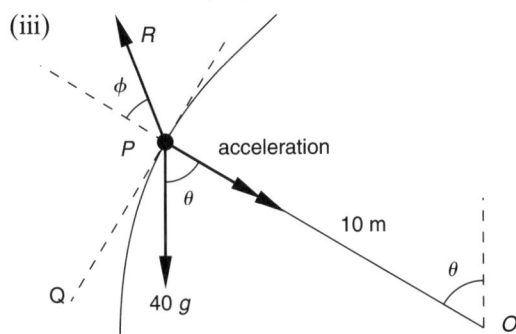

The angle that the weight makes with the radius is θ.

Since the acceleration is towards O (circular motion at constant speed) its tangential component is zero, therefore:
Resolving tangentially (along PQ):

$40g \sin \theta - R \sin \phi = 0$
$R \sin \phi = 40g \sin \theta$ (1)

(iv) Resolving radially:

$$40g \cos \theta - R \cos \phi = 40(10)\left(\dfrac{\pi}{30}\right)^2$$

$$40g \cos \theta - R \cos \phi = \dfrac{4\pi^2}{9}$$ (2)

(v) When $\theta = \dfrac{\pi}{6}$

Equation 1 $\Rightarrow$ $R \sin \phi = 20g$ (3)

Equation 2 $\Rightarrow$ $R \cos \phi = 40g \dfrac{\sqrt{3}}{2} - \dfrac{4\pi^2}{9}$

     (4)

From equation 3 ÷ equation 4:

$\tan \phi = 0.5849$
$\phi = 30.3°$

Substitute into equation 1: $R = 388$ N

10 Assume that the Moon moves with a constant speed in a circle around the Earth. Given that the Earth's radius is 6400 km, the distance between the Moon and the surface of the Earth is 384 000 km and that the acceleration due to gravity at the Earth's surface is 10 m s^{-2},

(i) show that the angular velocity of the Moon around the Earth is 2.62×10^{-6} radians per second, and

(ii) find the time, in days, taken for the Moon to make one revolution of the Earth. Give your answer to two decimal places.

(iii) What important assumption have you made about the Moon in your calculations?

(NICCEA, 1994)

• Let the mass of the Earth $= M_1$
Let the mass of the Moon $= M_2$

(i) Consider a mass m on the Earth's surface. By Newton's Law of Gravitation the force of attraction from the Earth is

$$\dfrac{GM_1 m}{(6400 \times 10^3)^2}$$

So by Newton's 2nd Law, applied towards the centre of the Earth:

$$\dfrac{GM_1 m}{(6400 \times 10^3)^2} = m \times 10$$

$$\Rightarrow GM_1 = 4.096 \times 10^{14}$$

Now consider the Earth and the Moon. The force of attraction is

$$\dfrac{GM_1 M_2}{(390\,400 \times 10^3)^2} = \dfrac{4096 \times 10^{11} M_2}{1.5241 \times 10^{17}}$$

$$= 0.002\,687 M_2$$

Distance between the Moon and the centre of the Earth is 390 400 km

So by Newton's 2nd Law, applied towards the centre of the Earth:

Force of attraction $= mr\omega^2$
$0.002\,687\,\cancel{M_2} = \cancel{M_2} \times 390\,400 \times 10^3 \omega^2$
$\omega^2 = 6.8827 \times 10^{-12}$
$\omega = 2.62 \times 10^{-6}$

(ii) $T = \dfrac{2\pi}{\omega} = 239\,4774\ \text{s} = 27.72\ \text{days}$

(iii) That the Moon can be regarded as a particle orbiting 384 000 km from the Earth's surface.

Exercises

1   A particle moves with constant angular speed around a circle of radius a and centre O. The only force acting on the particle is directed towards O and is of magnitude k/a^2 per unit mass, where k is a constant. Find, in terms of k and a, the time taken for the particle to make one complete revolution.
   (AEB, 1991)

2   (In this question take g as 9.8 m s^{-2}.)

   A child of mass 30 kg keeps herself amused by swinging on a 5 m rope, attached to an overhanging tree. She is holding on to the lower end of the rope and 'swinging' in a horizontal circle of radius 3 m.
   (a) Draw a diagram to show the forces acting on the girl.
   (b) Find the tension in the rope.
   (c) Show that the time she takes to complete a circle is approximately 4 seconds.
   (d) State any assumptions that you have made about the rope.
   (e) The girl's older brother then swings, on his own, on the rope in a horizontal circle of the same radius. Show that the tension in the rope is

   now $\dfrac{5mg}{4}$ where m is his mass. Find the time

   that it takes for him to complete one circle.
   (AEB Specimen Paper, 1996)

3

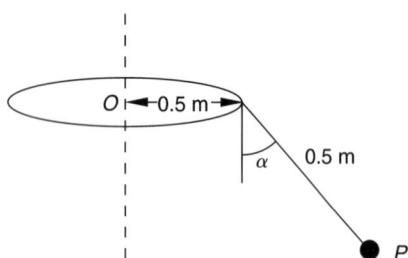

   A light, inextensible string of length 0.5 m is attached to a point on the circumference of a disc of radius 0.5 m. To the other end of the string is attached a small stone P of mass M kilograms. The disc is made to rotate in a horizontal plane, about a vertical axis through its centre O. At each instant, the string lies in a vertical plane through O. Given that the disc rotates at a constant rate of 2 rad s^{-1}, and that the string makes a constant angle α with the downward vertical, as shown in the figure, show that

   $\cot \alpha + \cos \alpha = 4.9$

   [Assume $g = 9.8$ m s^{-2}.]   (L, 1993)

4   [In this question you should assume $g = 9.8$ m s^{-2}.]

   A device in a fun-fair consists of a hollow circular cylinder of radius 3 m, with a horizontal floor and vertical sides. A small child stands inside the cylinder and against the vertical side. The cylinder is rotated about its vertical axis of symmetry.

   When the cylinder is rotating at a steady angular speed of 30 revs/min the floor of the cylinder is lowered, so that the child is in contact only with the vertical side.

   Given that the child does not slip, find, correct to two decimal places, the coefficient of friction between the child and the side.   (AEB, 1994)

5

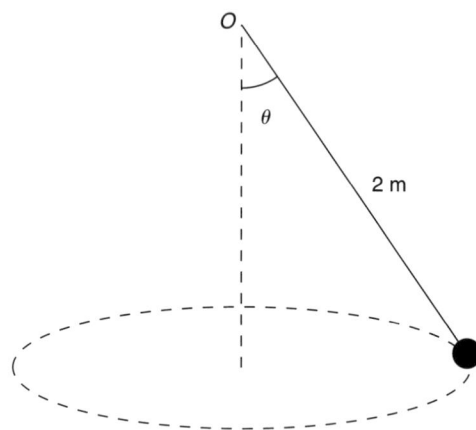

   In a simple model of a 'rotating swing', a particle of mass 30 kg is attached to one end of a light inextensible rope of length 2 m. The other end is attached to a fixed point O. The particle moves in a horizontal circle at a constant angular speed of 3 rad s^{-1}. The rope is inclined at a constant angle θ to the vertical. Find θ.

   In a more refined model, the rope is assumed to be elastic, with natural length 2 m and modulus of elasticity 10 000 N. The angular speed is the same as before. Find the angle θ in this case.

   [Assume $g = 9.81$ m s^{-2}.]
   (UCLES Specimen Paper, 1994)

6   A hollow circular cylinder of radius 0.4 metres is fixed with its axis horizontal. A body of mass 0.5 kg moves on the inside smooth surface of the cylinder and in a vertical plane perpendicular to the cylinder's axis. Find the least speed which the body must have at the lowest point of its path if it travels in complete circles.

   If the body was projected from the lowest point, A,

   inside the cylinder, with initial speed $\sqrt{\dfrac{3g}{2}}$ m s^{-1},

   where g is the magnitude of the acceleration due to gravity, to travel in the same vertical plane as before, show that it would leave the cylinder at a point such that the radius of the circle to that point makes an angle of $\cos^{-1}\left(\frac{7}{12}\right)$ with the upward vertical.

Find also the greatest vertical height above the lowest point A that the body would reach in its subsequent motion.

[Assume $g = 9.8$ m s^{-2}.]   (SEB, 1993)

7   A particle moves with constant speed u in a horizontal circle of radius a on the inside of a fixed smooth spherical shell of internal radius $2a$. Show that $u^2\sqrt{3} = ag$.   (AEB, 1994)

8   A spaceship of mass 10^4 kg is in a circular orbit 10^6 m above the surface of a planet whose diameter is 4×10^6 m. The mutual force of attraction between the spaceship and the planet can be written as $\dfrac{k}{r^2}$ where k is a constant with units N m^2 and r is the distance in metres between the centre of the planet and the spaceship.
   (a) The acceleration due to gravity on the planet's surface is 1.5 m s^{-2}. Show that $k = 6 \times 10^{16}$.
   (b) Find
        (i)   the period of the orbit
        (ii) the speed v m s^{-1} of the spaceship.   (OLE, 1993)

9   One end of a light inextensible string of length l is fixed at a point A and a particle P of mass m is attached to the other end. The particle moves in a horizontal circle with constant angular speed ω. Given that the centre of the circle is vertically below A and that the string remains taut with AP inclined at an angle α to the downward vertical, find $\cos \alpha$ in terms of l, g and ω.   (AEB, 1992)

Brief Solutions

1

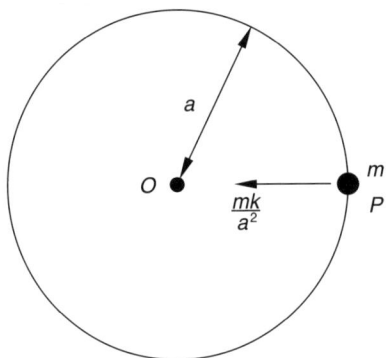

Resolve along PO:

$$\frac{mk}{a^2} = ma\omega^2 \quad\Rightarrow\quad \omega^2 = \frac{k}{a^3}$$

$$T = \frac{2\pi}{\omega} = 2\pi a\sqrt{\frac{a}{k}}$$

2   (a)

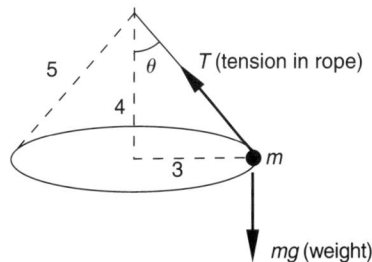

(b) r.v.: $T \cos \theta = mg$
              $\Rightarrow\quad T = 367.5$ N

(c) r.h.: $T \sin \theta = mr\omega^2$
              $\Rightarrow\quad \omega^2 = 2.45$

              $\Rightarrow\quad$ Period $= \dfrac{2\pi}{\sqrt{2.45}} = 4.01$ s (3 s.f.)

(d) It is light, inextensible and does not break.

(e) r.v.: $T_1 \cos \theta = mg \quad\Rightarrow\quad T_1 = \dfrac{5mg}{4}$

      r.h.: $T_1 \sin \theta = mr\omega^2 \quad\Rightarrow\quad$ Period $= 4.01$ s
      Period is independent of the mass

3

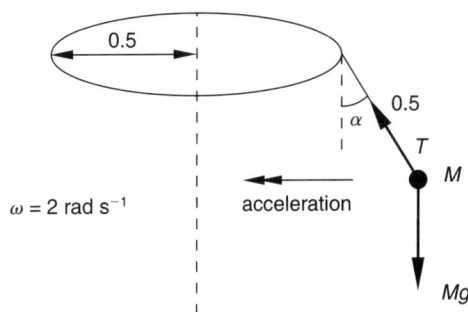

r.v.: $T \cos \alpha = Mg$
r.h.: $T \sin \alpha = M(0.5 + 0.5 \sin \alpha)2^2$
              $= 2M(1 + \sin \alpha)$

Divide: $\cot \alpha = \dfrac{9.8}{2(1 + \sin \alpha)}$

              $\Rightarrow\quad \cot \alpha + \cos \alpha = 4.9$

4

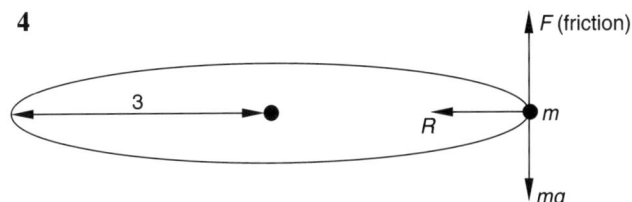

$$\omega = \frac{30 \times 2\pi}{60} = \pi \text{ rad s}^{-1}$$

r.v.: $F = mg$
r.h.: $R = m(3)\pi^2$

Not slipping $\therefore F \leq \mu R \quad\Rightarrow\quad \mu \geq \dfrac{F}{R} = 0.33$

$$\mu \geq 0.33$$

5

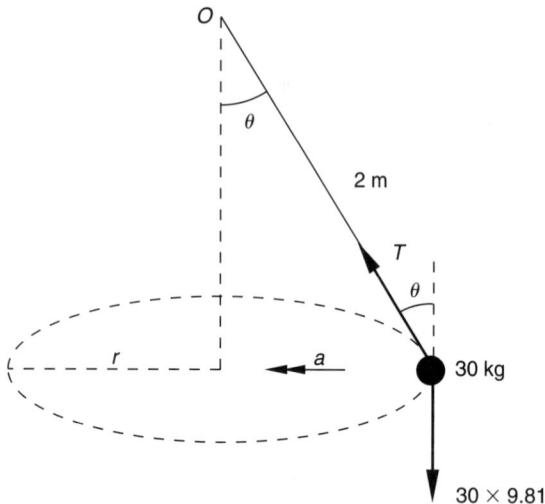

r.v.: $\quad T \cos \theta = 30 \times 9.81$

r.h.: $\quad T \sin \theta = 30 \times 2 \sin \theta \times 3^2$

$\qquad \Rightarrow \quad T = 540, \theta = 57°$

Now let length of rope $= 2 + x$ so

$$r = (2 + x) \sin \theta \text{ and } T = \frac{\lambda x}{l} = 5000x$$

r.v.: $\quad 5000x \cos \theta = 30 \times 9.81$

r.h.: $\quad 5000x \sin \theta = 30(2 + x) \sin \theta \times 9$

$\qquad \Rightarrow \quad x = 0.114 \text{ m}$

$\qquad \Rightarrow \quad \cos \theta = 0.5156 \quad \Rightarrow \quad \theta = 59°$

6

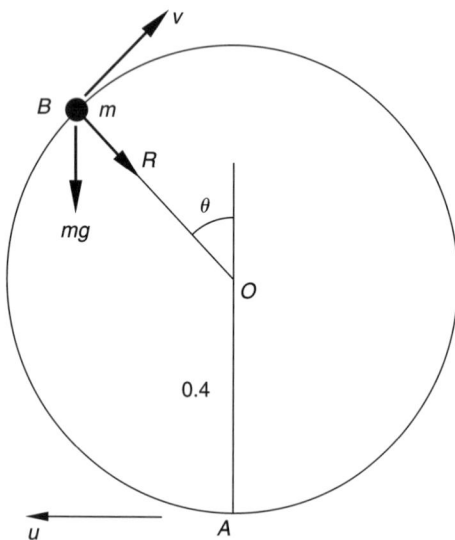

Consider a general situation first so that it can be used in both parts of the answer. Let speed initially be u, and v when angle with upward vertical is θ (still in contact)

Resolve along BO: $\quad R + mg \cos \theta = \dfrac{mv^2}{0.4}$ $\qquad$ (1)

Energy is conserved. Taking A as the potential energy datum,

$$\tfrac{1}{2}mu^2 = \tfrac{1}{2}mv^2 + mg\, 0.4(1 + \cos \theta)$$

$\Rightarrow \quad u^2 = v^2 + 0.8g(1 + \cos \theta)$ $\qquad$ (2)

In order to just perform complete circles, $R = 0$ when $\theta = 0$.

Equation 1 $\quad \Rightarrow \quad v^2 = 0.4g$

Equation 2 $\quad \Rightarrow \quad u^2 = 2g \quad \Rightarrow \quad u_{\min} = 4.43 \text{ m s}^{-1}$

Given $u^2 = \dfrac{3g}{2}$ so equation 2 gives

$$v^2 = g(0.7 - 0.8 \cos \theta).$$

This body leaves the cylinder when $R = 0$, so equation 1 gives

$0.4g \cos \theta = v^2 = g(0.7 - 0.8 \cos \theta)$

$\Rightarrow \quad \theta = \cos^{-1}\left(\tfrac{7}{12}\right)$

At the point of departure, the horizontal component of velocity is $v \cos \theta = 0.882$ (and this remains constant).

Equating the energy at the lowest point to the highest point:

$$\frac{1}{2}m\left(\frac{3g}{2}\right) = \frac{1}{2}m(0.882)^2 + mgh \quad \Rightarrow \quad h = 0.71 \text{ m}$$

7

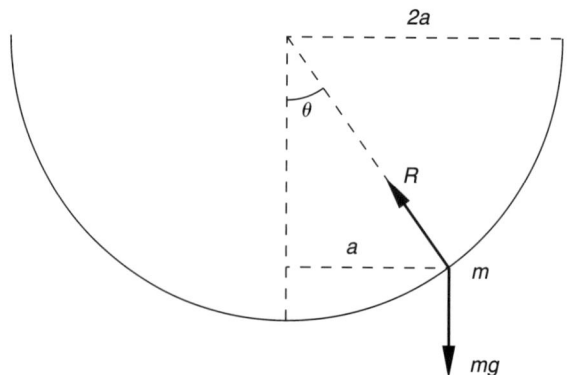

$\theta = 30°$

r.v.: $\qquad R \cos 30° = mg$

r.h.: $\qquad R \sin 30° = m\dfrac{u^2}{a}$

$\qquad \Rightarrow \quad \tan 30° = \dfrac{u^2}{ag} \quad \Rightarrow \quad u^2\sqrt{3} = ag$

8 (a) If the spaceship is on the surface

$$10^4 \times 1.5 = \frac{k}{(2 \times 10^6)^2} \quad \Rightarrow \quad k = 6 \times 10^{16}$$

(b) 10^6 m above the surface gives

$\qquad r = 10^6 + 2 \times 10^6 = 3 \times 10^6$

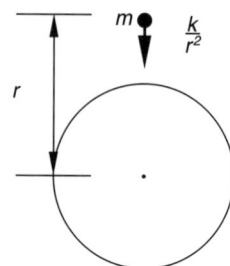

By Newton's 2nd Law:

$$\frac{k}{r^2} = mr\omega^2 \quad \Rightarrow \quad \omega^2 = 2.22 \times 10^{-7}$$

(i)   Period $T = \dfrac{2\pi}{\omega} = 3.70$ h

(ii) $v = r\omega = 1414$ m s^{-1}

9

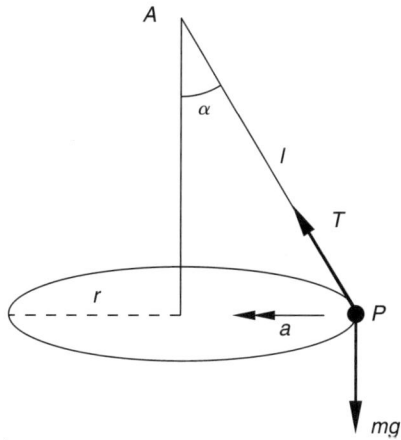

$r = l \sin \alpha$

r.v.: $T \cos \alpha = mg$

r.h.: $T \sin \alpha = m(l \sin \alpha)\omega^2$

$$\Rightarrow \quad T = ml\omega^2 \quad \Rightarrow \quad \cos \alpha = \frac{g}{l\omega^2}$$

25

Variable acceleration and SHM

$$\int \frac{x^4}{1} \frac{1}{1}$$

AS Level			A Level			Topic			
CORE	MODULAR	TRADITIONAL	CORE	MODULAR	TRADITIONAL		Date attempted	Date completed	Self-assessment
	✓	✓		✓	✓	**Derivation of differential equations**			
				✓	✓	**Solution of differential equations**			
		✓		✓	✓	**SHM:** $\ddot{x} = -\omega^2 x \Rightarrow x = a\cos(\omega t + \alpha)$			
		✓		✓	✓	**Simple pendulum**			

Fact Sheet

Equations of Motion

For straight-line motion

$$F = m\ddot{x} \quad \text{so} \quad \ddot{x} = \frac{F}{m}$$

where

$$\ddot{x} = \frac{d^2x}{dt^2} = \frac{dv}{dt} = \frac{dv}{dx}\frac{dx}{dt} = v\frac{dv}{dx} = \frac{d}{dx}(\tfrac{1}{2}v^2)$$

$$\left(x = \text{distance}, \quad t = \text{time} \quad \text{and} \quad v = \frac{dx}{dt} = \text{speed}\right).$$

Differential Equations

Decide which type of differential equation you have or which form of answer you require.

(i) $\dfrac{dv}{dt} = f_1(t) \quad \Rightarrow \quad \displaystyle\int dv = \int f_1(t)\, dt \quad \Rightarrow \quad v = F_1(t)$

(ii) $\dfrac{dv}{dt} = f_2(v) \quad \Rightarrow \quad \displaystyle\int \frac{dv}{f_2(v)} = \int dt \quad \Rightarrow \quad t = F_2(v)$

(iii) $v\dfrac{dv}{dx} = f_3(v) \quad \Rightarrow \quad \displaystyle\int \frac{v\, dv}{f_3(v)} = \int dx \quad \Rightarrow \quad x = F_3(v)$

(iv) $v\dfrac{dv}{dx} = f_4(x) \quad \Rightarrow \quad \displaystyle\int v\, dv = \int f_4(x)\, dx \quad \Rightarrow \quad \frac{v^2}{2} = F_4(x)$

Arbitrary Constants

Each general solution of a first-order differential equation includes an unknown, or arbitrary constant. It may be found either by substitution of the given values or by expressing the integrals as definite integrals.

Example

$$\frac{dv}{dt} = 5\omega \sin \omega t \quad \text{given that } v = 5 \text{ when } t = 0$$

Either

$$\int dv = \int 5\omega \sin \omega t\, dt \quad \Rightarrow \quad v = -5 \cos \omega t + C$$

$v = 5$ when $t = 0$ so $5 = -5 + C \quad \Rightarrow \quad C = 10, \quad v = 10 - 5 \cos \omega t$

or

$$\int_5^v dv = \int_0^t 5\omega \sin \omega t\, dt \quad \Rightarrow \quad [v]_5^v = [-5 \cos \omega t]_0^t$$

$$\Rightarrow \quad v - 5 = -5 \cos \omega t + 5 \quad \Rightarrow \quad v = 10 - 5 \cos \omega t$$

Simple Harmonic Motion

$\ddot{x} = -\omega^2 x$ has a general solution:

$$x = A \sin \omega t + B \cos \omega t \quad \text{period} \ \frac{2\pi}{\omega}$$

Note: ωt is measured in radians.

(i) If $t = 0$ when the particle passes through the centre of oscillation ($x = 0$), then $x = A \sin \omega t$, amplitude A.

(ii) If $t = 0$ when the particle has maximum displacement ($x = B$), then $x = B \cos \omega t$, amplitude B.

(iii) In other cases, $x = A \sin \omega t + B \cos \omega t$, or $x = C \cos (\omega t + \alpha)$, amplitude $\surd(A^2 + B^2) = C$.

Speed as a function of displacement

$$\ddot{x} = -\omega^2 x \quad \Rightarrow \quad v\frac{\mathrm{d}v}{\mathrm{d}x} = -\omega^2 x \quad \Rightarrow \quad v = \omega\surd(a^2 - x^2)$$

Notes

(i) Maximum displacement $x = \pm a$ occurs when $v = 0$.

(ii) Maximum speed $v = a\omega$ occurs when $x = 0$.

(iii) Maximum acceleration $\ddot{x} = \mp\omega^2 a$ occurs when $x = \pm a$.

(iv) The accelerating force always acts towards the centre of oscillation.

(v) For a particle on an elastic string the centre of oscillation is the position of static equilibrium.

Worked Examples

1 At time t seconds, where $t \geqslant 0$, the acceleration of a particle P, moving in a straight line, is $(2t - 8)$ m s^{-2}. Given that the initial speed of P is 12 m s^{-1}, find the times at which P comes instantaneously to rest. (L, 1993)

• Acceleration $= \dfrac{\mathrm{d}v}{\mathrm{d}t} = 2t - 8$

$$\therefore \int_{12}^{v} \mathrm{d}v = \int_{0}^{t} (2t - 8) \, \mathrm{d}t$$

$$[v]_{12}^{v} = [t^2 - 8t]_{0}^{t}$$

$$v - 12 = t^2 - 8t$$
$$v = t^2 - 8t + 12 = (t - 2)(t - 6)$$

The particle is instantaneously at rest when $t = 2$ s or 6 s.

2 A particle of mass 2 kg moves under the action of a constant force $(2\mathbf{i} + 4\mathbf{j})$ N. At time $t = 0$ the particle is stationary and at the point with position vector $(2\mathbf{i} + 5\mathbf{j})$ m. Find the position vector of the particle at time $t = 3$ seconds. (AEB, 1994)

• Force $= \dfrac{\mathbf{i}}{\mathbf{j}}\!\begin{pmatrix} 2 \\ 4 \end{pmatrix}$ Acceleration $= \dfrac{\text{force}}{\text{mass}} = \dfrac{\mathbf{i}}{\mathbf{j}}\!\begin{pmatrix} 1 \\ 2 \end{pmatrix}$

Velocity $= \int(\text{acceleration}) \, \mathrm{d}t$

$$\Rightarrow \quad \mathbf{v} = \dfrac{\mathbf{i}}{\mathbf{j}}\!\begin{pmatrix} t + c_1 \\ 2t + c_2 \end{pmatrix}$$

When $t = 0$, $\mathbf{v} = 0 \quad \Rightarrow \quad c_1 = c_2 = 0$

Displacement $= \int(\text{velocity}) \, \mathrm{d}t$

$$\mathbf{s} = \dfrac{\mathbf{i}}{\mathbf{j}}\!\begin{pmatrix} t^2/2 + c_3 \\ t^2 + c_4 \end{pmatrix}$$

When $t = 0$, $\mathbf{s} = \begin{pmatrix} 2 \\ 5 \end{pmatrix} \quad \Rightarrow \quad \mathbf{s} = \dfrac{\mathbf{i}}{\mathbf{j}}\!\begin{pmatrix} t^2/2 + 2 \\ t^2 + 5 \end{pmatrix}.$

When $t = 3$ the position vector of the particle is

$$\mathbf{s} = \dfrac{\mathbf{i}}{\mathbf{j}}\!\begin{pmatrix} 6.5 \\ 14 \end{pmatrix} \text{m or } (6.5\mathbf{i} + 14\mathbf{j}) \text{ m}$$

3 A particle starts with speed 20 m s^{-1} and moves in a straight line. The particle is subjected to a resistance which produces a retardation which is initially 5 m s^{-2} and which increases uniformly with the distance moved, having a value of 11 m s^{-2} when the particle has moved a distance of 12 m. Given that the particle has speed v m s^{-1} when it has moved a distance of x m, show that, while the particle is in motion,

$$v \frac{dv}{dx} = -\left(5 + \frac{x}{2}\right)$$

Hence, or otherwise, calculate the distance moved by the particle in coming to rest. (L)

• Retardation = 5 + λx.   [6.1]
When $x = 12$, retardation = 11 m s^{-2} so $\lambda = \frac{1}{2}$.

Acceleration $= v \frac{dv}{dx} = -\left(5 + \frac{x}{2}\right)$.

$$\int_{20}^{0} -v \, dv = \int_{0}^{x_1} \left(5 + \frac{x}{2}\right) dx$$

where x_1 is the distance moved in metres in coming to rest.

$$-\left[\frac{v^2}{2}\right]_{20}^{0} = \left[5x + \frac{x^2}{4}\right]_{0}^{x_1}$$

$$\frac{400}{2} = 5x_1 + \frac{x_1^2}{4}$$

$$x_1^2 + 20x_1 - 800 = 0$$
$$(x_1 - 20)(x_1 + 40) = 0$$

Required answer must be positive so the particle moves 20 m in coming to rest.

4 A particle moves with simple harmonic motion about a mean position O (i.e. acceleration $= -\omega^2 x$, where x is the displacement from O). When passing through two points 1.5 m and 2.0 m from O the particle has speeds 4 m s^{-1} and 3 m s^{-1} respectively.
(i) Find ω.
(ii) Find the amplitude of the motion.
(iii) Find the periodic time of the motion.
(NICCEA, 1993)

•

| 2 | 1 | 0 | 1 | 1½ |

$v = 3$m s^{-1} $v = 4$m s^{-1}

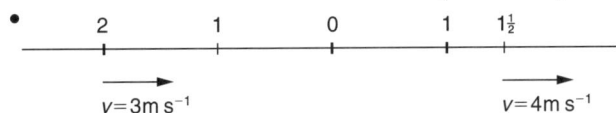

(i) Using $v^2 = \omega^2(a^2 - x^2)$ where a is the amplitude

$$16 = \omega^2(a^2 - 2.25) \tag{1}$$
$$9 = \omega^2(a^2 - 4) \tag{2}$$

Equation 1 − equation 2 gives

$$7 = \omega^2(-2.25 + 4) \implies \omega^2 = 4, \text{ i.e. } \omega = 2$$

(ii) Amplitude a
From equation 2:

$$9 = 4a^2 - 16 \implies a^2 = \tfrac{25}{4},$$
i.e. amplitude = 2.5 m

(iii) Periodic time $= \dfrac{2\pi}{\omega} = \pi$ seconds

5 An animal A runs in a straight line and has an acceleration of $0.05(20v - v^2)$ m s^{-2}, where v m s^{-1} is the speed of A. Show that, at time t seconds

$$\frac{dv}{dt}\left(\frac{1}{v} + \frac{1}{20 - v}\right) = 1$$

Given that at time $t = 0$, A passes through a point O with speed 4 m s^{-1}, show that at time t seconds the speed of A is given by

$$v = \frac{20e^t}{4 + e^t}$$

Hence, or otherwise, find the distance of the animal from O when $t = 9$. (AEB, 1994)

• Acceleration $= 0.05(20v - v^2) = \dfrac{dv}{dt}$

$$\frac{dv}{dt} \cdot \left(\frac{20}{20v - v^2}\right) = 1 \tag{1}$$

$$\frac{20}{v(20 - v)} = \frac{a}{v} + \frac{b}{20 - v} = \frac{a(20 - v) + bv}{v(20 - v)}$$

See Partial Fractions

Comparing numerators: $20 = 20a$, $0 = (-a + b)v$
Hence $a = b = 1$.

In equation 1:

$$\frac{dv}{dt}\left(\frac{1}{v} + \frac{1}{20 - v}\right) = 1$$

$$\therefore \int_{4}^{v}\left(\frac{1}{v} + \frac{1}{20 - v}\right) dv = \int_{0}^{t} dt \quad [6.1]$$

$$[\ln v - \ln(20 - v)]_{4}^{v} = t$$

$$\ln\left(\frac{v}{20 - v}\right) - \ln\left(\frac{4}{16}\right) = t$$

$$\ln\left(\frac{4v}{20 - v}\right) = t \implies \frac{4v}{20 - v} = e^t$$

$$4v = 20e^t - ve^t$$
$$v(4 + e^t) = 20e^t$$

Hence $v = \dfrac{20e^t}{4 + e^t}$

Then

$$v = \frac{ds}{dt} = \frac{20e^t}{4 + e^t}$$

$$\int_{0}^{s} ds = 20\int_{0}^{9} \frac{e^t}{4 + e^t} dt$$

$$s = [20\ln(4 + e^t)]_{0}^{9} \quad [6.1]$$

$$= 20\ln\left(\frac{4 + e^9}{5}\right)$$

$$= 148 \text{ m}$$

The animal is 148 m from O when $t = 9$.

6 A railway engine of mass 20 tonnes is working at a rate of 300 kW against constant resistances to motion of R newtons per tonne.

(i) If it can maintain a speed of 15 m s^{-1} on level ground find the value of R.

(ii) Find the acceleration of the engine when it is travelling at 12 m s^{-1}, if the power and resistances remain unchanged.

(iii) Travelling on the level track, the train reaches the bottom of a slope of inclination α where $\sin \alpha = \frac{1}{20}$. It then begins to climb the slope. Assuming that the power and resistances remain unchanged, show that, when the train is travelling at speed v m s^{-1}, the acceleration,

$\dfrac{\mathrm{d}v}{\mathrm{d}t}$, is given by $\dfrac{\mathrm{d}v}{\mathrm{d}t} = \dfrac{3(10 - v)}{2v}$

(iv) Hence, or otherwise, find the maximum speed which the engine can maintain on this incline.

(v) If the train is travelling at 5 m s^{-1} when it reaches the bottom of the slope, find the time it takes to reach a speed of 8 m s^{-1} up this slope.

[$g = 10$ m s^{-2}.]   (NICCEA, 1992)

• (i)

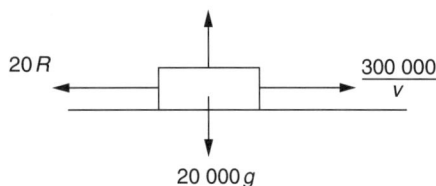

Power = 300 kW

Tractive force $= \dfrac{\text{power}}{\text{velocity}}$

$= \dfrac{300\,000}{v}$

Resistance = R N per tonne = $20R$

At a constant speed of 15 m s^{-1}

Tractive force = resistance

$\Rightarrow \quad \dfrac{300\,000}{15} = 20R$

$\Rightarrow \quad R = 1000$ N

(ii) When the speed is 12 m s^{-1} the tractive force is

$\dfrac{300\,000}{12} = 25\,000$ N

Accelerating force $= 25\,000 - 20R = 5000$ N

Acceleration $= \dfrac{\text{force}}{\text{mass}} = \dfrac{5000}{20\,000} = \frac{1}{4}$ m s^{-2}

(iii)

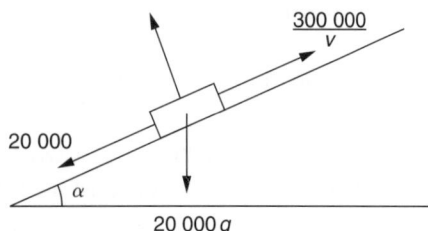

ΣForces up the slope

$= \dfrac{300\,000}{v} - 20\,000 - 20\,000g \sin \alpha$

$= \dfrac{300\,000}{v} - 20\,000 - 20\,000 \times 10 \times \dfrac{1}{20}$

$= \dfrac{300\,000}{v} - 30\,000 = \dfrac{30\,000}{v}(10 - v)$

Acceleration $= \dfrac{\mathrm{d}v}{\mathrm{d}t} = \dfrac{30\,000(10 - v)}{v \times 20\,000}$

$\Rightarrow \dfrac{\mathrm{d}v}{\mathrm{d}t} = \dfrac{3(10 - v)}{2v}$ (1)

(iv) Acceleration = 0 when $v = 10$.
Hence the maximum speed is 10 m s^{-2}.

(v) If the initial speed is 5 m s^{-1} then, from equation 1:

$\dfrac{2v}{10 - v} \dfrac{\mathrm{d}v}{\mathrm{d}t} = 3$

$\Rightarrow \quad 2 \displaystyle\int_5^8 \dfrac{v}{10 - v} \, \mathrm{d}v = \int_0^t 3 \, \mathrm{d}t$

$\Rightarrow \quad 2 \displaystyle\int_5^8 \left(\dfrac{v - 10}{10 - v} + \dfrac{10}{10 - v} \right) \mathrm{d}v = 3t$

$2[-v - 10 \ln (10 - v)]_5^8 = 3t$ 6.1

$2(-8 - 10 \ln 2 + 5 + 10 \ln 5) = 3t$

$t = \frac{2}{3}(10 \ln \frac{5}{2} - 3) = 4.1$ s

7 A particle moving on a straight line with speed v experiences a retardation of magnitude $b\mathrm{e}^{v/u}$, where b and u are constants. Given that the particle is travelling with speed u at time $t = 0$, show that the time t_1 for the speed to decrease to $u/2$ is given by

$bt_1 = u(\mathrm{e}^{-1/2} - \mathrm{e}^{-1})$

Find the further time t_2 for the particle to come to rest.

Deduce that $t_2/t_1 = \mathrm{e}^{1/2}$.

Find, in terms of b and u, an expression for the distance travelled in decelerating from speed u to rest.

• Retardation $= b\mathrm{e}^{v/u}$ so $\dfrac{\mathrm{d}v}{\mathrm{d}t} = -b\mathrm{e}^{v/u}$;

subject to $v = u$ when $t = 0$; $v = \dfrac{u}{2}$ when $t = t_1$.

$\displaystyle\int_u^{u/2} -\mathrm{e}^{-v/u} \, \mathrm{d}v = \int_0^{t_1} b \, \mathrm{d}t \quad \Rightarrow \quad [u\mathrm{e}^{-v/u}]_u^{u/2} = [bt]_0^{t_1}$

$u\mathrm{e}^{-1/2} - u\mathrm{e}^{-1} = bt_1$, thus $bt_1 = u(\mathrm{e}^{-1/2} - \mathrm{e}^{-1})$ (1)

Time taken to decrease speed to $\dfrac{u}{2}$ is given by

$$bt_1 = u(e^{-1/2} - e^{-1})$$

Changing the upper limits to $v = 0$, and $t = t_1 + t_2$, gives:

$$u(e^0 - e^{-1}) = b(t_2 + t_1)$$
$$b(t_2 + t_1) = u(1 - e^{-1}) \tag{2}$$

Equation 2 − equation 1 gives

$$t_2 = \frac{u}{b}(1 - e^{-1/2})$$

$$\frac{t_2}{t_1} = \frac{1 - e^{-1/2}}{e^{-1/2} - e^{-1}} = \frac{(1 - e^{-1/2})}{e^{-1/2}(1 - e^{-1/2})} = e^{1/2}$$

Retardation $= -v\dfrac{dv}{dx} = be^{v/u}$.

Separating and integrating by parts,

$$\int_u^0 -ve^{-v/u}\,dv = \int_0^{x_1} b\,dx$$

$$bx_1 = [vue^{-v/u}]_u^0 - \int_u^0 ue^{-v/u}\,dv$$

$$= [vue^{-v/u}]_u^0 + [u^2e^{-v/u}]_u^0$$

$$= u^2 - u^2e^{-1} - u^2e^{-1}$$

Hence $x_1 = \dfrac{u^2}{b}(1 - 2e^{-1})$.

8 A particle D moving along the x-axis has an acceleration in the positive x-direction of $k(26a^2x - 8x^3)$ where k and a are positive constants. Given that D has a speed of $a^2\sqrt{(6k)}$ when $x = a\sqrt{3}$ obtain an expression for the speed of D for any value of x. Determine the values of x at which D comes instantaneously to rest and show that the motion of D is confined to a finite region of the x-axis.   (AEB, 1983)

• $v\dfrac{dv}{dx} = k(26a^2x - 8x^3)$.

Separating and integrating,

$$\int_{a^2\sqrt{(6k)}}^v v\,dv = k\int_{a\sqrt{3}}^x (26a^2x - 8x^3)\,dx$$

$$\left[\frac{v^2}{2}\right]_{a^2\sqrt{(6k)}}^v = [k(13a^2x^2 - 2x^4)]_{a\sqrt{3}}^x$$

$$\frac{v^2}{2} - \frac{6}{2}ka^4 = 13ka^2x^2 - 2kx^4 - 39ka^4 + 18ka^4$$

$$\frac{v^2}{2} = 13ka^2x^2 - 2kx^4 - 18ka^4$$

$$v^2 = -2k(2x^4 - 13a^2x^2 + 18a^4)$$
$$v = \sqrt{[-2k(2x^2 - 9a^2)(x^2 - 2a^2)]}$$

$v = 0$ when $x^2 = \dfrac{9}{2}a^2 \Rightarrow x = \pm\dfrac{3a}{\sqrt{2}}$

and when $x^2 = 2a^2 \Rightarrow x = \pm\sqrt{2}a$.

Initially, $x = a\sqrt{3}$, also $v^2 \geqslant 0$ requires

$$(2x^2 - 9a^2)(x^2 - 2a^2) \leqslant 0$$

hence $\sqrt{2}a \leqslant x \leqslant \dfrac{3a}{\sqrt{2}}$.

9 A skydiver of mass m, falling vertically under gravity, experiences a resistive force of mkv, where her speed is v.
(i) By considering the equation of motion of the skydiver, show that her acceleration is $(g - kv)$.
(ii) Assuming that, at time $t = 0$, $v = 0$, show that her speed at time t is given by

$$v = \frac{g}{k}(1 - e^{-kt})$$

(iii) Given that $k = \tfrac{1}{5}\,s^{-1}$, what is the theoretical maximum speed that she can reach?
(iv) Given that $k = \tfrac{1}{5}\,s^{-1}$ show that she falls approximately 350 metres before her speed reaches 45 m s^{-1}.
    [$g = 10$ m s^{-2}]   (NICCEA, 1994)

• (i)

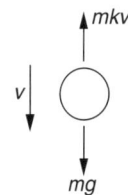

Accelerating force $= mg - mkv$
$\qquad\qquad\qquad = m \times$ acceleration
Hence acceleration $= (g - kv)$

(ii) Acceleration $= \dfrac{dv}{dt} = g - kv$

$$\Rightarrow \int_0^v \frac{dv}{g - kv} = \int_0^t dt$$

$$\Rightarrow \left[-\frac{1}{k}\ln|g - kv|\right]_0^v = [t]_0^t$$

$$\Rightarrow \frac{1}{k}\ln g - \frac{1}{k}\ln(g - kv) = t$$

$$\ln\left(\frac{g - kv}{g}\right) = -kt$$

$$\frac{g - kv}{g} = e^{-kt}$$

$$g - ge^{-kt} = kv$$

Hence, at time t,

$$v = \frac{g}{k}(1 - e^{-kt})$$

(iii) If $k = \frac{1}{5}$, $\quad v = 5g(1 - e^{-t/5})$

As $t \to \infty$, $\quad v \to 5g$

Maximum speed is 50 m s^{-1}.

(iv) Putting acceleration as $v\dfrac{dv}{ds}$

$$v\frac{dv}{ds} = g - kv \quad \Rightarrow \quad \int_0^v \frac{v}{g - kv}\,dv = \int_0^s ds$$

Substitute for k and g before integrating

$$\frac{v}{g - kv} = \frac{v}{10 - \frac{1}{5}v} = \frac{5v}{50 - v}$$

$$= 5\left(\frac{v - 50}{50 - v} + \frac{50}{50 - v}\right)$$

$$5\int_0^{45}\left(-1 + \frac{50}{50 - v}\right)dv = s$$

$$5[-v - 50\ln(50 - v)]_0^{45} = s \quad \boxed{6.1}$$

$$5[-45 - 50\ln 5 + 50\ln 50] = s$$

$$s = 350.6$$

Hence she falls approximately 350 m before her speed reaches 45 m s^{-1}.

10 A particle P has a position vector relative to a fixed origin O of $\mathbf{r} = 1\mathbf{i} + 2t\mathbf{j} + 3t^2\mathbf{k}$, where t is the time.

A force $\mathbf{F}$ is exerted on the particle where $\mathbf{F} = \dfrac{-3\mathbf{r}}{|\mathbf{r}|}$.

The work done by a force $\mathbf{F}$ displaced by $\delta\mathbf{r}$ is defined vectorially $\mathbf{F} \cdot \delta\mathbf{r}$.

Show that the work done by $\mathbf{F}$ in the time interval between $t = 1$ and $t = 2$ may be expressed as follows:

$$-3\int_1^2 \frac{18t^3 + 4t}{(9t^4 + 4t^2 + 1)^{1/2}}\,dt$$

Find the work done in this time interval.

• $\mathbf{r} = 1\mathbf{i} + 2t\mathbf{j} + 3t^2\mathbf{k}$, $\quad \mathbf{F} = \dfrac{-3(1\mathbf{i} + 2t\mathbf{j} + 3t^2\mathbf{k})}{\sqrt{(1 + 4t^2 + 9t^4)}}$

$$\frac{d\mathbf{r}}{dt} = 0\mathbf{i} + 2\mathbf{j} + 6t\mathbf{k}$$

Work done in time t is $\displaystyle\int \mathbf{F} \cdot \frac{d\mathbf{r}}{dt}\,dt$

$$\mathbf{F} \cdot \frac{d\mathbf{r}}{dt} = \frac{-3(4t + 18t^3)}{\sqrt{(1 + 4t^2 + 9t^4)}}$$

Work done between $t = 1$ and $t = 2$ is

$$-3\int_1^2 \frac{18t^3 + 4t}{\sqrt{(9t^4 + 4t^2 + 1)}}\,dt = -3[(9t^4 + 4t^2 + 1)^{1/2}]_1^2$$

$$= +3(\sqrt{14} - \sqrt{161})$$

$$= -26.8$$

11 Show that small oscillations of a simple pendulum of length l are simple harmonic with period $2\pi\sqrt{(l/g)}$. A pendulum clock beats seconds (i.e. one

half-period = 1 second) at a point where $g = 9.812$ m s^{-2}. Find the length of the pendulum correct to 3 significant figures. If the clock is moved to a place where $g = 9.921$ m s^{-2}, will the clock gain or lose? Find how much it will gain or lose during one day. To what length should the pendulum be altered if it is to register correctly? (SUJB)

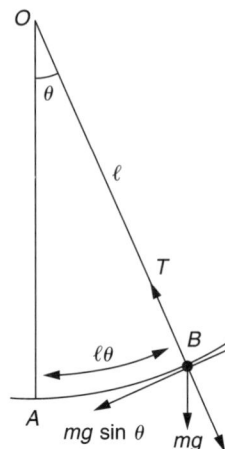

• Tangential component of force = $mg \sin\theta$ towards A, so

tangential acceleration = $g \sin\theta \approx g\theta$

where θ is small and measured in radians.

Arc length $AB = s = l\theta$, so $\theta = \dfrac{s}{l}$ and acceleration

towards A is $\dfrac{gs}{l}$, that is $-\ddot{s} = \dfrac{g}{l}s$

This is simple harmonic motion with period

$$2\pi\sqrt{\frac{l}{g}}.$$

For the clock, 1 period = 2 seconds, so

$$2\pi\sqrt{\frac{l}{g}} = 2 \quad \Rightarrow \quad l = \frac{g}{\pi^2}\text{ m}$$

When $g = 9.812$, $\quad l = 0.994\,16$ m.

When $g = 9.921$, $\quad$ period $= 2\pi\sqrt{\dfrac{0.994\,16}{9.921}}$

$$= 1.989\text{ s}.$$

This is less than 2 seconds so the clock will gain.

Number of 'beats' in 1 day is

$$\frac{24(60)(60)}{0.9945} = 86\,878.$$

Number of seconds in 1 day is

$$24(60)(60) = 86\,400$$

Thus clock gains 478 seconds = 7 min 58 s per day.

To register correctly, $l = \dfrac{9.921}{\pi^2} = 1.005$ m.

12 A scale pan is suspended from a fixed point A by a light elastic spring and a particle P of mass 0.3 kg is placed in the pan and attached to it with adhesive as shown.

The pan is pulled down from its equilibrium position and set in motion so that P moves in a vertical line through A with the base of the pan remaining horizontal.

Given that the motion of P is simple harmonic, with period $\dfrac{\pi}{5}$ s, and that the maximum and

minimum distances of P below A are 1.35 m and 0.85 m respectively, find
(a) the distance below A of the centre O of the oscillation, the amplitude of the oscillation and the maximum speed obtained by P
(b) the time to move directly upwards a distance of 0.125 m from O
(c) the maximum force, normal to the scale pan, that the adhesive has to exert on P
(d) the length of the spring, when, in the absence of adhesive, P would leave the pan. (Take g to be 10 m s^{-2}.)   (AEB)

• (a) Centre of oscillation is $\dfrac{1.35 + 0.85}{2}$ m below A, i.e. 1.1 m below A.

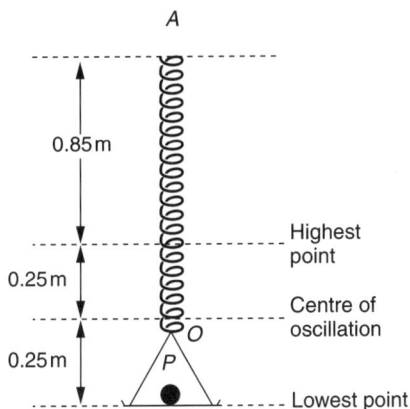

Amplitude $a = \dfrac{1.35 - 0.85}{2} = 0.25$ m.

Period $= \dfrac{2\pi}{\omega} = \dfrac{\pi}{5} \Rightarrow \omega = 10.$

Maximum speed $= \omega a = 2.5$ m s^{-1}.

(b) $x = 0.25 \sin 10t$, where x is the upward displacement from O, starting the timing at O.

When $x = 0.125$, $0.125 = 0.25(\sin 10t)$

$\sin 10t = \frac{1}{2} \Rightarrow 10t = \dfrac{\pi}{6}, \quad t = \dfrac{\pi}{60}$ s

Particle takes 0.0524 s to move upwards a distance of 0.125 m from O.

(c)

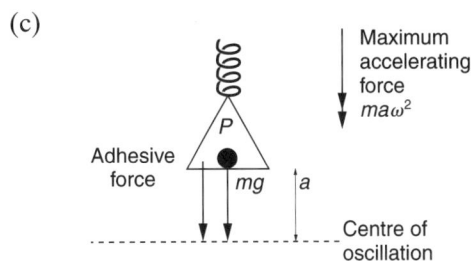

Maximum force that the glue has to exert is at the highest point, when a maximum acceleration downwards of $a\omega^2$ is experienced.

Accelerating force required $= 0.3(0.25)(100)$
$= 7.5$ N.

Weight of particle $= 3$ N, so adhesive has to exert a force of 4.5 N.

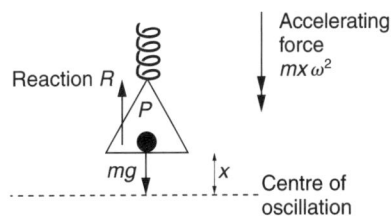

(d) If the particle is not glued it will leave the pan when $R = 0$, i.e. when the particle has an acceleration downwards of g, that is when

$-\omega^2 x = -10$
but $\omega^2 = 100$, so $x = +0.1$
$\Rightarrow$ length of spring $= 1$ m when P leaves the pan.

Exercises

1 A particle P moves so that, at time t seconds, its position vector $\mathbf{r}$ metres relative to a fixed origin O is given by

$\mathbf{r} = t^2\mathbf{i} + (3t - t^2)\mathbf{j} \quad t \geq 0$

where the unit vectors $\mathbf{i}$ and $\mathbf{j}$ are directed due east and due north respectively.

(a) Find the magnitude of the velocity, in m s^{-1}, of P when $t = 6$.

(b) Show that the acceleration of P is constant, and find the magnitude and bearing of this acceleration. (L, 1993)

2 At time t seconds a particle P has velocity

$$[(-4 \sin 2t)\mathbf{i} + (4 \cos 2t - 4)\mathbf{j}] \text{ m s}^{-1}$$

relative to a fixed origin O, where $\mathbf{i}$ and $\mathbf{j}$ are mutually perpendicular unit vectors.

(a) Find the magnitude of the acceleration of P.

(b) When $t = \dfrac{\pi}{8}$, find the angle that the

acceleration makes with the unit vector $\mathbf{i}$.

(c) Given that P passes through O at time $t = 0$, find the position vector of P relative to O at any subsequent time. (AEB, 1992)

3 A particle moves in a straight line with variable

acceleration $\dfrac{k}{1 + v}$ m s^{-2}, where k is a constant

and v m s^{-1} is the speed of the particle when it has travelled a distance x m. Find the distance moved by the particle as its speed increases from 0 to u m s^{-1}. (L)

4 One method of dyeing material is to immerse the cloth in a bath of water to which has been added d grams of concentrated dye. The material absorbs the dye at a rate equal to half the amount of dye remaining.

(a) If $\frac{3}{4}d$ grams of dye need to be absorbed to reach the desired colour, find the time taken to reach completion.

An alternative process is to keep the amount of dye present in the water constant at d grams by continuously adding dye throughout the process.

(b) Find the time now taken to complete the process.

5 [Assume $g = 10$ m s^{-2}.]

A particle of mass 3 kg is placed on a rough horizontal surface and a horizontal force of gradually increasing magnitude P is applied to it.

(a) (i) Draw a diagram and show the force of friction F acting on the particle together with all other forces.

(ii) If the coefficient of friction between the particle and the surface is μ, state the magnitude of the friction force when the particle remains in equilibrium, and determine its value when motion occurs.

(iii) Draw and label a sketch graph of F against P.

(b) After t seconds P is given by

$$P = \begin{cases} 2t & 0 \leqslant t \leqslant 7 \\ 0 & 7 < t \end{cases}$$

(i) Find the value of the coefficient of friction μ, if the body is in limiting equilibrium when $t = 4$.

(ii) Show that the distance travelled by the body in t seconds, where $4 \leqslant t \leqslant 7$, is

$$x = \tfrac{1}{9}(t^3 - 12t^2 + 48t - 64)$$

(iii) Find the distance travelled by the body while the force P is non-zero. (OLE, 1993)

6 A parachutist of mass m falls vertically under the action of the constant gravitational force and a

resisting force of magnitude $\dfrac{mv^2}{k}$, where v is the

speed of the parachutist and k is a positive constant. Given that the initial vertical speed of the parachutist is zero, show that after falling a distance x his speed is given by

$$v^2 = gk(1 - e^{-2x/k})$$

Explain why it is advisable to make k as small as possible, and suggest a way in which this may be achieved. (AEB, 1994)

7 A particle P, of mass m, moves along a straight line. A and B are two fixed points on this line at a distance $4a$ apart. The particle, P, moves under the action of two forces $\mathbf{F}_1$ and $\mathbf{F}_2$, where

$$\mathbf{F}_1 = \dfrac{3mg}{a}\,\overline{PA} \text{ and } \mathbf{F}_2 = \dfrac{mg}{a}\,\overline{PB}. \text{ (See figure.)}$$

(i) By considering the resultant force on the particle, show that P moves with Simple

Harmonic Motion, of period $\pi\sqrt{\dfrac{a}{g}}$, and that

the centre of the motion is at a distance a from A.

(ii) If the particle passes through the point A with a speed of $4\sqrt{2ag}$, find the amplitude of the motion.

(iii) What is the velocity of the particle at B?

(iv) How long will the particle take to travel from B to A? (NICCEA, 1992)

8 The stopping distance, x, is the distance travelled by a vehicle in the interval between the driver seeing an obstacle and the vehicle coming to rest. It is calculated on the assumptions that:

(a) a time interval T elapses between the obstacle being seen and the brakes being applied, during which time there is zero acceleration
(b) the brakes, when applied, immediately produce a retardation equal to that produced when the vehicle slides along the road. The coefficient of friction between the road and the tyres is a constant value μ.

Given that the initial speed is $u = 20$ m s^{-1}, $T = 2$ s, $\mu = 0.3$ and $g = 10$ m s^{-2}, find x in metres.

A more realistic model of the braking effect is that the retardation is dependent on the speed of the vehicle.

With the above values of u and T, and given that the retardation at speed v m s^{-1} is

$$\frac{660}{(v + 200)} \text{ m s}^{-2}$$

calculate the new value of x in metres. (AEB, 1983)

9 A car of mass M kg has a power of $16M$ watts and experiences a resistance of Mv where v m s^{-1} is the velocity at time t, and x is the displacement.

When $t = 0$, $x = 0$ and $v = 2$. Find

(a) an expression for $\dfrac{\mathrm{d}v}{\mathrm{d}t}$ and hence v in terms of t

(b) the maximum speed
(c) x in terms of v.

10 The speed v m s^{-1} of a particle moving on the x-axis is such that, at time t seconds

$$\frac{\mathrm{d}v}{\mathrm{d}t} = (3 - 5\mathrm{e}^{-t})$$

Given that $v = 1$ when $t = 0$ find v at any subsequent time.

Show also that the minimum value of v is approximately 0.53.

Given that the particle is of mass 0.2 kg, find the rate at which the force producing the motion is working at time t. (AEB, 1990)

Brief Solutions

1 $\mathbf{v} = 2t\mathbf{i} + (3 - 2t)\mathbf{j}$ $\mathbf{a} = 2\mathbf{i} - 2\mathbf{j}$
(a) When $t = 6$, $\mathbf{v} = 12\mathbf{i} - 9\mathbf{j}$
$|\mathbf{v}| = \sqrt{144 + 81} = 15$ m s^{-1}.
(b) $|\mathbf{a}| = 2\sqrt{2}$ m s^{-2} = constant
Direction of acceleration is S45°E or 135°.

2 $\mathbf{a} = -8\cos 2t\,\mathbf{i} - 8\sin 2t\,\mathbf{j}$ m s^{-2}
(a) $|\mathbf{a}| = 8\sqrt{\cos^2 2t + \sin^2 2t} = 8$ m s^{-2}

(b) $t = \dfrac{\pi}{8}$, $\mathbf{a} = -4\sqrt{2}\mathbf{i} - 4\sqrt{2}\mathbf{j} = -4\sqrt{2}\begin{pmatrix} 1 \\ 1 \end{pmatrix}$

Angle between $\mathbf{a}$ and $\mathbf{i}$ is given by

$$\mathbf{a} \cdot \mathbf{i} = |\mathbf{a}| \cos\theta \quad \Rightarrow \quad \frac{-4\sqrt{2}}{8} = \cos\theta$$

$\Rightarrow \quad \theta = 135°$.
(c) Position vector (integrate $\mathbf{v}$)
$\mathbf{r} = 2(\cos 2t - 1)\mathbf{i} + (2\sin 2t - 4t)\mathbf{j}$.

3 $v\dfrac{\mathrm{d}v}{\mathrm{d}x} = \dfrac{k}{1 + v} \quad \Rightarrow \quad \displaystyle\int v(1 + v)\,\mathrm{d}v = \int k\,\mathrm{d}x$

When $x = 0$, $v = 0$. Let $x = x_1$ when $v = u$; then

$$x_1 = \frac{u^2}{6k}(3 + 2u).$$

4 Let x be the amount of dye absorbed by the cloth.

(a) $\dfrac{\mathrm{d}x}{\mathrm{d}t} = \dfrac{(d - x)}{2}$ $\displaystyle\int_0^{3d/4} \frac{\mathrm{d}x}{d - x} = \int_0^{T_1} \frac{1}{2}\mathrm{d}t$

$T_1 = 4\ln 2 = 2.77$

(b) $\dfrac{\mathrm{d}x}{\mathrm{d}t} = \dfrac{d}{2}$ $T_2 = \frac{3}{2} = 1.5$

5 (a) (i)

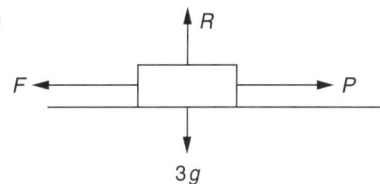

(ii) $F = P$ while $P < 3\mu g$ (equilibrium)
Motion occurs when $F = 3\mu g$ and $P > 3\mu g$.

(iii)

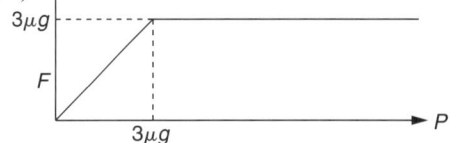

(b) (i) If $P = 8$ in limiting equilibrium, $3\mu g = 8$
$\Rightarrow \quad \mu = \frac{4}{15}$.
(ii) Accelerating force $= 2t - 8$ for $4 \leqslant t \leqslant 7$.

$$\text{Acceleration} = \frac{2t - 8}{3} = \frac{\mathrm{d}v}{\mathrm{d}t}$$

$$\Rightarrow \quad v = \left[\frac{t^2 - 8t}{3}\right]_4^t = \tfrac{1}{3}(t^2 - 8t + 16)$$

$$v = \frac{\mathrm{d}x}{\mathrm{d}t} \quad \Rightarrow \quad x = \tfrac{1}{3}\int_4^t (t^2 - 8t + 16)\,\mathrm{d}t$$

$$= \tfrac{1}{9}(t^3 - 12t^2 + 48t - 64)$$

(iii) When $t = 7$, $x = 3$ m.

6

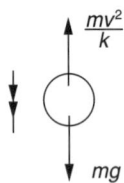

Accelerating force $= mg - \dfrac{mv^2}{k} = mv\dfrac{dv}{dx}$

$$\Rightarrow \int_0^x dx = \int_0^v \frac{kv}{kg - v^2}\,dv$$

$$\Rightarrow \quad x = \frac{k}{2}\ln\left(\frac{kg}{kg - v^2}\right)$$

$$v^2 = kg(1 - e^{-2x/k})$$

As $x \to \infty$, $v^2 \to kg$. If k is small, v is small.
This may be achieved by having a larger parachute.

7

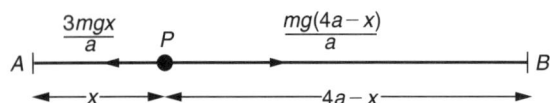

(i) Resultant force

$$F_2 - F_1 = \frac{mg}{a}(4a - x - 3x). \qquad \text{In the direction of } x \text{ increasing}$$

At the centre of oscillation $\sum F = 0$
$\sum F = 0$ at $x = a$.

Let $y = x - a$, i.e. displacement from centre of oscillation.

$$\ddot{y} = -\frac{4g}{a}y \quad \Rightarrow \quad \text{SHM}$$

$$\omega = 2\sqrt{\frac{g}{a}} \qquad \text{Period} = \frac{2\pi}{\omega} = \pi\sqrt{\frac{a}{g}}$$

(ii) If amplitude is b then $v^2 = \omega^2(b^2 - y^2)$.
   At A, $y = -a$ $v = 4\sqrt{2ag}$ $\Rightarrow$ $b = 3a$
(iii) At B, $y = 3a$ $\Rightarrow$ velocity $= 0$
(iv) Let the displacement from the centre of oscillation be $y = 3a\cos\omega t$

At B $y = 3a$ $\Rightarrow$ $t = 0$
At A $y = -a$ $\Rightarrow$ $\cos\omega t = -\frac{1}{3}$
     $\omega t = 1.9106$

Time from B to A is

$$\frac{1.9106}{\omega} = 0.9553\sqrt{\frac{a}{g}} \text{ seconds.}$$

8 (a) Thinking distance $= 40$ m,
      acceleration $= -3$ m s^{-2}.

   Total distance $= 40 + \dfrac{400}{6} = 106.7$ m.

   (b) With acceleration $= \dfrac{-660}{(v + 200)} = v\dfrac{dv}{dx}$.

   $$\int_{20}^0 - v(v + 200)\,dv = \int_0^{x_1} 660\,dx$$

   Total distance $= 40 + 64.6 = 104.6$ m.

9 (a) $\dfrac{dv}{dt} = \dfrac{16}{v} - v \quad \Rightarrow \quad \int_2^v \dfrac{v}{16 - v^2}\,dv = \int_0^t dt$

   $$t = \frac{1}{2}\ln\left(\frac{12}{16 - v^2}\right) \quad \Rightarrow \quad v = \sqrt{(16 - 12e^{-2t})}$$

   (b) As $t \to \infty$, $v \to 4$. Maximum speed $= 4$ m s^{-1}.

   (c) $v\dfrac{dv}{dx} = \dfrac{16 - v^2}{v} \quad \Rightarrow \quad \int_2^v \dfrac{v^2}{16 - v^2}\,dv = \int_0^x dx$

   $$\frac{v^2}{16 - v^2} = -1 + \frac{2}{4 - v} + \frac{2}{4 + v}$$

   $$x = \left[-v + 2\ln\left(\frac{4 + v}{4 - v}\right)\right]_2^v$$

   $$x = 2 - v + 2\ln\left(\frac{4 + v}{3(4 - v)}\right)$$

10 $\dfrac{dv}{dt} = 3 - 5e^{-t} \quad \Rightarrow \quad \int_1^v dv = \int_0^t (3 - 5e^{-t})\,dt$

   $$v - 1 = 3t + 5e^{-t} - 5 \quad \Rightarrow \quad v = 3t + 5e^{-t} - 4$$

   $$\frac{dv}{dt} = 0 \text{ when } t = 0.511$$

   $$\frac{d^2v}{dt^2} = 5e^{-t} \text{ which is positive}$$

   Hence, when $t = 0.511$, $v = 0.53$ is a minimum

   Rate of working $=$ force $\times$ velocity
   $\qquad\qquad\qquad\quad =$ mass $\times$ acceleration $\times$ velocity
   $\qquad\qquad\qquad\quad = 0.2(3 - 5e^{-t})(3t + 5e^{-t} - 4)$

26

Miscellaneous mechanics topics

AS Level			A Level			Topic			
CORE	MODULAR	TRADITIONAL	CORE	MODULAR	TRADITIONAL		Date attempted	Date completed	Self-assessment
				✓	✓	**Relative motion**			
				✓	✓	**Dimensions**			
				✓	✓	**Work done using vectors**			
				✓	✓	**Frameworks**			
				✓	✓	**Centres of mass by integration**			
				✓	✓	**Further differential equations**			

Fact Sheet

Relative Velocity

If $\mathbf{r}_A$ and $\mathbf{r}_B$ are the position vectors of A and B then $\dot{\mathbf{r}}_A = \mathbf{v}_A$ and $\dot{\mathbf{r}}_B = \mathbf{v}_B$ are the velocities of A and B.

The velocity of A relative to B is given by

$$_A\mathbf{v}_B = \mathbf{v}_A - \mathbf{v}_B$$

The velocity of B relative to A is given by

$$_B\mathbf{v}_A = \mathbf{v}_B - \mathbf{v}_A$$

Hence,

$$_A\mathbf{v}_B = -{_B\mathbf{v}_A}$$

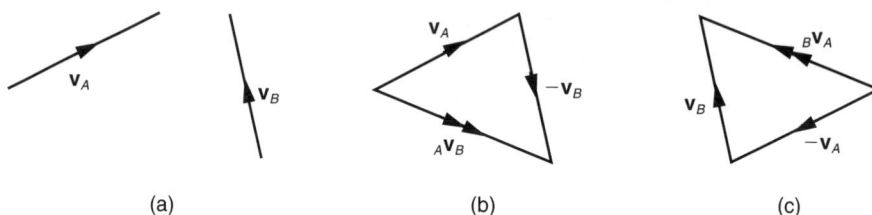

(a)   (b)   (c)

For three moving points $_A\mathbf{v}_B = {_A\mathbf{v}_C} - {_B\mathbf{v}_C}$ or $_A\mathbf{v}_B = {_A\mathbf{v}_C} + {_C\mathbf{v}_B}$.

Displacement

$\mathbf{r}_A - \mathbf{r}_B = \overline{BA}$, the displacement of A relative to B.
$\mathbf{r}_B - \mathbf{r}_A = \overline{AB}$, the displacement of B relative to A.

Distance between A and B is $|\overline{BA}| = |\overline{AB}|$.

Least distance occurs when $\overline{AB} \cdot {_A\mathbf{v}_B} = 0$ (or when $|\overline{AB}|^2$ is a minimum).

Collision will occur if $_A\mathbf{v}_B = k\overline{AB}$, provided $_A\mathbf{v}_B$ is a constant vector.

Graphical Method

This can be two distinct constructions, one for velocities and one for displacements, or one can be superimposed upon the other (usually more accurate and quicker).
(i) Choose scales for distances (to be used for all distances) and for velocities (to be used for all velocities).
(ii) Mark original positions of A and B on a diagram (L and M).
(iii) To find $_A\mathbf{v}_B$, construct the velocity vector triangle LCE. $_A\mathbf{v}_B = \overline{LE}$ in both magnitude and direction.

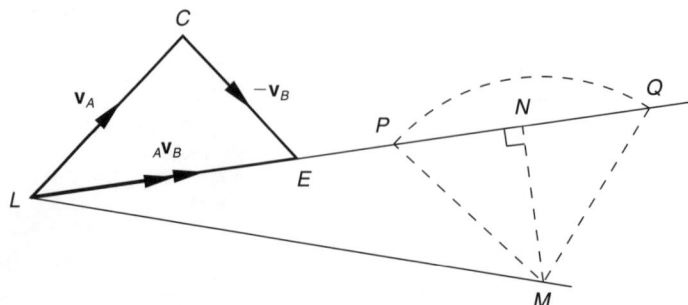

Least distance between particles is given by MN, where N is the foot of the perpendicular from M to LE (produced if necessary).

To find the time needed to reach any particular distance between the particles, find the one/two points P and Q on the line LE (produced if necessary), where $MP = MQ$ = required distance.

Time taken: $\dfrac{LP}{|_A\mathbf{v}_B|}$ or $\dfrac{LQ}{|_A\mathbf{v}_B|}$.

Collision occurs only if $MN = 0$, i.e. if LE is along LM. For collision, the velocity of A relative to B must be in the direction LM, the initial positions of A and B.

Dimensions

Any equation in mechanics must be dimensionally uniform.

Units are L for length
           M for mass
           T for time

$s = ut + \frac{1}{2}at^2$: s has dimension L
                 ut has dimensions $LT^{-1} \cdot T = L$
                 at^2 has dimensions $LT^{-2} \cdot T^2 = L$

Energy has units ML^2T^{-2}

Work Done by Vectors

The work done by a force moving from A to B is

$$\int_A^B \mathbf{F} \cdot d\mathbf{r} \text{ or } \int_A^B \mathbf{F} \cdot \frac{d\mathbf{r}}{dt} \, dt$$

Impulse $= \int \mathbf{F} \cdot dt$

Light Frameworks

(i) Each joint is in equilibrium.
(ii) The rods are usually light – neglect weight.
(iii) Each rod exerts equal and opposite forces at its ends on the joints. These are equal and opposite to those exerted by the joints on the rods.

     Arrows on force diagrams usually indicate the forces on the joints:
     —←——→— implies that the rod is under compression
     —→——←— implies that the rod is in tension.

Methods of Solution of Light Frameworks

(i) If possible find external forces first by taking moments and/or resolving.
(ii) Then resolve forces in two perpendicular directions at each joint and solve the resulting equations.

Differential Equations

See Chapter 12 for methods of solving differential equations arising from resisting and accelerating forces.

Worked Examples

1 (a) When attempting to find the equation for the path of a particle projected at speed u at an angle of elevation θ a student makes a mistake and produces this result.

$$y = x \tan \theta - \frac{xg}{2u^2 \cos^2 \theta}$$

where x and y represent horizontal and vertical displacements from the point of projection. Show, using dimensional considerations, that this formula cannot be correct.

(b) A particle of mass 5 kg has velocity given by

$$\mathbf{v} = (4 + 3t - t^2)\mathbf{i} + (3 + t^2)\mathbf{j}$$

where t is the time and $\mathbf{i}$ and $\mathbf{j}$ are perpendicular unit vectors in the directions of the x- and y-axes respectively of a Cartesian frame of reference with origin O.

(i) Find the initial speed and direction of the particle.

(ii) Find the force acting on the particle at time t.

(iii) What is the impulse of the force acting on the particle during the time interval $t = 0$ to $t = 2$?

(iv) If the particle is initially at the point A which has position vector $\mathbf{a} = 5\mathbf{i} + \mathbf{j}$ and at the point B at time $t = 3$ find the position vector of B.

(v) Find the time at which the acceleration is perpendicular to $\overline{CD}$ where C and D have position vectors $\mathbf{c} = \mathbf{i} - 4\mathbf{j}$ and $\mathbf{d} = 4\mathbf{i} - 2\mathbf{j}$ respectively. (NICCEA)

• (a) $y = x \tan \theta - \dfrac{gx}{2u^2 \cos^2 \theta}$

y has dimension L
$x \tan \theta$ has dimension L

$\dfrac{gx}{2u^2 \cos^2 \theta}$ has dimension $\dfrac{LT^{-2} \times L}{(LT^{-1})^2}$ which is non-dimensional.

Hence the formula cannot be correct.

(b) $\mathbf{v} = (4 + 3t - t^2)\mathbf{i} + (3 + t^2)\mathbf{j}$
(i) When $t = 0$, $\mathbf{v} = 4\mathbf{i} + 3\mathbf{j}$
speed = 5, direction is $\tan^{-1} \frac{3}{4}$

i.e. at angle 36.9° above the positive x-axis.

(ii) $\mathbf{f} = (3 - 2t)\mathbf{i} + 2t\mathbf{j}$ (acceleration)
Force = mass × acceleration
$= (15 - 10t)\mathbf{i} + 10t\mathbf{j}$

(iii) Momentum when $t = 0$ is $5\begin{pmatrix} 4 \\ 3 \end{pmatrix}$ and when $t = 2$ is $5\begin{pmatrix} 6 \\ 7 \end{pmatrix}$.

Impulse = change in momentum
$= 5\begin{pmatrix} 2 \\ 4 \end{pmatrix}$ or $10\mathbf{i} + 20\mathbf{j}$ or $\int_0^2 \mathbf{F} dt$

(iv) Since

$$\mathbf{v} = (4 + 3t - t^2)\mathbf{i} + (3 + t^2)\mathbf{j}$$

$$\mathbf{r} = \left(4t + \frac{3t^2}{2} - \frac{t^3}{3} + c\right)\mathbf{i} + \left(3t + \frac{t^3}{3} + d\right)\mathbf{j}$$

When $t = 0$, $\mathbf{r} = 5\mathbf{i} + \mathbf{j}$
Hence

$$\mathbf{r} = \left(4t + \frac{3t^2}{2} - \frac{t^2}{3} + 5\right)\mathbf{i} + \left(3t + \frac{t^3}{3} + 1\right)\mathbf{j}$$

When $t = 3$, $\mathbf{b} = 21.5\mathbf{i} + 19\mathbf{j}$

(v) $\mathbf{c} = \mathbf{i} - 4\mathbf{j}$ $\mathbf{d} = 4\mathbf{i} - 2\mathbf{j}$ $\overline{CD} = 3\mathbf{i} + 2\mathbf{j}$
$\mathbf{f} = (3 - 2t)\mathbf{i} + 2t\mathbf{j}$

This is perpendicular to CD when

$$\begin{pmatrix} 3 \\ 2 \end{pmatrix} \cdot \begin{pmatrix} 3 - 2t \\ 2t \end{pmatrix} = 0$$

i.e. $9 - 6t + 4t = 0 \implies t = 4\frac{1}{2}$

2 The position vector (in metres) at time t seconds, relative to a fixed origin O, of a particle P moving in a horizontal plane is given by

$$\mathbf{r} = \frac{2}{\pi}(\mathbf{i} \cos \pi t + \mathbf{j} \sin \pi t)$$

Describe as clearly as you can the path of P and determine the velocity of P at time t seconds.

Given that P is of mass 0.2 kg find the horizontal force acting on it at time t seconds.

When $t = 0$ a second particle Q, moving with constant velocity $(-6\mathbf{i} + 8\mathbf{j})$ m s^{-1} is at the point with position vector $\mathbf{b}$ metres. Find the position vector of Q at any subsequent time t in terms of $\mathbf{b}, \mathbf{i}, \mathbf{j}$, and t. Determine $\mathbf{b}$, given that the particles P and Q collide when $t = \frac{1}{2}$.

The mass of Q is 0.6 kg. On collision, the particles coalesce. Find the velocity of the composite particle immediately after collision. (WJEC, 1993)

• $\mathbf{r}_P = \dfrac{2}{\pi}(\mathbf{i} \cos \pi t + \mathbf{j} \sin \pi t)$

$$x^2 + y^2 = \frac{4}{\pi^2}(\cos^2 \pi t + \sin^2 \pi t) = \frac{4}{\pi^2}$$

Path is a circle of radius $\dfrac{2}{\pi}$ and centre $(0, 0)$.

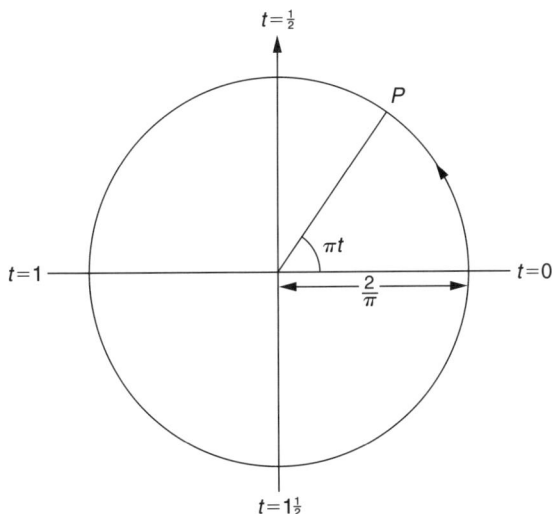

Velocity of P is

$$\mathbf{v}_P = \frac{2\pi}{\pi}(-\mathbf{i}\sin \pi t + \mathbf{j}\cos \pi t)$$

$$= 2(-\mathbf{i}\sin \pi t + \mathbf{j}\cos \pi t)$$

Acceleration of P is

$$\mathbf{a}_P = -2\pi(\mathbf{i}\cos \pi t + \mathbf{j}\sin \pi t)$$

Horizontal force is
$$\mathbf{F}_P = 0.2\mathbf{a}_P = -0.4\pi(\mathbf{i}\cos \pi t + \mathbf{j}\sin \pi t)$$

For Q, $\mathbf{r}_Q = \mathbf{b} + (-6t\mathbf{i} + 8t\mathbf{j})$

When particles collide, $\mathbf{r}_P = \mathbf{r}_Q$

$\mathbf{r}_P = \mathbf{r}_Q$ when $t = \frac{1}{2}$

$$\frac{2}{\pi}\left(\mathbf{i}\cos\frac{\pi}{2} + \mathbf{j}\sin\frac{\pi}{2}\right) = \mathbf{b} - 3\mathbf{i} + 4\mathbf{j}$$

$$\Rightarrow \quad \mathbf{b} = \frac{2}{\pi}\mathbf{j} + 3\mathbf{i} - 4\mathbf{j}$$

$$= 3\mathbf{i} + \mathbf{j}\left(\frac{2}{\pi} - 4\right)$$

Before collision, momentum of $P + Q$ is

$$0.2(-2\mathbf{i}) + 0.6(-6\mathbf{i} + 8\mathbf{j}) = -4\mathbf{i} + 4.8\mathbf{j}$$

After collision, momentum of $P + Q$ is $0.8\,\mathbf{v}$
$\therefore$ Combined velocity is $-5\mathbf{i} + 6\mathbf{j}$.

3 At noon, a ship, A, is travelling with a velocity of $4\mathbf{i} + 20\mathbf{j}$ km h^{-1} and a second ship, B, due north of it, is travelling with a velocity of $-3\mathbf{i} - 4\mathbf{j}$ km h^{-1} where $\mathbf{i}$ and $\mathbf{j}$ are unit vectors acting due East and North respectively.
(i) Find the velocity of A relative to B.
(ii) If the shortest distance between the vessels is 4.2 km find
    (a) the time, to the nearest minute, at which

they are closest together, and
(b) their original distance apart at noon.
(iii) If visibility at the time is 12 km show that they are within sight of each other for 54 minutes (to the nearest minute).
(iv) When the distance apart is a minimum what is the bearing of B from A?   (NICCEA, 1994)

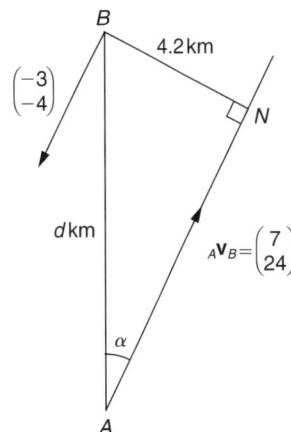

$$\mathbf{v}_A = \begin{pmatrix} 4 \\ 20 \end{pmatrix} \quad \mathbf{v}_B = \begin{pmatrix} -3 \\ -4 \end{pmatrix}$$

(i) $_A\mathbf{v}_B = \mathbf{v}_A - \mathbf{v}_B = \begin{pmatrix} 7 \\ 24 \end{pmatrix}$

$$|_A\mathbf{v}_B| = \sqrt{7^2 + 24^2} = 25$$

$$\tan \alpha = \tfrac{7}{24} \quad \sin \alpha = \tfrac{7}{25} \quad \cos \alpha = \tfrac{24}{25}$$

(ii) $AN = \dfrac{4.2}{\tan \alpha} = 14.4$ km, $AB = 15$ km

Time taken $= \dfrac{14.4}{25}$ h $= 35$ min

(a) The ships were nearest together at 12.35 p.m.
(b) Original distance apart 15 km.

(iii)

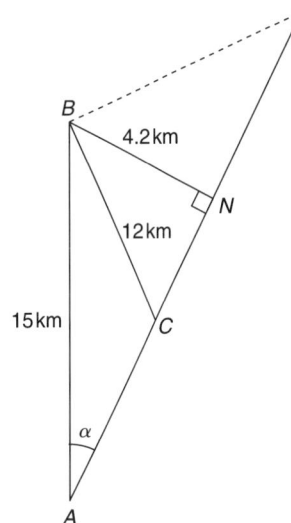

If BC is 12 km

$$\frac{\sin \alpha}{BC} = \frac{\sin C}{15} \implies \angle C = 20.49° \text{ or } 159.51°$$

$$\frac{BN}{CN} = \tan 20.5° \implies CN = 11.23 \text{ km}$$

Ships are visible for $\dfrac{2 \times 11.23 \text{ h}}{25} = 54$ min.

(iv) When the distance apart is a minimum, the bearing of B from A is

$$N(90 - \alpha)W \implies N74°W \text{ to the nearest degree.}$$

Alternatively – by vectors

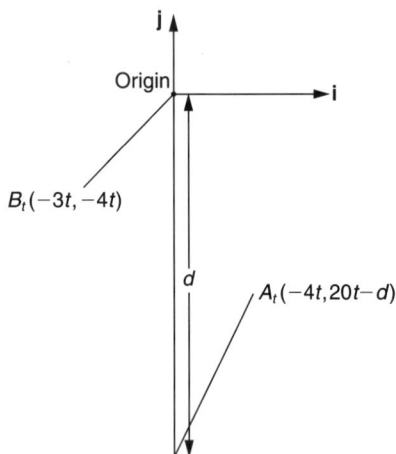

Displacement $BA = 7t\mathbf{i} + (24t - d)\mathbf{j}$

$$\begin{aligned}(\text{Distance } BA)^2 &= 49t^2 + 576t^2 - 48dt + d^2 \\ &= 625t^2 - 48dt + d^2 \\ &= (25t - \tfrac{24}{25}d)^2 + d^2\tfrac{49}{625}\end{aligned}$$

Minimum displacement occurs when $25t = \tfrac{24}{25}d$

and is $\left(\dfrac{7d}{25}\right)$.

$$\therefore 4.2 = \frac{7d}{25} \implies \text{ (b) } d = 15 \text{ km}$$

when (a) $t = 0.576$ h or 12.35 p.m.

(iii) When $BA = 12$ km

$$\begin{aligned}625t^2 - 720t + 225 &= 144 \\ \implies 625t^2 - 720t + 81 &= 0\end{aligned}$$

Roots t_1 and t_2:

$$t = \frac{720 \pm \sqrt{720^2 - 4.625.81}}{1250}$$

2.1a and 2.2a

$t_1 = 1.0256, t_2 = 0.1264$

They are within sight of each other for $t_1 - t_2$ hours, that is 53.952 min or 54 min

(iv) When $t = 0.576$
position of B is $(-1.73, -2.307)$
position of A is $(2.307, -3.465)$
Bearing of B from A is N74°W

4 Fred attempts to find a relationship between the speed and distance fallen for a parachutist. He makes a number of assumptions, one of which is that air resistance acts vertically upwards and is proportional to the square of the downward velocity.

(i) Show that the constant of proportionality for the air resistance has dimensions ML^{-1}.

Following his analysis he obtains the following expression

$$\text{speed} = \sqrt{u^2 + \frac{mg}{k}[1 - e^{-2kHF/m}]}$$

where

u is the initial horizontal velocity of the plane
m is the mass of the parachutist
g is the acceleration due to gravity
H is the distance fallen
and k is the constant of proportionality for the air resistance.

(ii) Using the method of dimensions check if Fred has made an error in his analysis and, if so, suggest the likely correct formula.

(NICCEA, 1993)

(i) Units should be MLT^{-2} (as for mg or any force)
Units of kv^2 are $k \times L^2T^{-2}$.
Hence k has units ML^{-1}.

(ii) (speed)^2 has units L^2T^{-2}.
Terms under the square root should have units L^2T^{-2}.
u^2 has units L^2T^{-2}.

$$\frac{mg}{k} \text{ has units } \frac{MLT^{-2}}{ML^{-1}} = L^2T^{-2}.$$

1 is non-dimensional.

$$\frac{2kH^2}{m} \text{ has units } \frac{ML^{-1}L^2}{M} = L.$$

$\therefore$ The exponential term is wrong: it should be non-dimensional, probably $e^{-2kH/m}$.

5 Two helicopter pilots are practising maintaining a straight course at a constant speed at a height of 100 metres above the surface of the sea. One pilot, flying due West at 100 kilometres per hour, flies over an anchored aircraft carrier. Fifteen minutes after this instant, the second pilot is 30 kilometres due South of the aircraft carrier and is supposed to be flying at 160 kilometres per hour in a direction 65° West of North. Unfortunately, although his speed is correct, he is a bit off course, with the potential tragic consequence that the two helicopters will collide unless evasive action is taken by one of the pilots.

Calculate,
(a) the time from the instant the first pilot flies over the aircraft carrier until the collision would happen if no evasive action were taken
(b) the angle by which the second pilot is off course. (SEB, 1993)

• Let the direction of H_2 be Nα°W

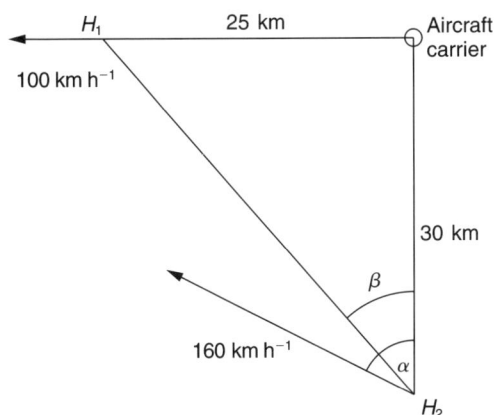

Helicopter 1 (H_1)   Helicopter 2 (H_2).
A collision will occur if the velocity of H_2 relative to H_1 is in the direction of H_1 relative to H_2 at any time
At time t_0:

H_1 is 25 km W of carrier
H_2 is 30 km S of carrier

i.e. bearing of H_1 from H_2 is N39.8°W.

Velocity of H_1 has components 100 W, 0 N.
Velocity of H_2 has components
$$160 \sin \alpha \text{ W, } 160 \cos \alpha \text{ N.}$$

For collision

$$\frac{160 \sin \alpha - 100}{160 \cos \alpha} = \tan (39.8) = \tfrac{5}{6}$$

$$\Rightarrow \quad \sin \alpha - \tfrac{5}{8} = \tfrac{5}{8} \cos \alpha$$
$$24 \sin \alpha - 20 \cos \alpha = 15 \tag{1}$$

$$24 \sin \alpha - 20 \cos \alpha = \sqrt{24^2 + 20^2} \sin (\alpha - \theta)$$

where $\tan \theta = \tfrac{20}{24}$ or $\tfrac{5}{6}$

$$\Rightarrow \quad 31.24 \sin (\alpha - \theta) = 15 \qquad \boxed{9.1}$$
$$\alpha - \theta = 28.7°$$
$$\alpha = 28.7° + 39.8°$$
$$= 68.5°$$

The direction of flight of the second helicopter is N68.5°W.

Time to collision is

$$\frac{30}{160 \cos \alpha} \times 60 \text{ min} + 15 \text{ min} = 30.7 \text{ min} + 15 \text{ min}$$
$$= 45.7 \text{ min}$$

and the pilot is 3.5° off course.
This could be done by vectors

6 A particle A has speed 6 m s^{-1} in the direction $\mathbf{i} + 2\mathbf{j} - 2\mathbf{k}$ and particle B has speed 7 m s^{-1} in the direction $2\mathbf{i} + 3\mathbf{j} - 6\mathbf{k}$. Find the position vectors of A and B relative to a fixed origin O at time t seconds, given that when $t = 0$ the position vectors are $(\mathbf{i} - \mathbf{j} + 2\mathbf{k})$ m and $(\mathbf{i} + 5\mathbf{j} + 4\mathbf{k})$ m respectively. Show that in the subsequent motion the minimum distance between A and B is $2\sqrt{5}$ m and find the time at which this position of minimum separation occurs.

When $t = 1$ a particle C is at the point with position vector $(\mathbf{i} + 4\mathbf{j} - 4\mathbf{k})$ m relative to O. Given that C moves with constant velocity $\mathbf{u}$ m s^{-1} and that B and C collide when $t = 3$, find $\mathbf{u}$. (AEB, 1993)

• *Particle A*
$|\mathbf{v}_A| = 6$ m s^{-1}
| direction vector of $\mathbf{v}_A| = \sqrt{1^2 + 2^2 + 2^2} = 3$
$$\therefore \mathbf{v}_A = (2\mathbf{i} + 4\mathbf{j} - 4\mathbf{k}) \text{ m s}^{-1}$$

Particle B
$|\mathbf{v}_B| = 7$ m s^{-1}
| direction vector of $\mathbf{v}_B| = \sqrt{2^2 + 3^2 + 6^2} = 7$
$$\therefore \mathbf{v}_B = (2\mathbf{i} + 3\mathbf{j} - 6\mathbf{k}) \text{ m s}^{-1}$$

Position vector of A is
$$\mathbf{r}_A = (\mathbf{i} - \mathbf{j} + 2\mathbf{k}) + t(2\mathbf{i} + 4\mathbf{j} - 4\mathbf{k})$$

Position vector of B is
$$\mathbf{r}_B = (\mathbf{i} + 5\mathbf{j} + 4\mathbf{k}) + t(2\mathbf{i} + 3\mathbf{j} - 6\mathbf{k})$$

Displacement vector is
$$\overline{AB} = \mathbf{r}_B - \mathbf{r}_A = (6\mathbf{j} + 2\mathbf{k}) + t(-\mathbf{j} - 2\mathbf{k})$$
$$(\text{Distance } AB)^2 = (6 - t)^2 + (2 - 2t)^2 \qquad \boxed{5.3}$$
$$= 5t^2 - 20t + 40$$
$$= 5(t^2 - 4t + 8)$$
$$= 5((t - 2)^2 + 4)$$
Minimum distance occurs when $t = 2$ s and has magnitude $\sqrt{20}$ or $2\sqrt{5}$ m.

At $t = 1$, position vector of C is
$$\mathbf{r}_C = (\mathbf{i} + 4\mathbf{j} - 4\mathbf{k})$$
At $t = 3$, position vector of C is
$$\mathbf{r}_C = (\mathbf{i} + 4\mathbf{j} - 4\mathbf{k}) + 2\mathbf{u}$$

If C collides with B when $t = 3$ then

$$(\mathbf{i} + 5\mathbf{j} + 4\mathbf{k}) + 3(2\mathbf{i} + 3\mathbf{j} - 6\mathbf{k})$$
$$= (\mathbf{i} + 4\mathbf{j} - 4\mathbf{k}) + 2(u_x\mathbf{i} + u_y\mathbf{j} + u_z\mathbf{k})$$

i component: $7 = 1 + 2u_x \Rightarrow u_x = 3$
j component: $14 = 4 + 2u_y \Rightarrow u_y = 5$
k component: $-14 = -4 + 2u_z \Rightarrow u_z = -5$

Hence $\mathbf{u} = 3\mathbf{i} + 5\mathbf{j} - 5\mathbf{k}$.

7 Prove, by integration, that the centre of mass of a uniform hemispherical shell of radius a is a distance $\frac{1}{2}a$ from the centre of its plane face.

The figure shows a uniform hemispherical shell of mass m resting with its curved surface in contact with a rough horizontal floor and a rough vertical wall.

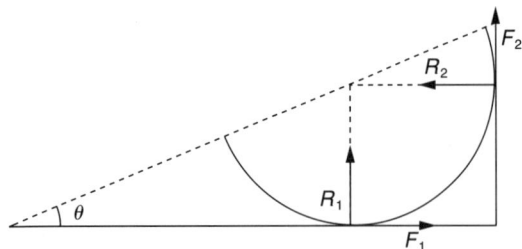

The reactions between the shell and the floor and wall have components R_1, F_1, R_2 and F_2 as indicated. Given that the coefficient of friction at both points of contact is μ and that equilibrium is limiting at both points of contact, show that

$$R_1 = \frac{mg}{1 + \mu^2}$$

Find, in terms of μ, an expression for $\sin \theta$.

(AEB, 1992)

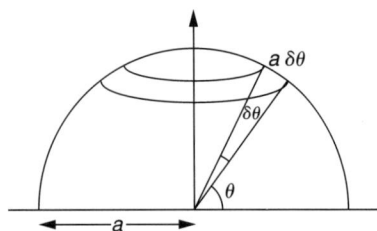

Using the elemental ring shown,
radius of the ring $= a \cos \theta$.

Surface area $= 2\pi a \cos \theta \, \delta s$ where δs is the arc length $a \, \delta \theta$.

Distance from plane face $= a \sin \theta$.

Let $\rho =$ mass/unit area.
Moment about the diameter of the plane face is

$$2\pi a^3 \cos \theta \sin \theta \, \rho \, \delta \theta$$

Total moment is

$$\int_0^{\pi/2} 2\pi a^3 \rho \sin \theta \cos \theta \, \mathrm{d}\theta = [\pi a^3 \rho \sin^2 \theta]_0^{\pi/2}$$

$$= \pi a^3 \rho$$

Total mass of the hemisphere $= 2\pi a^2 \rho$.

Hence distance of the centre of mass from the plane face is

$$\frac{\pi a^3 \rho}{2\pi a^2 \rho} = \frac{a}{2}$$

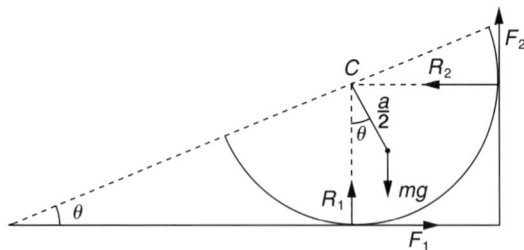

In limiting equilibrium

$$F_1 = \mu R_1$$
$$F_2 = \mu R_2$$

Sum of horizontal forces $= F_1 - R_2 = 0$
$$\Rightarrow R_2 = \mu R_1 \qquad (1)$$
Sum of vertical forces $= R_1 + F_2 - mg = 0$
$$\Rightarrow R_1 + \mu R_2 = mg \qquad (2)$$

Sum of moments about C: $= 0$

$$(F_1 + F_2)a = mg \frac{a}{2} \sin \theta \qquad (3)$$

From equations 1 and 2:

$$R_1 + \mu^2 R_1 = mg \Rightarrow R_1 = \frac{mg}{1 + \mu^2}, R_2 = \frac{\mu mg}{1 + \mu^2}$$

From equation 3:

$$\mu(R_1 + R_2) = \frac{mg}{2} \sin \theta$$

$$\Rightarrow \frac{\mu mg}{1 + \mu^2}(1 + \mu) = \frac{mg}{2} \sin \theta$$

$$\Rightarrow \sin \theta = \frac{2\mu(1 + \mu)}{(1 + \mu^2)}$$

8 (a) Show by integration that the centre of mass of a semi-circular lamina of radius r is at a distance

$\dfrac{4r}{3\pi}$ from the centre of the straight edge.

(b) The letter 'b' for a shop sign is to be cut from a sheet of plastic of uniform density. It consists of a rectangle 48 cm by 6 cm and a semi-circle of radius 12 cm, as shown in the figure.

   (i) Find the distances of the centre of mass G of this letter 'b' from OX and OY.
   (ii) The letter is to be held, with OY vertical, by 2 nails at P and Q positioned 3 cm from OY and 4 cm from the top and bottom of the letter as shown in the diagram. The nail at Q breaks, leaving the letter suspended from P. Find the angle which

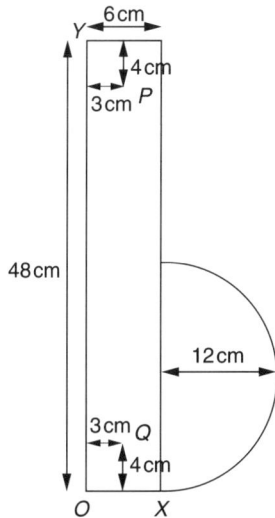

OY will then make with the vertical.

(NICCEA, 1992)

(a)

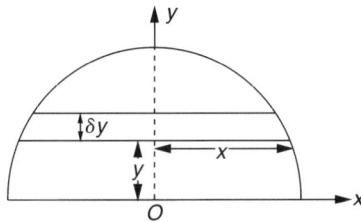

Consider an elemental strip distance y from the straight edge.

Radius of semi-circle $= r \implies x^2 + y^2 = r^2$. Mass of strip $= 2x\rho \, \delta y$ where ρ = mass/unit area.

Moment about straight edge is

$$\rho \int_0^r 2\sqrt{(r^2 - y^2)} y \, \mathrm{d}y = \frac{\pi r^2}{2} \rho \bar{y}$$

$$= \left[-\frac{2\rho}{3} (r^2 - y^2)^{3/2} \right]_0^r$$

$$\frac{\pi r^2 \rho}{2} \bar{y} = \frac{2\rho}{3} r^3 \implies \bar{y} = \frac{4r}{3\pi}$$

(b) (i)

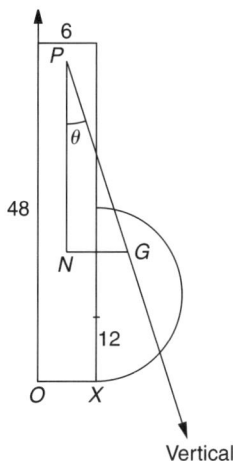

Mass of rectangle is 288ρ, centre of mass $(3, 24)$

Mass of semicircle is $\dfrac{144\pi}{2}\rho$,

centre of mass $\left(\dfrac{16}{\pi} + 6, 12 \right)$

For the letter 'b': mass $= 72\rho(4 + \pi)$, centre of mass $(\bar{x}, \bar{y})$

Taking turning moments about OY

$72\rho(4 + \pi)\bar{x}$

$$= 288\rho \times 3 + \frac{144\pi}{2}\rho \times \left(\frac{16}{\pi} + 6 \right)$$

$$= 72\rho \left(12 + \pi \left(\frac{16}{\pi} + 6 \right) \right)$$

$$= 72\rho(46.85)$$
$$\bar{x} = 6.56 \text{ cm}$$

$72\rho(4 + \pi)\bar{y} = 288\rho \times 24 + 72\pi\rho \times 12$
$(4 + \pi)\bar{y} = 96 + 12\pi = 133.7$
$\bar{y} = 18.72 \text{ cm}$

Distances of centre of gravity from OX and OY are 18.72 cm and 6.56 cm.

(ii) $\tan \theta = \dfrac{NG}{PN} = \dfrac{3.56}{25.28} \implies \theta = 8°$

i.e. OY makes an angle of $8°$ with the vertical.

9   A block of mass 0.15 kg is able to slide on a smooth, horizontal table. It is attached by a spring to a fixed point A and is connected to a damping device D by means of a rigid piston moving in a fixed pot, which contains a heavy oil.

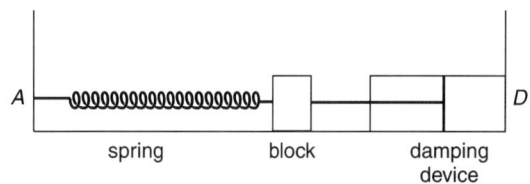

The stiffness of the spring is 0.2 N m^{-1} and the magnitude of the resistance due to the damping device is $0.4v$ newtons, where v metres per second is the speed of the block. The motion is started with the spring extended by 5 centimetres, by projecting the block horizontally with speed 14 centimetres per second towards A.

Let x metres denote the displacement of the block in the horizontal direction towards D, measured from the position of zero extension of the spring. Given that it is acceptable to use Hooke's Law for the tension exerted by the spring, show that a differential equation for the motion is

$$3\frac{d^2x}{dt^2} + 8\frac{dx}{dt} + 4x = 0$$

Find the particular solution of this equation satisfying the given initial conditions at $t = 0$.

Show that the block comes to rest momentarily after $\frac{3}{4}\ln 8$ seconds and describe the subsequent motion.

Suppose instead that the damping device had a resistance of $0.1v$ newtons. Write down the differential equation for this situation and describe briefly how the motion differs from that in the previous case. (SEB, 1993)

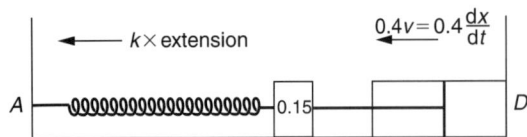

$k = 0.2$ N km^{-1} Tension = $k \times$ extension

$$k = \frac{\lambda}{\text{natural length}}$$

$x =$ extension in metres; initially $x = 0.05$.

Forces acting at any time = mass $\times$ acceleration

$$-kx - 0.4\frac{dx}{dt} = 0.15\frac{d^2x}{dt^2}$$

$$\Rightarrow \quad 3\frac{d^2x}{dt^2} + 8\frac{dx}{dt} + 4x = 0$$

Auxiliary equation is

$$3m^2 + 8m + 4 = 0$$
$$\Rightarrow \quad (3m + 2)(m + 2) = 0$$
$$\Rightarrow \quad m = -\tfrac{2}{3} \text{ or } -2$$

General solution is $x = Ae^{-2t} + Be^{-2t/3}$
$x = 0.05, \dot{x} = -0.14$ when $t = 0$

$$\Rightarrow \quad \begin{aligned} 0.05 &= A + B \\ -0.14 &= -2A - \tfrac{2}{3}B \end{aligned}$$

$$-0.04 = \tfrac{4}{3}B \quad \Rightarrow \quad B = -0.03, A = 0.08$$

Particular solution

$$x = 0.08e^{-2t} - 0.03e^{-2t/3}$$
$$\dot{x} = -0.16e^{-2t} + 0.02e^{-2t/3}$$

$\dot{x} = 0$ when $e^{-2t/3} = 8e^{-2t} \quad \Rightarrow \quad e^{4t/3} = 8$
$$\Rightarrow \quad \tfrac{4}{3}t = \ln 8$$
$$\Rightarrow \quad t = \tfrac{3}{4}\ln 8 \text{ s} = 1.56 \text{ s}$$

Therefore the block comes instantaneously to rest after $\frac{3}{4}\ln 8$ s, when $x = -0.0071$.

As t increases, x gradually increases to $x = 0$.

If the damping force is $0.1v$, the differential equation becomes

$$3\frac{d^2x}{dt^2} + 2\frac{dx}{dt} + 4x = 0$$

Auxiliary equation is $3m^2 + 2m + 4 = 0$ which has complex roots. Hence the block would have damped oscillations.

10 The figure shows a framework $ABCDEF$ consisting of nine smoothly jointed light rods. The rods AB, BC, CD and EF are each of length 8 m and are horizontal. The rods FB and EC are each of length 6 m and are vertical. The framework is simply supported in a vertical plane at A and D. Loads of 36 N and 18 N are attached at B and C respectively.

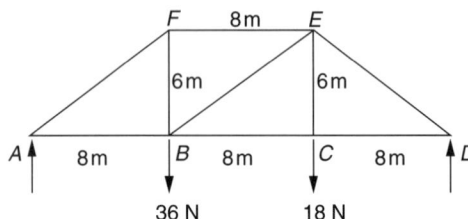

Calculate
(a) the reactions at A and D
(b) the forces acting in AF, AB and DE
(c) the forces acting in BF and BE.

An additional load of 18 N is now attached at C.
Find the force in BE. (AEB, 1984)

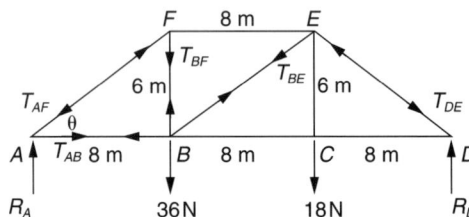

(a) For external forces

$$\sum F = 0 \quad \Rightarrow \quad R_A + R_D = 36 + 18 = 54 \qquad (1)$$

$\sum$ Moments about $A = 0$

$$36 \times 8 + 18 \times 16 = R_D \times 24$$
$$\Rightarrow \quad R_D = 24 \text{ N}, \qquad R_A = 30 \text{ N}$$

(b) From triangle ABF

$$\tan\theta = \tfrac{3}{4} \quad \sin\theta = \tfrac{3}{5} \quad \cos\theta = \tfrac{4}{5}$$

At A: $T_{AF}\sin\theta = R_A$
$$\Rightarrow \quad T_{AF} = 50 \text{ N} \qquad \text{Compression}$$
$T_{AF}\cos\theta = T_{AB}$
$$\Rightarrow \quad T_{AB} = 40 \text{ N} \qquad \text{Tension}$$
At D: $T_{DE}\sin\theta = R_D$
$$\Rightarrow \quad T_{DE} = 40 \text{ N} \qquad \text{Compression}$$
(c) At F: $T_{AF}\sin\theta = T_{BF}$
$$\Rightarrow \quad T_{BF} = 30 \text{ N} \qquad \text{Tension}$$
At B: $T_{BF} + T_{BE}\sin\theta = 36$
$$\Rightarrow \quad T_{BE} = 10 \text{ N} \qquad \text{Tension}$$

If a load of 18 N is added at C the forces would be symmetrical $\Rightarrow$ the force in $BE = 0$.

Exercises

1 At time $t = 0$ seconds, the position vectors (in metres) of two particles A and B are $7\mathbf{i} + 9\mathbf{j}$ and $5\mathbf{i} + 3\mathbf{j}$ respectively, and the particles are moving at all times with constant velocities (in m s^{-1}) of $4\mathbf{i} - 7\mathbf{j}$ and $9\mathbf{i} + 8\mathbf{j}$ respectively.
   (a) Write down the velocity of B relative to A.
   (b) Find the position vector of B relative to A at time t seconds.
   (c) Determine whether the particles will collide.
   (d) Find the cosine of the angle between their velocities. (WJEC, 1992)

2 The velocity $\mathbf{v}$ (in m s^{-1}) of a particle at time t seconds is given by

   $$\mathbf{v} = e^{4t}\mathbf{i} + ae^{-t}\mathbf{j}$$

   where a is a constant. Find the acceleration of the particle and determine the condition that has to be satisfied by a in order that the velocity and acceleration are perpendicular for some value of $t > 0$. (WJEC, 1992)

3 Two joggers, A and B, are each running with constant velocity on level parkland. At a certain instant, A and B have position vectors $(-60\mathbf{i} + 210\mathbf{j})$ m and $(30\mathbf{i} - 60\mathbf{j})$ m respectively, referred to a fixed origin O. Ninety seconds later, A and B meet at the point with position vector $(210\mathbf{i} + 120\mathbf{j})$ m.
   (a) Find, as a vector in terms of $\mathbf{i}$ and $\mathbf{j}$, the velocity of A relative to B.
   (b) Verify that the magnitude of the velocity of A relative to B is equal to the speed of A. (L, 1993)

4 Consider a coordinate system based on rectangular axes Ox, with unit vector $\mathbf{i}$, and Oy, with unit vector $\mathbf{j}$. In this question, distances are measured in metres and forces in newtons.
   (a) Find the work done when the point of application of the force $\mathbf{F} = 3\mathbf{i} + 2\mathbf{j}$ moves in a straight line from the point with position vector $2\mathbf{i} + \mathbf{j}$ to the point with position vector $6\mathbf{i} + 5\mathbf{j}$.
   (b) A particle is constrained to move along the x-axis under the action of a force $(2x - 3)\mathbf{i} + 2x\mathbf{j}$. Find the work done when the point of application of the force moves from $x = -2$ to $x = 6$. (SEB, 1993)

5 The drag force D (or resistance) on a car, due to its motion through the air, depends on the area perpendicular to motion, A, its velocity, v, and the density of the air, ρ.
   (i) Given that density is mass divided by volume, use the method of dimensions to show that

$D = \kappa\rho v^2 A$ where κ is a dimensionless constant. In a wind-tunnel test on a $\frac{1}{4}$ scale model car it was found that a drag of 9.3 N occurred at a wind speed of 12 m s^{-1}.
   (ii) Calculate the drag on the full scale car at a speed of 25 m s^{-1}. (CCEA, 1994)

6 A coastguard observes that a liner is cruising due West at 20 knots and a cargo ship is sailing at 15 knots on a bearing of 330°. At mid-day the liner is 8 nautical miles due North of the cargo ship. Find the velocity of the cargo ship relative to the liner.

   Given that they both continue to sail with the stated speeds and directions, calculate the time when they are nearest to each other. (SEB, 1994)

7 Two spots are moving on the screen of a microcomputer. At time t their position vectors are, respectively,

   $$\mathbf{r}_1 = t^2\mathbf{i} + t^3\mathbf{j} \text{ and } \mathbf{r}_2 = 2t\mathbf{i} - 2t^2\mathbf{j}$$

   Find the time $t > 0$ at which the directions of motion of the two spots are at right angles.

   Find the distance apart of the two spots at this time. (OLE, 1988)

8 (a) A *non-uniform* solid right circular cone has base radius 5 cm and height 15 cm. If the density at any point, ρ, is independent of distance from the point to the axis of the cone and is given by the expression

   $$\rho = 0.2\left(1 + \frac{x^2}{10}\right) \text{g cm}^{-3}$$

   where x is the distance from the vertex along the axis, show that
   (i) the mass is 0.3625π kg
   (ii) the centre of mass is approximately 12.41 cm from the vertex.

   (b) A composite body is formed by joining the cone above to a *uniform* solid hemisphere of radius 5 cm so that their bases coincide. If the mass of the hemisphere is 1.0 kg, find the centre of mass of the composite body. (The centre of mass of a uniform solid hemisphere of radius r is $\frac{3r}{8}$ from the centre along the axis.)

   (c) If AB is a diameter of the common circular base, what angle does AB make with the vertical if the composite body is freely suspended from the point A? (NICCEA, 1993)

9 The diagram overleaf shows a framework $ABCD$ of freely hinged light rods loaded at B, C and D with weights of magnitude 100 N, 150 N and 100 N respectively. The framework is in a vertical plane, smoothly hinged at the fixed point A and equilibrium

is maintained by a horizontal force of magnitude
P Newtons at D. The rods AD and BC are inclined

at $\dfrac{\pi}{4}$ to the horizontal, the rods AB

and CD are horizontal and rod BD is vertical. Find
(a) the value of P
(b) the magnitude and direction of the reaction at
    the hinge A
(c) the magnitude of the forces in each of the rods;
    stating in each case whether the rod is in
    tension or compression.   (AEB, 1994)

10 A target travelling with speed 10 m s^{-1} moves along
a fixed straight line which is inclined at 30° above
the horizontal and passes 6.25 m vertically above a
fixed point P.

(a) At the instant when the target is vertically
    above P a man fires a gun so as to hit the target.
    The bullet which is initially at P travels at a
    speed of 100 m s^{-1} at an angle of elevation of
    60°. The bullet's motion can be considered to
    be in a straight line during the time under
    consideration.

    Show that
    (i)   the bullet will not hit the target
    (ii) the approximate distance of closest
          approach of the bullet to the target is
          2.8 m.

(b) Instead of firing the gun, the man catapults a
    stone from P with speed $10\sqrt{3}$ m s^{-1} at an
    angle of elevation 60°. Effects of gravity now
    need to be considered.

    Show that the position vector of the target
    relative to P after t seconds is

    $5\sqrt{3}t\mathbf{i} + (6.25 + 5t)\mathbf{j}$ metres

    and show that the position vector of the stone
    relative to P at the same time is

    $5\sqrt{3}t\mathbf{i} + (15t - 5t^2)\mathbf{j}$ metres   (OLE, 1993)

Brief Solutions

1 Position vector of A at time t is
   $\mathbf{r}_A = (7\mathbf{i} + 9\mathbf{j}) + t(4\mathbf{i} - 7\mathbf{j})$
   Position vector of B at time t is
   $\mathbf{r}_B = (5\mathbf{i} + 3\mathbf{j}) + t(9\mathbf{i} + 8\mathbf{j})$
   (a) $_B\mathbf{v}_A = 5\mathbf{i} + 15\mathbf{j}$
   (b) $_B\mathbf{r}_A = (-2\mathbf{i} - 6\mathbf{j}) + t(5\mathbf{i} + 15\mathbf{j})$
       $= (-2 + 5t)\mathbf{i} + (-6 + 15t)\mathbf{j}$
   (c) Collision when $t = \frac{2}{5}$
   (d) Scalar product $\mathbf{v}_A \cdot \mathbf{v}_B = |\mathbf{v}_A||\mathbf{v}_B|\cos\theta$ where
       $\theta = $ angle between the velocities

       $\cos\theta = -0.206$

2 $\mathbf{v} = e^{4t}\mathbf{i} + ae^{-t}\mathbf{j}$ $\dfrac{d\mathbf{v}}{dt} = \mathbf{a} = 4e^{4t}\mathbf{i} - ae^{-t}\mathbf{j}$

   $\mathbf{a} \perp \mathbf{v}$ when $\mathbf{a} \cdot \mathbf{v} = 0$
   $\mathbf{a} \perp \mathbf{v} \Rightarrow 4e^{8t} - a^2e^{-2t} = 0 \Rightarrow a^2 = (2e^{5t})^2$
   $\Rightarrow a = \pm 2e^{5t}$

3 (a) Position vector of A is $\mathbf{r}_A = -60\mathbf{i} + 210\mathbf{j} + t\mathbf{v}_A$
       Position vector of B is $\mathbf{r}_B = 30\mathbf{i} - 60\mathbf{j} + t\mathbf{v}_B$
       Collision when $t = 90$
       $\Rightarrow \mathbf{r}_A = \mathbf{r}_B = 210\mathbf{i} + 120\mathbf{j}$
       $\Rightarrow \mathbf{v}_A = 3\mathbf{i} - \mathbf{j}, \mathbf{v}_B = 2\mathbf{i} + 2\mathbf{j}, _A\mathbf{v}_B = \mathbf{i} - 3\mathbf{j}$

   (b) $|\mathbf{v}_A| = \sqrt{10}$ m s^{-1} $|_A\mathbf{v}_B| = \sqrt{10}$ m s^{-1}

4 (a) Displacement $4\mathbf{i} + 4\mathbf{j}$, $\mathbf{F} = 3\mathbf{i} + 2\mathbf{j}$
       Work done $= \mathbf{F} \cdot \mathbf{d} = 20$ N m or 20 J.

   (b) Work done $= \displaystyle\int_{x_1}^{x_2} \mathbf{F} \cdot d\mathbf{r} = \int_{-2}^{6} (2x - 3)\,dx = 8$ J.

5 (i) ρ has dimensions ML^{-3}.
       v^2 has dimensions L^{2}T^{-2}.
       A has dimensions L^2.
       D has dimensions MLT^{-2}.
       $\kappa\rho v^2 A$ has dimensions
       ML$^{-3} \cdot$ L^{2}T$^{-2} \cdot$ L^2 = MLT^{-2}.
       Dimensionally correct.
   (ii) $D = 9.3, v = 12 \Rightarrow 144\kappa\rho A = 9.3$
       For full scale car
       $v_1 = 25, A_1 = 16A$ and $\kappa\rho A = \dfrac{9.3}{144}$

       $10\,000\kappa\rho A = $ drag $= \dfrac{10\,000 \times 9.3}{144} = 646$ N

6

   Velocity of cargo ship $= \mathbf{v}_C$.
   Velocity of liner $= \mathbf{v}_L$.

   $_C\mathbf{v}_L = \begin{matrix} \text{E} \\ \text{N} \end{matrix}\begin{pmatrix} -15\sin 30 \\ 15\cos 30 \end{pmatrix} - \begin{pmatrix} -20 \\ 0 \end{pmatrix} = \begin{pmatrix} 12.5 \\ 13 \end{pmatrix}$

Velocity of C relative to L is 18.03 knots on bearing 044°.

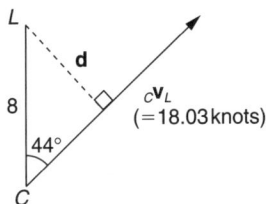

d = least distance = 8 sin 44°
$\qquad\qquad\qquad$ = 5.56 nautical miles
Cargo ship has travelled 8 cos 44 nautical miles at relative velocity of 18.03 knots.
Time 12.19 p.m. or 19.2 min after 12 noon.

7 $\quad \mathbf{r}_1 = t^2\mathbf{i} + t^3\mathbf{j} \qquad \mathbf{v}_1 = 2t\mathbf{i} + 3t^2\mathbf{j}$
$\quad \mathbf{r}_2 = 2t\mathbf{i} - 2t^2\mathbf{j} \qquad \mathbf{v}_2 = 2\mathbf{i} - 4t\mathbf{j}$
When directions of motion are perpendicular
$\mathbf{v}_1 \cdot \mathbf{v}_2 = 0$. That is

$$4t - 12t^3 = 0 \quad \Rightarrow \quad t = 0 \text{ or } \sqrt{\tfrac{1}{3}}$$

Reject $t = 0$ since $\mathbf{v}_1 = 0$ when $t = 0$

Positions at $t = \dfrac{1}{\sqrt{3}}$:

$$\mathbf{r}_1 = \frac{1}{3}\mathbf{i} + \frac{1}{3\sqrt{3}}\mathbf{j} \qquad \mathbf{r}_2 = \frac{2}{\sqrt{3}}\mathbf{i} - \frac{2}{3}\mathbf{j}$$

$$\mathbf{r}_2 - \mathbf{r}_1 = \left(\frac{2}{\sqrt{3}} - \frac{1}{3}\right)\mathbf{i} - \left(\frac{2}{3} + \frac{1}{3\sqrt{3}}\right)\mathbf{j}$$

$$|\mathbf{r}_2 - \mathbf{r}_1| = 1.19$$

8 (a)

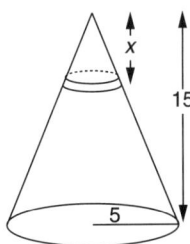

(i) Distance from the vertex
$\qquad$ = 3(radius of disc)

$$\text{Mass of disc} = \pi\left(\frac{x}{3}\right)^2 (0.2)\left(1 + \frac{x^2}{10}\right)\delta x \text{ grams}$$

$$\text{Total mass} = \frac{0.2\pi}{9}\int_0^{15} x^2\left(1 + \frac{x^2}{10}\right)dx$$

$$= 0.3625\pi \text{ kg}$$

(ii) Moment about the vertex is

$$\frac{0.2\pi}{9}\int x^3\left(1 + \frac{x^2}{10}\right)dx = 4.5\pi \text{ kg cm.}$$

$$\therefore 0.3625\pi\bar{x} = 4.5\pi$$
$$\Rightarrow \quad \bar{x} = 12.41 \text{ cm from the vertex.}$$

(b)

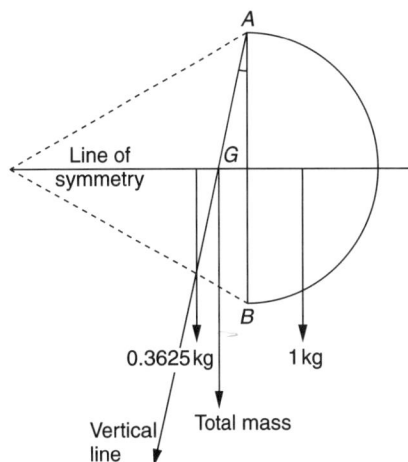

Moment about the vertex:

$$4.5\pi + 1\left(15 + \frac{15}{8}\right) = 2.139\bar{x}$$

$\bar{x} = 14.5$ cm from vertex, 0.5 cm from the common face.

(c) The angle between AB and the vertical is

$$\arctan\left(\frac{0.5}{5}\right) = 5.7°.$$

9

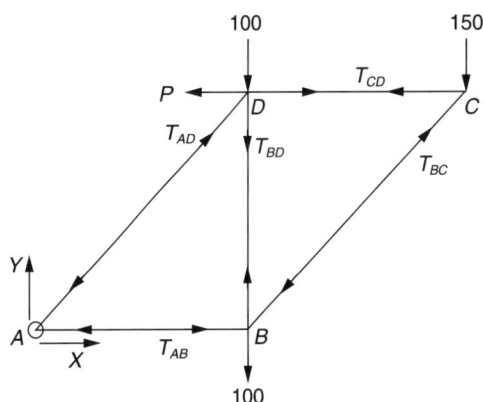

External forces
Resolving horizontally: $X = P$
Resolving vertically: $Y = 350$ N

(a) Moments about A:

$$Pd = 200d + 150(2d) \quad \Rightarrow \quad P = 500 \text{ N}$$

(b) Force at hinge $A = \sqrt{500^2 + 350^2}$
$\qquad\qquad\qquad\qquad = 610$ N at 35° above AB.

(c)

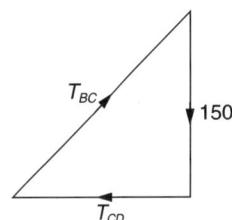

At C: $T_{CD} = 150$ N $\qquad$ tension
$\qquad\;\; T_{BC} = 150\sqrt{2}$ N $\qquad$ compression
At B: $T_{BD} = 100 + 150 = 250$ N $\;$ tension
$\qquad\;\; T_{AB} = 150$ N $\qquad$ compression

At D: $T_{AD} = (P - T_{CD})\sqrt{2} = 350\sqrt{2}$ N
compression
Use the forces at A as a check.

10 (a)

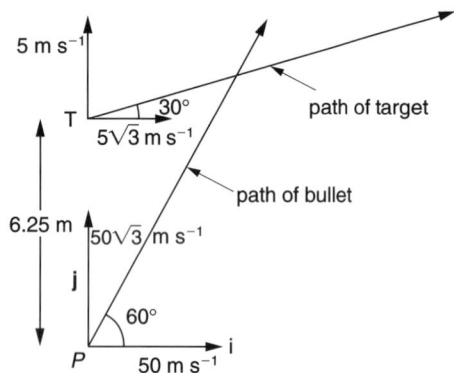

For the target: position vector at time t is
$5\sqrt{3}t\mathbf{i} + 5t\mathbf{j}$ relative to the point of projection.

Since the target is 6.25 m above P at the moment that the gun is fired:

position vector of the target relative to P is
$5\sqrt{3}t\mathbf{i} + (6.25 + 5t)\mathbf{j}$

position vector of the bullet relative to P is
$50t\mathbf{i} + 50\sqrt{3}t\mathbf{j}$

(i) For collision $5\sqrt{3}t = 50t$ which is not possible.

(ii) Displacement of the target from the bullet is

$$(5\sqrt{3}t - 50t)\mathbf{i} + ((5 - 50\sqrt{3})t + 6.25)\mathbf{j}$$
$$= -41.34t\mathbf{i} + (6.25 - 81.6t)\mathbf{j}$$

$$(\text{Distance})^2 = (-41.34t)^2 + (6.25 - 81.6t)^2$$
$$= 8368t^2 - 1020t + 39.06$$

$$\frac{\mathrm{d}}{\mathrm{d}t}(\text{distance})^2 = 16\,736t - 1020$$
$$= 0 \quad \text{when } t = 0.06094 \text{ s}$$

At this time $(\text{distance})^2 = 7.977$ m^2.

Minimum distance apart is
$\sqrt{7.977} = 2.8$ m.

(b) For the stone, the position vector relative to P is

$$5\sqrt{3}t\mathbf{i} + (15t - 5t^2)\mathbf{j}$$

Target as in (a)

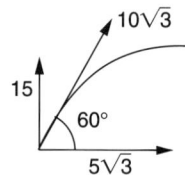

For collision:

$$5\sqrt{3}t\mathbf{i} + (15t - 5t^2)\mathbf{j} = 5\sqrt{3}t\mathbf{i} + (6.25 + 5t)\mathbf{j}$$
$$\Rightarrow \quad 5t^2 - 10t + 6.25 = 0$$
which has no real solutions.
Therefore the stone does not hit the target.

Index